COASTAL STABILIZATION

ADVANCED SERIES ON OCEAN ENGINEERING

Series Editor-in-Chief
Philip L- F Liu (*Cornell University*)

Advanced Series on Ocean Engineering – Volume 14

COASTAL STABILIZATION

RICHARD SILVESTER
JOHN R C HSU
The University of Western Australia

World Scientific
Singapore • New Jersey • London • Hong Kong

Published by

World Scientific Publishing Co. Pte. Ltd.

P O Box 128, Farrer Road, Singapore 912805

USA office: Suite 1B, 1060 Main Street, River Edge, NJ 07661

UK office: 57 Shelton Street, Covent Garden, London WC2H 9HE

Library of Congress Cataloging-in-Publication Data
Silvester, Richard, 1924–
    Coastal stabilization / Richard Silvester, John R. C. Hsu.
       p. cm. -- (Advanced series on ocean engineering ; vol. 14)
    Includes bibliographical references and index.
    ISBN 9810231377, -- ISBN 9810231547 (pbk.)
    1. Coastal engineering.   2. Marine sediments.   I. Hsu, John R. C.
II. Title.   III. Series.
TC209.S54   1997
627'.58--dc21   97-10527
                                     CIP

British Library Cataloguing-in-Publication Data
A catalogue record for this book is available from the British Library.

First published as "Coastal Stabilization — Innovative Concepts"
© 1993 by Prentice Hall, Inc.

First published 1997
Reprinted 1999

Printed in Singapore.

To
Marion and Lisa
for their patience and support

Contents

Preface

With the plethora of books being published on coastal and ocean engineering topics, the question may well be raised: "Why another one?" There is no doubt that the topic of the ocean and its continental margins is a large and deep one, so that there are many facets that can be covered by specialized treatises. The one with which this tome deals is that of sediment dynamics along the shorelines, even out to the edge of the continental shelf. It has been dealt with by geologists, geographers, and in more recent times by coastal engineers.

The study involves a knowledge of fluid mechanics, wave theory, sediment characteristics, and changeability of the sea. Because of these many variables it is very difficult to apply theory or mathematics rigorously to it. Attempts to do so by numerical analysis may contain one or two variables but the conclusions drawn must take cognizance of those omitted, where only rationalization can help in design. For these reasons topics involving sediment are not preferred for theses, since a nice package problem of finite dimensions is difficult to define.

Where physical models are utilized to solve some specific sedimentary problem on the coast there will always be some distortion entering the picture, similar to but more complex than in fluvial hydraulics. Conclusions are reached and reported soon after changes are effected in the prototype when sufficient time has not elapsed to verify the changes. It would be better for papers to await the outcome over say 5 years.

The output from the pens of coastal engineers is prodigious, particularly those from academic circles, through the proceedings of conferences held in many countries during any one year. Also, those involved in sedimentary problems must read journals ranging over a number of specialties in order to keep up with research around the world. There is need for a few workers who can read and digest this material in order to summarize and present it in a form useful to engineers for design. This is no easy task, as many judgments have to be made as to what should be included or omitted.

This book attempts to achieve this end by presenting the various theories involved in longshore drift and the fluid motions connected therewith. Despite the mathematical sophistication, further discussion is included of a descriptive nature which outlines the many variables that necessarily cannot be included in such theory. This is not meant to detract from the theoretical approach but to present a word of warning against its use without due thought for the problems as a whole. Nature works macroscopically and we should view coastal problems in the same manner.

Many new concepts are presented which are the outcome of papers published since 1974, when the senior author had a book *Coastal Engineering* published by Elsevier in Amsterdam. By bringing this information together in one volume it is hoped that the reader can obtain an overall view of the thinking process of the two authors. In this way, officers of the various governmental agencies can build up background knowledge which will help in the economical management of the coastline. At least the novel approaches should be priced at the same time as traditional solutions, taking into account costs of maintenance besides that of initial installation. University colleagues and practicing engineers might see a promising alternative to stabilizing eroded beaches, a method that has emerged from the engineering applications in coastal geomorphology.

One particular mode of coastal defense is stressed, by discussing it under many prototype conditions. This has been termed *headland control*, by which stabilization is provided by the insertion of fixed structures spaced along the coast, aided at times by reclamation. Bays are sculptured by Nature between these points which have a limiting indentation, when the coast can be considered as stable. No apology is afforded in emphasizing this procedure, as it is the authors' strong conviction that it is the breakthrough that has been awaited for some decades. Although relationships associated with it are only empirical at the moment, as are many phenomena connected with sediment movement in fluvial and marine situations, their verification has been as equally from prototype data as from model studies.

Richard Silvester
John R. C. Hsu

Introduction

The margin between the sea and the land is an extremely dynamic zone, for it is here that the motion of the sea interacts with the sediment or rock of the land. It seems superfluous to remark that the sea experiences large fluctuations of energy which it applies to the coast. Storms can wreak havoc sporadically on mobile material of the shoreline, while the more moderate swell waves effect sediment movement continually. The interaction of these two systems, which has been carried out over geologic time, causes net longshore transport along thousands of kilometers of coast. This results in denudation on some segments and massive silting on others, either as land mass or shoaling of the coastal margin in the form of continental shelves.

The shelves in their turn influence the waves generated in the deep water by twisting them almost normal to the coast by the time they steepen and break. This *refraction*, as it is called, is important in assessing the direction of waves encountering structures built in the nearshore zone. As man has constructed these of temporary or more permanent nature offshore, or on midparts of the shelf, he has become more interested in sediment movement in these greater depths. As he has reproduced the water motions in these locations more accurately, the further out he has proven sediment disturbance to take place. In fact it can be shown that the whole shelf is a highway for this drama of the sea.

In considering the energy available to transport sediment both in the surf zone, where the waves are broken, and offshore beyond the breaker line, the wave height and the wave period should be taken into account. These together determine the magnitude of velocities and accelerations of water-particle orbits as well as their amplitudes of excursion each wave cycle. The net motion per cycle, known as *mass transport*, also varies throughout the water column, but that at the bed is of greatest concern for sediment movement. The waves, in fact, exert a sweeping motion of particles in the direction of wave advance. The alongshore component of this net movement varies across the shelf as waves are refracted. There is a limit of reach of each wave component where water motion at the seabed becomes negligible, equal approximately to half the deepwater wave length. For the larger period waves this can be shown to extend across to the edge of the continental shelf where the profile assumes a greater gradient due to plain deposition, either from suspension or slippage from the shelf proper.

Continental shelves make up 4.5% of the ocean surface, which constitutes around 70% of the earth's surface. Of the remaining 30% one quarter is basic rock whereas sediment covers 75% of the land area. Of this latter zone 75% has been deposited under marine conditions, implying that it was at a shoreline when deposited. The other 25% is terrigenous or placed without contact with the sea. These measurements are based on surface samples, but it is likely that others taken from bore holes would prove that lower layers were deposited with marine debris. Thus it can be considered that the major portions of land masses have been constructed by wave action over geologic time which runs into thousands of millions of years.

The sequence envisaged is that the crust of the earth was distorted as it was formed and cooled. When the vapor condensed to form the sea, the major portion was submerged. That above the waterline suffered large variations in temperature, which helped break up the highest peaks that then fell into the sea. Later chemical breakdown aided in the formation of sediment which was dispersed by rain and runoff. Slowly the valleys in the rock formation were filled with coarse material and rivers became longer. As they reached flatter profiles their ability to transport large boulders decreased and only sand could be transported to the sea. Rivers still provide 95% or more of the sediment for waves to move along the coasts. As rivers lengthen, they are only able to carry fine sediment in suspension, which means a changed constitution of the material on the shelf downcoast of their mouths.

1.1 COASTAL ENVIRONMENT

The environment of greatest concern with respect to coastal margins is that of the motions of the sea itself. This is made up of waves, reaching from the surface to about half the deepwater wave length, tidal oscillations which influence the deepest waters of the ocean, and tsunamis caused by earthquake movements of the earth's crust. Both tides and tsunamis are equivalent to long waves which increase in height as they propagate over shoals such as continental shelves. They do not normally reach a point of instability and break, but there are locations where even this happens. Of course, tides are continuous with crests or high water occurring daily or twice daily. On the other hand, tsunamis occur very infrequently.

The waves that contain the largest energy are generated by strong winds within storm centers known as cyclones. These are repeated annually in certain zones of the ocean, particularly in latitudes 40° to 60°, where such centers travel constantly from west to east. Due to the circulation of air, anticlockwise in the northern hemisphere and clockwise in the southern hemisphere, waves are directed in an easterly direction toward equatorial regions. This results in west coasts receiving persistent swell, while eastern coasts experience waves from different directions during the year, dependent on the paths followed by cyclones passing seaward from the landmass.

It is essential for scientists and engineers dealing with marine matters to understand the generation of waves and to be able to forecast their characteristics from wind information. Waves are generally built up in deep water but then spread out across the ocean, or even oceans, to break on distant shores. In doing so they lose negligible energy, reducing in height only because the energy previously concentrated in the fetch is dispersed over much greater areas. This swell can be considered an efficient distributor of this large energy wind source to the shorelines of the world. These waves, either of storm character or swell, must traverse the continental shelf before arriving at the beach. Their transformation during this passage must be understood. Finally, waves break near the shore, so introducing new complexities in their application of energy to structures or natural features.

Thus, in studying sedimentary processes on a particular section of coast, types of wave and their duration over months or seasons of the year should be ascertained, known as a *wave climate*. The direction of approach is most important both for the design of structures and in the study of sediment movement. The assessment of the largest waves from the fiercest storms recorded in the region is fairly straightforward, but cognizance must be taken of whether a more intense cyclone could have or has occurred in historic time. Swell waves from distant storms vary in height and period throughout the day and are not computable from all their sources. Even measurements of them over a year or two are unlikely to provide the largest magnitudes over the lifetime of a structure. For the calculation of littoral drift, some mean heights, periods, and directions are normally used that can introduce errors of large magnitude. Thus, the compilation of a wave climate is necessarily qualitative rather than quantitative.

The level of the sea surface at the shoreline is important when calculating the reach of waves up a shoreline or on a structure. That due to tides is predictable from tide charts produced after measurements over a year or two. But even these vary throughout the year, so that it is difficult to judge whether the highest or spring tides should be used or some mean range when applied in a physical or mathematical model. Another water level change, the storm surge, occurs when a strong wind field approaches the coast. The stress applied to the water surface causes a current to flow toward the coast which cannot escape from a mildly sloped shelf zone. This results in a high water level over a short period which is maximum as the cyclone center passes over the coast. Besides wind shear, the buildup of water level is due to the low atmospheric pressure at the storm center. Such short-term rises are added to the tidal level at the time, which can result in catastrophic flood levels, as occurred on the Dutch coast in 1953. Prediction of this phenomenon is necessary to safeguard facilities and low-lying inland areas.

As waves break at the shoreline they may still be angled to the shore. They therefore contain a longshore component which generates a *littoral current*, as it is called, in the surf zone. In this turbulent region much sediment is thrown into suspension and is therefore carried alongshore by this current and is termed *littoral drift*. This may be accompanied by rip currents running normal to the shore, which have their own particular influence on dispersement of sediment.

Beyond the breaker line the oscillatory motion of water particles at the bed due to waves causes sediment to be put into pseudosuspension. The accompanying mass transport, which is maximum at the bed, moves this partially suspended material along the coast, where it oscillates see-saw fashion, essentially on one contour level of the bed profile. This action occurs with decreasing rates to the limit of wave influence. This depth limit increases with wave period but can encompass the complete shelf of oceanic margins, as noted earlier.

The sand of the beach and the dune area are also affected by wind. Material is blown from the berm toward the first dune and can in fact construct a foredune in front of it. If this is not eroded by storm waves over a number of years, it may become the larger first dune. The rate of transport by wind is thus important in the management of the coastal margin, but will not be dealt with in this tome.

The mechanism by which swell or storm waves move sediment normal to the beach involves both hydraulic and geomechanic principles. These two wave systems have opposite influences, the former accreting the berm, and the latter displacing it offshore in the form of a bar. In this sense storm waves place material, which was previously out of reach of water motions while in the berm, into circulation where swell or more moderate waves move it toward the beach. During this process sand is carried alongshore if the swell waves are arriving obliquely. It is this persistent obliquity which creates the problem of beach erosion or silting, because natural or man-induced interruptions to sediment supply create lateral movements in the resulting waterline, of short- and long-term duration.

The causes of these fluctuations must be understood by the coastal engineer if he is to succeed in any stabilization program. What happens on any section of coast is dictated by what has or could occur in the upcoast region from which sediment is being supplied. Equally, personnel in control of downcoast regions should be informed of plans that may impede transport of sand to their region of responsibility. The coast must therefore be viewed macroscopically or over reasonably long lengths. This is difficult sometimes due to the proliferation of local authorities in charge of small segments of coastline, each working to safeguard their territory, at the expense of adjoining zones.

Nature has had control of vast expanses of shoreline over geologic time, in fact all of it until man came into the picture relatively recently. It is pertinent to observe how she has maintained an equilibrium or relatively constant margin over centuries. The indications are that accretion has taken place over millions of years which has resulted in 75% of the land area of the globe. The mode of accumulation gives an insight into future growth or even denudation of these margins, remembering that climates change over time and hence so does the capacity of rivers to supply sediment for distribution along the coast.

Various geomorphological features are pronounced along the coastlines of the world and these should be explainable by the various processes in a qualitative fashion if not quantitatively at the present state of the art. For example, bays are sculptured between headlands which have predictable shapes, and barrier beaches are constructed across deep embayments in the coast. Both these features can be of major or minor proportions but explanations should be forthcoming for their occurrence.

As soon as man establishes himself on the coastal margin, by constructing facilities close to the water's edge, he cannot countenance the fluctuations in the waterline that Nature accepts as her mode of operation. He has, therefore, installed many types of structures to protect his possessions, some of which have exacerbated the problem. These attempts have been in ignorance, before the processes have been fully understood. This does not mean that new theories or novel ideas are readily accepted by the coastal fraternity, which is probably the most conservative of the engineering profession. There has been little progress made in marine structures over the past few decades. Some new approaches will be outlined which should be tried out in pilot studies, since verification in hydraulic models is difficult due to scaling effects, particularly those involving sediment movement.

One problem that is ubiquitous is that of silting at river mouths. Sediment moving along a shoreline arrives at such outlets, where water debouches to the sea, and immediately forms a shoal or even a sand spit rising above the waterline. This changes in position as waves continue to act on it, so making navigation a hazard. Depending on supply conditions, the outlet can be completely silted in spite of expensive attempts to keep the passage clear. Here again, observation of Nature should provide an answer to this worldwide problem.

Fluid motion of any sort provides many complex problems. It is a haven for mathematicians and source of work for thousands of scientists and engineers. Exact solutions are possible for some phenomena, but these are generally based on ideal conditions or must encompass empirical coefficients. When applied to prototype conditions, where many other energy inputs are present, further modifications to relationships must be entertained.

Mixing this fluid motion with the additional variable of sediment suspension or rolling at the bed makes the original fluid problem appear simple. The turbulence structure varies and the forces exerted on these mobile particles are difficult to theorize or verify physically. Even when some relationship is developed, it is almost impossible to check against prototype conditions. If successful it may explain some phenomenon for a specific location and for strict input conditions, but not for a site elsewhere with differing environmental variables. It is little wonder that many problems can be hypothesized and not dealt with rigorously by mathematics. This is where engineering begins and science ends since solutions must be found with a degree of intuition. The ocean margin is an exciting place for the thinker as well as the data collector.

1.2 STATE OF BEACH EROSION

As soon as man applies himself to the coast, by building facilities near the waterline, he becomes very conscious of the variabilities of Nature. The extreme application of energy from a wind system in a local storm situation, in the form of storm waves, causes the beach to temporally disappear. This is because so much water is poured onto the *berm*, that almost horizontal margin of sand between the sea and the land, that it subsides and is carried out to sea to accumulate in the form of an *offshore bar*. As noted already, this places this large volume of berm material, which was previously sagfeguarded from wave action, into direct contact with the waves. Milder waves after the storm, in the form of *swell*, then shift the bar back onto the beach, possibly over a few days but sometimes over some months, depending on timing of the strongest swell.

Thus, it was not until man exerted his influence on this dynamic margin that erosion was recognized and presumably had to be countered. However, in dealing with this problem, man has devised structures with little knowledge of the processes involved. Also, as ships have become larger, it has become necessary to dredge deeper and deeper channels into harbors, which interferes with longshore drift much more than physical structures such as groins and breakwaters. Even the seawalls that men built to protect one specific section of coast exerted an influence on beaches further along the shoreline. Although many of these solutions were not wholly satisfactory, they have been copied from one generation to another.

It has only been since the middle of the eighteenth century, when the industrial revolution took place, that man has been able to invest large sums in marine structures. This activity has expanded exponentially and with it the problems of beach erosion. The stage has been reached where some communities in the United States now disallow any further so-called protective devices on their shorelines. However, as man has created most of the problems, it behooves him to find solutions to them. The concept of "letting the lighthouse fall into the sea" is giving up too easily, as equally, shifting the lighthouse landward at great expense is not recognizing man's wonderful ability to find economical solutions when they are required.

1.3 ENGINEERING APPROACH

This is to stabilize the shorelines and keep them in their present positions, without long-term landward movement of the waterline at any place. This is not always possible, as any structure placed on the coast will accumulate material on the updrift side, which must necessarily cause erosion in the downdrift area. This is because swell waves persistently arrive from one oblique direction and constantly move sediment along the coast, which has continued over millions of years and is not likely to stop for many centuries to come. Recognition of this long-term trend should be incorporated into all maritime design programs.

1.3.1 Conventional Methods

Perhaps the first maritime structure to be devised was that of the seawall or rock revetment, in order to protect facilities built near the coast. These came within the ambit of the beach berm, which Nature demands during most storms, or certainly annually. Little was it recognized that waves would reflect from such edifices and so apply a double amount of energy to the seabed, especially adjacent to their faces. The longshore component of energy in this situation was almost doubled, so creating conditions for the collapse of these seawalls. Also downcoast of them reflected waves could add their energy to the incident waves and so accelerate the erosion in that area.

The next structure, the groin, was intended to impede the longshore sediment movement in the surf zone, the edge of which was generally the seaward limit of these impediments. Designers had little knowledge of the extent of this sediment motion so that erosion continued beyond their tips, resulting in deepening and increased slopes of the seabed offshore. In storms this resulted in more material being demanded for the bar, which was carried further seaward by rip currents formed at each groin. The groin field had to be filled again by the drift from upcoast, which resulted in greater erosion downcoast of the complex. Although these structures tended to make the shoreline more normal to the wave direction, which was helpful in decreasing littoral drift, the magnitude and spacing of them did not allow for much reduction. In spite of publications indicating their lack of success, they are still being installed worldwide.

Many other types of drift impediment have been invented of which the offshore breakwater has received the greatest attention in the past few decades. These are virtually seawalls constructed in deeper water than the previous tips of the groins. They have spaces between which permit waves to diffract behind them and so build sediment protuberances or salients which are meant to safeguard the shorelines at specific points. In spite of their expense they have proliferated in many countries. However, the oblique swell, that is present as before, reflects from their seaward faces and ultimately causes structural subsidence. Subsequent storm waves penetrate the gaps and also overtop the breakwaters, causing high water levels between them and the shoreline. This results in a strong rip current which carries accumulated material out to sea, where it is transported alongshore by the complex waves from the interaction of incident and reflected components.

1.3.2 New Concept

A new approach developed by the present authors over some three decades has evolved from seeing how Nature itself has sculptured the shorelines of the world. Sandy coasts are often separated by cliffs or headlands, and when swell waves arrive obliquely to the general coastline they form bays of specific shape. The sediment can still be moving along the coast, bypassing the rocky areas and being transported to the downcoast segments of the bays which are straighter than the curved section in the lee of the upcoast headlands. During this passage of littoral drift the bay is said to be in *dynamic equilibrium*, so long as the upcoast supply of sand remains steady. If it is reduced for any reason over a reasonable length of time, the bay will become more indented, or will recede in the curved portion. Should the supply cease altogether, the waterline will erode back to a limiting shape which is termed *static equilibrium*. For a given wave obliquity, the shape of this static equilibrium bay is predictable. This can then be compared to an existing plan form to check the stability of the coast, or the amount of erosion that can ensue should the drift from upcoast become zero.

This testing method can be applied to predict shorelines between new headlands which may be inserted in order to retain sand on certain segments of the coast where it is required. The bays so formed are much larger than the small compartments formed between groins, which are pseudo-bays. The swell approach direction is very persistent on any section of coast and hence its obliquity is reasonably constant. The new concept is a combination of the groin and the offshore breakwater, but the spacing alongshore and seaward is such as to create long lengths of curved beach. There is no better coastal defense than a beach since protective bars can be formed naturally during storms within the geometric domain of the embayment and dismantled later when they are not needed. This method of stabilization has been termed *headland control* and receives much attention in this book for reasons that will become apparent.

Waves

The major energy input of the sea to sediment transport is that due to waves and currents, which in turn is determined by the atmosphere in the form of winds and tides, respectively. These sources are continually available around the oceans of the world but more concentrated in some regions than others. The waves that are generated by winds within storm centers propagate outside them to travel vast distances, in some cases over two oceans, before being dissipated on a distant shoreline. These dispersing waves are known as swell which, although milder in nature than *storm waves*, are more persistent and therefore are important for predicting sediment phenomena. The direction of these predominant swells can be related to the shaping of the shorelines around the world (see Chapter 4).

This chapter is not meant to be a detailed study of wave phenomena, as this book concentrates on sediment motion. However, some knowledge of the generating process is needed in order to determine the wave climate for an area. The forecasting formulae for storm waves and swell have been developed since World War II, with a concerted effort in the 1970s. The water-particle orbital motions due to waves can be solved to several orders of accuracy, but the first order or linear theory is accepted in many cases, especially regarding sediment problems. As waves propagate across the continental shelf before arriving at a beach, they undergo various transformations such as refraction or diffraction, which influences their height. In the very shallower depths even the stable swell waves steepen and break, creating what is termed a *surf zone*. If waves strike an impermeable object such as a wall or breakwater, with or without breaking, they can reflect or overtop such a structure. Many of these phenomena will be outlined briefly, in order to provide background in the study of movable material.

Although some workers believe tidal currents are the major energy input to sediment motion in the ocean, these are restricted to deeper waters where wave motion is substantially reduced. Even there the currents at the bed, which are only a small fraction of those at the surface, have to be a certain value before they can be effective. However, the waves themselves can generate currents within the surf zone if arriving obliquely and are thus extremely important in the longshore transport of material. Another lateral current associated with waves is the net movement per wave cycle, in the direction of propagation for progressive waves. This is termed *mass transport*, which is maximum at the bed, but becomes negative farther up the water column, where only suspended sediment may exist. Other currents, emanating from storms, are due to wind stress on the ocean surface, which results in a seaward velocity at the bed close to the beach that is instrumental in removing berm material and forming it into an offshore bar. These landward currents at the surface also effect higher water levels in what are known as *storm surge*, which topic will not be discussed in this presentation.

2.1 WAVE GENERATION

No waves are produced unless there is a wind and hence the characteristics of this air motion should be looked at first. The developing waves can be considered to be in a storm area and are therefore termed *storm waves*, which are continually breaking. Mariners measure the force of the wind by the percentage of white caps on the sea surface, because waves are not growing if breaking is not occurring. Storms are caused by depressions in the atmospheric pressure, even though very modest in extent, which causes radial flow to this "low" region. Differential drag of the earth's surface, the sea in this case, causes this inflowing mass to have a tangential component which varies with distance from the center, and hence the phenomenon is called a *cyclone*. This is most marked in such features occurring in latitudes 5° to 35° which produce *tropical cyclones*. The ones in higher latitudes (25° to 60°) are broken into linear wind directions which change suddenly at fronts where precipitation is concentrated. These are termed *extra-tropical cyclones*, where centers can only be defined by *isobars*, lines of equal atmospheric pressure, drawn for surrounding regions.

Storm passages in the vicinity of, or passing across, a coast vary in frequency geographically, but are in the region of 1 to 12 times per annum. The strong winds from these may last but a few days, so that storm waves on a coast are of relatively short duration. The swell, emanating from storms a long distance away, may arrive almost continuously as there is bound to be a storm somewhere in the adjoining ocean margin. It will be seen that cyclones originate and move within specific latitudes at different times of the year, resulting in swell with rather persistent directions for different seasons of the year. Knowledge of their approach orientations in the vicinity of any coast is important to the coastal engineer concerned with sediment movement in any specific area.

2.1.1 Wind

For waves to be formed there must be a wind, but the speed of this air motion has to be defined at some specific height above the sea surface. This is because the velocity increases from some shear value to many times this, as measured some tens of meters from still water level (*SWL*). The effect on the water is twofold: that of current generation due to shear stress, and that of wave generation due to pressure pulsations. Wherever there is fluid motion there will be turbulence established by vortices within the fluid, or air body in this case. Turbulons of air are transported in the direction of the main wind vector, but also move upward and downward above the surface of the sea. These transient pressure applications to the water cause undulations in this surface which ultimately form waves. The associated current on the surface also influences water in close proximity to the surface to some minor depth.

The depression in the water surface due to the presence of a turbulon upon it is minute, but if it is maintained for a reasonable period with a speed that of the undulations it has created, so it will grow. This can be likened to a movable air pressure nozzle over a flume of water, as depicted in Fig. 2.1, where the magnitude of the air flow creates a certain wave length. If the nozzle is stationary no wave is created, and if it moves swiftly the depression is annulled. But should it move at the speed commensurate with the wave length or period in that depth of water, the wave will grow in height. One restriction applicable to this mechanism is that the wave length is very much smaller than the distance over which the eddy is applied on the water surface. Turbulence in the air is three-dimensional so that water depressions are random both spatially and temporally. Thus, waves are generated in many directions, with those traveling with the speed of the turbulon—or some component of it—growing in height. The critical angle of resonance with the wind vector is greater the smaller the wave period, the larger ones being in line with the wind direction.

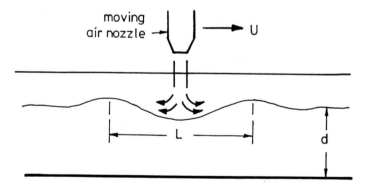

Figure 2.1   Equivalent wave generation by moving air jet

Within the boundary layer on the water surface, the wind gives a finite shear velocity, which is generally designated as U_*. However, velocities U above this level have subscripts in terms of distance in meters above *SWL* where the value is taken, for example U_{10} represents the velocity 10 meters above the sea surface. This is the value accepted for most marine activities, another common one being $U_{19.5}$ since many anemometers on large ships are located at this elevation. Wind speeds can be measured in meters per second, kilometers per hour, miles per hour, or knots (= nautical mile per hour), which are related as follows:

$$1 \text{ m/sec} = 3.6 \text{ km/hr} = 2.24 \text{ miles/hr} = 1.94 \text{ knots}$$

Wind profile. The wind profile accepted by Pierson (1964) was that proposed by Sheppard (1958) for open sea conditions, namely

$$U_y/U_{10} = 1 + (C_{10}^{1/2} \ln y/10)/k \tag{2.1}$$

where U_y is velocity for y meters above the sea surface, C_{10} is a drag coefficient that varies with U_{10}, and k is von Kárman's constant (= 0.4). The shear stress at the sea surface due to the wind is given by

$$\tau = \rho_a U_*^2 \tag{2.2}$$

where ρ_a is the density of the air and U_* is the shear velocity near the water surface. This can be expressed as

$$\tau = \rho_a C_y U_y^2 \tag{2.3}$$

where C_y is the resistance coefficient which varies with U_y. This is influenced by the roughness of the sea in terms of waves being produced which, as will be seen, is dependent on the distance over which the wind has blown, or the *fetch*.

The values of C_{10} as in Eq. (2.1) have been derived by Sheppard (1958) and Wilson (1960), and summarized by Wu (1969). Silvester (1974a) has graphed several results and suggested a curve

$$C_{10} = 0.00065 \, U_{10}^{1/2} \tag{2.4}$$

for U_{10} below 15 m/s with a constant value of $C_{10} = 0.0024$ for $U_{10} \geq 15$ m/s. Substitution in Eq. (2.1) gives relative values of U_{10} and $U_{19.5}$, as in Fig. 2.2. Values at other heights are obtainable from Eq. (2.1).

Wind shear. The currents generated by wind at the sea surface are due to shear and net movement forward of water particles in their orbital motion due to the waves. This latter is known as *mass transport*, which has a maximum value both at the surface and at the bed. Wu (1971) has shown that at the commencement of the wind blowing or for small fetches, the resulting current is mainly that due to shear stress. However, as *fetch* or *duration* increases, the mass transport becomes a greater proportion of the current. Also, in terms of the fraction of the wind velocity, the mass transport increases, while the wind drift decreases, as the fetch increases. The ratio between the total surface drift and the wind velocity decreases gradually as the fetch increases and approaches a constant value of about 3.5% at very long fetches.

Extra-tropical cyclones. It is worthwhile to examine the origin of winds that generate waves of great intensity, which occur in regions of low atmospheric pressure. Low in this sense implies only a reduction of 3.0% from the normal. A value of twice this amount would result in the severest storm possible. Air rushes in radially to this point

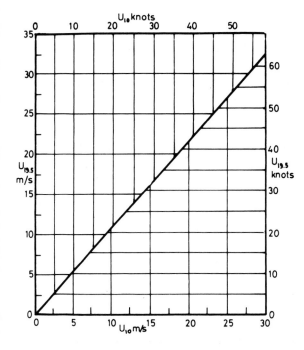

Figure 2.2 Matching wind velocities at 10 and 19.5 m above the sea surface

and rises but, due to the rotational speed of the earth, that nearest to the sea surface is sheared differentially by zones nearer the equator. This twists the radial inflow to cause a clockwise rotation in the Southern hemisphere and anti-clockwise in the Northern, which action is termed the *coriolis force*.

Weather charts are produced in which points of equal pressure are joined as *isobars*, the wind velocities then being predicted by their spacing. The closer they are the stronger the wind. The alignment of the isobars provides the vector of the wind at 600 m from the sea surface, which is beyond the frictional level of the earth. This is called the *geostrophic wind* whose velocity is given by

$$U_{gs}/g = (0.52/\sin\theta)\Delta p/\Delta n \qquad (2.5)$$

where θ = latitude in degrees, Δp = pressure below normal in millibars (34 millibars = 1 inch of mercury, normal atmosphere pressure = 1013 millibars), Δn = spacing of isobars in degrees of latitude, U_{gs} = wind velocity commensurate with g (acceleration due to gravity).

The wind velocity required for wave generation is the value near the sea surface, known as the *gradient wind*. It is derived by allowing for latitude and curvature of the isobars by

$$U_{gr} = U_{gs} \pm (U_{gr}^2/2\omega\, r \sin\theta) \qquad (2.6)$$

where U_{gr} = gradient wind velocity in knots, U_{gs} = geostrophic wind in knots, ω = angular velocity of the earth (= 0.262 rad/hr), r = radius of curvature of isobars in nautical miles (= 6,080 ft = 1,854 m = 1 sec longitude), and θ = latitude in degrees. The minus sign is used for low-pressure centers or *cyclones* and positive for higher pressure or *anticyclones*. For straight isobars $U_{gr} = U_{gs}$. Besides these variables the gradient wind varies with the difference in temperature between air and sea; colder air produces higher wind at the surface. This is difficult to apply as temperatures are not generally known at specific times so that only seasonal values can be assumed. The wind direction at the sea surface differs from that of the isobar spacing or the geostrophic wind. For extra-tropical cyclones it deflects 15° on average toward the center. This must be known when computing waves from a given storm sequence.

The air circulation in a low-pressure center consists of batches of warm and cool air which constitute *fronts* that rotate as depicted in Fig. 2.3 for the Southern hemisphere. At these cold, warm, or occluded fronts the isobars change direction swiftly, as do the winds. These fronts rotate at the speed of the gradient wind at that radius, so causing a curvature in the front, as exhibited in the figure. The center as a whole moves across the ocean, producing mobile fetches of varying lengths which makes wave forecasting from synoptic weather maps extremely difficult. An alternative is to accept measurements from anemometers on ships if there are sufficient numbers of these in the area of concern. Waves generated by one wind vector may pass into an area where the velocity or orientation changes. It is better to concentrate on the portion of the cyclone, that nearest the center, where high winds occur. Those moving in the same direction as the center as a whole will produce the highest waves. In the Southern hemisphere this is on the left of the center and opposite in the Northern hemisphere.

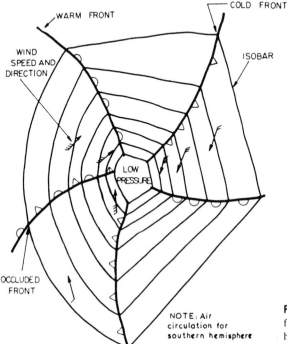

Figure 2.3 Typical isobar pattern for extra-tropical cyclone in Southern hemisphere

Anticyclones.   These are high-pressure centers with outward flowing air with a resultant rotation opposite to that of the cyclone mentioned above. These winds are milder and no cold or other fronts are involved, so that the widely spaced isobars can not be used for predicting wind velocities. These centers change in latitude throughout the year like the cyclones and move from west to east in the same manner. They provide the fine weather. During the hot summer period, the differential heating of the contiguous land mass causes air to rise and a sea breeze to form across the coast. These can have velocities of up to 20 knots and affect 80 km of the adjacent margin. They generate waves in a short choppy sea which can break on the coast obliquely and so transport sediment within a narrow surf zone. They may also act in concert with the longer period swell waves arriving at the coast, but their duration is much shorter.

Tropical cyclones.   These low-pressure centers, which occur between latitudes 5°–35° in both hemispheres, have unique wind structures that can produce some of the highest feasible velocities. They are known by different names geographically, being *hurricanes* on the east coast of the Americas, typhoons in the western Pacific, monsoons in the Indian Ocean, and plane tropical cyclones in Australia. The magnitude of the area covered by strong winds varies within any region and from one geographic location to another. The tropical version differs from the extra-tropical by the characteristics listed in Table 2.1.

TABLE 2.1   CHARACTERISTICS OF CYCLONIC WIND STRUCTURES

Tropical	Extra-tropical
Occur in low latitudes 5°–35°	Occur between 30°–60°
Isobars nearly circular	Isobars mainly straight
No cold fronts present	Many fronts circulate
Pressure gradients intensive	Moderate gradients exist
Wind speeds 20–60 m/s near center	Winds rarely exceed 15 m/s
Center is circulation calm zone (eye)	Center not readily defined
Precipitation intense overall	Rain moderate at cold fronts
Heat budget energy source (summer)	Pressure source (winter)
Center travels at 10 to 40 knots	Center travels about 30 knots
Normally N–S movement	Directions usually W–E

Like most natural phenomena, tropical cyclones have a life cycle of immature, mature, and waning. Even at their greatest intensity they differ in stature so that averages should be replaced by maxima in any specific class. The coastal engineer should examine historic intensities within the region of a port under study. Typical wind structures for the three major classes are shown in Fig. 2.4. Wind directions in (a) and (b) of this figure are shown for the Northern hemisphere where these particular versions occur; that in (c) is for the Southern hemisphere or more particularly the Australian version. Radial profiles of wind velocities are illustrated in Fig. 2.5, where it is seen that maximum speeds around the eye are of the same order even though the overall radii of wind action differs geographically (Syono 1963).

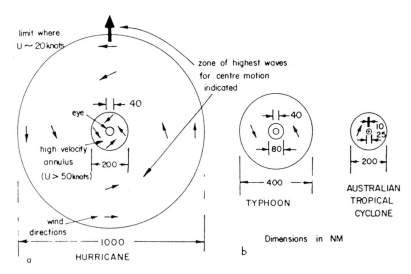

Figure 2.4   Plan of sea surface wind conditions in: (a) hurricane, (b) typhoon, (c) smaller tropical cyclone

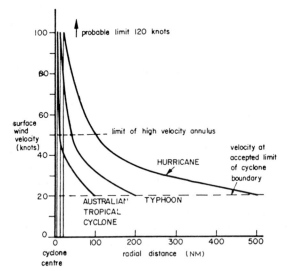

Figure 2.5   Radial wind distributions in tropical cyclones (Syono 1963)

The eye of a cyclone, where air is drawn up in a strong vortex motion, is visible as a black spot in the cloud structure from satellites. Within this eye surface winds are absent and the atmospheric pressure minimum, but at its edge winds suddenly rise to the maximum and then reduce radially as in Fig. 2.5. The diameter of the eye varies between classes of cyclone as noted in Fig. 2.4. Storm intensity is gauged by the product of eye radius (R) and difference in central and normal atmospheric pressure (Δp), to be discussed in Section 2.2.5. An important feature of these winds, as far as wave generation is concerned, is their direction. As seen in Fig. 2.4, they are 45° to the radius at the eye, but become tangential as the edge of the cyclone is reached. The centers themselves are in motion, which adds to the wind speed on one side, where larger waves are generated. The side differs in each hemisphere, and is important when the passage of a cyclone is being traced toward a particular section of coast. In this regard it is very difficult to predict the path of a tropical cyclone, even hours ahead, since slight changes in pressure around its boundary can change its motion (Renard and Levings 1969). The zone of high winds toward the coast also produces a storm surge, whereas on the other side of the tropical cyclone water is dragged from the coast to produce a transient low level.

Seasonal patterns.   The repetition of cyclonic centers over an ocean region, especially the paths followed, is of great importance in determining a wave climate for a particular section of coast. They should be studied for all seasons and can then be recorded in *cyclonicity charts*, which are either numbers of, or hours' duration of, centers occurring in 1° or 5° squares of latitude and longitude. These are obtained from regular synoptic weather charts, after which isopleths are drawn through squares of equal frequencies. Such a chart is depicted in Fig. 2.6, where the upper portion shows the number of centers per month, and the lower the direction of center motion. The general path of these tropical cyclones can be ascertained but, as mentioned above, the actual movement of any particular one is very unpredictable, so the worst case should always be accounted for. Other relevant data are normally recorded, such as lowest pressure experienced, radius of eye, and speed of travel.

Wind patterns are presented in certain atlases covering the oceans, which have resulted from storms and global circulations. Winter and summer conditions are generally given, but monthly values are also available (*Monthly Meteorological Charts*). Global wind vectors are shown in Fig. 2.7, plus the paths of major types of cyclone. In the same group of monthly charts are recorded the number of storm or swell observations from mariners within 5° squares of latitude/longitude over many years. Wave directions are provided in 45° bands, with increments of wave height categorized as in Fig. 2.8. Although only qualitative data on wave height, and none on wave period are available, the direction of swell waves can be extremely useful in sedimentological studies (considered in Chapter 4). Besides the above information, the zones of high winds or storm regions indicate the source of major swell. Such a presentation as in Fig. 2.9 highlights the concentration of cyclones moving from W to E in latitudes 40°–60° in both hemispheres. These produce almost continuous swell on the western margins of continents.

2.1.2 Mechanisms of Generation

Waves are undulations of the sea surface comprising vertical motions of water particles in such a sequence as to form crests and troughs that move across the surface. The difference in the heights of crest and trough are termed the *wave height*, and the horizontal distance between consecutive crests or troughs is known as *wave length*. The time between successive crests passing any point is the *wave period*. In deep water the water particles

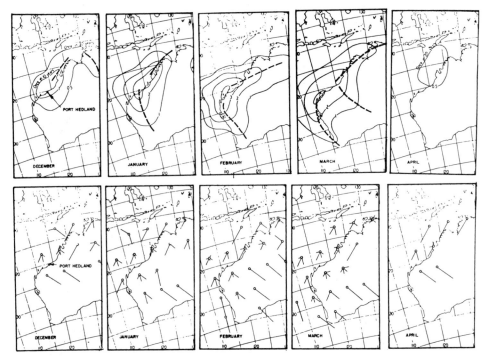

Figure 2.6 Cyclonicity chart for tropical cyclones covering 38 years over the northwest coast of Western Australia (modified from Brunt and Hogan 1956)

rotate in circular motion, the diameter of which at the sea surface is the wave height. The wave period (T) is related to the wave length in deep-water (L_o) by

$$L_o = gT^2/2\pi \qquad (2.7)$$

where g is acceleration due to gravity.

As a wind blows it simultaneously produces waves of many periods or lengths plus heights. Whereas the longer waves are slow in building up, the shorter period components increase swiftly to a point of instability when they break. This is aided by the longer waves, once they are formed, by reducing the lengths of the shorter waves at their crests (Longuet-Higgins and Stewart 1960, and Wu 1971) and breaking them on their forward face. This is one method of transferring energy from these readily generated short components to waves of greater length, thus increasing their height. As the fetch and duration of the wind increases so the total wave energy increases until saturation occurs, when the breaking absorbs all wind input, with wave heights and periods reaching their limit. This is termed a *fully arisen sea* and will be denoted by *FAS*. Wave characteristics are then specific to the steady wind velocity, not now being related to fetch or duration.

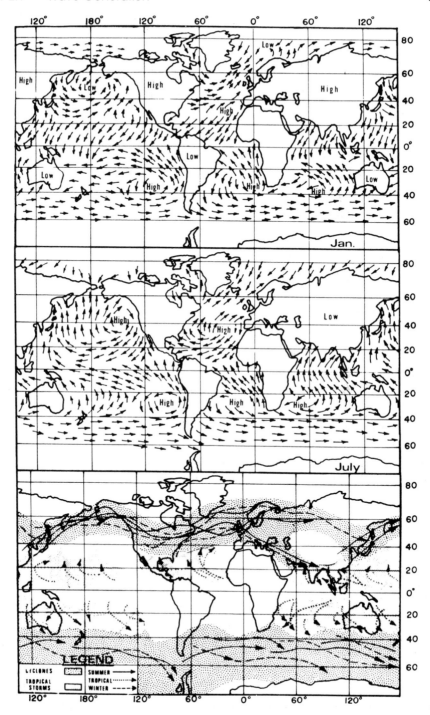

Figure 2.7   Global distribution of winds for months of January and July plus paths of tropical and extra-tropical cyclones

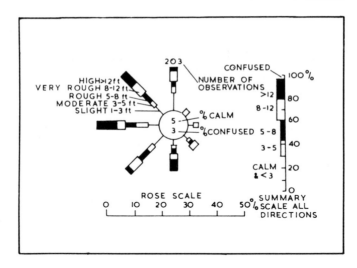

Figure 2.8 Typical wave rose in British Admiralty sea and swell charts

The four stages of wave generation, which are taking place simultaneously, have been discussed by Silvester (1974a) and many other researchers, so will be outlined very briefly here. These mechanisms are *resonance, shear flow, sheltering effect*, and *breaking*.

Phillips (1957) made his significant contribution to oceanography by proposing resonance from the random fluctuations of pressure on the sea surface as the origin of wave motion. These pressure pulses had to be applied for a reasonable time, while moving across the body of water at a speed corresponding to the celerity of the enlarging waves. Smaller turbulons of air closer to the surface travel slower than larger ones with their centers further from it. But waves of all periods could be formed concurrently, the smaller growing more swiftly than the larger, as already mentioned. These pulsations do not necessarily travel in the same direction as the general wind but can be angled to it, those most oblique produce the shorter period waves while the longest waves are generated nearly parallel to it. This resonance theory applies only to very small wave heights and so does not account for air-sea coupling. After this initiation other wave generating processes take over. The reader is referred to Kinsman (1965) for a detailed description of this phenomenon.

At the same time as Phillips published his theory, Miles (1957) proposed his shear flow model, based on a logarithmic wind-velocity profile. Although excluding turbulent fluctuations it involved pressures normal to the sea surface, since drag forces and drift currents were ignored. It assumes an initially disturbed surface which, by streamline flow of the air above it, produces regions of suction and pressure that are out of phase with the wave. This causes a force to be applied at the back of the crest and hence help the downward motion of the water surface. This aids wave buildup, but the theory becomes inapplicable once the streamlines break away from the wave crests. Some other mechanisms must then explain wave growth. Stewart (1961) has furthered this work.

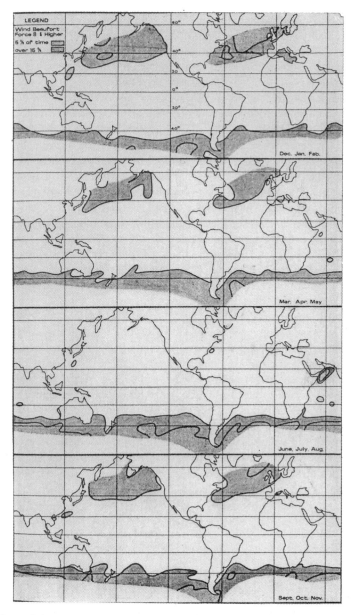

Figure 2.9 Global distribution of high wind velocity zones for quarters of the year

The previous two theories complement each other, since resonance from air turbulence provides initial undulations over a wide spectrum of periods, while shear flow selectively promotes the growth of shorter waves. Resonance causes a linear buildup of wave energy, whereas later coupling of the wind produces exponential growth. Waves with celerities close to the wind velocity (or $C/U \rightarrow 0.8$) may reach their limiting steepness (H/L) for the theory solely on resonance. The shorter waves ($C \ll U$) will pass through their linear growth to exponential via a transition before reaching this limit. Phillips and Katz (1961) have computed the times to the transition point for waves of various period. Longuet-Higgins et al. (1963) have measured waves by a buoy and concluded that 90% of the air pulsations were coupled to the waves and only 10% to general wind turbulence.

Waves continue to grow beyond the conditions predicted by Miles by a mechanism that might be termed a sheltering effect. This implies a suction at the front and pressure on the back of waves already formed, by which Jeffreys (1925) explained the complete generation process, and for which Stanton (1937) provided experimental evidence. Only pressures normal to the surface were considered and tangential shear ignored. Since the motion of water particles in the crest are in the wind direction, it would appear that any drag force applied to them would enhance orbital diameter and hence wave height. Hino (1966) combined the equations of Phillips and Miles without using any empirical factors. He showed that at a FAS the percentage of pressure to total drag on waves approaches 100%, whereas at the upwind end of the fetch it is only 50%. This is understandable as sea roughness increases along the fetch, so permitting a better grip of the wind on the waves. A steady state is reached when the energy input is consumed by wave breaking, turbulence, various other effects, and drift current.

Breaking is an integral part of wave generation which is considered by many researchers to be a dissipating mechanism. However, in the act of breaking, as depicted in Fig. 2.10, momentum could be transferred to orbital motion of surface water particles. This breaking on the crest of the longer wave is enhanced by the wind velocity, but as soon as the celerity of this carrying wave equals that of the wind velocity ($C/U = 1.0$), this process decreases and hence energy addition ceases. Longuet-Higgins (1969) has provided a theoretical basis for the transfer of energy from shorter to longer waves, noting that: "But when the short waves are forced to decay strongly by breaking on the forward slopes of the long waves the gain of energy by the latter is greatly increased. Calculations suggest that the mechanism is capable of imparting energy to sea waves at the rate observed."

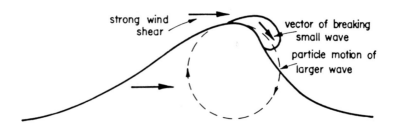

Figure 2.10 Possible momentum transfer from breaking short wave to longer-period wave

The recording of waves at some set point in the ocean can be analyzed to give the energy in each component generally measured in terms of frequency (f). These can then be graphed as a spectrum, as in Fig. 2.11, where energy is measured as square of the amplitude of the component with frequency f. The zones of major energy input

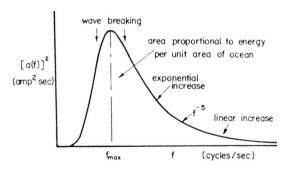

$[a(f)]^2$
$(amp^2\ sec)$

wave breaking

area proportional to energy
per unit area of ocean

exponential
increase

f^{-5}

linear increase

f_{max} f (cycles/sec)

Figure 2.11 Typical energy—frequency spectrum for ocean waves

2.1.3 Wave Characteristics

Waves that are still being generated or maintained by the wind are termed *storm waves*, while those that have left this zone and disperse across the oceans are called *swell*. Each have their distinct characteristics and have quite different influences on sediment motion at the coast.

Storm waves.   These undulations of the sea surface are very complex, consisting of waves of many lengths and heights, traveling in a variety of directions. The resulting major waves are therefore steep as they comprise enhanced small waves breaking at their crests. This action also makes the steeper waves asymmetrical or steep-fronted. These characteristics make for ready breaking of storm waves so that shoaling or contra currents can easily dissipate them. The multidirectional nature of storm waves is depicted in Fig. 2.12, where at the upwind end short waves are predominant, being quite angled to the wind. Further down the fetch the longer and larger waves are more aligned with the wind but not completely so. This brings out the importance of the fetch width in the generation process, because the oblique short waves have to be within the confines of the wind to reach instability and break before adding to the longer waves. A very narrow fetch would not permit this, except in a model flume or fiord where reflection occurs. The directional distribution of wave energy is schematized in Fig. 2.13, where average curves of $\cos^2\theta$ and $\cos^4\theta$ for short and long waves apply, respectively. As seen, some energy is directed almost at right angles to the wind vector either side.

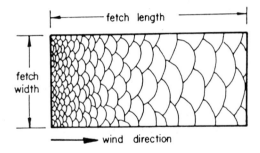

fetch length

fetch width

wind direction

Figure 2.12 Typical crest pattern of waves along a fetch

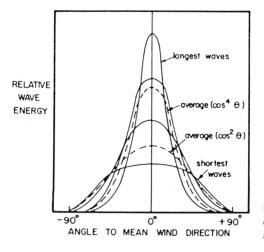

Figure 2.13 Directional energy distribution in waves of various periods for a specific wind velocity

When measuring waves the length of the record should be limited, in order that conditions for one specific wind speed are obtained for that location within the fetch. The heights of the wave trains could be graphed as a histogram, as in Fig. 2.14, for each incremental value in period (ΔT). If $\Delta T \rightarrow 0$ the curve represents the distribution of energy, which differs from the frequency distribution as in Fig. 2.11, because of the lack of a high-frequency tail. This energy distribution curve can be termed a *spectrum* as much as the frequency representation.

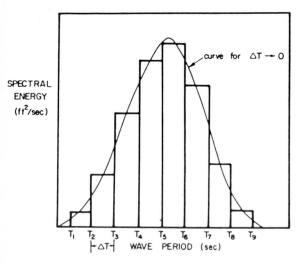

Figure 2.14 Histogram of wave energy for increments of wave period

As the wind blows along the fetch, as in the inset of Fig. 2.15, a steady spectrum will be reached at F_1 which is smaller than at F_2. That at F_3 will be larger, with that at F_4 the maximum for the specific wind velocity. This is the saturation limit or fully arisen sea so that the spectrum at F_5 is the same as for F_4. The shapes of these spectra under these fetch-controlled situations are similar. From the commencement of the wind

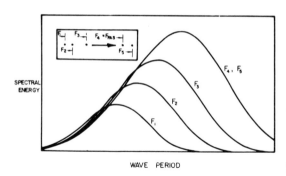

Figure 2.15 Energy-period spectrum for points along the length of a fetch

it takes time before waves with a given fetch to reach their limit. Thus, each fetch has some minimum period before a steady state is reached, so that the fetch curves in Fig. 2.15 can each have a duration (t) related to it. The fully arisen sea fetch F_{FAS} has a minimum duration t_{min} before it exists.

Swell waves.   Waves of any specific period within a fetch will have some direction where they are optimum, as seen in Fig. 2.13. Once generated these components continue in the same orientation as they pass out of the storm zone boundary. These swell waves propagate across the ocean in great circles, or straight lines on most map projections. In the region downwind of the fetch or *dispersal area*, as seen in Fig. 2.16, waves spread out circumferentially and radially from the end of the fetch and even to the sides somewhat. The bulk of the energy is concentrated within radii 30° to the wind vector. The proportion of the FAS energy reaching any point downwind is dictated by the angle contained between radii from the fetch boundary to the said point. The further distant this point the smaller the angle and hence the less energy available from the storm. Thus, the fetch width assumes importance in the size of swell waves available in the dispersal area, sometimes called the *decay area*, which is a misnomer, since wave energy is not being attenuated but only spread out. The greater the distance the less the number of waves that can interact and hence a reduction in the wave height.

Waves within the fetch could remain steady for some time once the minimum duration is reached at any given point; however, in the dispersal area swell waves will be varying continually as the longer waves reach any point first, followed by the middle periods with greater heights, and followed then by the smaller waves if they ever reach there. This is depicted by the wave spectra at different points within the dispersal area in Fig. 2.16. The area under the spectral curves represents the statistical wave height of $H_{1/3}$ or average of the third highest waves, used frequently in coastal engineering design. In the triangular region, adjacent to the downwind fetch extremity, all wave components can be traversing together so the sea is as complex as in the fetch itself, except that waves will not be breaking.

Swell waves are low in height compared to their length, or their wave steepness (H/L) is small. This causes them to be sinusoidal in shape and therefore very stable. As they pass across the oceans the orbital motion of the water particles associated with them

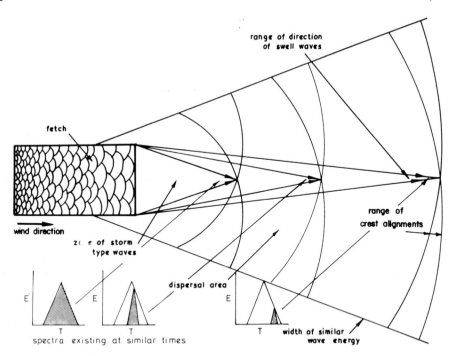

Figure 2.16   Wave spectra and propagation in the dispersal area

does not reach the bed, so they suffer very little loss of energy. Snodgrass et al. (1966) recorded waves at five points across the Pacific Ocean from the "roaring forties" (40°–50° S) to Alaska. This will be discussed further in Sections 3.2.2. One further feature of swell waves is the constancy of their direction at points on a coast. Even though fetches may spread over time along one latitude, the swell emanating therefrom is concentrated in a narrow band of orientations, especially when it is refracted or twisted due to shoaling depths across a continental shelf before reaching a shoreline.

2.2 WAVE FORECASTING

The prediction of ocean waves never really got off the ground, or off the water as it were, until the requirements of beach landings in World War II demanded this knowledge. Its development took place in two camps, one in the United States headed by Sverdrup and Munk (1947), who visually observed waves and related winds to the distinctly larger waves from which "significant height" and "significant period" evolved. It was fortuitous that later, as these water oscillations were studied statistically, the "significant" values applied to the average of the third highest waves ($H_{1/3}$).

The second group was based in the United Kingdom and led by Longuet-Higgins (1952, 1963) who analyzed wave records on a harmonic analyser from which he derived theories utilizing the sound wave theory of Rayleigh (1880) and electronic circuit noise as studied by Rice (1944–1945) and Eckart (1953). The UK data were put into predictive form by Darbyshire (1952).

After Longuet-Higgins (1952) derived statistics for a narrow spectrum, Neumann (1953) in New York derived theoretically an equation that represented a spectrum of waves, using the significant height and period as previously used. Pierson et al. (1955) then employed these spectra to derive tables and graphs which were the backbone of forecasting for the next decade. Many comparisons of these various formulae exhibited disparities that were difficult to explain, mainly because researchers were employing wind velocities at different heights above the sea surface. Pierson (1964) attended this matter by examining the wind profile in general. By adjusting the formulae to wind height measurements that were similar, he was able to reduce the divergencies.

At the same time Moskowitz (1964) undertook a comprehensive spectral analysis of data obtained from ship-borne wave recorders (Tucker 1956). This modified the spectral form previously used by Pierson et al. (1955) and for the FAS was in better agreement with other formulae. Pierson and Moskowitz (1964) then combined the wind profiles of Pierson (1964) and dimensionless spectral shape as suggested by Kitaigorodskii (1961). This produced the PM spectrum for the FAS, which is universally acknowledged as the best for this condition.

When the fetch or duration of the wind is insufficient to produce the maximum waves for a given velocity, it is said to be a *developing sea*. It is necessary to know the wave characteristics during these limiting conditions because an area of sea or lake may not be large enough for FAS conditions to exist, especially for higher wind velocities. The wave characteristics can readily be related to those for the FAS with fractions of the FAS fetch (F_{FAS}) or FAS duration (t_{FAS}) used to define them. This requires that F_{FAS} or t_{FAS} be known for any given wind velocity which can be put in terms of a dimensionless fetch which has a known value for FAS. The t_{FAS} derived from the time that the maximum waves along sections of fetch take to arrive at the point where FAS conditions exist. It is generally related to F_{FAS} which is then used for further proportioning of wave heights and periods.

One major study of developing waves was that by Hasselmann et al. (1973) who recorded waves along a fetch from Denmark with offshore winds. They found that the spectrum was peaked at the frequency appropriate to the fetch. This peakedness decreased as FAS was approached, so that at this condition it reverted to the PM spectrum. In the same year Toba (1973) presented a different spectral version from measurements, both in a model flume and in the field. Comparisons of these two formulae have been made by several workers. Another worker in this field is Mitsuyasu (1973, 1975a, 1975b) has produced papers on waves in limited fetch. He has developed forecasting formulae which differ slightly from those discussed above.

Many of the presentations are based on the frequency spectrum in which spectral energy is graphed as ordinate against wave frequency as abscissa. This introduces a frequency (f_m) at which maximum energy is concentrated. When formulae are transposed to period (reciprocal of frequency) the T_m so derived is not exactly the reciprocal of f_m because of the differentiation of the spectral curve and equating to zero for the peak value of f_m and T_m. The various formulae will be retained in frequency form to simply the mathematics. The area under the spectral curve is equal to $(H_{1/3}/4)^2$ in any form it is plotted.

The spectra for developing seas have constants that give peak values and spread from these that differ slightly from each other, so it is a matter of choosing those that fit best the data obtained from wave recorders. A choice will be made for the most ideal shape of the spectrum, which can then be simplified to give triangles, rising from a lower period (T_L) up to the peak energy point at T_{max} and then falling to the upper period value of T_U. Values of T_L/T_{max} and T_U/T_{max} then define the spectrum once T_{max} is derived from the forecasting formula.

The need to forecast design wave conditions for deep and shallow water is obvious, particularly at a place near the shore where sediment movement is paramount. Waves reaching any point may derive from two sources, that generated in deep water and then propagated shoreward, and that generated locally.

Design wave conditions for a project site, offshore or nearshore, can be determined from wave records or wave forecasting. The former approach requires local wave records to be analyzed either by statistical theory or using a wave energy spectrum, while the latter is based on empirical relationships developed elsewhere, from correlating wind conditions and waves. Meteorological information, such as synoptic weather charts, is utilized to provide wind data and hence wave conditions. Waves generated from tropical cyclones or hurricanes should also be considered.

The approach of wave forecasting, which is based on previous results of wave observations, through dimensional analysis, is commonly presented in graphical form for direct estimation of *significant wave height* and *significant wave period* under a given wind condition (wind speed, fetch, and duration). These are exemplified in Bretschneider (1958) and also collectively in *Shore Protection Manual* (SPM 1984). Three different models have been classified, these being (1) wave growth in fetch-limited conditions in deep water, (2) fully arisen sea in deep water, and (3) fetch-limited wave growth in shallow water. Although these three models remain in widespread use today, Vincent and Resio (1990) and Hurdle and Stive (1989) have found them to be inconsistent, and have proposed an alternative. On the other hand, although the term significant wave height has been used in both the statistical and spectral approach, Thompson and Vincent (1985) have commented on the discrepancies obtained for the three different models mentioned above, and suggested practical engineering methods for interrelating them in the shallow water via a wave energy spectrum.

In the following sections, wave statistics and empirical prediction methods are discussed, followed by wave spectra developed for the FAS in deep water. The TMA spectrum, which claimed to be useful in all water depths, including that in the surf zone (Hughes 1984), is examined. Waves from tropical cyclones are also considered, as well as waves in the dispersal area. Graphical presentations are provided for ready application.

2.2.1 Wave Statistics

The random nature of waves as recorded is illustrated in Fig. 2.17, where wave crests can occur both above the MSL line as drawn and below it. Equally troughs can appear

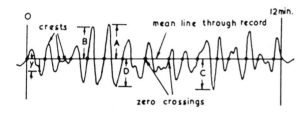

Figure 2.17 Wave record showing random wave profiles and notations used in wave statistics

above this mean line. It is not until this record of limited duration, generally 20 to 25 minutes, is analyzed statistically that any worthwhile information can be obtained. With wave generation, just as it is impossible to have a steady wind over an infinite duration, it is also impossible to treat a record of infinite length because conditions would have changed. But this almost spot measurement in time should give the same statistical result for any other produced by the same environmental input.

The first approach made in this field was by Sverdrup and Munk (1947) in the United States, who made visual observations, picking out the high waves which they termed *significant waves*. It was fortunate that these heights (H_s) correlated very well with the later statistical value of the highest third of the waves ($H_{1/3}$). Also the average period of water uprising through SWL (T_z = length of record/number of uprisings) approximated the significant period (T_s). The height at which wind speed was measured was not specified. Ten years later Munk (1957) stated: "I think the SMB method, at least in its earlier SM version, deserves retirement. It was first used in 1942 during the invasion of North Africa and published in 1947, and it is amazing that it should have survived so long. Its accomplishments were to organize multitudes of scattered data into a few dimensionally consistent empirical relations."

The next attempt at statistical analysis was by Longuet-Higgins (1952), who employed a harmonic analyzer on ship-borne wave records and applied theory applicable to sound waves and noise in electronic circuits, as mentioned earlier in this section. As crest-to-trough wave heights were not amenable to mathematical manipulation, measurements were made from SWL to the free surface (a). These were taken at points equidistant along the wave record of which there were N in number. A root mean square value (a_{rms}) was then calculated as:

$$a_{rms} = \left(\sum a^2/N \right)^{1/2} \tag{2.8}$$

Assuming a Rayleigh probability distribution, which is strictly applicable to swell waves only, values of $H_{1/3}$, $H_{1/10}$, and H_{ave} were related to a_{rms}.

Cartwright and Longuet-Higgins (1956) then examined the statistical distribution of crest heights above or below SWL, which permitted analysis of the FAS. This was found to depend on a_{rms} and $\mathcal{E}$, a measure of the spectral width, given by

$$\mathcal{E}^2 = 1 - (N_z/N_c)^{1/2} \tag{2.9}$$

where N_z is the number of uprisings through SWL, and N_c the number of crests in the limited length of record. For a single swell wave $N_z = N_c$ and therefore $\mathcal{E} = 0$. If on a single major wave many short waves were running then $N_c/N_z \rightarrow \infty$ and $\mathcal{E} \rightarrow 1.0$. This is a hypothetical situation, so that for FAS the value of ϵ tends to 0.8.

The ratios of $H_{1/n}/a_{rms}$ as noted earlier by Longuet-Higgins (1952) could be determined as in Table 2.2, where it is seen that $H_{1/3}/a_{rms}$ reaches 4.17 for FAS. A value of 4 is normally accepted so that:

$$\text{variance} = a_{rms}^2 = \sum a^2/N = (H_{1/3}/4)^2 \qquad (2.10)$$

It can also be seen in Table 2.2 that H_{ave} decreases from the swell condition to FAS which appears anomalous. Although higher major waves are experienced, many other smaller waves accompany them, which brings down the average. Thus, H_{ave} is of little engineering significance.

TABLE 2.2 WAVE HEIGHT PARAMETERS FROM SPECTRAL WIDTH (ϵ)

ϵ	0	0.2	0.4	0.6	0.8
$H_{1/3}/a_{rms}$	4.00	4.03	4.10	4.17	4.17
$H_{1/10}/a_{rms}$	5.09	5.09	5.24	5.45	5.73
H_{ave}/a_{rms}	2.51	2.51	2.45	2.26	1.63

Tucker (1957, 1961) has developed relationships for determining a_{rms} from a wave record as depicted in Fig. 2.17, in which A = height of the highest crest above SWL, B = second highest crest, C = lowest trough, and D = second lowest trough. By obtaining $H_1 = A + C$, and $H_1 = B + D$, the curves as in Fig. 2.18 can be used knowing N_z. Either curve should give the same value of a_{rms}. This result equates to a measurement of $H_{1/3}$ in the record by using Eq. (2.10).

2.2.2 Wave Forecasting for FAS

The first acceptable relationships for wave forecasting known as the *SMB* method resulted from Sverdrup and Munk (1947) and Bretschneider (1958), from which the initials SMB derived. Neumann (1953) derived an energy spectrum from visual observations of waves, from which the FAS was first recognized. From this the PNJ method evolved (Pierson et al. 1955). Later Pierson (1964) recognized the need to specify the height at which wind velocities were recorded, while at the same time Moskowitz et al. (1962) and Moskowitz (1964) analyzed 460 ship-borne wave records and found only 52 applied to FAS. From these two approaches, a spectral form was proposed for the FAS (Pierson and Moskowitz 1964) which became the highly acclaimed *PM* method of forecasting. Hasselmann et al. (1973) measured waves in a fetch limited sea to show peaking energies prior to the FAS being reached. These spectra are identical in shape, which Silvester and Vongvisessomjai (1978) put in period form as distinct from the frequency presentations normally used.

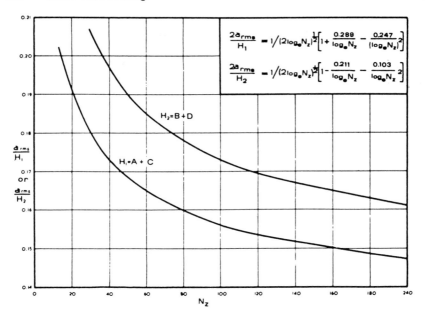

Figure 2.18 Derivation of a_{rms} from H_1, H_2, and N_Z, defined by Tucker (1957, 1961)

The PM spectrum for deepwater waves, when put in the frequency mode, is:

$$S_{H^2}(f) = \frac{\alpha g^2}{(2\pi)^4 f^5} \exp\left[-\beta\left(\frac{g}{2\pi U_{19.5} f}\right)^4\right] \tag{2.11}$$

where constants $\alpha = 0.0081$, and $\beta = 0.74$, $U_{19.5}$ is the wind velocity at 19.5 m above SWL, and g is acceleration due to gravity. Equation (2.11) is graphed in Fig. 2.19 for $U_{19.5}$ from 20 to 45 knots (i.e., from 10 to 23 m/s approximately). The changing frequency of the peaks should be noted, plus the substantial increase in area under the curve ($= (H_{1/3}/4)^2$) as the wind velocity grows. The distribution can be put in period form, for which the equation is

$$S_{H^2}(T) = \frac{\alpha g^2 T^3}{(2\pi)^4} \exp\left[-\beta\left(\frac{gT}{2\pi U_{19.5}}\right)^4\right] \tag{2.12}$$

Equation (2.12) is graphed in Fig. 2.20 for the same conditions as in Fig. 2.19, in which the ordinate is in m²/sec or ft²/sec. However, it seems that the spectra in period mode are more triangular in form, lacking the high-frequency tail.

The frequency f_{max} or T_{max} at the peak of these spectra can be obtained by differentiating Equations (2.11) and (2.12) and equating to zero, thus:

$$g/(2\pi U_{19.5} f_{max}) = (5/4\beta)^{1/4} = 1.14 \tag{2.13}$$

and

$$g T_{max}/(2\pi U_{19.5}) = (3/4\beta)^{1/4} = 1.00 \tag{2.14}$$

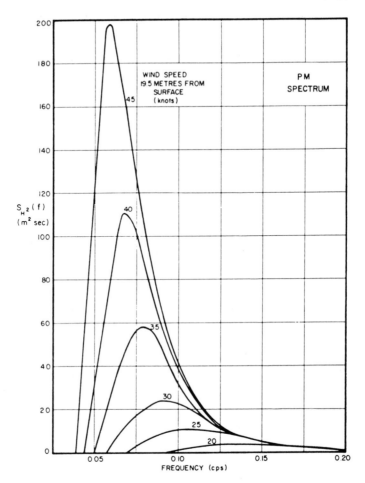

Figure 2.19 Pierson-Moskowitz (PM) frequency spectra for a range of wind velocities ($U_{19.5}$)

The round figure of 1.00 in Eq. (2.14) has no significance as it occurs only for $U_{19.5}$ in knots. Dividing Eq. (2.14) by (2.13) gives:

$$f_{max} T_{max} = 0.88 \qquad (2.15)$$

which constant of proportion should always be utilized if frequency spectra are to be transposed into period terms. This has not been observed in the literature by these authors as $T_{max} = 1/f_{max}$ is always employed, which gives slight deviations from measured data. Using the relationship given in Eq. (2.14) results in:

$$T_{max} = 2\pi U_{19.5}/g = (U_{19.5}/3) \qquad (2.16)$$

for T_{max} in seconds and $U_{19.5}$ in knots for the bracketed expression, which is an easy figure to remember, because for a 30 knots wind $T_{max} = 10s$.

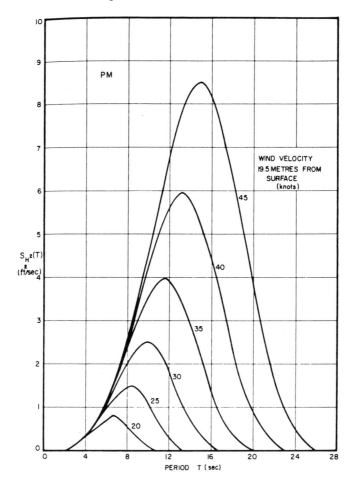

Figure 2.20 Pierson-Moskowitz (PM) period spectra for a range of wind velocities

By integrating Eq. (2.12) from 0 to ∞ the area under the spectral wave gives

$$\text{area} = (H_{1/3}/4)^2 = \alpha(U_{19.5})^4/4\beta g^2 \tag{2.17}$$

from which:

$$H_{1/3} = 2(\alpha/\beta)^{1/2}(U_{19.5})^2/g = 0.00564(U_{19.5})^2 \tag{2.18}$$

for $H_{1/3}$ in meters and $U_{19.5}$ in knots. Substitution of Eq. (2.16) into Eq. (2.18) gives

$$T_{\max} = 4.4(H_{1/3})^{1/2} \tag{2.19}$$

for $H_{1/3}$ in meters. As seen, Eq. (2.19) is independent of wind velocity but appears only for the FAS condition. The constant has been accepted as a little lower than the 4.4 shown.

The variables in Eq. (2.12) can be nondimensionalized by introducing wave age $(C/U_{19.5})$, so that:

$$S_{H^2}(T)g/U_{19.5}^3 = (\alpha/2\pi)(C/U_{19.5})^3 \exp[-\beta(C/U_{19.5})^4] \qquad (2.20)$$

which is graphed in Fig. 2.21, where a triangular shape with the same area under the curve, or equivalent $H_{1/3}$, is drawn. The apexes of both the original curve and the triangle are located at $C/U_{19.5} = 1.00$. From the figure the upper (T_U) and lower (T_L) limits of the triangular base are apparent, with the following ratios:

$$T_U/T_{\max} = 1.62; \quad T_L/T_{\max} = 0.35; \quad (T_U - T_L)/T_{\max} = 1.27 \qquad (2.21)$$

Thus, the bounds of periods in this equivalent FAS spectrum can be defined with negligible energy outside them. As wind velocity increases so does $T_{\max}$ and hence T_U and T_L, as also the total width $T_U - T_L$.

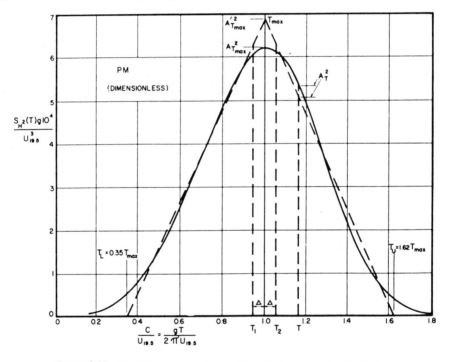

Figure 2.21   The PM period spectrum, with equivalent triangular distribution

The amplitude (A_T) of any wave band of specific period can be formed, which will be found useful in computing wave heights in the dispersal area in Section 2.2.6. The peak amplitude squared $(A_T)_{\max}^2$ at $T_{\max}$ has been shown by Silvester (1974a) to be:

$$(A_T)_{\max} = 0.0155(T_{\max})^{3/2} \quad \text{or} \quad (H_T)_{\max} = 0.031(T_{\max})^{3/2} \qquad (2.22)$$

for $(H_T)_{\max}$ in m/s$^{1/2}$. Knowing that $T_{\max} \doteq U_{19.5}/3$, then

$$(H_T)_{\max} = 0.0059(U_{19.5})^{3/2} \qquad (2.23)$$

for $H_{1/3}$ in m and $U_{19.5}$ in knots. The equivalent triangle has a peak whose value is:

$$(A'_T)_{max} = 0.0163(T_{max})^{3/2} \text{ or } (H'_T)_{max} = 0.0325(T_{max})^{3/2} \qquad (2.24)$$

which when converted becomes:

$$(H'_T)_{max} = 0.00616(U_{19.5})^{3/2} \qquad (2.25)$$

for $H'_{T\,max}$ in m and $U_{19.5}$ in knots. All of which naturally are slightly greater than those in Eqs. (2.21) and (2.22). Values for any intermediate period T are given by

$$(A_T/(A'_T)_{max})^2 = (T_U - T)/(T_U - T_{max}) \text{ for } T > T_{max} \qquad (2.26)$$

and

$$(A_T/(A'_T)_{max})^2 = (T - T_L)(T_{max} - T_L) \text{ for } T < T_{max} \qquad (2.27)$$

It is important to know the fetch for FAS to learn if sufficient is available for wave saturation. This has been computed by Silvester and Vongvisessomjai (1978) by considering the JONSWAP spectrum (Hasselmann et al. 1973) plus the PM spectrum for FAS (Pierson and Moskowitz 1964). The difference between these is that JONSWAP indicates a peak enhancement factor added to the right of Eq. (2.11) of:

$$\gamma \exp\left[-(f - f_m)^2/(2\sigma^2 f_m^2)\right] \qquad (2.28)$$

where γ is illustrated in Fig. 2.22, and the value of σ varies with that of σ_a and σ_b either side of the peak frequency f_{max}. As illustrated in the figure, factor γ is the ratio of peak energy (S_J) of JONSWAP to that of PM spectrum (S_{PM}) occurring at that particular fetch as dictated by α. The widths (σ) of this essentially triangular addition apply to midheight so the additional spectral area approximates $S_{PM}(\gamma - 1)(\sigma_a + \sigma_b)$.

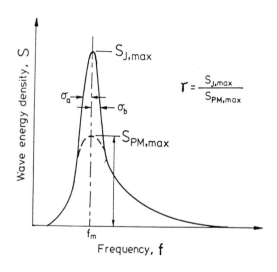

Figure 2.22 Definition sketch of JONSWAP spectrum

The variables in these spectra can be put in the form of three dimensionless relationships:

$$f_{\max} U_{10}/g = 3.5(g F/U_{10}^2)^{-0.33}(= \nu) \tag{2.29}$$

$$g H_{1/3}/U_{10}^2 = 0.0016(g F/U_{10}^2)^{0.5}(= 4\epsilon^{1/2}) \tag{2.30}$$

and

$$\alpha = 0.076(g F/U_{10}^2)^{-0.22} \tag{2.31}$$

This entails shifting from $U_{19.5}$ to U_{10} by $U_{19.5} = 1.07 \, U_{10}$. The extra terms ν and ϵ will be discussed later. This latter term ϵ should not be confused with that used in Eq. (2.9).

Silvester and Vongvisessomjai (1970, 1971) used a fetch ratio F/F_{FAS} for conditions other than FAS conditions, but any comparison with JONSWAP must be based on the dimensionless ratio $g F/U_{10}^2$ for FAS. This implies that beyond F_{FAS} wave growth is exceptionally weak. Bretschneider (1959) accepted that $g F_{FAS}/U_{10}^2 = 6 \times 10^5$ due to the fact that $g H_{1/3}/U_{10}^2$ increased very slightly from 10^4 to 6×10^5. To these authors it would appear that a value closer to 10^4 might have been more reasonable. Mitsuyasu (1975a, 1975b) reported a FAS value of 1.33×10^4.

The exponents in Eq. (2.29), (2.30) and (2.31) were those adopted by Hasselmann et al. (1973), others having been presented by Mitsuyasu (1975a, 1975b) and Liu (1971), but will not be discussed here. Choosing $f_{\max}$, because the exponent of 0.33 was common among all workers, the FAS values of $f_{\max} U_{10}/g$ for the PM spectrum, obtainable from Eq. (2.13), is 0.13 which requires $g F_{FAS}/U_{10}^2$ in Eq. (2.18) to be 1.95×10^4, for the specific constant and exponent used. When this is substituted into Eq. (2.30), $g H_{1/3}/U_{10}^2$ is 0.223, which is comparable to 0.24 obtained from Eq. (2.18). Substitution in Eq. (2.31) results in $\alpha = 0.0086$ which is commensurate with 0.0081, the accepted value for FAS. Thus, the value of $g F_{FAS}/U_{10}^2$ of around 20,000 could be used with confidence.

Comparison with Hurdle and Stive (1989) for limiting fetch in deep water was obtained from a value of duration limit t_{lim} substituted into related fetch relationships. This resulted for $U_{10} = 15$ m/s their value of $F_{FAS} = 531$ km compared to that from the above analysis of 447 km, a difference of around 10%. Using $U_{10} = 17.5$ m/s, or around 35 knots, the respective fetches are 584 and 609 km, with the alternate greater with a difference of 4%.

The shape of the spectra, particularly those based on period rather than frequency, are essentially the same throughout development because of the extra peak in small fetch conditions, as found by Hasselmann et al. (1973). No additional energy is available so that the area in this peak will be at the expense of the higher and lower sides of the bulk of the spectrum. It would appear that this peak rises uniformly from the existing high frequency bands already present so no extraction of spectral area should come from this side of the $f_{\max}$ value. Hasselmann et al. (1976) derived a shape parameter $\lambda = \epsilon \nu^4/\alpha$, referring to the variables in Eqs. (2.29) and (2.30). They noted that this parameter varied with the fetch parameter to about a 1/10th power, which was "not discernible within the scatter of the JONSWAP data." If in Eq. (2.29) $f_{\max} U_{10}/g^2$ varies with ν, then from Eq. (2.13) it is seen to vary with $(\beta)^{1/4}$, and with $g H_{1/3}/U_{10}^2$ varying with $4\epsilon^{1/2}$, then Eq. (2.18) shows it to vary with $(\alpha/\beta)^{1/2}$. The shape factor λ thus varies with $(\alpha/\beta)(\beta^{1/4})^4/\alpha$, or independent of all variables. Thus, the spectral shape in a developing sea should be the same as at FAS.

2.2.3 Average Spectra

Values of $H_{1/3}$ and T_{max} can become very high if winds around 25 m/s develop a FAS. Fortunately this is not generally the case, for when they occur in some meteorological complex they are normally associated with short fetches or short durations or both. Moskowitz (1964) in his selection of wave records with their related fetches refers to Walden (1963) and stated: "The fetch required for lighter winds (up to about 30 knots) to produce fully developed seas occurs frequently in most areas. Wind speeds greater that 30 knots are rarely associated with fetches great enough to produce fully developed seas."

   Using the dimensionless relationships of Eqs. (2.29) and (2.30), and accepting $gF_{FAS}/U_{10}^2 = 20,000$, values for $H_{1/3}$ and T_{max} ($= 0.88/f_{max}$) can be derived for acceptable limitations on fetch for higher wind velocities as in Table 2.3. There it is seen that fetches have been reduced in decrements of 50 kms as U_{10} increases from 15 to 30 m/s from 550 km at $U_{10} = 15$. The optimum $H_{1/3}$ for $U_{10} = 30$ is seen to be 9.7 m rather than 20.7 for a FAS. Equally, T_{max} is 12.6 s instead of 20.6 s predicted for this condition. In fact, T_{max} appears to have reached a limit of around 12 s for waves in any storm situation. Silvester (1974a) had previously suggested limits of $H_{1/3} = 10$ m and $T_{max} = 13$ s using different F_{FAS} values.

TABLE 2.3   HEIGHTS AND PERIODS OF WAVES FROM PROBABLE MAXIMUM FETCHES

U_{10} m/s	15	20	25	30
F_{FAS} kms	459	816	1275	1835
F_{act} kms	550	500	450	400
$F_{act}/F_{FAS}\%$	100	61	35	22
$(H_{FAS})_{1/3}$ m	5.2	9.2	14.4	20.7
$H/H_{1/3}\%$	100	78	59	47
$H_{1/3}$ m	5.2	7.2	8.5	9.7
$(T_{FAS})_{max}$ s	10.3	13.8	17.2	20.6
$T/T_{FAS}\%$	100	85	71	61
T_{max} s	10.3	11.7	12.2	12.6
$H_{1/3}$ (Scott)	5.5	7.6	10.0	12.7
$H_{1/3}$ (Silvester)	5.8	7.6	9.6	11.4

   Scott (1968) has analyzed wave data from a number of sources covering the Irish Sea and the Atlantic Ocean. He suggested a relationship which can be converted to:

$$H_{1/3} = 0.073 \, (U_{10})^{3/2} + 1.52 \qquad (2.32)$$

with $H_{1/3}$ in meters and U_{10} in m/s. These give values as listed in Table 2.3. Silvester (1974a) has suggested:

$$H_{1/3} = 0.384\ U_{10} \tag{2.33}$$

for similar dimensions as in Eq. (2.32), which give the results also listed in Table 2.3. It is seen that the empirical choice of fetches matches the observed data of Scott (1968) reasonably well. From also deriving T_{max} from the spectra inspected by Scott (1968), Silvester (1974a) plotted T_{max} versus $H_{1/3}$ which matched curves from FAS to limited fetch covering the various sources.

Mayeçon (1969) analyzed visual wave records from 500 ships in the North Atlantic over the years 1953 to 1961. The main ocean area was subdivided into 11 zones, but for the present purposes can be combined into 4 latitudinal bands. From the probability curves covering each zone, the following annual values of wave height were obtained:

Latitude (degrees)	20–30	30–40	40–50	50–60
$H_{1/3}$ (m)	7.9	10.7	12.2	13.7

Thom (1971) has applied an extreme value distribution to wave height measured at 3 h and 1 h intervals on 12 vessels over 7 to 10 years located between latitudes 30° to 50° in the north Atlantic and Pacific Oceans. The average $H_{1/3}$ for the 12 stations with mean recurrence interval of 2 years was 10 m. For intervals of 10, 25, and 50 years these averages were 13, 15, and 17 m respectively.

2.2.4 Wave Forecasting for Finite Depths

Many forecasting relationships have been developed, notably in the form of wave spectra and empirical formulae. For wave generation in shallow water by wind, wave growth is limited by water depth. Separate formulations have been proposed for this purpose, which are termed Pierson-Newmann-James method (*PNJ*, Pierson et al. 1955), and Sverdrup-Munk-Bretschneider method (*SMB*, Bretschneider 1958, 1977). Toba (1973) and Mit-suyasu (1973, 1975) have also proposed working formulae for this.

Based on wave energy approach and supported by numerous empirical data, *Shore Protection Manual* (SPM 1984) has given a simplified method for estimating wave conditions (H_{m0}, T_m) in deep water, as well as for shallow water, for a given wind speed and fetch or duration, where H_{m0} is the energy-based significant wave height (IAHR-PIANC 1986), and T_m is the wave period at the spectral peak (with $T_{1/3} = 0.95\ T_m$). It has been suggested that this is justifiable if cost and time are limited. Together with the requirement for water of finite depth, three sets of predictive formulae and several graphical presentations are produced in terms of dimensional parameters for ready application. A total of ten figures are given in SPM (1984), with each for a specific water depth from 1.5 m to 15 m in increments of 1.5 m. However, it should be noted that no wave breaking is assumed in these water depths, even at 1.5 m. Commenting on their use, Vincent and Resio (1990) state that: "The curves may be used to obtain a quick estimate, but the use of the wave equation is the preferred approach." The relationships presented above are not applicable to broken waves or wave in the surf zone.

Wave prediction formulae suggested by SPM (1984) for the fetch-limited cases under nearly constant wind speed U_A in the depth considered are in terms of dimensionless fetch parameter gF/U_A^2, from which wave height parameter gH_{m0}/U_A^2 and wave period parameter gT_m/U_A are computed, in addition to limiting duration parameter $gt_{\lim}/U_A$. The reference wind speed U_A is the adjusted wind speed (using procedures outlined in SPM 1984) given by

$$U_A = 0.71(R_T U_{10})^{1.23} \tag{2.34}$$

based on U_{10} in m/s, the wind speed recorded at 10 m height, with coefficient R_T as a factor for adjusting air-sea temperature difference (with default value of 1.1). If storm duration t is less than $t_{\lim}$, then a modified fetch length F' should be calculated and subsequently replace the F in the original parameter gF/U_A^2. On the other hand, if storm duration t is greater than $t_{\lim}$, then the equations for FAS should be applied.

Although the parametric relationships contained in SPM (1984) are still in widespread use today, Hurdle and Stive (1989) comment that: "The Manual presents different equations for three cases, deep water developing seas, deep water fully developed seas and shallow water. Unfortunately the equations presented do not match at the transition points between each of the cases. Further in very shallow water the relationship between critical duration and fetch is also questionable." They also find that "step changes of predicted properties with small changes in the input parameters." An alternative single set of equations is proposed (Hurdle and Stive 1989) to replace the three separate sets given in SPM (1984), hence, a decision about which set of equations should be applied can be removed. The set of revised equations are given by

$$gH_s/U_A^2 = 0.25\tanh[0.6(gd/U_A^2)^{0.75}]$$
$$\tanh^{1/2}\{4.3 \times 10^{-5}(gF/U_A^2)\tanh^{-2}[0.6(gd/U_A^2)^{0.75}]\}$$

$$gT_p/U_A = 8.3\tanh[0.76(gd/U_A^2)^{0.375}] \tag{2.35}$$
$$\tanh^{1/3}\{4.1 \times 10^{-5}(gF/U_A^2)\tanh^{-3}[0.76(gd/U_A^2)^{0.375}]\} \tag{2.36}$$

and

$$gt_{\lim}/U_A = 65.9\ (gF/U_A^2)^{0.667} \tag{2.37}$$

applicable to both deep and shallow water.

Equation (2.35) gives the significant wave height, whereas Eq. (2.36) provides the wave period at the peak spectral energy. The prediction of wave height and period from these revised equations not only matches with the results of SPM (1984) asymptotically but also exhibits a smooth transition between the shallow and deep water conditions, where those of the original SPM (1984) fail, according to Hurdle and Stive (1989). On the relationship of limiting duration proposed in Eq. (2.37), Hurdle and Stive (1989) indicate that: "A slightly modified (more conservative) version of the expression for the limiting duration in deep water is given for situations where the sea state may be limited by the duration of a storm rather than the fetch. It is recommended that this expression should be applied for both deep and shallow water as it is both easier to apply and gives more reliable results." Equations (2.35)–(2.37) are gaphed as Fig. 2.23 for ready application, in which t_e is equivalent to the t_{lim} in Eq. (2.37), denoting *effective* duration.

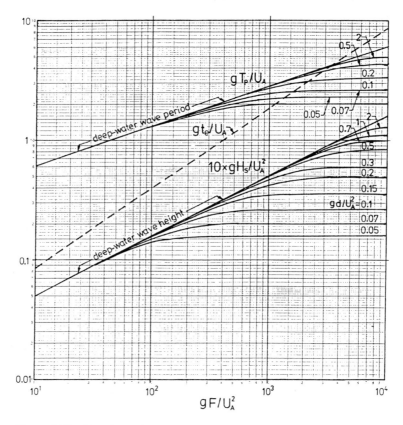

Figure 2.23 Wave forecasting relationships for finite depth based on dimensionless gF/U_A^2 (Hurdle and Stive 1989)

For predicting waves in finite depth of water, from given values of uniform wind speed U_A (in m/s), selected fetch F (in meters), and duration t (in second), Eq. (2.37) should be applied first to calculate the limiting duration (t_{lim}) specified by the given wind velocity and fetch. If the given duration (t) is sufficiently long compared to the limiting duration (t_{lim}) calculated, then wave height and period can be estimated directly from Eqs. (2.35) and (2.36) respectively. In the case of $t < t_{lim}$, the wave conditions to be determined are duration limited. An *effective fetch* (say F_e) should then be calculated from Eq. (2.37) based on the given duration (t), from which the new fetch value replaces that given in Eqs. (2.35) and (2.36).

Recent developments in the wind wave spectrum have produced a frequency spectrum which can be applied from deep to shallow water, even in the surf zone (Vincent 1984). This new form of shallow water spectrum, in self-similar form, is termed the

"TMA spectrum" (Bouws et al. 1985), which is extended from deep-water similarity principles of Kitaigorodskii et al. (1975). This spectrum contains all the historical developments on wave spectra, including Phillips (1958), Pierson and Moskowitz (1964), Hasselmann et al. (1973), and Kitaigorodskii et al. (1975). The TMA spectrum was named after Bouws et al. (1985) by "combining the first three letters of the three data sets used for field verification (Texel, MARSEN, and ARSLOE)."

The TMA spectrum has the final form of

$$S_{TMA}(f, d) = S_P(f)\phi_{PM}(f, f_m)\phi_J(f, f_m, \gamma, \sigma_a, \sigma_b)\phi_K(2\pi f, d) \qquad (2.38)$$

It consists of the product of equilibrium range $[S_P(f)]$ of Phillips (1958) and the shape function $[\phi_{PM}]$ of PM spectrum (Pierson and Moskowitz 1964) for FAS conditions in deep water, which is the PM frequency spectrum given in Eq. (2.11). Multiplying the first two parts on the RHS of Eq. (2.38) with JONSWAP shape function $[\phi_J]$, which was given by Hasselmann et al. (1973), becomes the JONSWAP spectrum for developing seas in deep water. Finally the shallow-water self similar shape function $[\phi_K]$ of Kitaigorodskii et al. (1975) is attached, which eventually forms the TMA spectrum for developing sea in depth-limited conditions (Vincent 1982, 1984, 1985; Hughes 1984; Vincent and Hughes 1985; Bouws et al. 1985; Hughes and Miller 1987). Thus, the complete form of TMA spectrum can be expressed as:

$$S_{TMA}(f) = \alpha g^2 (2\pi)^{-4} f^{-5} \exp[-5/4(f/f_m)^{-4}] \times$$
$$\exp\{ln(\gamma) \exp[-(f - f_m)^2/2\sigma^2 f_m^2]\} \times \phi_K(2\pi f, d) \qquad (2.39)$$

Hughes (1984) indicated that wave length in finite depth water can be sufficiently calculated by linear wave theory. A shallow-water limit of wave celerity $C = (gd)^{1/2}$ was also adopted in the derivation of the TMA spectrum, in which the parametric constants α and γ are allowed to redefine empirically for all water depths using the following relationships:

$$\alpha = 0.0078 \ \kappa^{0.49} \qquad (2.40)$$

$$\gamma = 2.47 \ \kappa^{0.39} \qquad (2.41)$$

where

$$\kappa = (U_{10}^2/g)k_m = (U_{10}^2/g)(2\pi/L_m) \qquad (2.42)$$

in which $k_m = 2\pi/L_m$ is the wave number of the wave at peak frequency f_m, and L_m is the linear wave length at depth d.

Although TMA was claimed by Hughes (1984) as suitable for developing wind waves in all water depths, from deep to shallow (even in the surf zone), Goda (1990) has commented that: "The TMA spectral model is intended for use in wave hindcasting and forecasting in water of finite depth. Various mechanisms of wave attenuation by bottom friction, percolation, breaking, and others are supposed to be included in the function ϕ implicitly. Use of the TMA model in general wave transformation problems in shallow water should be made with caution, because the model is essentially for wind waves at the growing stage."

A graphical example of the TMA spectra in variable depth is drawn by the authors as in Fig. 2.24, using Eq. (2.39) for a uniform wind at $U_A = 20$ m/s over a 500-km fetch on water depths indicated in the figure. Comparison is also made with the JONSWAP spectrum at $d = 50$ m. Unlike other single peak spectra, it is seen that the TMA has the ability of presenting a second peak or a humped feature in spectral shape in shallower depths, which has been observed in both prototype and model conditions for wave growth (Mitsuyasu 1968, 1969; Kamronrithisorn 1978) and even in long-traveled swell (Goda 1983, 1990). It is hoped that, when this method matures in the future, it might serve as a better tool for predicting wind waves in shallow waters. However, no graphical presentation is produced here for application.

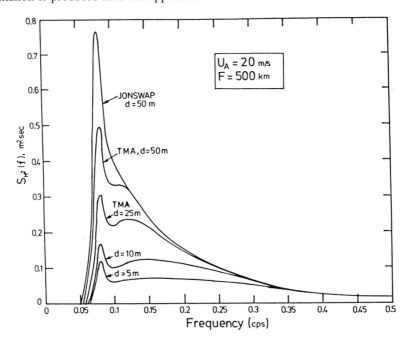

Figure 2.24 TMA frequency spectra for varying water depth

2.2.5 Waves from Tropical Cyclones

As discussed in Section 2.1.1, the wind structure in these cyclones is virtually circular even though the vectors are angled differently from the eye to the outer periphery. This makes the determination of a fetch very difficult for normal prediction formulae to be used. There are many variables involved in the task of forecasting maximum waves in these features, which are not readily verifiable. For example, the radius of the eye can only be ascertained with certainty if it passes directly over a station where personnel are available to record it. Even the pressure at the center can only be extrapolated from values at various radii outside it. The speed of the cyclone can be determined over hours but it can vary very swiftly between such assessments.

The first requirement is to compute the maximum wind speed around the eye at radius R. Bretschneider (1957) used a relationship:

$$U_{max} = 0.868[73(\Delta p)^{1/2} - R(0.575f)] \tag{2.43}$$

where Δp is the difference between the normal atmospheric pressure (= 29.7 inches of mercury, or 1013 mbar, i.e., 1 in. Hg = 34 mbar) and that at the center, R is in nautical miles, and f is the Coriolis parameter varying from 0.2 to 0.3 for latitudes 5° to 30°. Since the second term in Eq. (2.43) is only 5% that of the first, it could be ignored, yielding:

$$U_{max} = 60(\Delta p)^{1/2} \tag{2.44}$$

Of the data presented by Holliday (1969), the equation given by Kraft (1961) appears most suitable:

$$U_{max} = 80(\Delta p)^{1/2} \tag{2.45}$$

for Δp in inches of mercury. This gives a higher U_{max} than Eq. (2.44), it being the maximum sustained wind at the edge of the eye. If a reasonable fetch width is considered, the wind velocity at its outer edge will be around half this U_{max} so that the average could be closer to that of Eq. (2.44), which is preferred. This maximum applies to height 35 m above the sea surface, so that:

$$U_{10} = 0.865\ U_{max} \tag{2.46}$$

Equations (2.44) and (2.46) are graphed in Fig. 2.25.

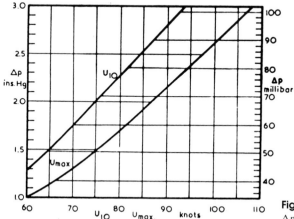

Figure 2.25   Values of U_{max} and U_{10} for Δp in Hg inches or millibars

Young (1988a, 1988b) has developed a parametric model for predicting waves from tropical cyclones, which requires the determination of an equivalent fetch for use in the JONSWAP (Hasselmann et al. 1973) relationship:

$$gH_s/U_{10}^2 = 0.0016(gF/U_{10}^2)^{0.5} \tag{2.47}$$

or its new scaling version:

$$g(H_s)_{max}/U_{max}^2 = 0.0016(gF/U_{max}^2)^{0.5} \tag{2.48}$$

An effective radius (R') is defined as

$$R' = 22{,}500\log_{10} R - 70{,}800 \tag{2.49}$$

where both R' and R are in meters. This is then substituted, together with U_{max} and V_f, into the following equations:

$$F/R' = -0.002175U_{max}^2 + 0.01506U_{max}V_f - 0.1223V_f^2 + 0.219U_{max}$$
$$+ 0.6737V_f + 0.7980 \tag{2.50}$$

where V_f is the forward speed of the cyclone in m/s. To find an equivalent fetch F, Eq. (2.50) is plotted in Fig. 2.26. Inserting F into Eq. (2.48) provides $(H_s)_{max}$. It can be noted that V_f can be too small or too great for a given U_{max} to produce a large F or large $(H_s)_{max}$. If it equals the group wave velocity of the largest waves, these will remain in the high-intensity wind field longer. For $U_{max} \geq 20$ m/s a $V_f = 5$ m/s gives the highest F/R' and hence largest H_s from Eq. (2.48), making $V_f/U_{max} = 0.25$. The spectral peak period T_{max} can be computed from:

$$gT_{max}/U_{max} = 0.25(gF/U_{max}^2)^{0.33} \tag{2.51}$$

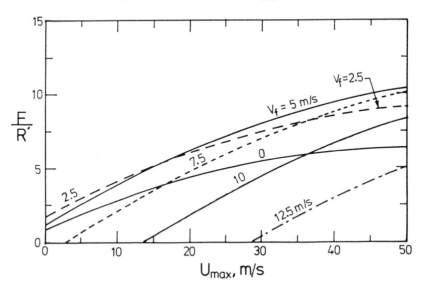

Figure 2.26   Values of F/R' versus U_{max} for a range of V_f

The waves predicted above are being generated around the eye, particularly those propagating in the direction of center movement. At greater radii they are reduced in a manner as given by the *Shore Protection Manual* (1984), from which Fig. 2.27 has been derived. This gives height ratio H_r/H_R for a range of r/R and comprises 3 curves

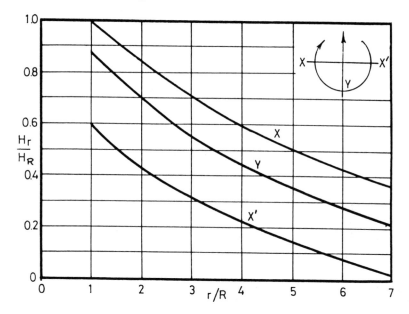

Figure 2.27 Ratio of wave height at radius r to that at eye radius (R) for alignments parallel and normal to the cyclone vector

representing axes as in the inset. Young (1988a) supplied many plan forms of H_r/H_R for various U_{max} and V_f values, but Fig. 2.27 should suffice for most purposes.

A plan and radial wind distribution in a hurricane is shown in Fig. 2.28, even though the air rotation is shown for the southern hemisphere. It should be compared with Figs. 2.4 and 2.5 for tropical cyclones in other geographical regions.

2.2.6 Waves in Dispersal Areas

As noted in Fig. 2.16 and also in the inset of Fig. 2.29, the swell arriving at some point on the centerline of the fetch must come from within the fan enclosed by angle θ. All other waves will bypass this point. The further from the fetch (D) or the smaller the fetch width (W), the smaller this angle and hence the less energy arriving. The ratio of the energy along this axis $(H_{1/3})^2$ to that at the downwind end of the fetch is shown dotted in Fig. 2.29 against D/W. This can be compared to the energy reduction as recorded by Snodgrass et al. (1966) for waves traversing the Pacific Ocean, as given by the full lines.

The curve in Fig. 2.29 is based on a $\cos^4 \theta$ energy distribution which applies to waves where $C/U \geq 1.0$. The shorter period components have a less peaked angular distribution of $\cos^2 \theta$. Integration across the distribution from $-90°$ to $+90°$ results in the curve of Fig. 2.30. By drawing the two arcs from the side boundaries of the fetch to the point in question, as in Fig. 2.29, measure the half angles $(\theta/2)$ either side of the axis, which are then marked on Fig. 2.30 to give two percentages of the energy $(H_{1/3})^2$. The difference between these is the percentage of the energy at the end of the fetch. If a point is angled α to one side of the centerline the angles from each edge of the fetch

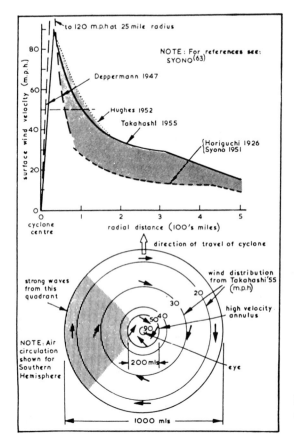

Figure 2.28 Fetches with wind distribution in tropical cyclones (Syono 1963)

will differ, as seen in Fig. 2.31, dictated by α and D/W. As α in Fig. 2.31 increases, it becomes the angle θ in Fig. 2.30 from which θ_1 and θ_2 are measured either side to give the energy differential. Where this occurs on the curved segments in Fig. 2.30, these percentages will decrease, becoming minimal around 30° from the fetch centerline. The $\cos^2 \theta$ trace should be used for T_L to $T_{\max}/2$ and $\cos^4 \theta$ for $T_{\max}/2$ to T_U (Silvester and Vongvisessomjai 1971).

Besides this circumferential spreading the waves disperse radially, the higher period waves arriving before the median and lower period components of the spectrum. To derive the band width present at any time it is necessary to know the duration time (t_e) in excess of that required for FAS. Since 95% of the FAS value of $H_{1/3}$ and $T_{\max}$ is produced in 80% of t_{FAS} (Silvester and Vongvisessomjai 1970), it is more reasonable to assume t_e after $t_{95} = 0.80_{FAS}$ even though $(H_{FAS})_{1/3}$ is used for the height calculation.

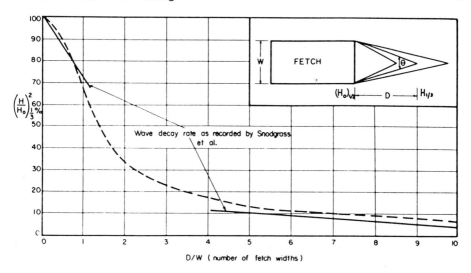

Figure 2.29   Energy reduction with distance directly downwind of a fetch

Each wave component travels at its group velocity, which for deep water is $C_o/2 = gT/4\pi$. At a given distance (S) from the downwind end of the fetch, at some time t after t_{95}, the period T_1 of the train just arriving is given by

$$t = S/1.52T_1 \qquad (2.52)$$

where t is time in hours; S is distance in nautical miles (1 NM = 1854 m), and T_1 is wave period in seconds. The period T_1' of the train just departing, which is a faster train travelling the same distance over a smaller time $t - t_e$, is given by

$$t - t_e = S/1.52T_1' \qquad (2.53)$$

Thus the band width present is $T_1' - T_1$, depicted in Fig. 2.32, encompassing a certain proportion of the energy existing at the end of the fetch. This may or may not be FAS, as illustrated by equations for $(H_{T\,\text{max}})^2$ and $(H_{T\,\text{max}}')^2$ in the figure. To facilitate the calculation of this proportion, the cumulative area curve is produced in Fig. 2.32, the utilization of which requires T_1 and T_1' to be expressed as a fraction of either T_{max} or T_U. If T_1' exceeds T_U $(= 1.62T_{\text{max}})$, then only the triangular portion of the equivalent spectrum with a base $T_U - T_1$ is used.

The spectrum of Fig. 2.32 has a dimensionless ordinate $(H/H_{T\,\text{max}}')^2$ so that when $T_1' - T_1 \to 0$, due to $t_e/t \to 0$, or S being large (making t large in comparison to t_e), the height of the single period swell wave can be determined. Equations (2.51) and (2.52) have been graphed in Fig. 2.33 for ready assessment of T_1 and T_1'. Should T_1 so derived exceed T_U, then no waves have yet arrived at the site. If T_1' is less than T_L then all energy of engineering significance has already passed.

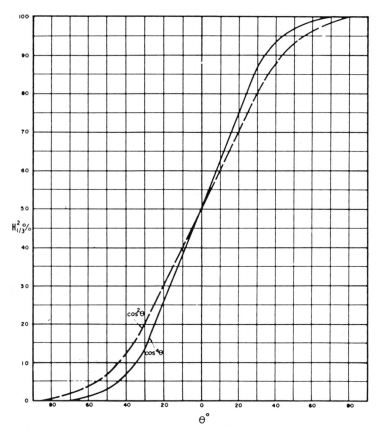

Figure 2.30 Angular distribution of wave energy according to $\cos^2\theta$ or $\cos^4\theta$

2.3 WAVE KINEMATICS

As has been noted, the swell wave with its small steepness (H/L) is sinusoidal in form, for which the mathematics of profiles and orbital motions is quite simple, known as *linear theory*. However, as it reaches shoaler water its crest increases at the expense of the trough so that these characteristics become nonlinear, or have added terms in exponential series that describe them. Orders of accuracy or number of terms included have risen proportionately, but the second order will suffice for this discussion about sediment processes. Most waves to be discussed are termed *progressive* in that the crests proceed in order across the ocean. Should these reflect normally to a structure then a progressive wave moves opposite to the incident which creates a *standing wave*. These have completely different profiles and orbital motions to those of the progressive wave. When reflected obliquely two wave crests are angled to each other, thus establishing a *short-crested system*, whose profiles and orbital motions vary in three dimensions, rather than two for progressive or standing waves. This short-crestedness occurs in the wave

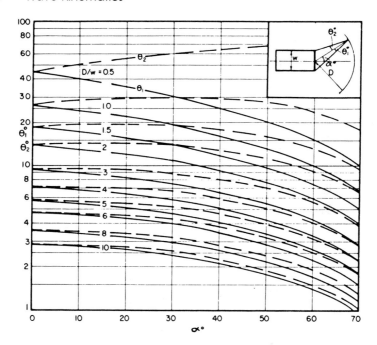

Figure 2.31 Fan angles to points downwind of a fetch

generation process and many other occasions in nature, leading to the strong conclusion by these authors that they are more important for study and analysis than the two-dimensional waves generally employed in theory and physical models. This is especially the case in their influence on sedimentary bed material.

Waves come in many periods from capillary ($T < 0.1$ s) to tidal ($12 < T < 24$ hr) but those of concern here are termed *gravity waves* ($1 < T < 30$ s), with a concentration on periods from 5 to 20 s. Generally speaking, wave theories can be broadly grouped into small and finite amplitude, based on the mathematical complexity and the depth over which the waves propagate. A linear theory (Airy 1845, Stokes 1847, Fenton 1985) is used for deep water ($d/L_o > 0.5$), cnoidal for transitional depths ($0.16 < d/L_o < 0.5$) (Korteweg and De Vries 1895, Keulegan and Patterson 1940, Svendsen 1974, Fenton 1979), and solitary wave theory for shallow ($d/L_o < 0.1$) (Boussinesq 1872, McCowan 1891, Grimshaw 1971, Fenton 1972). Other theories have been developed, such as tro-choidal (Gerstner 1802), stream-function (Chappelear 1961a, Dean 1965, 1990, Chaplin 1980), vocoidal (Swart and Loubser 1979a, 1979b) and numerical solutions (Schwartz 1974, Cokelet 1977).

The three primary variables H, L, and d, referred to as the *geometric trio* (Carter 1988), are put into various parameters such as wave steepness (H/L), relative height (H/d), and relative depth (d/L), which cover many complex environments. Ursell (1953) developed HL^2/d^3 to distinguish between cnoidal and Stokes wave theory. Other parameters are often used such as wave number $k = 2\pi/L$, and wave angular frequency

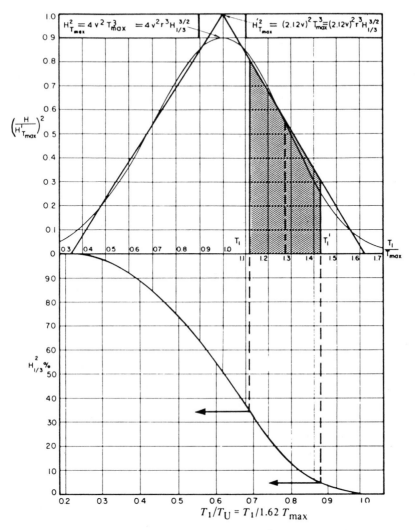

Figure 2.32 Dimensionless period spectrum and $(H_{1/3})^2$ as percentage of that at the end of the fetch

$\sigma = 2\pi/T = 2\pi f$, but these will not be used in the following presentation. Wave profiles are given in terms of water surface elevation above SWL along wave length L. Velocity components u, v, w denote horizontal, vertical and normal to propagation directions. For more detailed background on theories of oscillatory gravity waves, the reader is referred to Lamb (1932), Stoker (1957), Wehausen and Laitone (1960), Wiegel (1964), Kinsman (1965), Ippen (1966), Silvester (1974a), Whitham (1974), Le Méhauté (1976), Phillips (1977), Horikawa (1978), LeBlond and Mysak (1978), Mei (1983), Dean and Dalrymple (1984), Massel (1989), and Herbich (1990). Only the resultant expressions will be presented here, mostly in graphical form.

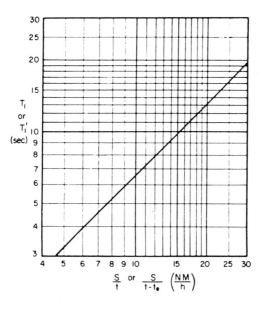

Figure 2.33 Periods of waves arriving at points distance S from the fetch, t hours after $0.80t_{FAS} = t_{95}$

2.3.1 Progressive Waves

Numerous wave theories have been presented for various depths for which their validity must be ensured. Comparisons have been conducted on theoretical grounds (Dean 1970) and by experiments (Le Méhauté et al. 1968). A detailed summary and reanalysis was given by Silvester (1974a), who stated: "The former are based upon boundary criteria, whilst the latter compare velocities and amplitudes of water-particle motions against those predicted by the equations. Only the experimental verification is a basic measure of accuracy, since the theoretical test can only be comparative."

Linear wave theory.   The wave celerity C can be given by:

$$C \equiv L/T = (gT/2\pi)\tanh 2\pi d/L = [(gL/2\pi)\tanh 2\pi d/L]^{1/2} \qquad (2.54)$$

and wave length by:

$$L = (gT^2/2\pi)\tanh 2\pi d/L \qquad (2.55)$$

where C = wave celerity, L = wave length, T = wave period, d = water depth, and g = acceleration due to gravity.

Wave celerity is determined by both T and d, but when $d/L \geq 0.5$ the hyperbolic term ($\tanh 2\pi d/L$) becomes unity, so that for deep water (with subscripts o)

$$C_o = gT/2\pi \qquad \text{or} \qquad (gL_o/2\pi)^{1/2} \qquad (2.56)$$

thus excluding the depth term. The gravity term requires specific dimensions, so that

$$C_o = 1.56T \,(\text{m/s}) = 5.12T \,(\text{ft/s}) = 3.03T \,(\text{knots}) \qquad (2.57)$$

Thus

$$C/C_o = L/L_o = \tanh 2\pi d/L \tag{2.58}$$

In shallow water as $d/L \to 0$, $\tanh 2\pi d/L \to 2\pi d/L$, celerity becomes dependent only on depth:

$$C = (gd)^{1/2} \tag{2.59}$$

This reduction as the beach is approached is fortunate for mankind, not only for his bathing propensity but also for protection of his beaches against wave attack.

From Eq. (2.58) it can be seen that:

$$(d/L_o)/(d/L) = \tanh 2\pi d/L \tag{2.60}$$

Thus, if d/L_o is known so is d/L. In any equations these terms are interchangeable, but it is preferable to work with d/L_o since L_o is directly available from wave period. In the appendix to this volume are listed values of d/L_o and d/L together with many other parameters useful in wave mechanics, even though programmable calculators and microcomputers are now available.

Water-particle motions can be viewed from two concepts, as seen in Fig. 2.34; the first is to note the changes taking place at fixed points in a water column, and the second is to travel with the particle and record temporal variations. The former is termed the *Eulerian* presentation and the latter *Lagrangian*, in which displacements from the mean position do not change the kinetic variables. Velocities and accelerations are given best by the Eulerian solution, while amplitudes of motions should employ Lagrangian. Close to the bed the two approaches give similar results. Mean positions of water particles in the Lagrangian solution are denoted by x_o, y_o and their amplitudes of motion by x, y. This subscript o should not be confused with that denoting deep water for other variables. In the Eulerian form positions are given just in x and y.

As variations occur over time the ratio t/T is used as fractions of the wave period, as is x/L for spatial variations along the wave length. Hence, in the circular functions *cos* and *sin* ratios are used alternatively. Since Stokes (1847) put Airy (1845) into acceptable form and extended his theory to higher orders, the following equations might be termed Stokes I. Velocities (Eulerian) in transitional depths are given by:

$$u = \frac{\pi H}{T} \left[\frac{\cosh 2\pi (y + d)/L}{\sinh 2\pi d/L} \right] \cos 2\pi (x/L - t/T) \tag{2.61}$$

$$\text{dimension [depth factor]} \qquad \text{phase}$$

and

$$v = \frac{\pi H}{T} \left[\frac{\sinh 2\pi (y + d)/L}{\sinh 2\pi d/L} \right] \sin 2\pi (x/L - t/T) \tag{2.62}$$

In Eq. (2.61) it is seen that the dimension of u varies directly with H and inversely with T. The depth (y) from the surface influences magnitude as much as the depth (d) itself. At the bed ($y = -d$) the vertical velocity (v) becomes zero, as dictated by the physical nature of the motion. The change in velocities with time is given by the phase

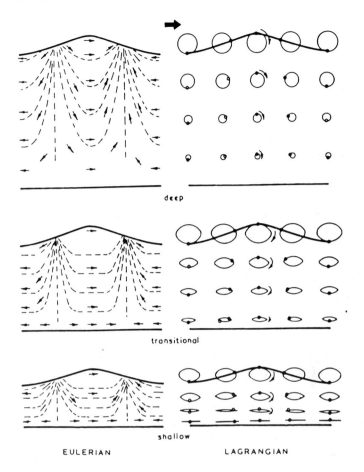

Figure 2.34   Water-particle motions in progressive waves

term, varying between zero and unity. In most problems the coastal engineer will want only the maximum value, when this term becomes unity. A subscript *max* could be used to highlight the exclusion of temporal and spatial variation. When the horizontal velocity is maximum the vertical is zero, and vice versa. This is obvious from the Lagrangian presentation in Fig. 2.34, where orbits are seen to be either circular or elliptical.

In deep water, Eqs. (2.61) and (2.62) revert to:

$$u = \frac{\pi H}{T} \exp(2\pi y/L_0) \cos 2\pi (x/L_0 - t/T) \qquad (2.63)$$

and

$$v = \frac{\pi H}{T} \exp(2\pi y/L_0) \sin 2\pi (x/L_0 - t/T) \qquad (2.64)$$

It is seen that the vertical and horizontal velocities are the same, resulting in a constant circular motion that varies exponentially with depth. When $d/L_0 = 0.5$ the resultant velocity $u^2 + v^2 = \exp(-2\pi) = 0.0019$ or 0.2% of that at the surface.

In shallow water Eqs. (2.61) and (2.62) become:

$$u = \frac{H}{2}\sqrt{\frac{g}{d}}\cos 2\pi (x/L - t/T) \tag{2.65}$$

$$v = \frac{\pi H(y+d)}{Td}\sin 2\pi (x/L - t/T) \tag{2.66}$$

It is seen that u is independent of y and T, indicating that it is uniform throughout the depth and determined by shallow water celerity ($\sqrt{gd}$). The vertical velocity varies linearly with depth and inversely with T, being zero at the bed when $y = -d$.

Orbital amplitudes (Lagrangian) are given by:

$$x = -\frac{H}{2}\left[\frac{\sinh 2\pi (y_0 + d)/L}{\sinh 2\pi d/L}\right]\sin 2\pi (x_0/L - t/T) \tag{2.67}$$

and

$$y = \frac{H}{2}\left[\frac{\sinh 2\pi (y_0 + d)/L}{\sinh 2\pi d/L}\right]\cos 2\pi (x_0/L - t/T) \tag{2.68}$$

Displacements are either side of mean position x_0, y_0 and form an elliptical orbit with major axis of length $H(\sinh 2\pi d/L)^{-1}\cosh 2\pi (y_0+d)/L$ and focal distance $H/\sinh 2\pi d/L$. The values of x and y are both influenced by location y_o and water depth d.

In deep water Eqs. (2.67) and (2.68) revert to:

$$x = \frac{H}{2}\exp(2\pi y_0/L_0)\sin 2\pi (x_0/L_0 - t/T) \tag{2.69}$$

$$y = \frac{H}{2}\exp(2\pi y_0/L_0)\cos 2\pi (x_0/L_0 - t/T) \tag{2.70}$$

The radius of the circular orbits is thus seen to reduce exponentially as did the velocity. In shallow water the displacements become

$$x = \frac{HT}{4\pi}\sqrt{\frac{g}{d}}\sin 2\pi (x_0/L - t/T) \tag{2.71}$$

$$y = \frac{H(y_0 + d)}{2d}\cos 2\pi (x_0/L - t/T) \tag{2.72}$$

In Eq. (2.71) it is seen that x varies with T but not y_0, giving uniform motion throughout the depth, approximating to $H/2\sqrt{2\pi d/L_o}$. The vertical oscillation (y) varies with y_0, becoming zero at the bed, the ellipses becoming flatter through the water column.

By taking maxima values, or eliminating the circular functions, and putting hyperbolic functions to one side, Eq. (2.61) becomes for u_{max}:

$$\frac{u_{max}}{\pi H} = \frac{\cosh 2\pi (y + d)/L}{\sinh 2\pi d/L} = X \tag{2.73}$$

In this way X can be graphed against d/L or d/L_0 with y/d as a variable as in Fig. 2.35. The value of u_{max} can be derived simply by knowing H and T. The same principle can be used for v_{max}, du/dt, dv/dt, x_{max}, and y_{max}. Also included are curves for ratios v_{max}/u_{max} and $(dv_{max}/dt)(du_{max}/dt)$, which can be useful in calculating forces on immersed objects.

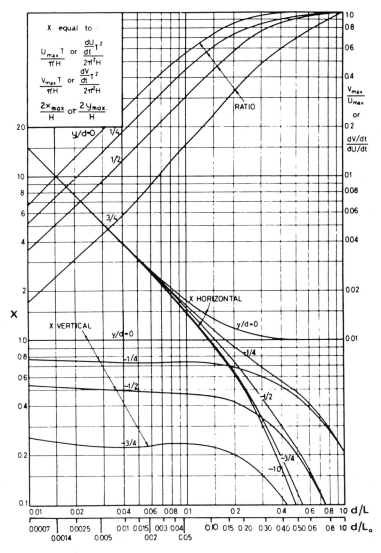

Figure 2.35 Velocity, acceleration, and amplitude parameters for water particles in progressive waves—linear theory

It becomes apparent that

$$u_{\max} = 2\pi x_{\max}/T \tag{2.74}$$

so that when maximum values of either amplitude or velocity are known the other is readily assessable for a given T.

Finite height theory.   Linear theory strictly applies to swell type waves with very modest height and very large length, the former of which increases as shallower depths are reached and the latter decreases so increasing wave steepness, and thus making it inapplicable. It can be used for approximations which serve many engineering applications. However, it is useful to outline changes that occur in these finite height conditions by presenting the second-order modifications.

The surface profile is given by:

$$y_s = \frac{H}{2}\cos 2\pi(x/L - t/T)$$

$$+ \left[\frac{\pi H^2}{4L}\left(1 + \frac{3}{2\sinh^2 2\pi d/L}\right)\cosh 2\pi d/L\right]\cos 4\pi(x/L - t/T) \tag{2.75}$$

The first term represents the sinusoidal form and the second represents an elevation (ΔH) of the mean water level (*MWL*) above the still water level (*SWL*). This gives a steepening of the crest and flattening of the trough, as depicted in the inset of Fig. 2.36. The MWL is an imaginary surface midway between the crest and trough of the wave with an amplitude $a = H/2$. Of more interest is the height of crest above SWL (a_c) or the depth to the trough (a_t) so that $a_c + a_t = H$.

As seen in the lower section of Fig. 2.36, the wave steepness (H/L) exceeds the deep-water value (H_o/L_o) as d/L decreases, until instability is reached at a theoretical value of:

$$(H/L)_b = 0.142\tanh 2\pi d/L \tag{2.76}$$

with a more realistic constant of 0.12 (Danel 1952), both of which are drawn in the figure.

The upper half of Fig. 2.36 gives curves for a_c/H, for the same d/L abscissa, for varying values of H/L. The limit of the Stokian theory is demarcated, when cnoidal must take over (Laitone 1962). The breaking limit is indicated for various nearshore slopes as well as for a horizontal bed when $H/d \approx 0.73$.

Beyond $d/L = 0.5$ the curves become horizontal so that a_c/H is determined strictly by H/L, which also represents H_o/L_o. These values are given by

$$a_c/H = 0.5(1 + 1.57H/L) \tag{2.77}$$

$$a_t/H = 0.5(1 - 1.57H/L) \tag{2.78}$$

For example, a wave with $H_o/L_o(= H/L) = 0.1$ gives $a_c/H = 0.58$.

For horizontal velocity (Eulerian) in transitional depths an extra term is added to Eq. (2.61) as follows:

$$+\frac{3}{4}\left(\frac{\pi H}{T}\right)\left(\frac{\pi H}{L}\right)\left[\frac{\cosh 4\pi(y + d)/L}{\sinh^4 2\pi d/L}\right]\cos 4\pi(x/L - t/T) \tag{2.79}$$

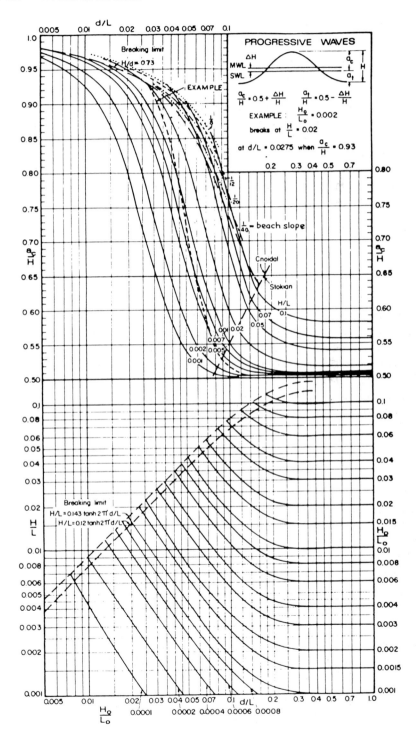

Figure 2.36 Steepening and vertical asymmetry of progressive waves

This term fluctuates at twice the frequency of the first in Eq. (2.61), so adding and subtracting values to those of the linear theory. A differential results, giving higher forward velocity at the crest than at the trough. The time for the half orbit $(T/2)$ remains the same so that the water particle travels farther in the crest than it does backwards in the trough. The addition of the term $\pi H /L$ in Eq. (2.79) indicates the importance of the wave steepness in this modification to orbital velocity.

The cnoidal theory on which most of Fig. 2.36 is based will not be presented here. The reader is referred to Keller (1948), Laitone (1959), Chappelear (1962), Fenton (1979), plus Isobe (1985) for details.

Empirical orbits throughout depth.   Since no theory of waves predicts orbital particle motions accurately for all depths, it is logical to employ empirical relationships. That of Goda (1974) has been checked against data from these major experimental sources and shown to be acceptable over most of the range of two predominant dimensionless ratios of d/L and H/d (Silvester 1974b). Design data are presented from the surface to the bed over the full workable range for near-breaking waves. Further research is required to verify these important phenomena of velocities, accelerations, and amplitudes of water particles.

As has been shown in Section 2.2.2, the FAS produces breaking waves of some limiting height in terms of $H_{1/3}$ and period T_{max} where this energy is concentrated. As far as orbital motions are concerned, for the design of marine structures or sediment movement on the floor these incipient breakers are the optimum that should be considered. These reduce in shallow water as seen in the graph of $H_{1/3}/(T_{max})^2$ versus d/L_o as in Fig. 2.37. These may be in terms of $H_{1/3}$, $H_{1/10}$, or H_{max}, but the frequency at which these occur must also be taken into account, especially with respect to sediment motion. In this sense the waves may be considered as monochromatic with energy equal to the area

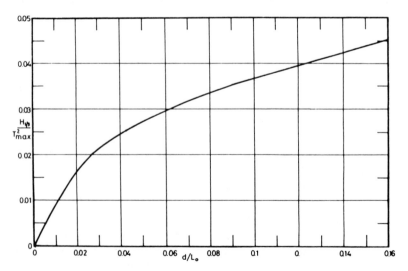

Figure 2.37 $H_{1/3}/(T_{max})^2$ versus d/L_o for storm waves

under the spectrum or $(H_{1/3}/4)^2$ with period T_{max}. Higher waves of $H_{1/10} = 1.275H_{1/3}$ (equal to the highest $H_{1/3}$ in 10 years), or $H_{max} = 1.86H_{1/3}$ (equal to the highest in each 1,000 waves).

Although wave celerity and profile can be determined adequately by certain theories, none have been found suitable to predict orbital motions of water particles. Dean (1974), in comparing his stream function with other theories against experimental data of Le Méhauté et al. (1968), observed a standard deviation in horizontal velocities throughout the depth of 0.17 for his stream function and of 0.235 for the Goda (1974) modification of the Airy theory. However, even this check was made for only eight velocity profiles and then mainly for very shallow conditions of $d/L \approx 0.05$.

This lack of theory should be accepted as a fact of life for the present and thorough testing carried out in order to derive an acceptable empirical formula. Goda (1974) has presented a relationship for U_{max} which has been compared to data of various workers, the final results of which will be given below. Once U_{max} has been determined reasonably accurately, other maxima such as $(dU/dt)_{max}$ and x_{max} can be derived simply from linear theory as previously noted. Values of U_{max} should be examined under the crest at SWL, its distribution from SWL to the bed, and from SWL to the ocean surface. These are found to vary both with d/L and H/d. Above SWL the distribution requires a concurrent evaluation of crest height (a_c) above this datum.

Silvester (1974b) examined the experimental data of Goda (1974) and concluded that for $d/L > 0.2$ the linear theory predicted U_{max} at SWL accurately, and that for $d/L < 0.3$ this can be expressed as

$$U_{max}/(gd)^{1/2} = 2H/3d \tag{2.80}$$

This is exhibited in Fig. 2.38, where the data straddle the empirical line of Eq. (2.80) quite well. In the same publication Goda (1974) presented a modification of the Airy solution for SWL which can be put in the form

$$\frac{U_{max}}{\sqrt{gd}} = \frac{(1 + \alpha\sqrt{H/d})^{1/2}}{(\tanh 2\pi d/L)^{1/2}} \left(\frac{\pi d}{2L}\right)^{1/2} \frac{H}{d} \tag{2.81}$$

where α is a factor dependent on d/L as illustrated in Fig. 2.39. The $U_{max}/(gd)^{1/2}$ for SWL remains sensibly constant at $2H/3d$ for $d/L < 0.3$ but rises above this for higher values of d/L, which are necessarily controlled by the limiting H/d at these greater depths; as noted in Fig. 2.39, a curve that arises from tests by Goda (1974).

The optimum values of Eq. (2.81) for all possible conditions of d/L and H/d are presented as a curve in Fig. 2.40, together with Eq. (2.80) and data from Iwagaki and Sakai (1970) and Le Méhauté et al. (1968). (Raw data were supplied to the senior author from Professor Sakai of Kyoto University, Japan, which had been used in several publications. These will be referred to as Sakai's data.) It is possible, with further experimentation, which requires a degree of sophistication for the necessary accuracy, that a slightly higher value than $2H/3d$ may be required for $d/L < 0.3$. Goda (1974) acknowledged: "in the actual sea ... the orbital velocities under wave crests may be larger than those expected from experimental velocities in a laboratory flume." As noted

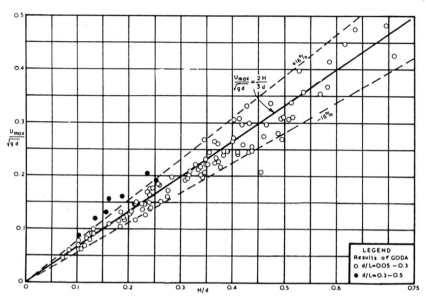

Figure 2.38   Dimensionless $U_{\max}$ versus H/d from data by Goda (1974)

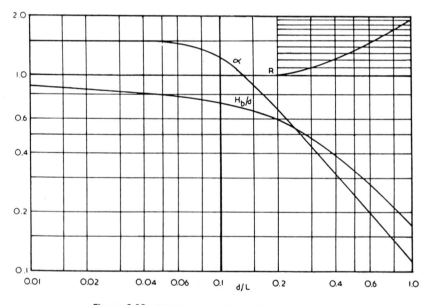

Figure 2.39   Variables of α, H_b/d, and R versus d/L

already, the linear equation serves adequately for $d/L \geq 0.3$, which can be expressed in the following manner:

$$U_{\max}/(gd)^{1/2} = 1.255(d/L)^{1/2}(H/d) = R(2H/3d) \qquad (2.82)$$

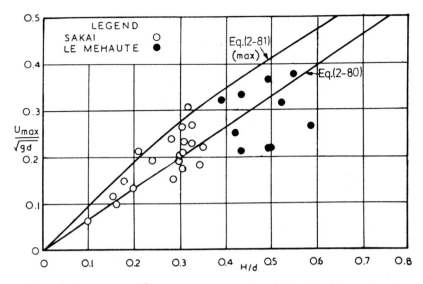

Figure 2.40 $U_{\max}/(gd)^{1/2}$ versus H/d, showing Eq. (2.80) and optimum values of Eq. (2.81) when $d/L > 0.3$, see text for data sources

If $U_{\max}/(gd)^{1/2}$ is computed from Eq. (2.80), this SWL value must be multiplied by a factor R when $d/L \geq 0.3$, a curve of which is included in Fig. 2.39.

For vertical distribution from SWL to the bed, the empirical equation of Goda (1974) takes the form:

$$U_{\max} = K \frac{\pi H}{T} \frac{\cosh 2\pi z/L}{\sinh 2\pi d/L} \tag{2.83}$$

where

$$K = \left[1 + \alpha(H/d)^{1/2}(z/d)^3\right]^{1/2} \tag{2.84}$$

Fig. 2.41 defines variables, where it is seen that z is a vertical location above the bed. From Eqs. (2.83) and (2.84) it can be shown that $U_{\max}$ at any depth is given by

$$\frac{U_z}{U_{SWL}} = \left[\frac{1 + \alpha\sqrt{H/d}(z/d)^3}{1 + \alpha\sqrt{H/d}}\right]^{1/2} \frac{\cosh 2\pi z/L}{\sinh 2\pi d/L} \tag{2.85}$$

The *RHS* of Eq. (2.85) for any given z/d varies with both d/L and H/d. However, for $d/L < 0.2$ this ratio becomes progressively more important. As seen in Fig. 2.39, the limiting value of H/d in these shoaler depths is greater then 0.6 and the factor α quickly approaches its value of 1.5.

The vertical distribution from SWL to the wave surface was studied by Iwagaki et al. (1974), for crests in shallow water breaking waves. Experimental curves were grouped according to type of breaker, namely, spilling, plunging (3 categories) and surging. For the deeper water conditions emphasized here only spilling breakers are considered for

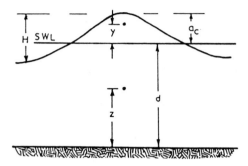

Figure 2.41 Definition sketch of variables, for empirical water-particle orbits

U_{max} from SWL to limiting crest height (a_c) above it. Values at $y/a_c = 0.75$ and 0.5 are also noted since values at 0.25 were essentially the same as at SWL.

Velocity ratios U_y/U_{SWL} are presented in Fig. 2.42 over a workable range of d/L, where it is seen that for deep water the surface values can reach 400% of those at SWL. The column through which this distribution operates is limited by a_c, data for which was obtained from Goda (1974) and can be expressed as:

$$a_c/d = 0.835(H/d)^{1.275} \tag{2.86}$$

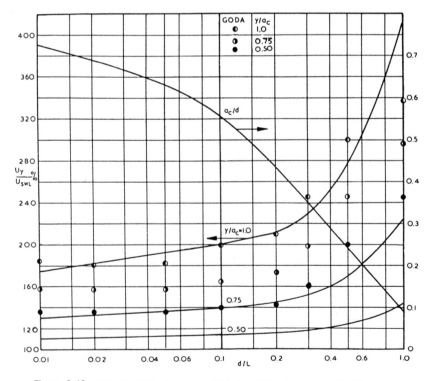

Figure 2.42 Velocity distribution above SWL for spilling breakers and limiting values of a_c/d

Since the connection between limiting H/d and d/L has been established as in Fig. 2.39, the relationship between a_c/d and d/L can be obtained by the curve in Fig. 2.42.

Equation (2.85) from Goda (1974) also predicts for this upper zone and is plotted in Fig. 2.42 for the surface and $y/a_c = 0.75$ and 0.5. It is seen that the surface-plotted points correspond reasonably well with the curve of experimental values, although the peak does not reach its value of 410%. The Goda predictions for $y/a_c = 0.75$ and 0.5 are substantially above the experimental curves. This implies that the distribution is more uniform or almost triangular. This needs further verification.

The overall distribution from bed to surface combines all the above discussion, for which near-breaking wave conditions only will be considered. Equation (2.80) can be used for SWL with the R factor entering for $d/L > 0.2$. The percentage of these values at other depths is shown in Fig. 2.43 in terms of $U_{max}/(gd)^{1/2}(2H/3d)$, the denominator of which is that at SWL. It is seen that up to $d/L = 0.15$ the vertical distribution is essentially the same, but in deeper water the energy is concentrated above SWL for near-breaking waves. Few experimental data would be available to verify these curves as limiting conditions of stability are implied. Table 2.4 is derived from Fig. 2.43 for ready use in design.

The acceleration $(dU/dt)_{max}$ from linear theory is given by:

$$(dU/dt)_{max} = 2\pi U_{max}/T \tag{2.87}$$

and maximum horizontal amplitude of motion computed from:

$$x_{max} = U_{max}T/2\pi \tag{2.88}$$

Maximum vertical velocity V_{max} as fraction of U_{max} is a function of d/L which for linear theory is:

$$\text{vertical/horizontal} = \tanh 2\pi z/L \tag{2.89}$$

In Table 2.4 percentages of $V_{max}/(gd)^{1/2}(2H/3d)$ are listed from the water surface to the bed over the complete range of d/L. Ratios of vertical accelerations and amplitudes of motion can be similarly obtained from the same percentages.

2.3.2 Standing Waves

These occur, as noted previously, by two progressive waves moving opposite to each other, so that at each half period the crests make contact which doubles the wave height, or adds the heights if they differ. The profile by linear theory for all depths of water is:

$$y_s = (H/2)\cos 2\pi x/L \cos 2\pi t/T \tag{2.90}$$

The velocities (Eulerian) are:

$$u = \frac{\pi H}{T}\left[\frac{\cosh 2\pi(y+d)/L}{\sinh 2\pi d/L}\right]\sin 2\pi x/L \sin 2\pi t/T \tag{2.91}$$

and

$$v = \frac{\pi H}{T}\left[\frac{\sinh 2\pi(y+d)/L}{\sinh 2\pi d/L}\right]\cos 2\pi x/L \sin 2\pi t/T \tag{2.92}$$

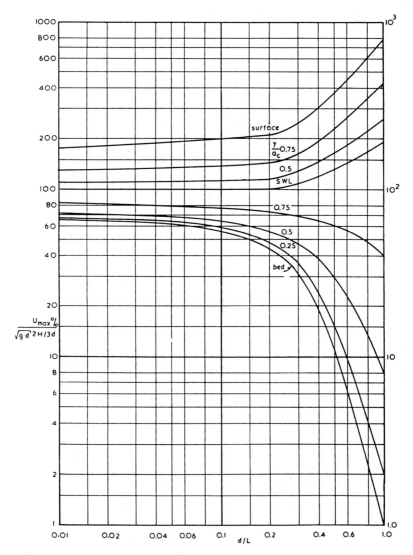

Figure 2.43   Values of $U_{max}/(gd)^{1/2}(2H/3d)$ from surface to bed for d/L from 0.01 to 1.0

The nodes (only horizontal motion) occur where $\cos 2\pi x/L = 0$ or $x/L = 1/4$ and $3/4$, and the antinode (only vertical motion) when $x/L = 0$ (at the wall) and $1/2$. These are presented in Fig. 2.44 for various depth notations.

Deepwater values thus become:

$$u = \frac{\pi H}{T} \exp(2\pi y/L_0) \sin 2\pi x/L_0 \sin 2\pi t/T \qquad (2.93)$$

TABLE 2-4 VARIATIONS OF DIMENSIONLESS PARAMETERS WITH d/L

Percentages of following parameters	index	d/L 0.01	0.02	0.04	0.06	0.08	0.10	0.15	0.20	0.30	0.40	0.50
	d/L_o	0.0007	0.0025	0.01	0.022	0.037	0.056	0.11	0.17	0.286	0.395	0.498
(a_c/d) max		0.73	0.69	0.65	0.62	0.58	0.555	0.49	0.435	0.35	0.29	0.24
(H/d) max		0.88	0.85	0.81	0.78	0.75	0.72	0.66	0.60	0.48	0.39	0.33
$U_{max}/\sqrt{gd\dfrac{2H}{3d}}$	$\dfrac{y}{a_c}$ = 1.0	174	182	190	194	197	200	206	211	248	305	371
	0.75	130	133	136	137	138	140	142	145	163	194	228
	0.5	110	111	112	113	113.5	114	115	116	130	146	166
$T(dU/dt)_{max}/2\pi\sqrt{gd\dfrac{2H}{3d}}$; $2\pi x_{max}/T\sqrt{gd\dfrac{2H}{3d}}$	$\dfrac{z}{d}$ = 1.0	100	100	100	100	100	100	100	100	110	121	134
	0.75	82.5	82	81	80	79	78	75	74	70	65	61
	0.50	71.5	71	69.5	68	66	64.5	60	56	47	38	29
	0.25	67.5	67	65.5	64	61.5	59.5	53	48	36	24	15
	0	66.5	65.5	64.5	62	60	56	50	44	31	19	11
$V_{max}/\sqrt{gd\dfrac{2H}{3d}}$	$\dfrac{y}{a_c}$ = 1.0	11	21	39	54	67	75	89	95	109	121	134
	0.75	10	19	36	50	62	71	86	93	108	120	134
	0.5	9	17	32	45	58	67	82	91	107	120	134
$T(dv/dt)_{max}/2\pi\sqrt{gd\dfrac{2H}{3d}}$; $2\pi y_{max}/T\sqrt{gd\dfrac{2H}{3d}}$	$\dfrac{z}{d}$ = 1.0	7	13	25	36	47	56	74	85	104	120	134
	0.75	5	10	18	27	36	44	61	72	91	109	126
	0.50	4	6	12	18	24	30	44	57	82	103	121
	0.25	2	3	6	9	12	16	23	32	53	73	92

66

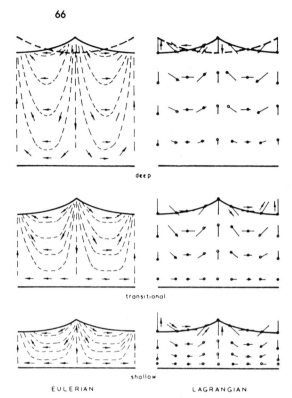

Figure 2.44 Water particle motions in standing waves

and

$$v = \frac{\pi H}{T} \exp(2\pi y/L_0) \cos 2\pi x/L_0 \sin 2\pi t/T \qquad (2.94)$$

At any depth the horizontal and vertical values are equal. As depicted in Fig. 2.44, the orbits are rectilinear, so that resultant velocities $[= (u^2 + v^2)^{1/2}]$ differ throughout the length of the wave.

In the shallow water the velocities become

$$u = \frac{HL}{2Td} \sin 2\pi x/L \sin 2\pi t/T \qquad (2.95)$$

and

$$v = \frac{H}{Td}(y + d) \cos 2\pi x/L \sin 2\pi t/T \qquad (2.96)$$

Thus, in shallow water u is independent of depth (y), whereas v varies linearly with it, being zero at the bed. At the antinodal point at the floor no particle motion takes place.

The water particles are displaced from their initial location (x_o, y_o) by

$$x = -\frac{H}{2} \left[\frac{\cosh 2\pi (y_0 + d)/L}{\sinh 2\pi d/L} \right] \sin 2\pi x_0/L \cos 2\pi t/T \qquad (2.97)$$

and

$$y = \frac{H}{2} \left[\frac{\sinh 2\pi (y_0 + d)/L}{\sinh 2\pi d/L} \right] \cos 2\pi x_0/L \cos 2\pi t/T \qquad (2.98)$$

They are thus rectilinear and simple harmonic with resulting amplitude of:

$$(x^2 + y^2) = \frac{H}{2\sqrt{2}} \frac{(\cos 4\pi x_0/L + \cosh 4\pi y_0/L)^{1/2}}{\sinh 2\pi d/L} \qquad (2.99)$$

and inclined at angle θ to the horizontal such that

$$\tan \theta = \tan 2\pi x_o/L - \tanh 2\pi (y_o + d)/L \qquad (2.100)$$

For deep water these amplitudes become:

$$x = -\frac{H}{2} \exp(2\pi y_0/L_0) \left[\sin 2\pi x_0/L_0 \cos 2\pi t/T \right] \qquad (2.101)$$

and

$$y = \frac{H}{2} \exp(2\pi y_0/L_0) \left[\cos 2\pi x_0/L_0 \cos 2\pi t/T \right] \qquad (2.102)$$

They decrease with depth, giving x and y as 4% of the SWL value when $y_o/L = 0.5$. Shallow water amplitudes are given by:

$$x = \frac{HL}{4\pi d} \sin 2\pi x_0/L \cos 2\pi t/T \qquad (2.103)$$

and

$$y = \frac{HL}{2d}(y_0 + d) \cos 2\pi x_0/L \cos 2\pi t/T \qquad (2.104)$$

Horizontal displacement (x) is independent of y_o, with y varying directly with it, being zero at the bed. The picture of particle motion given in Fig. 2.44 is for opposing waves of equal height.

Wave theories for standing waves of finite height have been developed to second or higher order of approximation (Miché 1944, Kishi 1957, Tadjbakhsh and Keller 1960, Goda 1967). However, there are a few differences in form, especially in the use of $\sin 2\pi x/L$ or $\cos 2\pi x/L$ for the resulting wave height and other subsequent wave variables, in combination with $\sin 2\pi t/T$ or $\cos 2\pi t/T$ for time. For example, the second order theory (Miché 1944) for waves of finite height gives a new profile as follows:

$$y_s = \frac{H}{2} \sin 2\pi x/L \sin 2\pi t/T$$

$$+ \frac{\pi H^2}{4L} \coth 2\pi d/L \left[\sin^2 2\pi t/T - \frac{3 \cos 4\pi t/T + \tanh^2 2\pi d/L}{4 \sinh^2 2\pi d/L} \right] \cos 4\pi x/L \qquad (2.105)$$

The first term is similar to that for the linear theory (see Eq. 2.90), but with different combinations of trigonometric functions in space and time, whereas the second represents an incremental increase in depth above SWL. Unlike the progressive wave, this ΔH deviation is maximum at the crest and zero at the nodes, as seen in the inset of Fig. 2.45. The magnitude of this rise is

$$\Delta H = \frac{\pi H^2}{4L}\left[1 + \frac{3}{4\sinh^2 2\pi d/L} - \frac{1}{4\cosh^2 2\pi d/L}\right]\coth 2\pi d/L \qquad (2.106)$$

As ΔH increases so does the wave height until a maximum is reached when breaking occurs (Miché 1944, Danel 1952) given by:

$$(H/L)_{\max} = 0.22\tanh 2\pi d/L \qquad (2.107)$$

A fourth-order theory (Goda 1967) reduces this breaking limit for $d/L < 0.5$ as seen in Fig. 2.45, thus giving $a_c/H = 0.75$, or the crest constituting 75% of the total wave height. The crest height (a_c) above SWL is obtained from:

$$a_c = 0.5H + \Delta H \qquad (2.108)$$

Should one of the opposing progressive waves be smaller in height than the other, as in partial reflection from a wall, the resulting nodes and antinodes will be located in the same position but the water-particle orbits will not be rectilinear in character. As seen in Fig. 2.46, different degrees of reflection produce elliptical orbits which become flatter as complete reflection is approached. It is seen in the figure that there is a slight vertical oscillation at the nodes, whereas in a complete clapotis it is zero. Figure 2.46 was produced from photographs presented by Wallet and Ruellan (1950).

2.3.3 Short-Crested Waves

When two progressive waves are angled to each other, the crests form a diamond pattern, termed by French as *clapotis gaufré*. Although in prototype conditions the heights and periods of these two interacting waves may differ, for the present discussion it will be presumed they are equal. Several theories have been developed for short-crested wave systems (Fuchs 1952, Chappelear 1961b, Hamada 1965, Hsu 1979, Hsu et al. 1979, Roberts 1983). However, only the third-order approximation of Hsu (1979) will be included here. First, amplitude of the surface profiles of these complex waves will be presented, with crest heights (a_c) proportional to wave height (H_{sc}). It is important to know the celerity (C_s) of the island crests for various angles (2θ) between the wave components. Then orbital velocities in the horizontal and vertical direction will be discussed for all positions throughout the diamond pattern and down the water column. The masss transport due to these orbits will be treated in Section 2.3.4.

From the definition sketch shown in Fig. 2.47, It is seen that at crest intersections a higher than normal mound of water will form, so creating a new crest system of island type. These propagate parallel to the vector C_s shown, with successive crests spread a distance L apart, which is greater than the wave length of the two component waves. A new dimension is introduced, the distance (L') between synchronous crests which is finite,

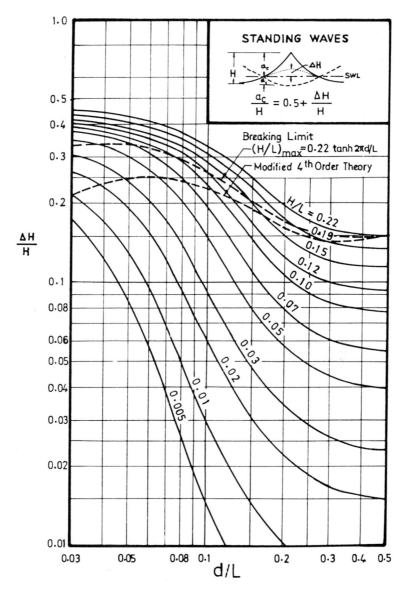

Figure 2.45 Vertical asymmetry of standing waves

so introducing the term *short-crested wave*. Crests of progressive waves are assumed to be infinite in theory and in modeling.

The surface undulations in this system are quite complex, as seen in the contours displayed in Fig. 2.48 for progressive waves angled 120° to each other. It is seen that the crests rise to 70% of the wave height above SWL, while the trough is depressed to around 20% of this value. There is a zero or SWL contour of diamond shape encompassing

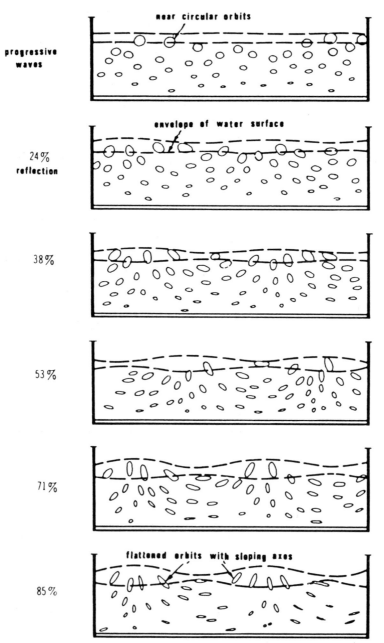

Figure 2.46 Wave profiles and water-particle orbits in partial standing waves

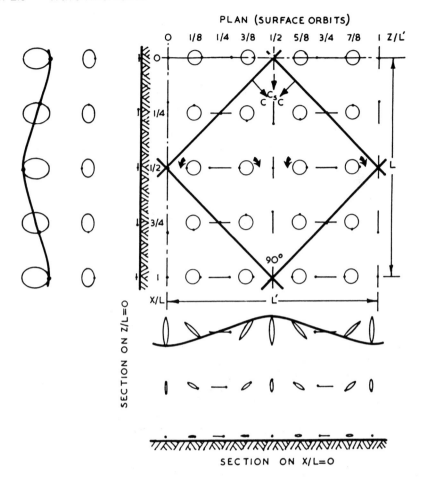

Figure 2.47   Orbital motions of water particles within a short-crested wave system

the trough. The crest height as a proportion of short-crested height (a_c/H_{sc}) is plotted in Fig. 2.49 against d/L for varying values of H/d for the component waves, with the breaking limit shown of:

$$(H_{sc}/L)_b = 0.142 \tanh 2\pi d/L \qquad (2.109)$$

in which L is the wave length of the short-crested wave in the direction of propagation. It is seen that for very small depths a_c/H_{sc} can rise to 0.77.

The celerity of the short-crested island crests can be related to that of the incident or reflected wave as in Fig. 2.50, where it is seen to vary with θ, the angle between the short-crested orthogonal and each component crest. The celerity of the incident wave is $C = (gT/2\pi)\tanh 2\pi d/L_A$, (Eq. 2.54), where L_A refers to the wave length of the incident component.

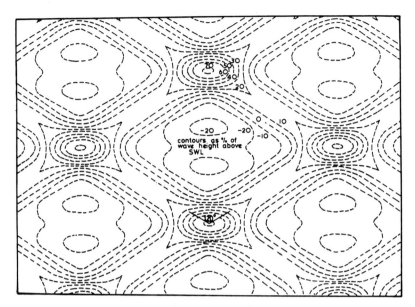

Figure 2.48 Water surface contours produced by two progressive waves angled 120°
to each other

Reverting back to Fig. 2.47, it is seen that the water particle motions are three-dimensional, with components vertical, parallel, plus normal to the wall. These proportions vary along wall alignments spaced at different Z/L' distances from it. At $Z/L' = 0$, 1/2, 1, ... the orbital planes are parallel to the wall, even though those drawn in section of Fig. 2.47 on $X/L = 0$ are flattened ellipses for clarity, they should be straight lines. As noted in the section of Fig. 2.47 on $Z/L' = 0$, the surface particles may have their major axes of the elliptical orbit vertical, but this becomes horizontal further down the water column, being rectilinear at the bed.

Halfway between the above alignments, at $Z/L' = 1/4$, 3/4, 5/4, ... the orbits are in fact rectilinear at all depths and parallel to the bed. They are oriented normal to the wall. This motion can be likened to the horizontal orbits in the standing waves at the nodal points. As noted previously, this only applies when the component waves are of equal height, otherwise there will be some vertical motion at these nodes, both for the two- and three-dimensional cases.

For locations along paths at $Z/L' = 1/8$, 3/8, 5/8, 7/8, ... the orbital motions are ellipses, in planes that are angled to the horizontal. Their slopes vary throughout the depth, becoming zero at the bed. The slopes also vary at each successive $L'/8$ interval, being directed upward toward the path of the island crests (see section on $X/L = 0$ in Fig. 2.47). For the angle $\theta = 45°$ as in the figure, these ellipses when viewed from above would be circular at the bed, but for any other angle would be elliptical. The direction of rotation in these virtual vortices also alternates at successive $L'/8$ alignments, it being in the direction of crest propagation on the near side to such alignments. Along each alignment parallel to the wall this rotation is constant.

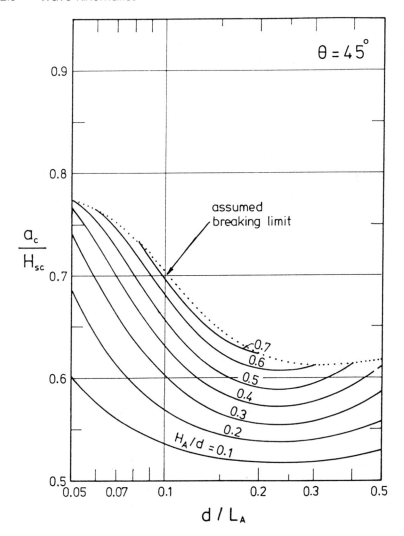

Figure 2.49 Ratio of a_c/H_{sc} for short-crested wave for equal component heights

Details of water motion from the surface to the bed is available from Hsu (1990), but those occurring at the bed are most relevant with respect to sediment motion. For this the maximum velocity within the orbit for any given direction is important, which can vary depending on the order of accuracy used in the computation. Values of first and third order will be discussed here.

In Fig. 2.51 are shown ratios of maximum horizontal velocities (U) at the bed parallel to the wall to similar maxima for the incident wave should it propagate along this wall $U(90°)$. Each curve is designated also by the position of the orbit across the L' line normal to the wall in terms of Z/L'. Thus, $U_1/U_1(90°) - 0$ designates linear

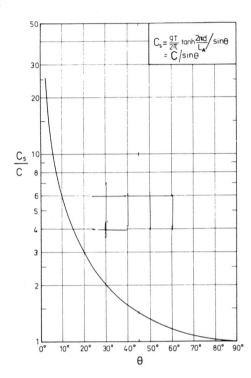

$$C_s = \frac{gT}{2\pi} \tanh\frac{2\pi d}{L_A} \Big/ \sin\theta$$
$$= C / \sin\theta$$

Figure 2.50 Ratio C_s/C for various half angles (θ) between waves

theory for this ratio at alignments $Z/L' = 0, 1/2, 1, \ldots$, while $U_3/U_3(90°) - 0$ refers to the same at locations for third-order theory (Hsu and Silvester 1989). It is seen that this latter ratio reaches 2 for an obliquity of the wave crest at $80°$. Even at $45°$ this ratio is 1.4 and naturally reduces to zero for the standing wave when $\theta = 0°$, since no velocity parallel to wall is possible. As θ approaches $90°$ a Mach-stem effect is introduced where a crest normal to the face propagates along it. This is of little significance as its action is similar to that of the the the island crest itself in terms of the moving sediment.

Halfway between the above alignments (i.e., at $Z/L' = 1/4, 3/4, 5/4, \ldots$) the vertical motion of the water surface is practically zero, so that only horizontal water-particle motions occur. As noted earlier, these points can be likened to the nodes in the standing wave case and these are the nodes for the short-crested system. Although these orbits are normal to the wall they can suffer a net movement along the wall or mass transport, to be discussed later. These ratios $(V_1/U_1(90°) - 1/4)$ approach 2 for $\theta = 0°$ or the standing wave and become zero for $\theta = 90°$.

At alignments halfway again between the above two, $Z/L' = 1/8, 3/8, 5/8, \ldots$ the elliptical orbits become planar with the bottom next to the bed, having both U and V components. They are shown as $U_1/U_1(90°) - 1/8$, where it is seen to reach 1.36 for $\theta = 80°$, decreasing to unity at $\theta = 90°$, and becoming zero for the standing wave at $\theta = 0°$. The value of $V_1/U_1(90°) - 1/8$ is maximum (1.4) at $\theta = 0°$. At $\theta = 45°$ the maxima of U_1 and V_1 are the same at unity, implying that at the bed circular motion occurs with uniform velocities around the orbit. However, for all other angles U and V

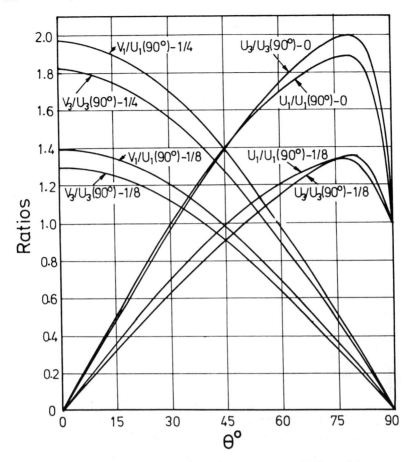

Figure 2.51 Maximum bed velocities at specific alignments (Z/L') parallel to a reflecting wall for various obliquities (θ) of wave crest to wall. Velocity U is parallel to
the wall and V normal to it. $U(90°)$ represents bed velocity of incident wave. Subscripts
1 and 3 refer to linear and third-order theories. $Z/L' = 0$, 1/8 or 1/4

components differ, which indicates that these orbits at the bed are elliptical. The major
axes of these ellipses will be angled to the wall, being closer to the normal for $\theta < 45°$
and more obtuse for $\theta > 45°$.

Verification of these orbital motions has been carried out by Hsu (1979) using
a basin shown in Fig. 2.52, where it is seen that a tunnel with perspex ceiling was
constructed beneath the basin floor. A general view of the setup is shown in Fig. 2.53.
Waves of very limited duration were propagated to the reflecting wall at 45°, where
motions of polystyrene beads ($SG = 1.03$ and $D = 1$ to 2 mm) were photographed
every 1/24 s by cine-camera. The perspex ceiling had line grids grooved into it at 100
mm in each direction for ready acquisition of bead position. Each second frame of the
film was projected on a screen from which accurate measurements of orbits could be
taken. Each test had to be compiled in 15 s by which time several wave cycles were

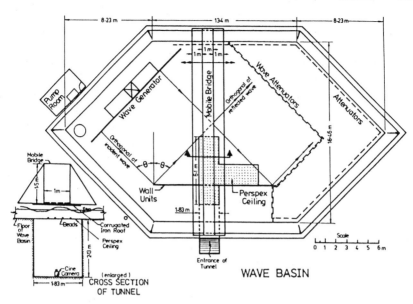

Figure 2.52 Plan view of an outdoor wave basin with details of the tunnel construction beneath it (Hsu 1979)

Figure 2.53 General view of wave basin as used by Hsu (1979)

recorded, besides positions after each wave cycle, or the mass transport which will be discussed later.

Orbits as measured (Hsu 1979) for $\theta = 45°$ and $d/L = 0.2182$ and $H/L = 0.0298$ (incident wave length) are recorded in Fig. 2.54 at specific Z/L' alignments from the reflecting wall. These could not be at ideal values of 0.125, 0.25, and 0.5 due to bead mobility during placement. However, the general picture is obtained of parallel to the wall at $Z/L' = 0.02$ and 0.51, normal to the wall at $Z/L' = 0.25$, and almost circular at $Z/L' = 0.125$ and 0.375. The square dots are bead locations after each wave period T and therefore indicate mass transport velocity. It can be noted in the $Z/L' = 0.23$ alignment that although orbital motion is normal to the wall there is a mass transport parallel to it, as will be shown theoretically later.

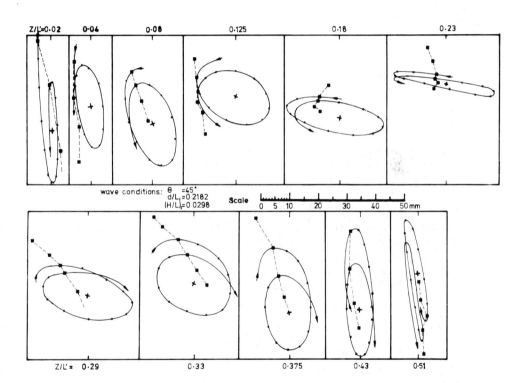

Figure 2.54 Orbital paths of beads recorded at specific distances from the reflecting wall. Circular dots at $T/12$ intervals and squared dots after each T

A more comprehensive series of orbits is available in Fig. 2.55, where seven cycles are depicted. The first two orbits probably occurred prior to complete reflection as the mass transport trace is normal to the wall prior to becoming parallel to it. The similarity of the orbits is outstanding even though they are displaced each wave period by this net movement.

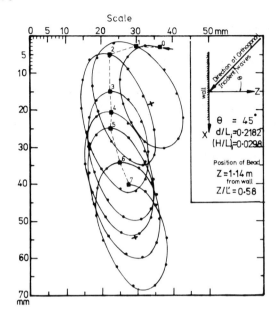

Figure 2.55 Multiple bed orbits for point $Z/L' = 0.58$ showing mass transport

2.3.4 Mass Transport

Waves in deep water have circular water-particle orbits, which in shoaler conditions become ellipses, with these flattening toward the bed until plain oscillatory motion exists at this boundary.   However, the particles do not arrive exactly back at their original position after each wave period, but are displaced either forward or backward from it. This cyclic net motion is termed *mass transport* and is very important with respect to sediment transport. The theory for it assumes that across a plane normal to the orthogonal there must be a balance in volume in either direction, since bodies of fluid are not carried laterally due to this phenomenon.

There are three wave systems to be examined, namely progressive, standing, and short-crested waves, each of which have completely different distributions of velocities, if they can be called such, throughout the water column.   One assumption that might be questioned is that laminar conditions exist at the bed, which is very unlikely since water motions are quite active there when the mass transport is quite strong. However, the theory has been proven correct in model conditions (Russell and Osorio 1958) even though these introduce their own distortions of being in enclosed bodies of water.

The following is not meant to be a dissertation on the topic but only an outline to show the reader its importance in moving sediment toward the coast. This is fortunate for man for if wave action tended to transport material seaward the oceans would be filled with all the sediment weathered from rocks and most of the landmass would be rocky. Civilization in fact could not have evolved in its present form because of the vegetation needed for man to survive.

Progressive waves.   The oscillatory motions of water particles at the bed suffer a greater speed forward, as the wave crest passes any point, than backward during the passage of the trough. Since the direction of each is the same this means that the particle will end up closer to shore after each wave cycle. Net movement increases with wave height squared, so that mass transport is enhanced toward the beach. This does not imply that storm waves with many waves of different periods and large heights improve this landward transport because it has almost the opposite effect, only swell type waves now being considered. This again is important with regard to sediment transport because these waves are more persistent than storm sequences.

Within the boundary layer for this oscillatory motion the mass transport is distributed throughout its thickness δ as in Fig. 2.56A, where a maximum occurs at 0.3δ, becoming zero at the bed. This is 1.1 times the value at the upper edge of he boundary layer. This mass transport velocity U_{BP} at the edge of the boundary layer for progressive waves is given by

$$U_{BP} = \frac{5\pi^2 H^2}{4TL \sinh^2 2\pi d/L} \tag{2.110}$$

from Longuet-Higgins (1953), or

$$gXT^2/H^2 = \frac{77.4}{\sinh^2 2\pi d/L \tanh 2\pi d/L} \tag{2.111}$$

where X is a unit of length forward per wave period T. For peak values at 0.3δ multiply the above constants by 1.1. Equation (2.111) is drawn in Fig. 2.57, which shows the increase in mass transport at the bed as d/L_o decreases.

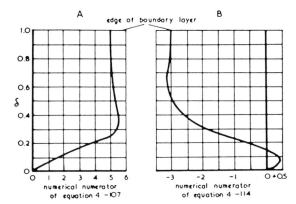

Figure 2.56 Distribution of mass transport throughout boundary layer: A. of a progressive wave; B. of a standing wave

Its distribution throughout the water column is given Fig. 2.58, where X is expressed in feet per wave period. For $d/L_o = 0.5$, the value at the bed is practically zero and then rises to negative values, contra to wave propagation, before becoming positive at the surface, which applies down to $d/L_o = 0.10$. At $d/L_o = 0.01$ the bed value is large in the direction of crest movement but has a negative maximum up to the surface.

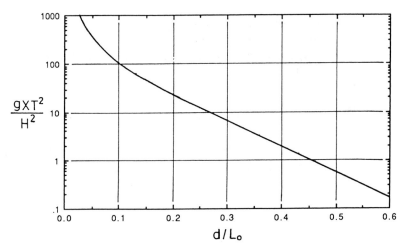

Figure 2.57 Forward displacement due to mass transport in bottom boundary layer in progressive waves

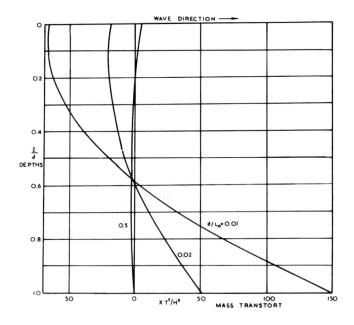

Figure 2.58 Distribution of mass transport velocity throughout water depth

Standing waves.   The distribution of mass transport within the boundary layer for a standing wave is shown in Fig. 2.56B, where a change from positive to negative occurs at 0.078δ. This implies that sediment suspended slightly will move in one direction, whereas that rolled across the bed could move in the opposite direction. Since in most prototype situations sand will be in the upper region of the boundary layer, and even further, the negative direction as shown in the figure would apply.

Noda (1968) has computed the mass transport at the upper edge of the boundary layer as:

$$U_{BS} = \frac{3\pi^3 H^2}{8TL \sinh^2 2\pi d/L} \sin 4\pi x/L \qquad (2.112)$$

where x is measured along the wave length with origin at the node. This strong net movement establishes cells of circulation in $L/4$ compartments, as seen in Fig. 2.59, whose vertical dimensions are dictated by d/L, encompassing the whole depth when it is large and reducing to about half depth when small. Sediment will collect mainly at the antinodes, but small accumulations may occur at the nodes, as shown in the figure, due to the opposite rotating cells not shown to scale. The prototype effects of these standing waves are discussed in Section 7.1.

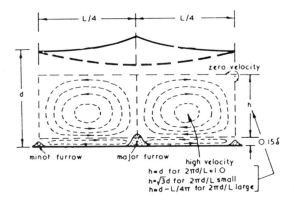

Figure 2.59   Mass transport cells in standing waves

Short-crested waves.   The analysis of mass transport in these complex waves has been carried out by Hsu (1979, 1990), Hsu et al. (1980), Mei et al. (1972), Tanaka et al. (1972) and Dore (1974). Its distribution throughout the boundary layer varies with the angle between component waves (2θ), distances normal to the wall (Z/L'), having also vertical and two horizontal components. These will not be presented here, but the bed values are shown in Fig. 2.60 as ratios to the same values for the incident waves were they to propagate along the wall. For example, one is expressed as $U_{M2}/U_{M2}(90°) - 0$, denoting mass transport within the boundary to the second order for the direction parallel to the wall or orthogonal of island crests, the $90°$ refers to the incident waves as noted above, the 0 represents alignments parallel to the wall at $Z/L' = 0, 1/2, 1, \ldots$

As seen in the figure, this ratio reaches 3.6 for $\theta = 80°$ decreasing to unity at $\theta = 90°$ and zero for $\theta = 0°$. At $\theta = 45°$ this ratio is 2.2, or the mass transport at these alignments is over twice that for the incident wave should it be dissipated by breaking rather than be reflected obliquely. For $\theta = 60°$ and $30°$ the ratios are 3 and 1.4 respectively, which is likely to be the range of obliquity for waves against breakwaters, as will be discussed in Chapter 7. At the nodes, or $Z/L' = 1/4, 3/4, 5/4, \ldots$, the ratio reaches 0.6 at $\theta = 38°$. This occurs even though the water-particle motions are only

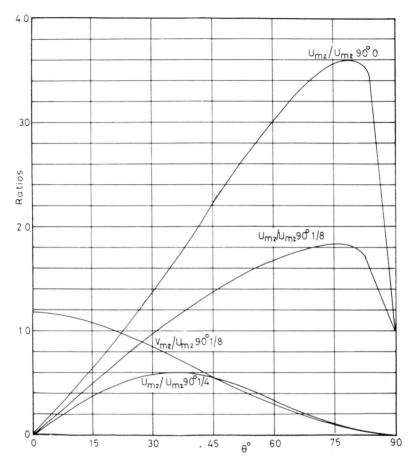

Figure 2.60 Ratio of mass transport of short-crested system (U_{M2}) to that for the incident wave ($U_{M2}90°$) for various Z/L' alignments. Angles of incident crests to the reflecting wall are shown

normal to the wall. This mass transport was displayed in Figs. 2.54 and 2.55, where the square dots represented positions at the end of each cycle, or the net movement over a wave period. At alignments of $Z/L' = 1/8, 3/8, 5/8, \ldots$, the movement parallel to the wall $(U_{M2}/U_{M2}(90°) - 1/8)$ approaches a maximum of 1.83 at $\theta = 76°$ reducing to unity at $\theta = 90°$ and to zero at $\theta = 0°$. Transport normal to the wall, with direction toward $Z/L' = 0, 1/2, 1, \ldots$ alignments, is given by $V_{M2}/U_{M2}(90°) - 1/8$, which rises to 1.18 for the standing wave ($\theta = 0°$) and decreases to zero at $\theta = 90°$. The U_{M2} and V_{M2} values are equal at $\theta = 27°$.

The mass-transport distribution is repeated in each $Z/L' = 1/4$ segment, as seen in Fig. 2.61, where the ratio of mass transport for the short-crested wave over that for progressive (U_{sc}/U_{pr}) is drawn for $\theta = 30°, 45°$ and $60°$. It is maximum along alignments $Z/L' = 0, 1/2, 1$, which is fed by ejection from the $L'/8$ regions either

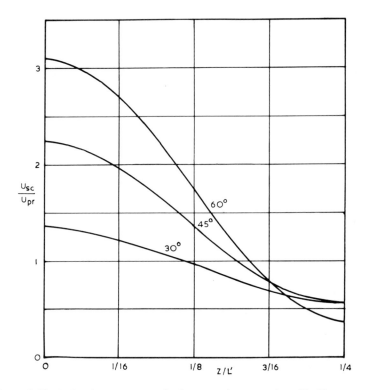

Figure 2.61 Ratio of mass transport in short-crested wave to that of incident wave (U_{sc}/U_{pr}) for various half angles between waves

side (Silvester 1985). But even at $Z/L' = 1/4$, where water-particle orbits are normal to the wall there is a net movement parallel to it, being half that for a progressive wave parallel to it. Thus, maximum orbital velocities and mass transport occur along the same alignments. The vortex motions at $Z/L' = 1/8$, 3/8, 5/8, ... produce much macroturbulence that is conducive to sediment suspension, which will be discussed in Chapter 7.

2.4 WAVE TRANSFORMATIONS

The variations in wave characteristics as they propagate from deep to shallow water are generally studied theoretically using linear theory and monochromatic waves (Le Méhauté and Wang 1980). This has been found satisfactory for most depths of interest but can become erroneous very close to the beach, where breaking also occurs. Water particles that are circular in deep water become elliptical in shoaler water with greater net movement shorewards per wave cycle. In very shallow water, orbits become almost rectilinear.

When waves arrive obliquely, with bed contours essentially parallel to the coast, they will be twisted almost normal to it by the time they break. This action is known as *refraction*, which also has an effect on the wave height. Should part of a wave crest be intercepted by some object which dissipates or reflects its energy, the remainder will bypass and twist sharply to the leeward side, a process termed *diffraction*. This continuance of the wave is accompanied by energy spreading into the shadow region with varying height along this circular curving crest. The action of reflection against objects varies with the wave height and period as also the type of face impacted. The instability of waves causing breaking depends on the slope of the bottom and the wave steepness. The waves, whether already broken or not, can run up a beach or structure and even overtop it. Some of these aspects will be outlined in this section.

2.4.1 Shoaling

Where waves can be considered arriving normal to a coast, with its bed contours parallel to it, they will become higher and lessen in length since their total energy is accepted as remaining the same. There could be some losses due to bottom friction and other factors but these can generally be ignored, particularly for studying sedimentary processes.

The assumption is therefore made that the kinetic and potential energy of an almost sinusoidal wave are equal. Integration of this over a wave length results in a total wave energy:

$$E_T = E_K + E_P = wH^2L/8 \tag{2.113}$$

taken per unit length of wave crest. The energy per unit area of ocean thus becomes:

$$E = E_T/L = wH^2/8 \tag{2.114}$$

where w is the unit weight of sea water, E_K is kinetic energy, and E_P is potential energy due to the presence of the wave.

It can be shown that power in a wave can be expressed as:

$$P = wH^2C_g/8 \tag{2.115}$$

where C_g is the *group wave velocity*, which is half the wave celerity in deep water and becomes equal to it in shallow water. This power transmitted forward by a wave in one depth will be equal to that being transmitted at any other depth. Thus:

$$wH_0^2C_{go}/8 = wH^2C_g/8 \tag{2.116}$$

or shoaling coefficient

$$K_s = \frac{H}{H_o} = \sqrt{C_{go}/C_g} = \sqrt{C_o/2C_g} \tag{2.117}$$

Since

$$\frac{C_g}{C} = \frac{1}{2}\left[1 + \frac{4\pi d/L}{\sinh 4\pi d/L}\right] \tag{2.118}$$

this ratio varies only with d/L so that H/H_o also varies only with d/L and can therefore be listed with it, as in the appendix. It is generally written as H/H_o' to distinguish it as normal approach without other influences. It is graphed in Fig. 2.62, together with values of C_g/C and $\tanh 2\pi d/L$.

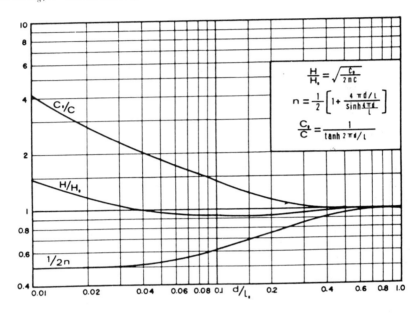

$$\frac{H}{H_o} = \sqrt{\frac{C_o}{2nC}}$$

$$n = \frac{1}{2}\left[1 + \frac{4\pi d/L}{\sinh\frac{4\pi d}{L}}\right]$$

$$\frac{C_o}{C} = \frac{1}{\tanh 2\pi d/L}$$

Figure 2.62 Variables involved in wave shoaling for linear wave theory

Shoaling of finite height waves has been derived by Stiassnie and Peregrine (1980). Using the cnoidal wave theory Shuto (1974) and other researchers have developed approximate forms, one of which is given in Fig. 2.63 (Goda 1985). Isobe (1985) has confirmed that these curves agree reasonably well with experimental data. Yasuda et al. (1982) have considered shoaling quantities to breaking.

2.4.2 Refraction

As noted previously, $C/C_o = \tanh 2\pi d/L$, so that a wave traveling obliquely to a bed contour will have part of its crest say in deep water with another section in shallower water, as depicted in Fig. 2.64 for a step change in depth. Omitting any reflection or loss of energy over this mild change in depth, it is seen that the crest must change in orientation, because the path followed by any point in it, which must be normal to it or along its orthogonal, must satisfy the equation:

$$\sin\alpha / \sin\alpha_o = L/L_o = C_d/C_o = \tanh 2\pi d/L \qquad (2.119)$$

where C_d equals C previously. Equation (2.119) is that applicable to light waves as in Snell's law.

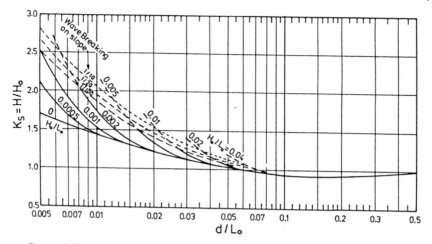

Figure 2.63 Shoaling coefficient H/H_o versus depth ratio d/L_o for various H_o/L_o (Goda 1985)

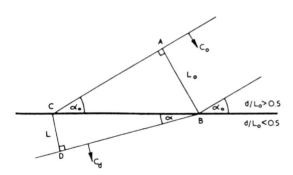

Figure 2.64 Refraction of a wave over a step

Where the change in depth is gradual (not stepped) the wave crest will be curved instead of angled. Thus, for a beach with uniformly parallel contours, as depicted in Fig. 2.65, the depths can be expressed as celerity C or d/L_o. Beyond $d/L_o > 0.5$ the wave celerity can be accepted as constant, but for smaller depths it reduces, so twisting the crests and therefore the orthogonal, as seen in the figure. The obliquity at the limiting depth of $d/L_o = 0.5$ is given as α_o, the angle between wave crest and bed contour or between the orthogonal and a normal to the bed contour. At some intermediate depth this angle has reduced to α as given by Eq. (2.119), which is plotted in Fig. 2.66 by a curve of d/L_o versus C_d/C_o, which then gives α for any given α_o and C_d/C_o. The inset provides an enlargement for C_d/C_o approaching unity.

Again for parallel contours, if the approach angle (α_1) is known from some intermediate depth rather than in deep water (α_o), the obliquity at some smaller depth contour (α_2) can be obtained from Fig. 2.66 using $C_d/C_o = (C_2/C_o)/(C_1/C_o)$. For example,

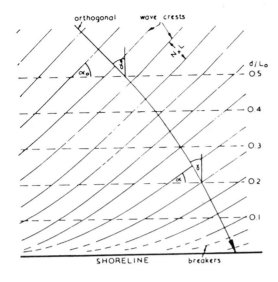

Figure 2.65 Oblique waves refracting across a uniformly sloped shelf

when

$$\alpha_o = 30°, \text{ at } d_1/L_o = 0.17, C_d/C_o = 0.85 \quad \text{giving } \alpha_1 = 25°$$

$$\alpha_o = 30°, \text{ at } d_2/L_o = 0.05, C_d/C_o = 0.52 \quad \text{giving } \alpha_2 = 15°$$

taking $\alpha_1 = \alpha_o = 25°$, $(C_2/C_o)/(C_1/C_o) = 0.52/0.85 = 0.62 = C_d/C_o$ giving $\alpha_2 = 15°$
Changes in orthogonal direction should be made midway between the contours used.

For nonparallel bed contours, as in Fig. 2.67, each successive contour is angled θ_1, θ_2, and so on to the deepwater contour (C_o). An approach angle α_o across C_o must cross C_1 with an approach angle of $\alpha_o - \theta_1$. If this causes the new orthogonal to approach C_1 normally no refraction will take place. But if $\alpha_o - \theta_1$ is finite, then α_1 is given by

$$\frac{\sin\alpha_1}{\sin(\alpha_o - \theta_1)} = \frac{C_1}{C_o} \tag{2.120}$$

which should be read in Fig. 2.66 as $\alpha_o = \alpha_o - \theta_1$ and $\alpha = \alpha_1$. Note that α_o represents any approach angle from deeper water. The use of θ_1 between contours C_o and C_1 is mainly by way of explanation, for in practice the angle between the contours can be measured by a protractor. In Fig. 2.67 the protractor base line is placed along C_1 with the 90° line through turning point A (on the bisector of the angle between C_o and C_1) to mark line B, which is normal to C_1. Then place the protractor center on A, still with the 90° line through B, and measure the approach angle $(= \alpha_o - \theta_1)$ and after using Fig. 2.66 mark α_1 to the bisector of $\theta_1 + \theta_2$. The approach angle to C_2 is $\alpha_1 + \theta_1 + \theta_2$, as seen by the extension of the orthogonal to D and the dotted line parallel to C_1. The traverse angle α_2 is then given by:

$$\frac{\sin\alpha_2}{\sin(\alpha_1 + \theta_1 + \theta_2)} = \frac{\sin\alpha_1/\sin\alpha_o}{\sin(\alpha_1 + \theta_1 + \theta_2)} = \frac{C_2/C_o}{C_1/C_o} = \frac{C_d}{C_o} \tag{2.121}$$

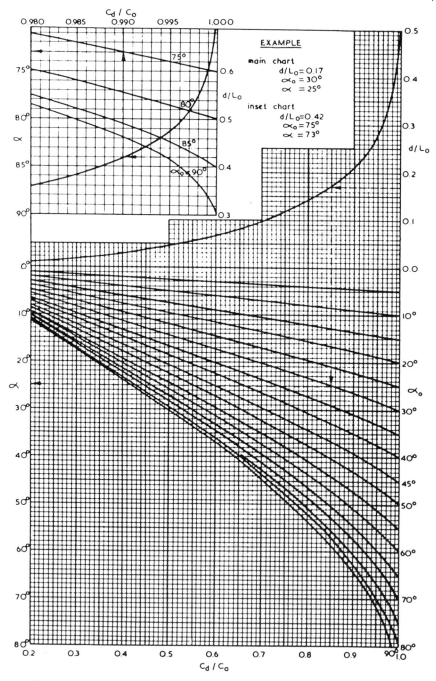

Figure 2.66 Refraction relationship between d/L_o and C_d/C_o (upper), and α_o and α (lower)

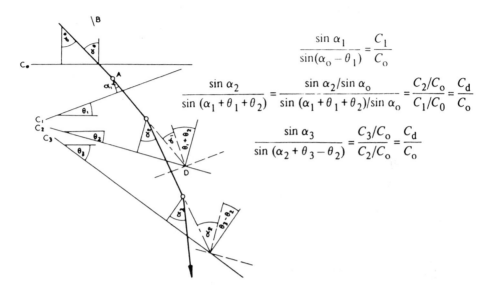

$$\frac{\sin \alpha_1}{\sin(\alpha_o - \theta_1)} = \frac{C_1}{C_o}$$

$$\frac{\sin \alpha_2}{\sin (\alpha_1 + \theta_1 + \theta_2)} = \frac{\sin \alpha_2/\sin \alpha_o}{\sin (\alpha_1 + \theta_1 + \theta_2)/\sin \alpha_o} = \frac{C_2/C_o}{C_1/C_0} = \frac{C_d}{C_o}$$

$$\frac{\sin \alpha_3}{\sin (\alpha_2 + \theta_3 - \theta_2)} = \frac{C_3/C_o}{C_2/C_o} = \frac{C_d}{C_o}$$

Figure 2.67   Definition sketch for refraction across straight contours angled to each other

In Fig. 2.66 the values of d_2/L_o and d_1/L_o give respective values of C_2/C_o and C_1/C_o, the ratio of which gives C_d/C_o, to be used with $\alpha_o = \alpha_1 + \theta_1 + \theta_2$ to find α_2. It is possible for the orthogonal to change quadrants but the principle remains the same.

Most contours can be straightened where it is envisaged the orthogonal will pass through them so that contour curvature will not be discussed, but can be seen in Silvester (1966, 1974a). The same method can be used for tracing an orthogonal seaward instead of landward. In this case Fig. 2.66 is entered from angle α but the C_d/C_o is computed as before, by taking the ratio $(C_2/C_o)/(C_1/C_o)$ as though entering from deeper water. The value of α_o is thus determined and the orthogonal continued seaward. If the respective contours are parallel the angle α_o is drawn directly, but if they are oblique to each other the angle between them is added or subtracted from α_o to obtain the correct alignment.

In Section 2.4.1 the *shoaling coefficient* $K_s = (C_o/2C_g)^{0.5}$ was derived (see Eq. 2.117). Consider now a wave train approaching a shoreline with parallel contours as in Fig. 2.68, where two orthogonals are spaced b_o apart in deepwater or b_s apart as measured along the contour. This spacing b_s remains the same at any shallower contour even though the orthogonal angle has changed. This implies that the normal distance between them has changed since:

$$b_s = b/\cos \alpha = b_o/\cos \alpha_o, \text{ or } b/b_o = \cos \alpha/\cos \alpha_o \qquad (2.122)$$

The power transmitted forward between the two orthogonals is assumed to be constant, so that:

$$w H_o^2 b_o C_{go}/8 = w H^2 b C_g/8 \qquad (2.123)$$

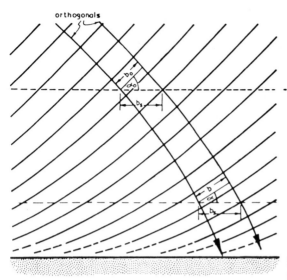

orthogonals

Figure 2.68 Orthogonal spacing over a uniformly sloped shelf

or

$$\frac{H}{H_o} = \left(\frac{C_{go}}{C_g}\right)^{0.5}\left(\frac{b_o}{b}\right)^{0.5} = \left(\frac{C_{go}}{C_g}\right)^{0.5}\left(\frac{\cos\alpha_o}{\cos\alpha}\right)^{0.5} = K_s K_r \qquad (2.124)$$

where K_r is known as *refraction coefficient*. Since α can be determined for any d/L once α_o is known, then both K_s and K_r are functions of d/L or d/L_o. Thus, the wave height can be predicted from Table 2.5 from a knowledge of H_o, α_o, and T (or d/L_o). In the table it will be noticed that for $\alpha_o = 0$, or normal approach, the height ratio decreases to 0.913 at $d/L_o = 0.15$ before rising toward the beach. This is due to C/C_g in Eq. 2.118 (designated in Fig. 2.62 as $1/2n$) decreasing faster than C_o/C rising in intermediate depths, thus reducing H/H_o initially, as seen in Fig. 2.62.

Tracing wave rays or orthogonals by the above methods can become tedious at times, but nearshore obliquities can readily be obtained without tracing orthogonals across the complete continental shelf if bottom contours are reasonably uniform. The angle of the chosen contour to the deepwater value needs to be known for the correct assessment of orthogonal orientation. Munk and Arthur (1952) have derived a method useful for computer programs (Wilson 1966). Refraction can also occur when waves cross currents (Bretherton and Garrett 1969, Stiassnie and Peregrine 1980). Refraction of wave spectra has been carried out by Longuet-Higgins (1957) and Karlsson (1969). In prototype conditions, wave refraction is often combined with the process of diffraction, which has been an area of significant advancement in numerical calculation in recent years (Berkhoff 1972, 1976; Smith and Sprinks 1975; Radder 1979; Lozano and Liu 1980; Hashimoto 1982; Berkhoff et al. 1982; Kirby and Dalrymple 1983; Booij 1983; Liu 1983; Mei 1983; Liu and Tsai 1984). This topic is not treated here.

TABLE 2-5 RATIO OF WAVE HEIGHT AT SPECIFIC DEPTH RATIO TO DEEPWATER HEIGHT FOR WAVES REFRACTING ACROSS A UNIFORMLY SLOPED SHELF

d/L_o	α_o 0	5	10	15	20	25	30	35	40	45	50	55	60	65	70	75	80	85
0.005	1.692	1.689	1.680	1.664	1.642	1.614	1.578	1.536	1.486	1.429	1.363	1.307	1.223	1.124	1.013	0.877	0.722	0.504
0.01	1.439	1.435	1.427	1.413	1.395	1.372	1.342	1.307	1.267	1.218	1.160	1.096	1.026	0.944	0.850	0.740	0.606	0.430
0.04	1.063	1.062	1.057	1.048	1.038	1.023	1.004	0.980	0.953	0.923	0.883	0.840	0.789	0.753	0.656	0.576	0.473	0.335
0.06	0.993	0.992	0.987	0.982	0.973	0.959	0.944	0.926	0.903	0.874	0.841	0.801	0.753	0.699	0.633	0.553	0.457	0.326
0.08	0.953	0.953	0.950	0.946	0.938	0.927	0.919	0.898	0.878	0.852	0.823	0.787	0.742	0.690	0.622	0.548	0.453	0.326
0.10	0.931	0.931	0.929	0.925	0.918	0.910	0.898	0.883	0.866	0.843	0.816	0.783	0.743	0.693	0.630	0.555	0.459	0.329
0.15	0.913	0.913	0.912	0.909	0.905	0.898	0.892	0.882	0.869	0.853	0.832	0.805	0.769	0.726	0.669	0.595	0.498	0.353
0.20	0.917	0.917	0.916	0.913	0.910	0.907	0.901	0.886	0.894	0.873	0.857	0.840	0.813	0.773	0.724	0.650	0.548	0.397
0.25	0.932	0.932	0.931	0.931	0.929	0.926	0.923	0.918	0.912	0.904	0.894	0.879	0.858	0.829	0.787	0.713	0.620	0.453
0.30	0.949	0.949	0.949	0.948	0.947	0.945	0.943	0.941	0.937	0.932	0.925	0.915	0.902	0.882	0.850	0.796	0.699	0.524
0.35	0.964	0.963	0.963	0.963	0.962	0.961	0.960	0.959	0.956	0.954	0.950	0.942	0.934	0.922	0.897	0.855	0.773	0.600
0.40	0.976	0.976	0.976	0.976	0.975	0.975	0.974	0.974	0.972	0.971	0.968	0.965	0.960	0.951	0.937	0.910	0.846	0.684
0.45	0.985	0.985	0.985	0.985	0.985	0.984	0.984	0.984	0.983	0.982	0.980	0.978	0.975	0.971	0.963	0.945	0.902	0.769
0.50	0.991	0.991	0.991	0.990	0.990	0.990	0.990	0.990	0.989	0.989	0.988	0.987	0.985	0.983	0.977	0.967	0.940	0.841

2.4.3 Diffraction

Waves striking a structure can be partially dissipated or reflected, but the remainder of their crests can bypass the tip and spread in a curved fashion into the shadow zone behind. Three examples are given in Fig. 2.69, where it is seen that these crests, form circles concentric on the tip known as the *diffraction point*. Energy is spread from the straight crest sections past the limiting orthogonals, shown dotted in the figure, to provide wave heights in the curved segments, which diminish the further from these orthogonals. Diffraction studies concentrate on the heights existing in this shadow zone, by determining a *diffraction coefficient* (K_d) which is the ratio of any height to that of the incident wave. As the reflected wave can also spread in circular fashion from the face, its energy can appear in this zone. This is not shown in Fig. 2.69 and is generally minor in character.

The case depicted in Fig. 2.69A, generally termed the semi-infinite breakwater, appears to be a misnomer because the inference is that the body of water beyond it is infinite in order to supply all the energy required for diffraction. Transfer of energy crestwise can be limited, as in the case of the breakwater gap in Fig. 2.69B, where wave heights in the straight segment of the waves will reduce with distance from the gap. The third case, in Fig. 2.69C, is that of an island-type structure where diffraction can take place around its extremities (Wada 1965). It is seen that the circular crests form a short-crested wave system which has been discussed in Section 2.3.3. These have particular significance in salient formation behind an offshore breakwater, as will be dealt with in Section 6.2.

Semi-infinite breakwater. The diffracting waves in Fig. 2.69A are shown as circular, but this may be distorted a little since in shoal conditions the wave height has some influence on the celerity. Since the heights will vary along the curved crests, to a minimum adjacent the leeside of the structure, wave speed can thus vary but this can normally be ignored.

The theory of water-wave diffraction emanates from that applied to optical diffraction (Penney and Price 1952). It is difficult to visualize the transfer of energy along a crest but the incident height exists in proximity to the limiting orthogonal and none adjacent to it. The closest hydraulic equivalent is that of a collapsing dam, as depicted in Fig. 2.70, where water at the original dam site remains at 4/9 of the original depth, which might be equated to the incident height (H_i). It will be shown that the height along the shadow line or limiting orthogonal remains essentially at half that of the incident wave, no matter the distance from the diffraction point. The analogy of the dam burst does not imply that water in the form of a current is passing along the crest into the shadow zone. Lim (1968) showed that this may occur for the first 0.2 of a wave length behind a breakwater but further than this no current was measurable.

The general case for the semi-infinite breakwater is illustrated in Fig. 2.71, where a train of waves is arriving at angle θ (orthogonal to structure). Waves may be reflected at an equal angle and be diffracted through $360° - 2\theta$ before adding to the diffracted energy of the incident wave. In this process these waves add energy to the incident waves so

Sec. 2.4   Wave Transformations

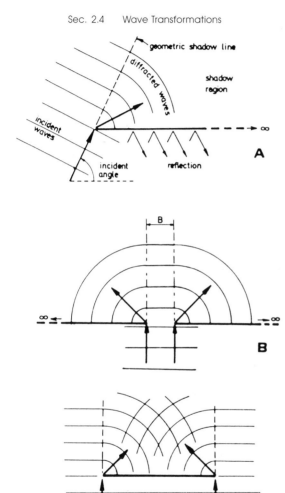

Figure 2.69 Wave diffraction: A. behind a semi-infinite breakwater, B. through a breakwater gap, and C. behind an island or offshore breakwater

increasing heights just beyond the tip of the breakwater (Fan et al. 1967). This region will not be treated in the following discussion. Despite the trend to design breakwaters for the fullest dissipation of waves, diffraction theory is normally based on 100% reflection (Silvester and Lim 1968). The location of some point P within the shadow region can be defined by angle α from the shadow line and R/L from the diffraction point, or

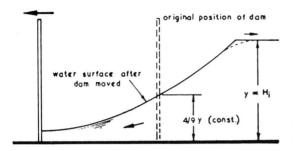

Figure 2.70 Analogy of wave diffraction to the dam break problem

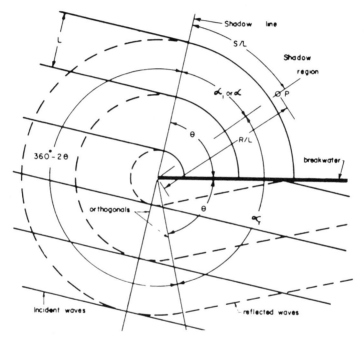

Figure 2.71 Definition sketch for semi-infinite breakwater

circumferential distance S/L and R/L. This latter system can be reduced to S/L alone with little loss of accuracy.

Silvester and Lim (1968) separated the incident and reflected wave components with insignificant difference from the exact solutions provided by Larras (1966) and Wiegel (1962). It can be seen from Fig. 2.71 that the reflected wave must be diffracted to point P through an angle of $360° - 2\theta + \alpha$. The final result for $K_d = H/H_i$ is plotted in Fig. 2.72 for various values of R/L. It is noteworthy in this figure that for $\alpha = 0$, or along the shadow line, $K_d = 0.5$ irrespective of R/L. This may be added to by the reflected component diffracting through $360° - 2\theta$ if it is in phase with the incident wave. Normally the width of breakwaters is small compared to wave length so waves are likely to be partially in phase, but for natural headlands with distances from reflecting wall to diffraction point perhaps equaling half a wave length the reflected may discount the

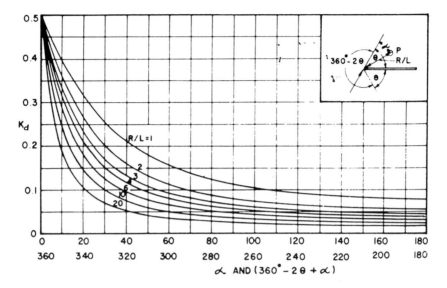

Figure 2.72 Diffraction coefficient for all incident angles with or without reflection.
For $\theta < 60°$ and $R/L < 3$ add 0.1 to K_d

incident wave. Note that Fig. 2.72 applies only for $\theta > 60°$ and $R/L > 3$, for smaller values add 0.1 to K_d.

The reduction in height from the shadow line will be similar no matter how far a wave has progressed from the diffraction point. This concept has been presented by Silvester and Lim (1968), resulting in the plot of K_d versus S/L as in Fig. 2.73, where comparison is made with tests by Putnam and Arthur (1948). The agreement with the zero reflection line is good, since their waves diffracted from a rectangular corner. Thus, in any plan of a harbor, once the shadow line is drawn, the circular arc from it to the point in question can be measured directly or computed ($\pi R \alpha° / 180° L$) in order to find S/L for applying Fig. 2.73. Partial reflection can be applied linearly between the two curves.

Breakwater gap. The definition sketch for the case shown in Fig. 2.69B is illustrated with greater detail in Fig. 2.74, where wave guides were used by Silvester (1978, 1981) to obviate wave reflection. It is understandable that with structures either side of a limited gap, reflection can play a significant part in supplying energy into the shadow region. It will increase wave heights at the entrance and for some wave lengths from the two diffraction points.

Solutions to the breakwater gap problem have been provided by Morse and Rubenstein (1938), Lacombe (1952), and Larras (1966), but that most readily applicable is Penney and Price (1952). This was used by Blue and Johnson (1949), who also modified it to extract the influence of partial reflection plus other scaling effects. Penney and Price (1952) showed K_d to be proportional to $B/(yL)^{1/2}$ along the incident orthogonal passing through the center of the gap (see Fig. 2.74). This is equivalent to $B/(LR)^{1/2}$

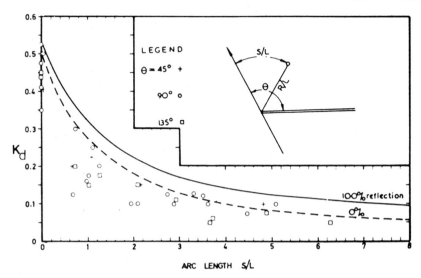

Figure 2.73 Diffraction coefficient using arc length measurement (S) from shadow line. Data from Putnam and Arthur (1948) for no reflection. For $\theta < 60°$ and $R/L < 3$ add 0.1 to K_d

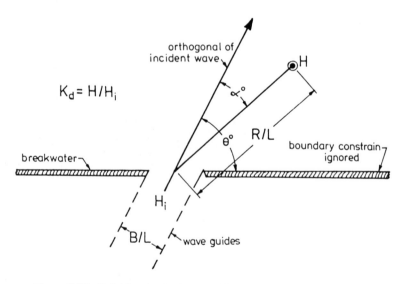

Figure 2.74 Definition sketch of waves diffracting through a breakwater gap

or $(B/L)/(R/L)^{1/2}$. This suggests that $K_d(R/L)^{1/2}/(B/L)$ could be a reasonable dimensionless parameter for comparison with other outstanding variables of θ and α, as in the figure.

Tests were carried out by Silvester (1978) which were compared to the theory of Penney and Price (1952) and shown to not follow strictly the theory applicable to sound

and light waves. There appears to be some energy spreading when differential heights
are developed along water wave crests, which is not so for other types of fluctuation.
The resulting curves, displayed in Fig. 2.75, should be conservative but closer to the
experimental results obtained and those of Mobarek (1962). Curves of $K_d(R/L)^{1/2}/(B/L)$
versus α are presented for various B/L values. These apply only for $R/B \geq 2$ and R/L
from 0.5 to 4.0. Hatching indicates a transition occurring between $K_d(R/L)^{1/2}/(B/L) =$
0.5 to 0.6, for which high values on the LHS should be low values on the RHS, and
vice versa. To aid design Table 2.6 is provided, in which values in the R/L column
relate to the extremities of the hatched areas in Fig. 2.75. Experimental results obtained
for $\alpha = 0°$ were significantly below 1.0 (produced by the theory for large R/L), which
was also found by Blue and Johnson (1949). These more realistic values have been
listed in the table. When $B/L \geq 5$ each breakwater can be considered the same as the
semi-infinite case discussed earlier.

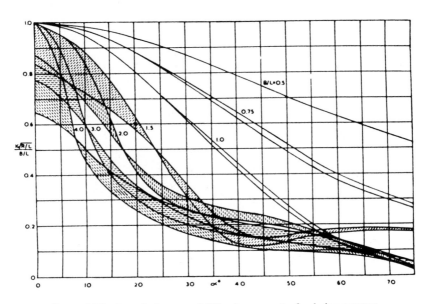

Figure 2.75   Smoothed curves of diffraction parameter for design purposes

Combined diffraction and refraction has been studied by some workers, which was
noted briefly in Section 2.4.2, but will not be discussed here. However, this procedure
takes place in the lee of headlands which are emphasized in this tome. A fine example
is presented at Crowdy Head in Fig. 2.76, which is on the coast of New South Wales,
Australia (Reader's Digest 1983). Here the major swell waves are approaching from
SSE, to arrive normal to the southern beach and diffracting around the headland and then
refracting to arrive normal to the bayed beach to the north. This action will be discussed
in greater detail in Chapter 4.

TABLE 2.6 VALUES OF $10 K_d \sqrt{R/L}/(B/L)$ VERSUS B/L AND $\alpha°$

B/L	R/L	$\alpha = 0°$	$10°$	$20°$	$30°$	$40°$	$50°$	$60°$	$70°$	$80°$
0.5	0.5	10.0	9.85	9.38	8.70	7.8	7.03	6.23	5.53	4.95
0.75	≥ 1	10.0	9.68	8.82	7.70	6.45	5.15	4.00	3.15	2.51
		10.0	9.68	8.82	7.58	6.2	4.89	3.75	2.96	2.42
1.0	2	10.0	9.47	8.06	6.26	4.45	2.75	1.47	0.72	0.40
	≥ 3	10.0	9.47	8.06	6.15	4.20	2.50	1.30	0.53	0.20
1.5	3	8.25	7.20	5.75	3.60	1.75	1.60	1.85	1.85	1.75
	≥ 4	8.85	8.85	6.25	3.05	1.25	1.35	1.65	1.75	1.65
2.0	3	8.25	6.90	4.40	3.05	2.55	2.10	1.40	0.70	0.22
	≥ 4	8.50	8.40	4.15	2.60	2.20	1.85	1.30	0.60	0.12
3.0	3	7.75	5.95	3.58	2.60	2.10	1.75	1.40	0.80	0.34
	≥ 6	8.00	6.10	3.00	2.05	1.60	1.30	1.05	0.50	0.12
4.0	4	6.5	5.00	3.00	2.00	1.60	1.30	1.05	0.72	0.28
	≥ 8	7.00	4.55	2.60	1.70	1.20	1.00	0.85	0.48	0.12

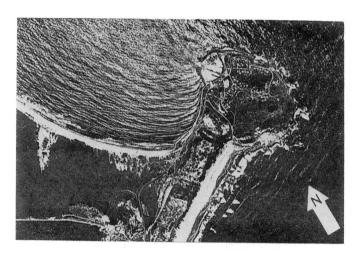

Figure 2.76 Combined diffraction and refraction around Crowdy Head, N.S.W., Australia into leeward bay (Reader's Digest, 1983)

REFERENCES

AIRY, G. B. 1845. Tides and waves. *Encyclopaedia Metropolitana,* V, Article 192, 241–396.

BERKHOFF, J. C. W. 1972. Computation of combined refraction-diffraction. *Proc. 13th Inter. Conf. Coastal Eng., ASCE 1*: 471–90.

———. 1976. Mathematical models for simple harmonic linear water waves—wave diffraction and refraction. *Delft Hydrau. Lab.,* Pub. No. 163.

BERKHOFF, J. C. W., BOOY, N., and RADDER, A. C. 1982. Verification of numerical wave propagation models for simple harmonic linear water waves. *Coastal Eng. 6*: 255–79.

BLUE, F. L., and JOHNSON, J. W. 1949. Diffraction of water waves passing through a breakwater gap. *Trans. Am. Geophys. Un. 30*: 705–18.

BOOIJ, N. 1983. A note on the accuracy of the mild-slope equation. *Coastal Eng. 7*: 191–203.

BOUSSINESQ, J. 1872. Theorie des ondes et des remous qui se propagent le long d'un canal rectangulaire horizontal, en communiquant au liquide contenu dans ce canal de vitesses sensiblement parreilles de la surface au fond. *Jour. Mathematiques Pures et Applicquees 17*: 55–108.

BOUWS, E., GÜNTHER, H., ROSENTHAL, W., and VINCENT, C. L. 1985. Similarity of the wind wave spectrum in finite depth water. 1. spectral form. *J. Geophys. Res. 90*: (C1): 975–86.

BRETHERTON, F. P., and GARRETT, C. J. 1969. Wave trains in inhomogeneous moving media. *Proc. Roy. Soc. London, Ser. A 302*: 529–54.

BRETSCHNEIDER, C. L. 1957. Hurricane design wave practice. *Proc. Waterways, Harbor Div., ASCE* 83(WW2): paper 1238.

———. 1958. Revisions in wave forecasting: deep and shallow water. *Proc. 6th Inter. Conf. Coastal Eng., ASCE 1*: 30–67.

———. 1959. Wave variability and wave spectra for wind generated gravity waves. *Beach Erosion Board*, Tech. Memo. No. 18.

———. 1977. On the determination of the design ocean wave spectrum. Look Lab., Hawaii Univ., Tech. Rep., 7, no. 1, 1–23.

BRUNT, A. T. and HOGAN, T. 1956. The occurrence of tropical cyclones in Australian region. *Trop. Cycl. Symp.*, Brisbane, 1–14.

CARTER, R. W. G. 1988. *Coastal Environments: An Introduction to the Physical, Ecological and Cultural Systems of Coastlines*. London: Academic Press.

CARTWRIGHT, D. E., and LONGUET-HIGGINS, M. S. 1956. The statistical distribution of the maxima of a random function. *Proc. Roy. Soc. London, Ser. A 237*: 212–232.

CHAPLIN, J. R. 1980. Development of stream-function theory. *Coastal Eng. 3*: 179–205.

CHAPPELEAR, J. E. 1961a. Direct numerical calculation of wave properties. *J. Geophys. Res. 66*: 501–08.

———. 1961b. On the description of short-crested waves. U.S. Army Corps Engrs., *Beach Erosion Board*, Tech. Memo. No. 125.

———. 1962. Shallow water waves. *J. Geophys. Res. 67*: 4693–704.

COKELET, E. D. 1977. Steep gravity waves in water of arbitrary uniform depth. *Phi. Trans. Royal Soc., Ser. A 286*: 183–260.

DANEL, P. 1952. On the limiting clapotis. *Gravity Waves*, Natl. Bur. Stds., Circ. 521, 35–38.

DARBYSHIRE, J. 1952. The generation of waves by wind. *Proc. Roy. Soc., Ser. A 275*: 299–328.

DEAN, R. G. 1965. Stream function wave theory: validity and application. *Proc., ASCE Specialty Conf. Coastal Eng.*, 269–300.

———. 1970. Relative validities of water wave theories. *J. Waterways and Harbors Div., ASCE* 96(WW1): 105–19.

———. 1974. Evaluation and development of water wave theories for engineering applications, Vols. I & II. *U.S. Corps of Engineers*, Coastal Eng. Res. Center, Special Rep. No. 1.

———. 1990. Stream function wave theory and applications. *Handbook of Coastal and Ocean Engineering, Vol. 1*, ed. J. B. Herbich. Houston, Texas: Gulf Publishing Co., 63–94.

DEAN, R. G., and DALRYMPLE, R. A. 1984. *Water Waves Mechanics for Engineers and Scientists*. Englewood Cliffs, N.J.: Prentice Hall.

DORE, B. D. 1974. The mass transport velocity due to interacting wave trains. *Meccanica 9*: 172–78.

ECKART, C. 1953. The theory of noise in continuous media. *J. Acoust. Soc. Am. 25*: 195–99.

FAN, S. H., CUMMING, J. E., and WIEGEL, R. L. 1967. Computed solutions of wave diffraction by semi-infinite breakwater. *Univ. Calif., Berkeley*, Tech. Rep. HEL-1-8.

FENTON, J. D. 1972. A ninth-order solution for the solitary wave. *J. Fluid Mech. 53*: 257–71.

————. 1979. A high-order cnoidal wave theory. *J. Fluid Mech. 94*: 129–61.

————. 1985. A fifth-order Stokes theory for steady waves. *J. Waterway, Port, Coastal and Ocean Eng., ASCE 111*(2): 216–34.

FUCHS, R. A. 1952. On the theory of short-crested oscillatory waves. *Gravity Waves*, U.S. Natl. Bur. Stds., Circ. 521, 187–200.

GERSTNER, F. 1802. *Theorie der Wellen*. Abhandlungen der koniglichen bomischen Gesellschaft der Wissenschafte, Prague.

GODA, Y. 1967. The fourth order approximation to pressure of standing waves. *Coastal Eng. in Japan 10*: 1–11.

————. 1974. Wave forces on a vertical cylinder: Experiments and a proposed method of wave force computation. *Port and Harbour Res. Inst.*, Tech. Rep. 8.

————. 1983. Analysis of wave grouping and spectra of long-travelled swell. *Port and Harbour Res. Inst.*, Rep. 22, No.1, 3–41.

————. 1985. *Random Seas and Design of Maritime Structures*. Tokyo: University of Tokyo Press.

————. 1990. Random waves and spectra. *Handbook of Coastal and Ocean Engineering, Vol. 1*, ed. J. B. Herbich. Houston, Texas: Gulf Publishing Co., 175–212.

GRIMSHAW, R. 1971. The solitary wave in water of variable depth. *J. Fluid Mech. 46*: 644–52.

HAMADA, T. 1965. The secondary interactions of surface waves. *Port and Harbour Res. Inst.*, Japan, Rep. No. 10.

HASHIMOTO, H. 1982. Numerical solution of the parabolic equation for wave refraction and diffraction. *Proc. 29th Japan Conf. Coastal Eng.*, JSCE, 115–19. (In Japanese)

HASSELMANN, K., et al. 1973. Measurements of wind-wave growth and swell decay during the Joint North Sea Wave Project (JONSWAP). *Deut. Hydrogr. Inst.*, Hamburg, Rep. No. 12.

HASSELMANN, K., ROSS, D. B., MÜLLER, P., and SELL, W. 1976. A parametric wave prediction model. *J. Phyical Ocean. 6*: 200–208.

HERBICH, J. B., ed. 1990. *Handbook of Coastal and Ocean Engineering, Vol. 1*. Houston, Texas: Gulf Publishing Co.

HINO, M. 1966. A theory on the fetch graph, the roughness of the sea and the energy transfer between wind and wave. *Coastal Eng. in Japan 9*: 1–26, 1966; also *Proc. 10th Inter. Conf. Coastal Eng., ASCE 1*: 18–37, 1966.

HOLLIDAY, C. 1969. On the maximum sustained winds occurring in Atlantic hurricanes. *U.S. Dept. Comm.*, Tech. Mem. WBTM-SR-45.

HORIKAWA, K. 1978. *Coastal Engineering: An Introduction to Ocean Engineering*. Tokyo: University of Tokyo Press.

HSU, J. R. C. 1979. Short-crested water waves. *Ph.D. thesis*, University of Western Australia.

————. 1990. Short-crested waves. *Handbook of Coastal and Ocean Engineering, Vol. 1*, ed. J. B. Herbich. Houston, Texas: Gulf Publishing Co., 95–174.

HSU, J. R. C., and SILVESTER, R. 1989. Model test results of scour along breakwaters. *J. Waterway, Port, Coastal and Ocean Eng., ASCE 115*(1): 66–85.

HSU, J. R. C., SILVESTER, R., and TSUCHIYA, Y. 1979. Third-order approximation to short-crested waves. *J. Fluid Mech. 90*: 179–96.

HSU, J. R. C., TSUCHIYA, Y., and SILVESTER, R. 1980. Boundary-layer velocities and mass tansport in short-crested waves. *J. Fluid Mech. 99*: 321–42.

HUGHES, S. A. 1984. The TMA shallow-water spectrum description and applications. U.S. Army Corps of Engrs., *Coastal Eng. Res. Center*, Waterways Expt. Station, Vicksburg, Miss., Tech. Rep. CERC-84-7.

HUGHES, S. A., and MILLER, H. C. 1987. Transformation of significant wave heights. *J. Waterway, Port, Coastal and Ocean Eng., ASCE 113*(6): 588–605.

HURDLE, D. P., and STIVE, R. J. H. 1989. Revision of SPM 1984 wave hindcast model to avoid inconsistencies in engineering applications. *Coastal Eng. 12*: 339–51.

IAHR-PIANC. 1986. List of sea state parameters. Joint publ. by *IAHR* Section on Maritime Hydraulics and *PIANC*, Supplement to Bulletin No. 52.

IPPEN, A. T., ed. 1966. *Estuary and Coastline Hydrodynamics.* New York: McGraw-Hill, Inc.

ISOBE, M. 1985. Calculation and application of first-order cnoidal wave theory. *Coastal Eng. 9*: 309–25.

IWAGAKI, Y., and SAKAI, T. 1970. Horizontal water particle velocity of finite amplitude waves. *Proc. 12th Inter. Conf. Coastal Eng., ASCE 1*: 309–25.

IWAGAKI, Y., SAKAI, T., TSUKIOKA, K., and SAWAI, N. 1974. Relationship between vertical distribution of water particle velocity and type of breakers on beaches. *Coastal Eng. in Japan 17*: 51–58.

JEFFREYS, H. 1925. On the formation of water waves by winds. *Proc. Royal Soc. London, Ser. A 107*: 189–206.

KAMRONRITHISORN, P. 1978. Determination of JONSWAP spectral parameters. *Master Eng. thesis,* Asian Inst. of Tech., Bangkok.

KARLSSON, T. 1969. Refraction of continuous ocean wave spectra. *J. Waterways and Harbors Div., ASCE 95*(WW4): 437–48.

KELLER, J. B. 1948. The solitary wave and periodic waves in shallow water. *Comm. Appl. Math. 1*: 323–39.

KEULEGAN, G. H., and PATTERSON, G. W. 1940. Mathematical theory of irrotational translation waves. *J. Res. National Bureau Standard,* U.S. Dept. Commerce, 24, Res. Paper RP-1272, 47–101.

KINSMAN, B. 1965. *Wind Waves.* Englewood Cliffs, N.J.: Prentice Hall.

KIRBY, J. T., and DALRYMPLE, R. A. 1983. A parabolic equation for the combined refraction-diffraction of Stokes waves by mildly varying topography. *J. Fluid Mech. 136*: 453–66.

KISHI, T. 1957. Clapotis in shallow water. *J. Public Works Res. Inst.,* Japan, 2, Paper 5, 1–10.

KITAIGORODSKII, S. A. 1961. Application of the theory of similarity to the analysis of wind-generated wave motion as a stochastic process. *Izv. Akad. Nauk., S.S.S.R. Ser Geofiz. 1*: 105–17. (English transl. 1: 73–80)

KITAIGORODSKII, S. A., KRASITSKII, V. P., and ZASLAVSKII, M. M. 1975. On Phillips' theory of equilibrium range in the spectra of wind-generated gravity waves. *J. Physical Ocean. 5*: 410–20.

KORTEWEG, D. J., and DE VRIES, G. 1895. On the change of form of long-waves advancing in a rectangular canal, and on a new type of long stationary waves. *Phil. Magazine,* 5th Ser. *39*: 422–43.

KRAFT, R. H. 1961. The hurricane central pressure and highest winds. *Mariner's Weather Log. 5*(5).

LACOMBE, H. 1952. The diffraction of a swell. A practical approximate solution and its justification. *Gravity Waves*, U.S. Natl. Bur. Stds., Circ. 521, 129–40.

LAITONE, E. V. 1959. Water waves, IV; shallow water waves. *Univ. of California,* Berkeley, Inst. Eng. Res., Tech. Rep. No. 82–11.

———. 1962. Limiting conditions for cnoidal and Stokes waves. *J. Geophys. Res. 67*: 1555–64.

LAMB, SIR H. 1932. *Hydrodynamics.* New York: Dover Publications.

LARRAS, J. 1966. Diffraction de la houle par les obstacles rectilignes semi-indéfinis sous incidence oblique. *Cah. Océanogr. 18*: 661–67.

LEBLOND, P. H., and MYSAK, L. A. 1978. *Waves in the Ocean.* Amsterdam: Elsevier.

LE MÉHAUTÉ, B. 1976. *An Introduction to Hydrodynamics and Water Waves.* New York: Springer-Verlag.

LE MÉHAUTÉ, B., and WANG, J. D. 1980. Transformation of monochromatic waves from deep to shallow water. U.S. Army Corps of Engrs., *Coastal Eng. Res. Center*, Tech. Rep. 80–2.

LE MÉHAUTÉ, B., DIVOKY, D. M., and LIN, A. C. 1968. Shallow water waves: a comparison of theories and experiments. *Proc. 11th Inter. Conf. Coastal Eng., ASCE 1*: 86–107.

LIM, T. K. 1968. Wave diffraction. *Master Eng. thesis*, Asian Inst. of Tech., Bangkok.

LIU, P. C. 1971. Normalized and equilibrium spectra of wind wave in Lake Michigan. *J. Phys. Ocean. 1*: 249–59.

LIU, P. L.-F. 1983. Wave-current interactions on a slowly varying topography. *J. Geophys. Res. 88*: 4421–26.

LIU, P. L.-F., and TSAI, T. K. 1984. Refraction-diffraction model for weakly nonlinear water waves. *J. Fluid Mech. 141*: 265–74.

LONGUET-HIGGINS, M. S. 1952. On the statistical distribution of the heights of sea waves. *J. Marine Res. 11*: 245–66.

———. 1953. Mass transport in water waves. *Phil. Trans. Roy. Soc., Ser. A 245*: 535–81.

———. 1957. On the transformation of a continuous spectrum by refraction. *Proc. Camb. Phil. Soc. 53*(I), 226–29.

———. 1963. The effect of non-linearities in statistical distributions in the theory of sea waves. *J. Fluid Mech. 17*: 459–80.

———. 1969. A non-linear mechanism for the generation of sea waves. *Proc. R.Soc., Ser A 311*: 371–89.

LONGUET-HIGGINS, M. S., and STEWART, R. W. 1960. Changes in form of short gravity waves on long waves and tidal currents. *J. Fluid Mech. 8*: 565–83.

LONGUET-HIGGINS, M. S., CARTWRIGHT, D. E., and SMITH, N. D. 1963. Observations of the directional spectrum of sea waves using the motions of a floating buoy. *Proc. Conf. Ocean Wave Spectra,* 111–31.

LOZANO, C. J., and LIU, P. L-F. 1980. Refraction-diffraction model for linear surface water waves. *J. Fluid Mech. 101*: 705–20.

MASSEL, S. R. 1989. *Hydrodynamics of Coastal Zones.* Amsterdam: Elsevier.

MAYEÇON, R. 1969. Etude statistique des observations de vagues. *Cah. Oceanogr. 21*: 487–501.

McCOWAN, J. 1891. On the solitary waves. *Phil. Magazine,* 5th ser. *32*: 45–48.

MEI, C. C. 1983. *The Applied Dynamics of Ocean Surface Waves.* New York: Wiley-Interscience.

MEI, C. C., LIU, P. L.-F., and CARTER, T. G. 1972. Mass transport in water waves. *Ralph M. Parsons Lab., M.I.T.*, Rep. No. 146.

MICHÉ, R. 1944. Mouvements undulatoires des mers en profondeur constante ou décroissant 4. *Ann. Ponts Chaussées 114*, 25–78; 131–64; 270–92; 369–406.

MILES, J. W. 1957. On the generation of surface waves by shear flows. *J. Fluid Mech., 3*: 185–204.

MITSUYASU, H. 1968. On the growth of the spectrum of wind-generated waves I. *Res. Inst. Applied Mechanics*, Kyushu Univ., Japan, Rep. 16, No. 55, 459–82.

———. 1969. On the growth of the spectrum of wind-generated waves II. *Res. Inst. Applied Mechanics*, Kyushu Univ., Japan, Rep. 17, No. 59, 235–48.

———. 1973. On the growth of duration-limited wave spectra. *Res. Inst. Applied Mechanics*, Kyushu Univ., Japan, Rep. 20, No. 66.

———. 1975a. On the growth of duration-limited wave spectra. *Res. Inst. Applied Mechanics*, Kyushu Univ., Japan, Rep. 23, 31–60.

———. 1975b. The one-dimensional wave spectra at limited fetch. *Res. Inst. Applied Mechanics*, Kyushu Univ., Japan, Rep. 23, No. 72.

MOBAREK, I. E. 1962. Effects of bottom slope on wave diffraction. *Univ. Calif. Berkeley,* Tech. Rep. HEL-1-1.

Monthly Meteorological Charts, H.M.S.O. London, or *Atlas of Sea and Swell Charts*, U.S. Govt. Printing Office, Washington, D.C. (data listed for various oceans).

MORSE, P. M. and RUBENSTEIN, P. J. 1938. The diffraction of waves by ribbons and slits. *Phys. Rev. 54*: 895–98.

MOSKOWITZ, L. 1964. Estimates of power spectrum for fully developed seas for speeds of 20 to 40 knots. *J. Geophys. Res. 69*: 5161–79.

MOSKOWITZ, L., PIERSON, W. J., and MEHR, E. 1962. Wave spectra estimated from wave records obtained by O.W.S. Weather Explorer and O.W.S. Weather Reporter, 1, 2. New York Univ., Tech. Rep. 3.

MUNK, W. H. 1957. Letter to the Editor. *Trans Am. Geophys. Un. 38*: 778.

MUNK, W. H., and ARTHUR, R. S. 1952. Wave intensity along a refracted ray. *Gravity Waves*, U.S. Nat. Bur. Stand., Circ. 521, 95–109.

NEUMANN, G. 1953. On ocean wave spectra and a new method of forecasting wind generated sea. U.S. Army Corps of Engrs., *Beach Erosion Board*, Tech. Mem. 43.

NODA, H. 1968. A study on mass transport in boundary layers in standing waves. *Proc. 11th Inter. Conf. Coastal Eng., ASCE 1*: 227–47.

PENNEY, W. G., and PRICE, A. T. 1952. The diffraction theory of sea waves and shelter afforded by breakwater. *Phil. Trans. Roy. Soc., Ser. A 224*: 236–53.

PHILLIPS, O. M. 1957. On the generation of waves of turbulent winds. *J. Fluid Mech. 2*: 417–45.

———. 1958. The equilibrium range in the spectrum of wind generated waves. *J. Fluid Mech. 4*: 426–34.

———. *The Dynamics of the Upper Ocean.* 2nd ed. Cambridge, UK: Cambridge University Press.

PHILLIPS, O. M., and KATZ, E. J. 1961. The low frequency components of the spectrum of wind-generated waves. *J. Fluid Mech. 19*: 57–69.

PIERSON, W. J. 1964. The interpretation of wave spectrums in terms of the wind profile instead of the wind measured at a constant height. *J. Geophys. Res. 69*: 5191–5203.

PIERSON, W. J., NEUMANN, G., and JAMES, R. W. 1955. Practical methods for observing and forecasting ocean waves. *U.S. Hydrogr. Office,* Publ. 603.

PIERSON, W. J., and MOSKOWITZ, L. 1964. A proposed spectral form for fully developed wind seas based on the similarity theory of S. A. Kitaigorodskii. *J. Geophys. Res. 69*: 5181–90.

PUTMAN, J. A., and ARTHUR, R. S. 1948. Diffraction of water waves by breakwaters. *Trans. Am. Geophys. Union 29*: 481–90.

RADDER, A. C. 1979. On the parabolic equation method for water-wave propagation. *J. Fluid Mech. 95*: 159–76.

RAYLEIGH, LORD. 1880. On the resultant of a large number of vibrations of the same pitch and of arbitrary phase. *Phil. Mag. 10*: 73–78.

READER'S DIGEST. 1983. *Guide to the Australian Coast.* Sydney, Australia: Reader's Digest Services Pty Ltd.

RENARD, R. J., and LEVINGS, W. H. 1969. The navy's numerical hurricane and typhoon forecast scheme: application to 1967 Atlantic storm data. *J. Appl. Meteorol. 8*: 719–25.

RICE, S. O. 1944–1945. Mathematical analysis of random noise. *Bell System Tech. J. 23*: 282–332; *24,* 46–156.

ROBERTS, A. J. 1983. Highly nonlinear short-crested waves. *J. Fluid Mech. 135*: 301–21.

RUSSELL, R. C. H., and OSORIO, J. D. C. 1958. An experimental investigation of drift profiles in a closed channel. *Proc. 6th Inter. Conf. Coastal Eng., ASCE* 171–193.

SCHWARTZ, L. W. 1974. Computer extension and analytical continuation of Stokes expansion for gravity waves. *J. Fluid Mech. 62*: 553–78.

SCOTT, J. R. 1968. Some average sea spectra. *Q. Trans. R. Inst. Mar. Archit. 110*: 233–39.

SHEPPARD, P. A. 1958. Transfer across the earth's surface and through the air above. *Q. J. R. Meteorol. 84*: 205–24.

Shore Protection Manual. 1984. 4th ed., U.S. Army Corps Engrs., *Coastal Eng. Res. Center,* U.S. Govt. Printing Office, Washington, D.C.

SHUTO, N. 1974. Nonlinear long waves in channel of variable section. *Coastal Eng. in Japan 17*: 1–12.

SILVESTER, R. 1966. An aid to constructing wave-refraction diagram. *Trans. Inst. Engr. Aust.* CE 8, 123–27.

———. 1974a. *Coastal Engineering, 1.* Amsterdam: Elsevier.

———. 1974b. Water particle orbits in deep to shallow water waves. *Proc. 5th Austral. Conf. Hyd. and Fluid Mech.,* 310–16.

———. 1978. Diffraction through a breakwater gap. *Proc. 4th Inter. Conf. Coastal Eng., ASCE,* 128–31.

———. 1981. Diffraction through a breakwater gap. *Trans. Instn. Engrs. Aust.,* CE23(2): 114–17.

———. 1985. Sediment by-passing across coastal inlets by natural means. *Coastal Eng. 9*: 327–46.

SILVESTER, R., and LIM, T. K. 1968. Application of wave diffraction data. *Proc. 11th Inter. Conf. Coastal Eng., ASCE 1*: 248–70.

SILVESTER, R., and VONGVISESSOMJAI, S. 1970. Energy distribution curves of developing and fully arisen seas. *J. Hydr. Res., IAHR 8*: 493–521.

———. 1971. Computation of storm waves and swell. *Proc. Instn. Civil Engrs. 48*: 259–83.

————. 1978. Spectral growth of waves to the fully arisen sea. *Proc. Inter. Conf. on Water Res. Eng.*, A.I.T., Bangkok, *1*: 375–94.

SMITH, R., and SPRINKS, T. 1975. Scattering of surface waves by a conical island. *J. Fluid Mech. 72*: 373–84.

SNODGRASS, F. E., GROVES, G. W., HASSELMAN, K. F., MILLER, G. R., MUNK, W. H., and POWERS, W. H. 1966. Propagation of ocean swell across the Pacific. *Phil. Trans. Roy. Soc., Ser. A 259*: 431–97.

STANTON, T. 1937. The growth of waves on water due to the action of wind. *Proc. Roy. Soc., Ser. A 137*: 283–93.

STEWART, R. W. 1961. The wave drag of wind over water. *J. Fluid Mech. 10*: 189–94.

STIASSNIE, M., and PEREGRINE, D. H. 1980. Shoaling of finite-amplitude surface waves on water of slowly-varying depth. *J. Fluid Mech. 97*: 783–805.

STOKER, J. J. 1957. *Water Waves*. New York: Interscience.

STOKES, G. G. 1847. On the theory of oscillatory waves. *Math. and Phys. Papers,* London: Cambridge University Press, *1*: 314–26.

SVENDSEN, IB A. 1974. *Cnoidal Waves over a Gently Sloping Bottom.* Inst. of Hydrodynamics and Hydraul. Eng., *Tech. Univ. of Demark, Ser. Paper 6.*

SVERDRUP, H. V., and MUNK, W. H. 1947. Wind sea and swell theory of relations for forecasting. *U.S. Hydrogr. Office*, Publ. 601.

SWART, D. H., and LOUBSER, C. C. 1979a. Vocoidal water wave theory, vol. 1, derivation. Stellenbosch, South Africa, *CSIR* Res. Report 357.

————. 1979b. Vocoidal water wave theory, vol. 2, verification. Stellenbosch, South Africa, *CSIR* Res. Report 360.

SYONO, S. 1963. Structure of typhoons. *Proc. Inter. Seminar Tropical Cyclones, 1962, Tokyo,* Meteorol. Agency Tech. Rep. 21, 121–31.

TADJBAKHSH, I., and KELLER, J. B. 1960. Standing surface waves of finite amplitude. *J. Fluid Mech. 8*: 442–51.

TANAKA, N., IRIE, I., and OZASA, H. 1972. A study on the velocity distribution and mass transport caused by diagonal, partial standing waves. *Rep., Port and Harbour Res. Inst.* Japan *11*(3): 112–40. (In Japanese)

THOM, H. C. S. 1971. Asymptotic extreme value distributions of wave heights in the open ocean. *J. Mar. Res. 29*: 19–27.

THOMPSON, E. F., and VINCENT, C. L. 1985. Significant wave height for shallow water design. *J. Waterway, Port, Coastal and Ocean Eng., ASCE, 111*(5): 828–42.

TOBA, Y. 1973. Local balance in the air-sea boundary process. III: On the spectrum of wind waves. *J. Oceanogr. Soc. Japan 19*: 209–20.

TUCKER, M. J. 1956. A ship-borne wave recorder. *Trans. Inst. Nav. Archit. 98*: 236–50.

————. 1957. The analysis of finite length records of fluctuating signals. *Brit. J. Appl. Phys. 8*: 137–42.

————. 1961. Simple measurement of wave records. *Proc. Conf. Wave Recording, Civil Eng., Natl. Inst. Oceanogr.,* 22–23.

URSELL, F. 1953. The long wave paradox in the theory of gravity waves. *Proc. Cambr. Philos. Soc. 49*: 685–94.

VINCENT, C. L. 1982. Depth-limited significant wave height: a spectral approach. U.S Army Corps of Engrs., *Coastal Eng. Res. Center*, Waterways Expt. Station, Vicksburg, Miss., Tech. Rep. 82–3.

———. 1984. Shallow water waves: a spectral approach. *Proc. 19th Inter. Conf. Coastal Eng., ASCE 1*: 370–82.

———. 1985. Depth-controlled wave height. *J. Waterway, Port, Coastal and Ocean Eng., ASCE 111*(3): 459–75.

VINCENT, C. L., and HUGHES, S. A. 1985. Wind wave growth in shallow water. *J. Waterway, Port, Coastal and Ocean Eng., ASCE 111*(4): 765–70.

VINCENT, C. L., and RESIO, D. T. 1990. Wave forecasting and hindcasting in deep and shallow water. *Handbook of Coastal and Ocean Engineering*, Vol. 1, ed. J. B. Herbich., Houston, Texas: Gulf Publishing Co., 213–48.

WADA, A. 1965. On a method of solution of diffraction problems. *Coastal Eng. in Japan 8*: 1–19.

WALDEN, H. 1963. Comparison of one-dimensional wave spectra recorded in the German Bight with various theoretical spectra. *Proc. Conf. Ocean Wave Specrtra 1963*, 67–91.

WALLET, A., and RUELLAN, F. 1950. Trajectories of particles within a partial clapotis. *La Houille Blanche 5*: 483–89.

WEHAUSEN, J. V., and LAITONE, E. V. 1960. Surface Waves. *Handbuch der Physik,* ed. W. Flügge, Berlin: Springer-Verlag, 9: 446–78.

WHITHAM, G. B. 1974. *Linear and Nonlinear Waves.* New York: Wiley-Interscience.

WIEGEL, R. L. 1962. Diffraction of waves by semi-infinite breakwaters. *Proc. ASCE 88*(HY1): 27–44.

———. 1964. *Oceanographical Engineering.* Englewood Cliffs, N.J: Prentice Hall.

WILSON, B. W. 1960. Note on surface wind stress over water at low and high wind speeds. *J. Geophys. Res. 65*: 3377–82.

WILSON, W. S. 1966. A method for calculating and plotting surface wave rays. U.S. Army Corps of Engrs., *Coastal Eng. Res. Center,* Tech. Mem. No. 17.

WU, J. 1969. Wind stress and surface roughness at the sea interface. *J. Geophys. Res. 74*: 444–55.

———. 1971. Observations on long waves sweeping through short waves. *Tellus 23*: 364–70.

YASUDA, T., GOTO, S., and TSUCHIYA, Y. 1982. On the relation between changes in integral quantities of shoaling waves and breaking inception. *Proc. 18th Inter. Conf. Coastal Eng., ASCE 1*: 22–37.

YOUNG, I. R. 1988a. A parametric model for tropical cyclone waves. *Uni. College, Aust. Defense Force Acad.,* Uni. of New South Wales, Res. Pap. 28.

———. 1988b. Parametric hurricane wave prediction model. *J. Waterway, Port, Coastal and Ocean Eng., ASCE 114*(5): 637–52.

Beach Processes

A major distinction can be made between storm waves that are still being generated when a cyclone passes near or across the coast, and swell waves that arrive from distant storms. Their action both offshore and within the surf zone is opposite in character, which results in significant transverse motion of sediment between the beach and the offshore bar, as well as transport alongshore. Thus, any study of processes occurring along the shoreline requires an assessment of the characteristics of these two wave conditions, or what may be termed a *wave climate*.

Although in the past great attention has been devoted to littoral drift in the surf zone, there is the wider zone beyond the breakers where sediment transport occurs, out to the ultimate reach of the longest waves of engineering significance. Since it has been shown in Chapter 2 that the optimum waves in storms and hence the resulting swell have a period about 13 seconds, disturbance of the bed by waves can extend down to 130 meters. This may be accepted as the depth along the continental shelf and hence it is this width that constitutes the highway for longshore sediment transport (Silvester 1974).

Whereas the longitudinal component of wave energy decreases as waves refract, almost normal to shore as they approach the beach, the breaking of waves at a slight angle generates a littoral current, which maximizes transport along the coast in the surf zone. This capacity to move material varies greatly over time due to the changing profile of the beach from a swell-built parabolic cross section to an undulating storm surface with one or more bars parallel to the beach. The process of storm waves removing the beach berm and submerging it in the form of a bar can be likened to putting it into circulation, for the waves to act upon. The milder waves or swell sweep this material back to reform the beach while at the same time transporting it downcoast, when they arrive at an angle to the coast.

Once the bar is fully formed and is breaking the majority of incoming storm waves, and subsequently the swell waves, the surf zone is at its widest and the breaker heights greatest. It is at this time that the littoral current plus littoral drift are optimized, decreasing in magnitude as the bar is removed, and the profile reverts back to the swell-built curve of increasing depths, when the surf zone is of minimal width. While the sand is stored in the beach berm, the waves can only activate the beach face or a small fraction of the total volume and hence littoral drift becomes negligible.

Although the bar is being moved shoreward, after the peak of the storm, breaker heights are greater and their asymmetry produces a stronger current which is acting over a large width of surf zone. All these factors make for changing transport capacity which are generally not taken into account when assessing average littoral drift over a year. This is no doubt the reason for the large fluctuations in predictive formulae. Besides lateral movement during storm sequences, longer-term variations occur due to supply conditions from rivers, impediments to transport by breakwaters and groins, and differential application of energy along the coast.

Tidal ranges around the bulk of the world's shorelines are 2 m or less, which will not affect the discussion on beach processes. Larger tides tend to smooth out the profile and result in wider surf zones that oscillate transversally. The resulting offshore bar is steeper in character but is smoothed more readily with a falling tide. Hence, measurements of profiles on such beaches will differ from those with negligible tides. Where large tidal oscillations occur there is a greater propensity for rises in water level from storm surge.

A distinction can be made between oceanic shorelines with persistent swell and enclosed seas with a preponderance of storm wave activity. A third class, of partially enclosed seas, introduces coasts where some parts suffer swell action and others on which locally generated waves are predominant. Tidal ranges can also vary around their periphery.

Estuaries where fresh water issues into the ocean encompass stratification, with river outflow at the surface maintaining a salt water wedge at the mouth or upstream from it. This produces a landward current beneath the wedge at the bed whilst a seaward flow occurs upstream from its toe. Siltation occurs at this null point whose position varies with the stage of the tide. The action of swell waves in sweeping sediment shoreward causes estuaries to continually silt up.

Most coastal problems are involved with sand but some shorelines consist of cohesive soils such as silt or mud. This material cannot be blown into dunes and hence such shorelines are generally swampy or at a level of high tide only. The muddy shore consists mainly of a vertical solid face with a very mild offshore profile, consisting of very mobile flocks suspended above the bottom. Where such depositions are wide the waves are attenuated greatly and hence do not reach the shoreline itself. There are many problems of channel siltation resulting from mass transport of this fluid bed into these depressions.

Similar to sandy beaches, shingle beaches are also permeable, thus being regarded as a natural defense against the onslaught of storm waves. The major differences between these two materials is that the shingle foreshore has steeper slope, bigger prevailing waves in front, and diminishing swash due to its high permeability. At some locations, their direction of littoral drift may be opposite to each other due to different wave conditions required for their movement.

3.1 ORIGIN OF BEACHES AND DUNES

The margin between the sea and vegetated land can comprise an almost horizontal surface of sand or cobbles. Coastlines can also contain cohesive soils or mud, but these and the shingle versions will be discussed separately from sandy shorelines in Sections 3.9 and 3.10. A typical cross section of such a segment is shown in Fig. 3.1, where the berm is backed by a foredune, followed by a first dune and multiple dunes beyond. This occurs where the landmass is building seaward due to continual accretion caused by deposition of sand where littoral drift is virtually continuous.

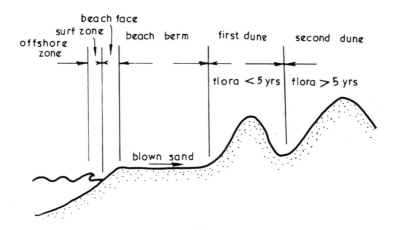

Figure 3.1   Definition sketch of a typical cross section of a dune system

Despite the inferred stability of such a coast the beach berm and sometimes the foredune may disappear due to storm waves removing them seaward and forming an offshore bar. This transient erosion, which may last for only a few days but could extend to some months, cannot be eliminated by man. All he can do is to plan his infrastructure on the coastal margin to be out of harm's way during these sequences. Longer-term erosion, caused by the imbalance of sand removed from some segment to that replacing it, is also under human control by various means which influence the longshore movement of sediment.

It is always a source of wonder to the authors why this margin of pure sand, very dynamic in nature, should exist between the sea, with its changing wave conditions, and the land, with its great variety of plants, bushes, and trees. Why does nothing grow on the beach? As noted already, storm events on the coast lead to the removal of part or all the beach berm out to sea, together with any seeds that would have been deposited on it since the last storm. As the bar is being moved back to shore by the milder waves after the storm, the seeds contained therein act like very fine sediment.

As seen in Fig. 3.2, when the crest of a swell wave passes an offshore region all sandlike particles are in suspension and hence moved landward. But before the trough arrives the orbital motion of water has a downward and zero horizontal motion which tends to ground sediment particles, particularly the heavier or larger sizes. Finer components remain in suspension and are then acted upon by the seaward segment of the water-particle orbit. In these shallow-water conditions this lasts longer than the crest motion landward. In this way the finer or lighter elements of the sediment complex are sorted seaward, while the coarser fractions are moved landward. Thus the lightweight seeds are lost to the beach berm and hence no plant growth can take place.

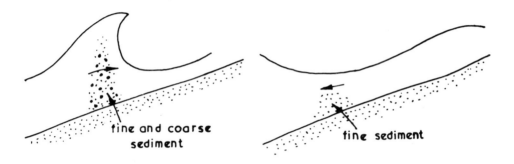

Figure 3.2   Effect of wave sorting on various sizes of particles

At the landward edge of the berm, which may have not been demanded for the past year or two, salt-tolerant species may be established. It is this green margin that should be used in measuring the width of a berm. As sand is blown landward by sea breezes or other winds its velocity adjacent to the beach surface is impeded by plants already growing. This aids in deposition so that a dune or foredune commences to form, depending on rates of sand supply and plant growth so vegetation may survive or be buried by an increasing height of the dune. This foredune could be usurped by waves during the less frequent fiercer storms, resulting in a wider beach and hence greater windblown volume to reestablish the foredune.

A delicate balance exists between erosive and accretive mechanisms. On a shoreline that is stable the sand continually fed to a foredune turns it into a first dune of considerable height. This prevents the accumulating sand from smothering vegetation in the swales and on the crests of other dunes landward of it. An unstable beach, on the other hand, with periodic removal of beach, suffers a surfeit of blown sand that can bury plants not only in the foredune but also those established on the first and second dunes and thereon. Such areas are called *blowouts*.

Although sand can be a source of pleasure in the form of beautiful beaches or dunes, its mobility under wave action and currents, or by wind in the dry condition, can be of great concern to engineers and geologists. Lack of sand in one place is generally not connected with the problems of surplus in another, but in fact they are all related. It is the longshore transport of material that provides most coastal engineering problems.

3.1.1 Coastal Terminology

Although hypotheses on the genetic origin and evolution of coastal landforms differ among geologists, terminology that defines various coastal features along a beach profile is basically identical. A typical beach profile with related terms, as shown in Fig. 3.3 (*Shore Protection Manual* 1984), has been accepted by geographers, geologists, geomorphologists, and coastal engineers alike. As seen in this figure, *shoreline* is generally defined as the line "where land and water meet," which migrates to and fro with the tide, so referred to as high water shoreline or low water shoreline, etc. The exact state of the shoreline is dependent upon the state of tide, wave conditions, and slope of the beach. *Beach* (or *shore*) is the zone from mean low water line to the inner edge of the landward limit of effective wave action; normally to the foot of a coastal cliff or to the line of permanent vegetation. A beach includes *foreshore* and *backshore*, the former lying between the low and high water marks where waves uprush and backwash, and the latter extending from the high water shoreline to the landward limit of waves during the most severe storms.

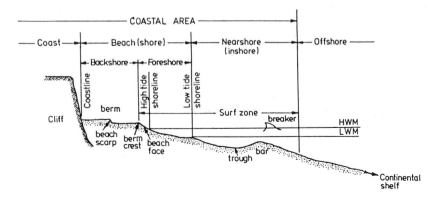

Figure 3.3   Definition sketch of a typical beach profile (*Shore Protection Manual* 1984)

The term *coast*, which has often been mistaken as beach, is defined as "a strip of land of indefinite width (maybe several miles) that extends from the shoreline inland to the first major change in terrain features" (*Shore Protection Manual* 1984). Therefore, coast includes only the broad zone of land directly landward from the foot of a cliff or the edge of permanent vegetation, but not any region containing water. However, a common term of *coastal area* could include coast, beach (or shore), and inshore (which is the zone of variable width extending from the low water line through the breaker line). The definitions for *nearshore* and *offshore* are also given in Fig. 3.3. A *coastline* can be interpreted in two different ways. Technically speaking, coastline is the line that forms the boundary between the coast and the beach (see Fig. 3.3), but it has been commonly taken as the line that forms the boundary between the land and the water, which is equivalent to *shoreline*. The total length of the world coastline is estimated at 500,000 km (Bird 1984), of which only 20% are sandy beaches (The Institution of Civil Engineers 1985).

3.1.2 Sand on Beaches

Sand consists of tiny grains of quartz, which are weathered from granite or other igneous rocks by mechanical or chemical means. It is composed of silicon (Si) and oxygen (O) bounded chemically to form silica (SiO_2). A sand grain is generally of glassy appearance but can contain other minerals to give it distinct coloration. There are pink beaches in Bermuda, purple beaches in Western Australia, black beaches in Hawaii, and white coral debris in most tropical areas. The most beautiful beach consists perhaps of white sand that has a calcium carbonate skin added to it, after a lengthy stay in warmish seawater.

Sand can occur in a variety of sizes, as seen in Table 3.1, where the whole range of sediment dimensions is given for convenience. The relationship between particle size in millimeter and microns (μ m) is straightforward, while the ϕ-scale is a logarithmic scale to base 2 of the particle diameter D in millimeters, such that $\phi = -\log_2 D = -3.322 \log_{10} D$. Theoretically, the size of beach sediment can be determined by a visual comparison with a standard, by sieve or settling tube analysis (*Shore Protection Manual* 1984). The coarser the material the steeper the beach formed from it, as witnessed by the mean slope of beach face (Shepard 1963) in Table 3.1. The majority of sand on the coast was originally discharged there by rivers, delivering the weathered components from their catchments, with some from eroding cliffs and other sources. Geologically young countries, generally with short steep rivers, transport mainly coarse material such as cobbles. The older and larger continents discharge sand, whilst the very long rivers, that necessarily pass through relatively flat expanses, deliver suspended mud and silt. The coastal margins of the world can be related to the sediment supply from these major inputs (McGill 1958).

TABLE 3.1 CLASSIFICATION OF MATERIAL ON BEACHES

Wentworth description	Particle diameter D (mm)	Microns (μ m)	ϕ scale $\phi = -\log_2 D$	Mean slope of beach face
Clay	$\leq 1/256$	≤ 3.9	> 8	
Silt	$1/256 \sim 1/16$	$3.9 \sim 62.5$	$8-4$	
Sand: very fine	$1/16 \sim 1/8$	$62.5 \sim 125$	$4-3$	$1°$
fine	$1/8 \sim 1/4$	$125 \sim 250$	$3-2$	$3°$
medium	$1/4 \sim 1/2$	$250 \sim 500$	$2-1$	$5°$
coarse	$1/2 \sim 1$	$500 \sim 1000$	$1-0$	$7°$
very coarse	$1 \sim 2$	≥ 1000	$0 \sim -1$	$9°$
Granule	$2 \sim 4$		$-1 \sim -2$	$11°$
Pebble	$4 \sim 64$		$-2 \sim -6$	$17°$
Cobble	$64 \sim 256$		$-6 \sim -8$	$24°$
Boulder	≥ 256		< -8	

This natural process of gravity fed water and sediment delivery to the coast has been observed by McIntyre (1977): "The basic nature of a shore is determined by the geology and topography of the coastline and adjacent land, and by the physical processes operating on them, particularly the actions of rivers and waves which create, supply, and distribute sedimentary material." Besides rivers, some material transported and deposited on the coast is gained from the collapse of sea cliffs and also faunal debris and chemical precipitation in warm seas. These make a small percentage of the total taken globally but can be significant on a local scale.

Sand is very durable since in submerged condition it can travel 100 miles with only 0.01% to 0.05% loss in mass. Aeolian transport, however, can effect 5% to 10% loss over a similar distance (McIntyre 1977). The finer the grain the less likely the loss from abrasion. Sand particles are noncohesive and when wet behave like a fluid. When subjected to rising groundwater, a state of liquefaction can be produced which prevents the sand from supporting an applied load. Even mild slopes existing on the seabed can slump due to such action of waves (Polous 1988).

The vegetation on the foredune, first dune, and previously formed dunes will differ, due to the time available over which it has existed. The greater the height of a dune the more sparse its plant cover because of the greater distance to the water table. The swales between will have denser growth due to the proximity of water. A dune system can thus be readily perceived from aerial photographs, as seen in Figs. 3.4 and 3.5, in which the orientations depict previous shorelines since dunes are quickly stabilized by vegetation. Those further inland will have more mature plants, even trees, while those nearer the sea margin are covered in more modest bush. Most sedimentary land areas will have been deposited under marine conditions, with ground undulations of dune character that may be quite unrelated to the current wind conditions over them.

As mentioned previously, the grain size of sediment on a beach is closely related to the wave energy acting on it, hence introducing the phenomenon of sorting of sediments by waves. Sorting is a statistical term which defines the degree of uniformity of grain size distribution for sediments present at a location, or places alongshore or across a beach profile. As there are constant adjustments of these profiles to suit the changing wave conditions, their morphology and sediment distribution will reflect the degree of exposure of a beach to these external conditions. From the relationship between the approximate range of beach slope against sand grain size, as given in Fig. 3.6 (see also Table 3.1), it can be said that finer materials are usually found on gentler slopes. The scattering apparent in this figure is attributable to variations in wave steepness and wave period.

3.1.3 Blowouts

An abundance of sand is available on wide beach berms, as exist on cuspate forelands or tombolos in the lee of islands or offshore reefs. Much material is available to bury developing vegetation, which zone is termed a *blowout*. This would appear to be a misnomer and perhaps should be replaced with a word purporting burial rather than removal. Although it is acknowledged that destruction of plants by fire or earth-moving

Figure 3.4   Ancient dune lines parallel to the coast with current and ancient blowouts
exhibited

equipment can initiate erosion by wind, such instances are very localized and can be
remedied quickly by Nature or by man. However, the large tracts of moving sand
covering existing growth, including sometimes fully grown trees, are mainly caused by
aeolian forces acting on an adjacent wide beach.

As seen in Fig. 3.7, excess sand accumulation can occur at the apex of a cuspate
foreland. Also as observed in Fig. 3.8, material can be blown along a beach by incessant
sea breezes to accumulate on a section of coast. In this figure the shape of the resultant
mass is indicative of the wind direction. Normal dune accretion together with current
and ancient blowouts are exhibited in Fig. 3.4, where parallel lines indicate swales and
crests of dunes and broken curved crests are previous burial systems.

McArthur and Bartle (1980) have examined such features along the coast of Western
Australia. Their dating of deposits have shown that this action took place many thousands
of years ago, certainly before man was present in any significant numbers. Thus, it is
a short-term view to blame man's intervention in destroying vegetation as the major
cause of these so-called blowouts. Nature acts on a much more mammoth scale and so
reasons for these significant aberrations must be sought in much longer mechanisms. By
checking ages of samples from these ancient features, the same authors have traced their

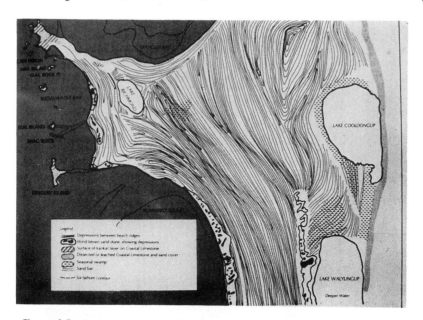

Figure 3.5 Section of Western Australian coast with dune system marked by lines as this land accreted

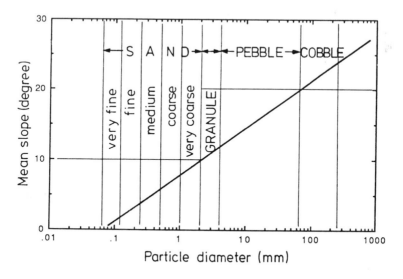

Figure 3.6 Relationship between beach slope and sand grain size

boundaries, as indicated in Figs. 3.9 and 3.10. These exhibit zones that are still mobile, as well as others that have been partially stabilized by vegetation. The orientation of these mounds indicates the direction of the persistent winds. The scale of these accretions is in terms of many kilometers crosswise from the coast and tens of kilometers along the

Figure 3.7 Excess sand accumulation at the apex of a cuspate foreland by waves

shoreline. Dune stabilization which forms part of coastal management programs will be discussed in Section 5.2.

3.2 WAVE CLIMATE

The sweeping of sediment shoreward by swell, and not the opposite, has been the reason for accumulation of land as continents, in spite of the sporadic demand of beach material during storm sequences. Hence any study of coastal geomorphology must involve this predominant source of energy, that due to the waves. In the past this has been discounted by geologists and engineers, but its importance in the total sequence of sediment movement from weathered rock, to river transport, to wave dispersal along the coast, finally to some sink or trap, is now receiving the attention it deserves.

Geomorphology is a fascinating study since it encompasses the whole globe over the complete geologic time scale. There are tremendous forces involved in transporting prodigious volumes of material many thousands of kilometers along continental margins. What more obvious source for this energy than that transferred from cyclonic winds to the surface of the sea. The storm waves so generated are then spread across the oceans, with negligible energy loss, to arrive as swell at distant shorelines. There they disturb and suspend the sediments and move them alongshore, but sensibly in one direction. The accumulation of vast areas of deposition, either as land or continental shelves, in

persistant sea
breeze

Figure 3.8   Accumulation of wind-blown
sand by sea breezes

this manner may appear too ambitious for any unique energy source, but it should be remembered that it has been acting over some 2000 m years (Stoddart 1969).

Even though waves have been discussed fully in Chapter 2, there are certain characteristics relating to sediment movement which require emphasis here. The two categories, of storm waves and swell, have their own particular influence on beaches. These should be understood before the phenomenon of littoral drift can be discussed adequately. The long-term changes effected on a shoreline by the persistent swell then become clear.

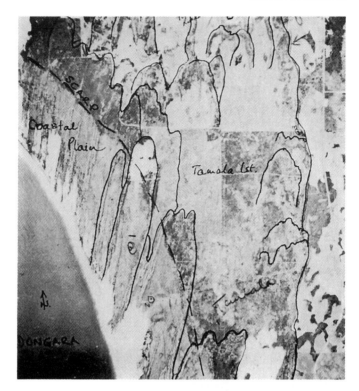

Figure 3.9 Blowout alignments dictated by persistent southerly winds on Western Australian coast (McArthur and Bartle, unpublished data, 1980)

3.2.1 Storm Waves

When a cyclone or strong wind system is in the vicinity of a coast the waves are breaking as they continue to receive energy. These can be termed *storm waves* which may still be building up, or have reached a fully arisen state. Even when the fetch itself is not adjacent to the shoreline, but is a short distance from it, the sea surface will be undulating in a complex manner without the short period components that promote the breaking process.

As noted in Chapter 2, the time to reach the steady state, plus the distance over which the wind is required to blow, vary with the wind speed. Higher velocities produce larger waves which comprise many wave trains of differing period and height. These are not aligned with each other, the smaller period components being more angled to the wind direction while the larger and longer waves are almost parallel to the wind vector. This randomly fluctuating surface can be measured by a wave recorder and the results plotted in the form of a spectrum. This is normally drawn as wave energy $(H/2)^2$ against frequency (f) but can also be plotted with period (T) as the abscissa. Each spectrum represents the condition at some position along the fetch (F) after a specific duration (t), as seen in Fig. 3.11. These increase in size along the generation area (measured

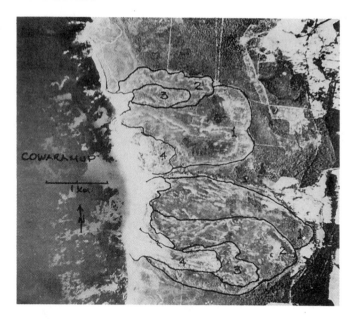

Figure 3.10 Blowouts distinguished by age on Western Australian coast (McArthur and Bartle 1980)

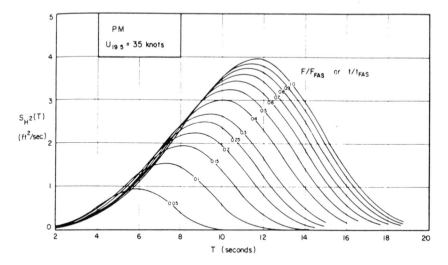

Figure 3.11 Wave spectra for differing fetches in a storm

parallel to the wind direction), until at some minimum fetch ($F_{\min}$) with concomitant minimum duration ($t_{\min}$) for the given wind speed, no further growth takes place. This condition is known as the *fully arisen sea* and is termed FAS, so that its respective fetch is F_{FAS}.

The multidirectional nature of these storm waves due to the generation process is enhanced by the changing orientations and positions of fetches as the cyclone center, and its related circulating cold fronts, passes near or across the coast. This is illustrated in Fig. 3.12 where wave energy vectors account for fetch and duration as cyclonic winds approach a coast at different angles. Throughout a storm sequence such waves can arrive from a wide fan of directions and tend to cancel each other with respect to longshore component of energy when breaking near the beach (Mason et al. 1984). This will be seen to have importance in the placement of beach material more or less directly offshore rather than alongshore, despite the vast amount of energy being applied to the shoreline.

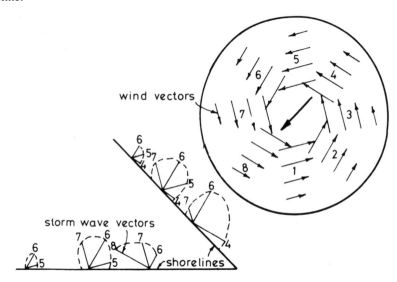

Figure 3.12   Storm wave energy vectors as a northern hemisphere cyclone approaches and crosses a coast

Any particular segment of coast receives very little storm wave energy throughout a year. Individual storms may last for three days, two weeks at the most, and may be repeated on differing scales but a few times only over twelve months. Thus these high-energy sequences are of short duration, which restricts their influence on longshore sediment motion. For the remainder of the year, possibly 40 weeks or more, the same site may receive oceanic swell from distant storms.

An exception to the above conclusion regarding little longshore drift from storm-type waves is along coasts of inland seas. Since the decay area for waves to spread either radially or circumferentially from fetches is insufficient, waves will contain a number of components angled to each other and are therefore of modified storm character. This is where knowledge of cyclone paths and durations is of importance. Their repetition throughout the year and over the years can define the location of fetches and directions of waves reaching a specific point on the coast. Cyclonicity charts as discussed in Section 2.1.1 and presented in Fig. 2.6 are extremely useful in producing a wave climate.

With knowledge of the wind circulation around such centers (anticlockwise in the northern hemisphere and clockwise in the southern hemisphere) the fetches for waves can be determined, with larger waves produced in those where the wind direction equates the path followed by the storms. If the resultant of these storm wave vectors is oblique to a coast, then longshore drift can result even in the absence of swell activity.

In obtaining such a resultant of wave energy in a relatively small enclosed sea, where the dimensions are only a fraction of most meteorological features, historic wind data should be obtained around its periphery. These would consist of wind velocities of various durations (reduced to consistent heights above the sea surface) in a range of directions, and can be summed and arranged in order to obtain a single wind rose for the complete body of water. The point on the coast under study then has arcs drawn in various compass directions from which fetch lengths are derived for waves generated along those alignments. Limitations with respect to wind duration are not likely to enter this picture but only fetch length, which can vary along these several arcs.

In this event significant wave height ($H_{1/3}$) varies with wind velocity (U) and the square root of fetch (F)—that is, $H_{1/3} = f(UF^{1/2})$, whereas for the significant wave period $T = f(U^{1/3}F^{1/3})$, (Silvester and Vongvisessomjai 1978). If wave energy is required ($H_{1/3}^2$), the vector component should consist of $\Sigma(tU^2F)$, where t is the duration of each U in that direction with fetch F. If the energy per wave was considered more relevant ($H_{1/3}^2L$), where L is the wave length of the peak wave in the spectrum (varying with T^2), the vector component becomes $\Sigma(tU^{8/3}F^{5/3})$. By computing the summations for each direction, a vector resultant will indicate the obliquity of wave energy in some form on that location of the coast. As far as sediment movement is concerned it is only the quadrant in which it occurs that becomes important in determining the direction of net movement. As will be illustrated in Chapter 4, there are many physical features on the coast that can indicate this.

3.2.2 Swell

The storm waves generated within fetches surrounding a low-pressure center will continue to propagate in the direction that they were initiated. They therefore pass through the boundaries of the wind fields at varying angles to the wind vector, being more aligned for periods closer to the peak component in the FAS spectrum. At some point downwind of the fetch, as seen in Fig. 2.16, the only waves to arrive are contained within a small fan encompassed by the width of the fetch. This limits the wave energy available (as noted in Section 2.2.6) and hence the heights are substantially reduced.

Besides this restriction by circumferential spreading there is also radial separation of wave components due to their different speeds across the ocean, known as group-wave velocity. This is half the celerity of the wave in this deep water which varies directly with wave period. Thus, the higher period waves in the spectrum arrive at any point first, followed by the median (also the highest waves), with the smallest components trailing behind. The resulting spectra at various distances from the fetch at a particular time were illustrated in Fig. 2.16, which involves a narrow band of periods present. This radial spreading of energy also reduces the wave energy available at any instant

122 Beach Processes Chap. 3

so that swell wave heights become 5–10% of those when first generated, as seen in Fig. 2.29.

Also, as seen in Fig. 2.16, the waves passing through the point some distance from the fetch are closely aligned. Even if the fetch were slightly displaced sideways, waves from it would arrive essentially from the same direction. A significant feature of storm waves being generated in the oceans is that they emanate from cyclones passing from west to east between latitudes 40° to 60° in both hemispheres, as depicted in Fig. 3.13 (Davies 1964). These paths oscillate north and south with the seasons, some even extending to the 30° latitude for short periods. As seen by the wind circulations in each hemisphere, the largest storm waves, and hence the predominant swell, travel eastward and toward the equator. It is the west coasts of continents, therefore, that receive persistent swell from that quadrant. On the east coasts swell arrives from the tail end of these cyclones moving away from the coast. This is complemented by swell and storm waves produced by tropical cyclones between 10° and 25° latitude in either hemisphere, which tend to move westward and away from the equator. Thus, wave climates for east and west continental margins can differ significantly. Even where storm centers are in proximity to a coast before crossing it, the swell waves will be concentrated in a narrow band of directions whereas the deepwater storm waves are spread over a wide fan, as illustrated in Fig. 3.14.

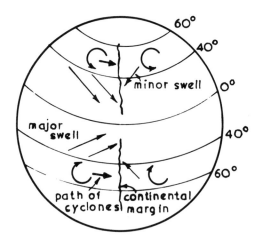

Figure 3.13 Passage of cyclones within 40°–60° latitudes (Davies 1964)

Another characteristic of swell waves is their larger duration. It takes longer for waves of all periods to disperse through a point away from the fetch than it does for them to exist as storm waves within it. In a study of swell waves generated near New Zealand to the shores of Alaska, Snodgrass et al. (1966) found that: "A typical event begins at 30 mc/s and ends at 80 mc/s, lasting for 2 days at Tutuila and for a week at Yakutat; the progressive lengthening being attributable to dispersive stretching of the wave train." They also observed: "Once or twice a week a wave train associated with a severe southern storm leads to an identifiable 'event' that can be traced across the entire

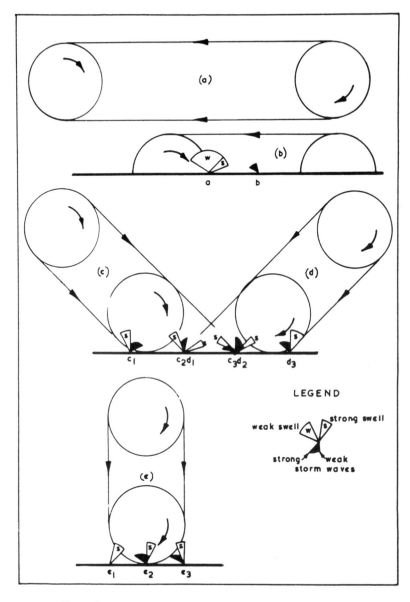

Figure 3.14 Wave incidence from cyclones in proximity to a coast

ocean." They presented spectra which have been modified to read in seconds rather than frequency in Fig. 3.15, for distances of zero, 1,000, and 10,000 km from downwind end of the fetch. This displays very little attenuation as waves propagate further from this source of energy. Greater reduction occurs in low period components, no doubt, due to their dispersion at greater angles to the wind direction.

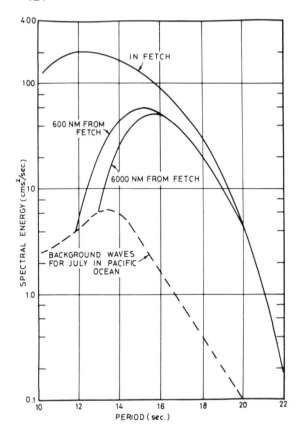

Figure 3.15 Typical spectra recorded at various locations across the southern Pacific Ocean (Snodgrass et al. 1966)

With the distribution of high wind zones, as shown in Fig. 2.9, it is apparent that swell could be arriving continually on most oceanic margins, but especially the west coasts. But even in enclosed seas such as the Mediterranean, Baltic, Sea of Japan, Caspian, etc., extra-tropical cyclones will have repetitive paths and frequencies dependent on their latitude. Points on the coast at greater latitudes will experience fiercer storm waves whilst those closer to the equator will receive milder components, partially dispersed from the generating areas. The west coasts of these inland seas will suffer from greater wave energy than the east coasts because the latter receive waves from fetches moving away from them. There will be greater periods of calm sea than on oceanic margins.

Another significant feature of swell is its continual change in height and period. As seen by the spectra in the dispersal area of Fig. 2.16, the band of waves arriving at any instance is varying continually, with each containing different energy content. Swell characteristics thus change hourly which makes it difficult to assess some mean height or period, as is done for most calculations of annual littoral drift.

In summary, it can be stated that swell waves are small in height, generally long in period, of long duration, persistent in direction, and varying in energy over time. Because of these characteristics they will influence the direction of net sediment movement along the coast. Bird (1984) observed that the effect of oceanic swell is to modify the configuration of the coasts, through which irregular beach outlines in plan may be simplified into a succession of curved sandy beaches between protruding rocky headlands, to be discussed in Chapter 4. However, it will be shown that the swell arriving soon after each storm will be the predominant input for this longshore one-way trend, since they act when sand is more available to them.

Thus, in determining a wave climate for a specific section of coast it is not only the local winds that must be considered but the waves arriving from distant storms as swell. This is aided by the publication by authorities in many countries of wind, plus sea and swell charts (*Monthly Meteorological Charts of Oceans* and *Atlas of Sea & Swell Charts*) which provide roses of either category of wave, with percentages of time in 45° bands around 360°, as was illustrated in Fig. 2.8. These are drawn for each 5° square of latitude and longitude, with adjacent squares matching since cyclonic patterns cover larger areas of the ocean at any one time. These data have been gathered over many years from mariners when a ship has been located within that square at some specific time of the year. Thus roses are presented for each month which can be integrated quarterly (seasonally) or annually. As indicated in Fig. 2.8 the heights of waves are only specified roughly and no periods stated. A more detailed picture can be gained from Hogben and Lumb (1967), available in an updated form (Hogben et al. 1986, Hogben 1988) with tables of height and period for each location on an annual basis. In these publications data have been summed for several 5° squares where the statistics are similar.

Whereas such sea and swell charts give only qualitative data for height and period, the directional nature of ocean waves can be extremely useful. When roses are inspected over each ocean a wave pattern is observed which is necessarily consistent with the wind pattern observed over them as depicted in Fig. 2.7. Extra-tropical cyclones are continuous throughout the year, as seen in the figure, whereas the tropical centers occur during the summer for the specific hemisphere. The pronounced west to east motion of the 40° to 60° cyclones is evident, whereas the tropical storms have more of a westerly movement and toward the poles. The actual paths of these intense high-wind events are extremely variable, being activated by small pressure differences around their boundaries. However, cyclonicity charts, as illustrated in Fig. 2.6, can provide a general picture of their movement around any continental margin.

Besides swell waves, which are predominant in sediment motion and scouring problems of seawalls and breakwaters, the coastal engineer needs to know the heights, periods, and directions of the largest storm waves, for the design of marine structures. In the sea and swell charts previously noted are recorded monthly wind roses for the same 5° squares. These should be used in conjunction with data on paths followed by storm centers in the vicinity of the coast under study. Wave hindcasting is difficult at any time, taking into account changing speeds and directions of fetches and variable wind speeds within them. It may be preferred to use some maximum wind velocities with their durations to determine some optimum significant height and period for use in design. This is a good insurance policy even if more sophisticated data and procedures are employed. Once a period at the peak of the spectrum (T_{max}) is derived and hence a d/L_o ratio, there is a specific $H_{1/3}$ for storm waves in shoaling depths, as given in Fig. 2.37 and for deeper regions in Fig. 2.39.

3.3 SEDIMENT MOVEMENT BEYOND BREAKER LINE

Waves approaching a coast exert their influence on a sedimentary bed by placing particles into suspension or partial suspension due to the oscillatory motion of water particles at that level. The concentration of sand is greatest adjacent to the floor and it is there, within the boundary layer of the waves (only one or two centimeters) that currents for lateral movement must be taken into account. As discussed in Chapter 2, the net transport of water or mass transport is optimum at this level, whereas tidal currents are minimal. Even though the latter are not zero at the bed their influence could be much less, changing in direction as they do throughout the tidal cycle. They could be very much smaller than the net movement per wave cycle which moves sediment in the direction of wave propagation. This net movement decreases as orbital motions do, as d/L_o increases.

This sweeping motion shoreward commences farther seaward for long period components, but increases for them all as shallower water is reached. During their propagation across the continental shelf waves are refracted, if angled to the bed contours, so that the longshore component of this net movement is reduced as the beach is approached. Even so, the longshore component $(xgT^2 \sin\alpha/H^2)$ (Silvester 1974) is increased closer inshore, as indicated in Fig. 3.16, where x is progress per wave period (T). Thus, while tidal currents decrease adjacent to the coast the longshore component of mass transport due to the waves assumes importance.

This sweeping of sediment shoreward is very fortuitous for mankind, for if the reverse were the case all sediment supplied to the coast from weathering of rock would have been spread across the ocean floors. Very few sedimentary expanses would have been available for vegetation so that evolution of flora and fauna would have been drastically impeded. This lack of deposition throughout the oceans was termed by Menard (1961) as the "great paradox of marine sedimentation." It is no longer a paradox when the predominant influence of mass transport is understood.

Another factor not generally taken into account in the offshore region is the differential refraction that takes place across the continental shelf. Waves of varying period arriving simultaneously, even from the same deepwater direction, will commence twisting at different depths and therefore will be angled to each other. This produces a wave system termed *short crested*, as the intersection of angled crests produces island mounds of water whose distance apart are finite, and not infinite as assumed in two-dimensional progressive waves. The orbital motions of water particles are complex, as discussed in Section 2.3.3, and conducive to suspending material from the bed and transporting it horizontally, influenced by the increased mass transport within this system.

The angle of obliquity to a coast with contours parallel to it differs in any depth with the wave period. This is illustrated in Fig. 3.17 where angles in a 20 m depth for $\alpha_o = 45°$ are determined by period T. Differential refraction of waves occurs which causes them to be angled to each other as they cross the continental shelf, particularly around

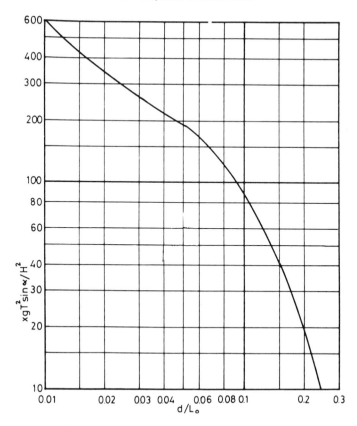

Figure 3.16 Dimensionless longshore component of mass transport at the bed as waves traverse the continental shelf for $\alpha_o = 45°$, based on laminar theory of Longuet-Higgins (1953)

$d/L_o \approx 0.25$. Thus, these short-crested systems appear to these authors more important than progressive waves, especially in modeling or analyzing sediment movement offshore.

As soon as bed oscillations at the sea floor are sufficient to disturb the sediment it will form into ripples or dunes whose crests are aligned with those of a progressive wave system. These bed undulations are symmetrical in cross section with heights and spacing dictated by the maximum orbital movement of the water particles at the bed, which, in turn, depend on the height and period of the waves in a given depth of water. They reach an equilibrium profile after which horizontal transport becomes minimal. However, a change in wave characteristics demands new spacings, so producing excessive lateral movement during the transition. A similar pulse of activity occurs when wave direction alters since the bed dunes must be realigned. As noted above, swell waves are continually changing in height and period. Hence, sediment movement could be much greater in the prototype than is recorded in flume studies with unidirectional monochromatic waves, even discounting the scaling errors of modeled waves. Waves angled to each other,

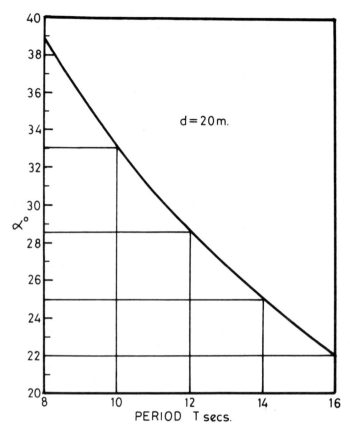

Figure 3.17 Refraction angle for waves in 20 m depth for $\alpha_o = 45°$

which are considered by the authors to be the norm rather than the exception, will also enhance suspension of material, as will be discussed in Chapters 5 and 7. The importance of ripple formation in transport of sediment beyond the breaker line has been noted by Vongvisessomjai et al. (1986) as follows: "It is found from the study that the sediment transport rates are strongly controlled by the rates of growth and migration of ripples."

The motion shoreward is in the direction of wave propagation which is oblique to the shore. As the beach is approached the bed steepens so that gravity plays its part, moving sediment particles normal to the contours. The resultant action, therefore, is of a "saw-tooth variety," with sediment transported alongshore in pulses as changes in wave characteristics effect new spacings and orientations of bed dunes.

The point to be made is that significant movement of sediment is taking place seaward of the breaker zone and hence beyond the influence of most shoreline structures. If the volume of material being removed from any section of coast is being replaced by an equal amount from upcoast, the bed level will remain steady or "in regime." However, if there is a dearth of supply the bed will be denuded and the area deepened. This will result in steepening of the beach profile, which may slump, but in any case will demand more sand to construct the offshore bar during a subsequent storm. Some of this volume will remain offshore to make up for the previous scouring and hence the waterline recedes and erosion ensues. It is a worthwhile insurance policy to invest in annual or biennial hydrographic surveys of the offshore area if scouring is anticipated. Even so, by the time they display erosion the volumes so removed will require substantial remedial measures to rectify.

3.4 LATERAL MOVEMENT IN SURF ZONE

As waves proceed shoreward the decreasing depths cause the waves to shorten in length and hence to increase in height. Their steepness (ratio of height to length) reaches a maximum limit when they are forced to break, after which they become a bore that travels to the beach. This act of breaking produces two distinct zones of action as far as sediment movement is concerned. There is the *surf zone* between breakers and beach and an *offshore zone* seaward of the breaking waves (Fig. 3.3), the sediment movement in which has been discussed previously.

The extent of the surf zone is clearly exhibited in the prototype, although its width can vary substantially, with waves breaking offshore at distances dictated by wave heights and periods. As will be shown later it also varies with the type of wave—storm wave or swell—since these two classifications have distinctly different, almost opposite effects on the beach profile. Storms remove the beach berm to place it offshore while the milder or swell waves replace it back onshore. During the former the surf is wide and for the latter condition it is very narrow.

3.4.1 The Swell-Built Profile

Consider now the action of swell waves on a beach, which are almost monochromatic, with undulating wave height due to the interaction of components of slightly varying period. This system of waves is predominant on oceanic margins, as it applies throughout the year and persistently from a small fan of directions in one seaward quadrant. As they proceed across the continental shelf they refract, becoming almost normal to the coast before breaking. The mass transport increases up to this point so that much sediment is contained in the bores which surge up the beach face. During their passage across the surf zone more material is suspended, thus loading the uprush water with sand.

The broken wave swashes up the beach face with its water percolating through it, so long as there is adequate time between each wave. As seen in Fig. 3.18, this water soaks down to the water table some distance below the face, eventually to be returned back to the sea. The downrush, when a trough arrives, is smaller than the uprush due to this percolation and therefore cannot carry much of the sediment load back down the beach face. Also, the hydraulic jump associated with this flow reversal is small. The result is that swell waves, with several seconds between each crest, will leave material on the face and hence the beach accretes.

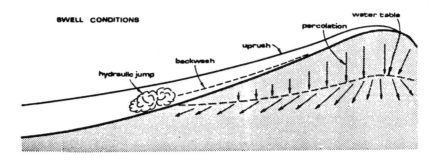

Figure 3.18   Hydraulic conditions in accreting beach

The widening of the berm continues only so long as material is available offshore to be fed into the breaking waves. A parabolic profile results (Dean 1977), with a small hump on it near the breaker line, on which particles oscillate back and forth without any net movement to or from the beach. They can move "see-saw" fashion along the beach but this has little to do with the beach profile itself.

The slope of the beach face in the finally formed berm depends on size of sediment in the area. Finer material produces milder slopes than course, which also dictates the reach of run-up and the resulting height of the berm. When first placed on the beach, sand is very loose and walking or running along it causes sinking of the feet into it. Such an observation can indicate how recently material was deposited. As waves continue to swash across the face it becomes well compacted, sufficient sometimes to take wheeled vehicles.

During this buildup by swell waves, sorting of sand grains takes place, by the process as discussed with Fig. 3.2. During the DUCK-82 beach experiments Richmond and Sallenger (1984) confirmed: "It is also demonstrated that coarse sediment may move onshore while finer material may simultaneously move offshore."

3.4.2 The Storm-Built Profile

Now consider what happens to this swell-built profile when storm waves arrive. These are very steep and contain much water above the mean sea surface. They comprise waves of many periods, or constitute a wide spectrum, with heights appropriate to each. A crest arrives almost every second instead of every few seconds and hence large volumes of water are thrown over the beach face, which is quickly saturated. This situation is depicted in Fig. 3.19, where it is seen that the water table has become almost coincident with the face itself. The downrush now nearly equals the uprush so that much sand is dragged down the face into the hydraulic jump, which is increased in size. This is one mechanism by which the berm is eroded and its contents placed into suspension.

Another factor, of equal if not greater importance, is the flow of the excess ground-water returning to the sea. At the waterline, where the hydraulic jump is also located, it is moving vertically, thus causing liquefaction or a "quicksand" effect (Longuet-Higgins 1983). This aids suspension so helping to undermine the toe of the beach face, which

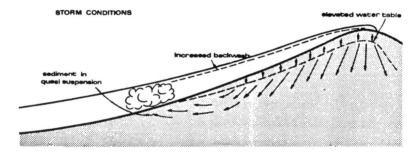

Figure 3.19 Swell-built beach profile suffering storm waves

progressively retreats landward, as seen in Figs. 3.20A and B, from tests conducted in a large wave tank by Dette and Uliczka (1987). Steep faces of approximately 50° are created on this retreating berm, from which incoming storm waves reflect extremely well.

Partial standing waves are established with a period equal to the *surfbeat*, as batches of large waves periodically meet this face, which can be steeper than the angle of repose of dry sand due to the damp conditions. The reader would have seen remnants of a steep scarp left at the back of a berm, indicating how much beach was demanded during the previous storm sequence. This scarp may well be adjacent to the frontal dune, with roots exposed of vegetation established there. The disappearance of a berm can take place very swiftly. Katoh et al. (1987) have reported: "During the erosional process, the shoreline rapidly recesses in one or two days."

During a storm the high wind velocities exert a shear on the water surface, which together with the mass transport of the waves, generate a landward current in the order of 3.5% of the wind velocity (Wu 1975). This forces a return flow at the bed which is carrying the bulk of material being eroded from the beach. This seaward current necessarily decreases as it reaches greater depths, but will also be influenced by the partial standing wave oscillation already noted. The rate of cross-shore sediment transport is swift at commencement, but reduces with time as the eroding forces weaken, due to the bar becoming well formed.

From the tests recorded in Fig. 3.20C, it can be seen that the associated bar is not recognizable until 1.27 hours after commencement of the model waves. It is more pronounced at 2.70 hours and by 3.85 has formed a distinct trough landward of it. The erosion of the berm, as noted in Fig. 3.20B, from the original 1:4 slope is extremely swift (see 0.33 hours line), but reduces as the bar reaches equilibrium conditions. At this stage wave dissipation over the bar is optimized and hence wave impact on the steep beach face is substantially reduced. However, if the duration of the storm is great, continued reflection of even broken waves will supply more material offshore, filling the swale between beach and bar to almost a horizontal plane.

The processes described above imply that sediment from the berm is transported to and over the bar where it falls down the seaward face in the presence of excessive wave action. It will tend to be deposited at the angle of repose for the material concerned. Sallenger and Howd (1989) have reported: "It is also interesting to note that as the bar

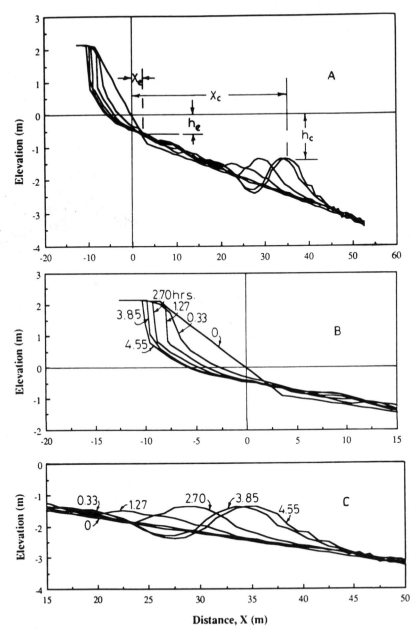

Figure 3.20 Progressive berm erosion and formation of a bar: A. overall view, B. detail of berm recession, and, C. detail of bar development over time (Dette and Uliczka 1987)

migrated offshore during both storms the slope of the seaward face of the bar increased, resulting in an increase in relative breaker height." Their paper was analyzing the hypothesis that bar location was related to the point of breaking. Although this may be so for the swell-built profile with a minor bar in position, for storm conditions it is not so. As Sallenger et al. (1985) have concluded: "The bar was in no way related to breaker zone processes; through most of the storm the bar-crest distance offshore was typically only 10% of the surf zone width." This resulted in the statement by Sallenger and Howd (1989): "We conclude that the offshore migration of nearshore bars is not necessarily associated with break-point processes."

The actual location of the bar offshore for any given wave conditions has interested many workers. There is a strong belief that it is related to a partial standing wave system produced by reflection of incoming waves, or groups of high waves producing a surf beat frequency or period. Sediment, particularly that placed in suspension, would accumulate at the antinode of these systems. Dally (1987), who conducted flume tests with surf beat optimized, proved that bar position was unrelated to it. He believed undertow was the predominant mechanism.

Since waves associated with surf beat would certainly be of shallow-water type, their celerity and hence wave length would be dictated by depth. Dolan (1983) has assumed the bar crest depth, but it seems more appropriate to assume the trough or some proportion of it. A conclusion of Sallenger et al. (1985) is relevant here: "During the storm, the bar acted to maintain its form, the ratio of trough depth to crest depth remained roughly constant." This value averaged at 1.24. "Analysis of the bar position in terms of a standing wave motion showed that the causative wave period must have been longer than that of incident waves, probably on the order of a minute. Wave data showed significant energy in the infragravity band at these periods although no definite link has been made." Mason et al. (1984) have also concluded: "However additional analysis of the wave and current data is required before the relationship between infragravity waves and bar response can be confirmed."

Wright et al. (1979) have observed some Australian beaches over a period of time and concluded that during initial bar formation and its passage seaward incident wave periods are dominant in establishing the standing wave oscillation. As a storm reaches its peak and the bar is in its final position infragravity periods associated with surf beat assume importance. They state: "The characteristic periods of inshore resonance vary from incident-wave period for highly reflective beaches through to periods in the surf-beat region for highly dissipative systems. Characteristic periods do not vary on a continuum, but shift from incident wave period through successively lower subharmonics to surf beat as dissipation increases. This progression of modes implies firstly that beaches do not vary between reflective and dissipative extremes on a continuum, but that they tend to shift to different states as wave climate etc. change (as the surf zone or trough width changes)."

During this migration of the bar seaward no rip currents are formed as the excessive transport of water toward the beach at the surface, due to stress of the wind, causes a reverse flow at the bed over the complete bar form. However, as milder waves commence to move the bar shoreward the higher water level in the trough region requires an outlet.

Weak spots in the bar provide this passage which deepens and widens as the rip current develops. On average the spacing of these is about double the distance of the bar crest from shore, but as the bar moves shoreward this ratio becomes 3 to 6 (Wright et al. 1979). At this stage waves entering the trough region through these gaps in the bar diffract from successive openings and create short-crested wave systems which result in *beach cusps*. In fact, sizable protrusions of the beach can occur due to the alternate longshore drift caused by these diffracting waves, which is accentuated as the bar approaches the beach. There is a tendency for rip currents to change location if the swell waves are angled to the coast since the bar gaps are accreted on the upcoast side and eroded on the downward extremity of the opening.

When a portion of beach, at the rear of the berm, is not removed seaward over the course of say 2 years, plants may germinate and develop. It is this line of vegetation, probably helping to form a foredune, that can be taken as the edge of the active berm. The wider any beach berm the more severe the waves encountered on that section of coast. This is displayed in Fig. 3.21, where the more exposed beaches have greater width.

Figure 3.21   Aerial photograph of Cloudy Bay, Bruny Island, Tasmania, Australia, showing greater berm width along more exposed section of coast (Reader's Digest 1983)

It must be expected that the whole berm at any beach will be demanded by storm waves at least once every 2 years, or maybe every year. Of course, there can be more intense cyclones at less frequent intervals that may erode not only the berm but also the portion of the first or foredune. However, the beach will be wider than normal after such events, because of the greater volume of bar to be spread across this flat feature.

It becomes obvious that the first storm of winter will have the most dramatic effect, since the accreted profile from the previous swell season will be either partially or fully removed to form the protective bar. A subsequent storm could effect greater erosion under either of two conditions. If it is accompanied by a higher sea level due to a higher tide or storm surge the bar will need to be elevated to break the storm waves. Should the second storm be of longer than normal duration the attenuated waves still arriving at the steep beach face could continue to nibble at it and spread sand in the swale between bar and beach. In either case erosion is increased.

It has been noted already that waves within any one storm arrive from many directions, due both to the generating process and the changing orientations of fetches as it crosses the coast. Obliquity of waves to the shoreline is more or less cancelled out as the net longshore current is practically zero. During the DUCK-82 experiments Mason et al. (1984) observed two storms in both of which southward currents occurred first but turned northward during the course of the storm. Another factor in discounting littoral drift during storms is their short duration. It seems acceptable, therefore, that beach material is removed directly offshore during the course of a storm, there virtually to be put "into circulation" for swell waves to act upon when returning it back to the beach.

The changes from swell-built to storm profile have been termed by Chapman and Smith (1980) as the "swept prism." They showed that on a certain section of the Australian coast this extended to a depth of 8 m which was 300 m from the waterline. This is but one example of the outside limit of the first offshore bar that is ultimately transported back to the beach.

During this removal and return process further sorting of sediment takes place. The first material to be deposited by the outgoing current is coarsest, followed seaward by the medium-sized grains, with the finest components being placed further offshore. This hypothesized deposition is depicted in Fig. 3.22, which requires verification in the field. When this sand is placed back on the beach the fine sediment is placed at the rear of the berm, the median is then transferred to the center of it, whilst the coarsest material is last

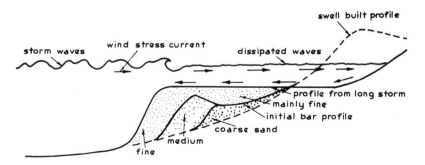

Figure 3.22 Hypothesized sediment size distribution within the offshore bar formed by a storm

to be brought ashore which is placed in the final beach face of the full berm. Vertical stratification can also occur due to the uprush decreasing in velocity up the beach face, so concentrating smaller particles near the surface and coarser material at lower levels. Problems have been encountered in relating sand sizes between sites alongshore, which may be due to differing locations on the berm where samples have been taken.

The natural sequence of storm waves forming a protective bar is beautiful to behold, even if menacing at the time. It is a preservation procedure by nature for all the sediment previously pushed toward the coast in the accretion of land. It is a mechanism that man cannot hope to emulate due to the massive volumes of material being shifted offshore at short notice. Nor does he have to think of copying this action because it is already provided for him. Suggestions of constructing a permanent offshore bar by placement of rubble stone mounds is discussed in Section 5.3.4.

3.4.3 Bar Characteristics from Large Wave Tank Tests

Larson (1988) has cited some 240 references on beach profile changes and found only a countable few reported tests carried out in large wave tanks, with the remaining majority from small-scale model experiments. Because there are no applicable scaling laws for interpreting the results of small-scale models, due mainly to scale distortion, it was concluded that: "data sets from laboratory experiments performed with small-scale facilities are of limited value for establishing quantitative understanding of profile change in nature" (Larson 1988). On the other hand, field data useful for quantifying beach profile change are extremely rare, thus calling for results obtained in large wave tanks in which wave height and period near prototype conditions are used.

Tests have been carried out in large wave tanks (LWT) in 1956–57 and 1962 at the Beach Erosion Board (the predecessor of Coastal Engineering Research Center), U.S. Army Corps of Engineers, and in 1982 at the Central Research Institute of Electric Power Industry, Japan. These test results have been summarized by Kajima et al. (1983a, 1983b), Kraus and Larson (1988), and Larson and Kraus (1989), and designated as CE (standing for Coastal Engineering Research Center) and CRIEPI results from Japan. From these tests a total of 32 data entries were obtained, of which 18 refer to bar profiles and 14 to berm dominated conditions, as designated by the previous authors. Also similar research has been conducted in 1986 by Dette and Uliczka (1987) at University of Hanover, Western Germany. These will now be referred to as: (CE) United States 8 tests, (CRIEPI or shortened to PI) Japanese 10 tests, and (UH) Germany 2 tests, in the various figures in this section. Some of the pertinent physical quantities are collected from the reports with others computed by the authors, of which a summary is provided for the test conditions and primary results as in Table 3.2.

In these tests, large waves with periods from 3 to 16 seconds and heights ranging from 0.74 to 1.8 m in flumes of $200 \times 5 \times 6$ m (deep) approximately have been employed for studying berm erosion and bar formation during storm conditions. The beach slopes ranged from 1:50 to 1:4, in which uniform sediment diameters from 0.22 to 0.47 mm were included. Some resultant beach profiles had a significant single bar, while others showed dual bars or mixed bar and berm features. To assess the optimum berm erosion

TABLE 3.2 DATA SUMMARY FOR THE 20 BAR PROFILES IN LWT TESTS

No. (1)	Case no. (2)	H (m) (3)	T (sec) (4)	d (m) (5)	D (mm) (6)	Beach slope (7)	H_o (m) (8)	w (m/s) (9)	X_c (m) (10)	h_c (m) (11)	V_{bar} (m³/m) (12)	$T/\sqrt{H_o}$ (sec) (13)
1	CE 100[a]	1.28	11.33[b]	4.57	0.22	1:15	1.081	0.031	39	1.16	18.5	10.89[b]
2	CE 300[a]	1.68	11.33[b]	4.27	0.22	1:15	1.402	0.033	41	1.31	30.1	9.57[b]
3	CE 400[a]	1.62	5.60[b]	4.42	0.22	1:15	1.717	0.031	44	1.52	28.6	4.27[b]
4	CE 500	1.52	3.75	4.57	0.22	1:15	1.645	0.031	56	1.44	33.3	2.92
5	CE 700[a]	1.62	16.00[b]	3.81	0.22	1:15	1.118	0.037	40	1.79	24.5	15.13[b]
6	CE 401[a]	1.62	5.60[b]	4.42	0.40	1:15	1.717	0.055	37	1.28	19.6	4.27[b]
7	CE 501	1.52	3.75	4.57	0.40	1:15	1.645	0.059	44	1.16	22.1	2.92
8	CE 901	1.34	7.87[b]	3.96	0.40	1:15	1.246	0.059	36	1.28	11.3	7.05[b]
9	PI 1-8	0.81	3.00	4.50	0.47	1:20	0.852	0.056	20	0.55	1.6	3.25
10	PI 2-1	1.80	6.00[b]	3.50	0.47	3:100	1.758	0.061	70	1.12	7.5	4.53[b]
11	PI 3-1	1.07	9.10[b]	4.50	0.27	1:20	1.040	0.031	36	0.91	12.4	8.92[b]
12	PI 3-2	1.05	6.00[b]	4.50	0.27	1:20	1.101	0.031	39	0.76	15.3	5.72[b]
13	PI 3-4	1.54	3.10	4.50	0.27	1:20	1.619	0.035	66	1.23	24.9	2.44
14	PI 4-2	0.97	4.50[b]	4.00	0.27	3:100	1.058	0.033	57	0.83	4.5	4.37[b]
15	PI 4-3	1.51	3.10	4.00	0.27	3:100	0.604	0.031	107	1.23	23.5	2.45
16	PI 5-2	0.74	3.10	3.50	0.27	1:50	0.799	0.035	73	0.61	6.5	3.47
17	PI 6-1[a]	1.66	5.00[b]	4.00	0.27	1:10	1.778	0.036	40	1.39	36.5	3.75[b]
18	PI 6-2[a]	1.12	7.50[b]	4.50	0.27	1:10	1.097	0.036	21	0.98	25.3	7.16[b]
19	UH-1[a]	1.50	6.00[b]	5.00	0.33	1:4	1.590	0.0447	14	1.18	48.7	4.76[b]
20	UH-2	1.50	6.00[b]	5.00	0.33	1:4/1:20	1.590	0.0447	34.5	1.3	11.0	4.76[b]

[a] These 8 cases were selected to correlate maximum beach erosion (relevant values shown in Table 3.3).

[b] These 14 cases, with $T/\sqrt{H_o} \geq 3.75$ s per column (13), were selected to correlate bar volumes (see explanation notes under subheading "equilibrium point" in text).

for practical engineering applications, consideration is given only to the equivalence between the volumes of the accreted bar and the eroded berm.

Various dimensionless parameters were calculated and used in the reanalysis. For wave conditions, parameters such as H_o/L_o, H_b/L_b, and H_b/gT^2 were tried, with surf similarity by $\tan\beta/(H_o/L_o)^{1/2}$, $\tan\beta/(H_b/L_b)^{1/2}$, or $(H_b/gT^2)/\tan\beta$, and for sediment to wave conditions by H_o/wT, H_b/wT, D/H_o, D/H_b, $(H_o/wT)/\tan\beta$, and H_b^2/gT^2D. In these parameters $\tan\beta$ is the offshore bed slope, w the fall velocity of the sediment, and D is the grain diameter in appropriate dimensions.

The topics to be discussed are:

1. Boundary between bar and berm profile
2. Rate of offshore sediment movement
3. Bar crest distance from beach
4. Bar crest depth from SWL
5. Volume of bar
6. Equilibrium point on the profile
7. Equivalence of erosion and bar volumes

Boundary between bar and berm profile. Previous graphs used H_o/wT with wave steepness H_o/L_o (Larson 1988, Larson and Kraus 1989), and plotting of D/H_b with H_b/gT^2 (Sunamura and Maruyama 1987; see also Larson and Kraus 1989) produced a boundary line inclined to the coordinates, as seen in Fig. 3.23, which was difficult to apply. However, as shown in Fig. 3.24, drawing of H_b/wT against $\tan\beta/(H_o/L_o)^{1/2}$ yields a horizontal line at H_b/wT about 4.1 for demarcating the bar and berm profiles. Data in each category of bar and berm spread uniformly either side of this demarcation line throughout the whole range of the abscissa, except for case 700 in the CE series, which had its water depth changed from 4.11 m to 3.81 m at its tenth hour of operation. The same relationship of $H_b/wT \approx 4.10$ was consistently observed when plotting H_b/wT

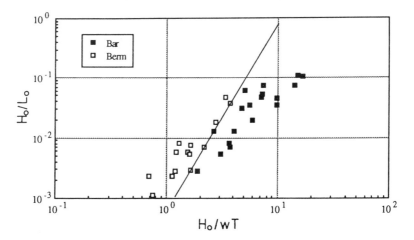

Figure 3.23 Demarcation of bar and berm profiles based on H_o/L_o and (H_o/wT) (Larson and Kraus 1989)

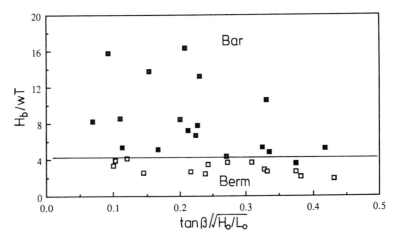

Figure 3.24 Demarcation as in Fig. 3.23, but with (H_b/wT) versus $\tan\beta/\sqrt{H_o/L_o}$

versus $\tan\beta/(H_b/L_b)^{1/2}$, H_o/L_o, H_b/L_b, and H_b/gT^2 respectively. It was found in Fig. 3.24 and other graphs (not shown) that either bar or berm profile can be produced over the full range of H_o/L_o (from 0.001 to 0.05) and H_b/L_b (from 0.02 to 0.09) tested in the LWT, depending on a matching value of H_b/wT. Hence, it is inadequate to define a storm condition by a value of wave steepness alone (such as $H_o/L_o > 0.025$ after Johnson 1956 and many others cited in Larson 1988 and Larson and Kraus 1989).

Rate of offshore sediment movement.   As a storm commences the insta-bility of the swell-built profile to the energy now being applied to it produces a swift transport of material from the berm face to offshore. When this offshore bar is fully developed the majority of waves will break over it. Larson and Kraus (1989) provide 12 data sets for peak transport over time, which resulted in an equation

$$q_m = q_{mo}/(1 + \alpha t) \qquad (3.1)$$

where

$$q_m = \text{peak transport rate}$$

$$q_{mo} = \text{peak transport rate at time } t = 0$$

$$\alpha = \text{decay coefficient of peak transport rate}$$

$$= \text{average value of 0.91 per hour.}$$

Kajima et al. (1983a, 1983b) had used an exponential decay rate, as did Sawaragi and Deguchi (1980). Although this is expected on theoretical grounds Larson and Kraus (1989) believed it approached zero too quickly, since transport continued for extensive periods. The same swift reduction in rate occurred when bar material was subsequently returned shoreward by milder waves, as is evident in all sedimentary processes when energy and load levels are disparate.

Equation (3.1) has been plotted for case CE-300 in Fig. 3.25, which exhibits the initial swift reduction in seaward movement of sediment from an eroding beach. The decay coefficient (α) could well vary in prototype situations but the shape of the curve would be similar. In the first 4 hours of this LWT test, transport had diminished to 1/5 of the initial rate, but it took another 50 hours before this reduced to 1/50 of it.

Bar crest distance from beach.   This is designated as X_c, and is measured from the original waterline to the apex of the bar, as illustrated in Fig. 3.26, of which the progressive berm erosion and formation of the bar were shown in Fig. 3.20. Larson and Kraus (1989) attempted to relate parameters containing this variable with wave and beach characteristics but much scatter resulted. The data have been reanalyzed more successfully, as seen below.

A dimensionless ratio X_c/L_o is plotted against deepwater wave steepness in the form $(H_o/L_o)/\tan\beta$, where $\tan\beta$ is the original offshore slope of the beach, as in Fig. 3.27, which results in the equation:

$$X_c/L_o = 0.022 + 1.508(H_o/L_o)/\tan\beta + 0.1404[(H_o/L_o)/\tan\beta]^2 \qquad (3.2)$$

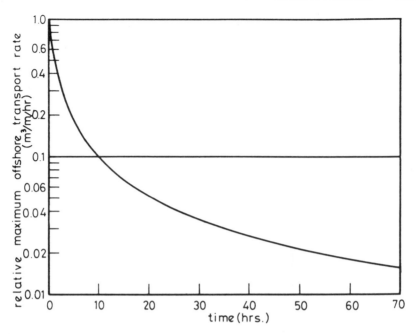

Figure 3.25 Rate of transport across the foreshore as derived by Larson and Kraus (1989)

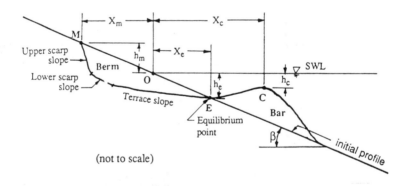

(not to scale)

Figure 3.26 Definition sketch of an idealized bar shape on a storm-built profile, showing optimum berm erosion and relevant variables analyzed

The first constant (0.022) could be assumed as zero, since for $H_o/L_o = 0$ there should be no bar formed and hence $X_c = 0$. Besides wave steepness the bed slope is also important in fixing the bar distance (X_c), although by varying (H_o/L_o) alone a relationship could not be formed for bar formation (Sallenger et al. 1985, Larson 1988, Larson and Kraus 1989). The different sources of data, as listed in Table 3.2 and the legend of Fig. 3.27, follow the curve quite well.

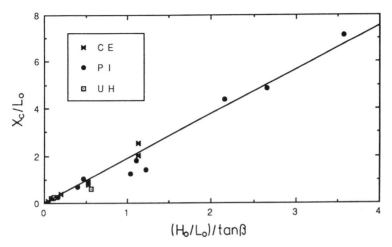

Figure 3.27 Dimensionless bar distance (X_c/L_o) versus wave and foreshore slope parameter $(H_o/L_o)/\tan\beta$

Bar crest depth from SWL.   As seen in Fig. 3.26, the crest of the bar is submerged to a depth of h_c below SWL. This can conveniently be related to X_c, wave characteristics and beach slope by three parameters. However, they should all give similar h_c no matter which is used. The first curve relates $h_c/L_o \tan\beta$ to X_c/L_o as in Fig. 3.28, to which the following equation applies:

$$h_c/L_o \tan\beta = 0.0269 + 0.391 X_c/L_o \qquad (3.3)$$

for which the first constant could be assumed as zero for reasons as previously given. The second relationship was dimensional as in Fig. 3.29, of which the

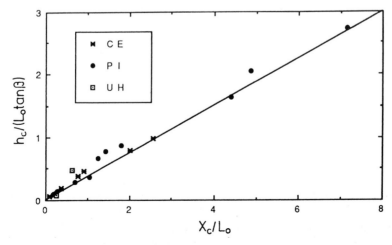

Figure 3.28 Dimensionless bar crest depth $h_c/(L_o \tan\beta)$ versus dimensionless bar distance (X_c/L_o)

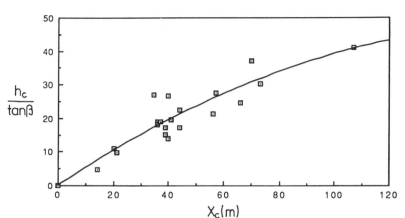

Figure 3.29 Bar crest depth $h_c/\tan\beta$ versus bar distance (X_c)

equation is:

$$h_c/\tan\beta = 0.6366 + 0.436 X_c \tag{3.4}$$

As seen by the values of $h_c/\tan\beta$ the constant 0.6366 is negligible and could be assumed as zero. The third parameter (h_c/H_o) plotted against $\tan\beta/(H_o/L_o)$, as in Fig. 3.30, has some degree of scatter, but the equation so derived is

$$h_c/H_o = 0.7529 + 0.0096\tan\beta/(H_o/L_o) + 0.0011[\tan\beta/(H_o/L_o)]^2 \tag{3.5}$$

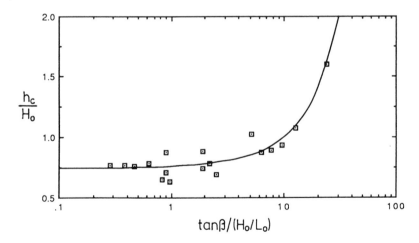

Figure 3.30 Dimensionless bar crest depth (h_c/H_o) versus $\tan\beta/(H_o/L_o)$

These equations should all give similar values of h_c for any given wave condition. Thus, assuming $H_o/L_o = 0.025$, $\tan\beta = 0.05$, and $L_o = 150$ m, then $H_o = 3.75$ m and $(H_o/L_o)/\tan\beta = 0.5$.

From Fig. 3.27 or Eq. (3.2), $X_c/L_o = 0.81$ or $X_c = 121.7$ m.
From Fig. 3.28 or Eq. (3.3), $h_c/L_o \tan \beta = 0.34$ or $h_c = 2.6$ m.
From Fig. 3.29 or Eq. (3.4) for $X_c = 121.7$, $h_c/\tan \beta = 53.7$ or $h_c = 2.7$ m.
From Fig. 3.30 or Eq. (3.5), for $\tan \beta/(H_o/L_o) = 2$, $h_c/H_o = 0.78$ or $h_c = 2.9$ m.
Minor differences in h_c exist because of the slight scatter existing in some graphs, but overall the agreement is reasonable.

Volume of bar.   Perhaps, more important than X_c or h_c, the volume contained in a bar per unit length of beach can be used for useful purposes. Data supplied by Larson and Kraus (1989) has been correlated as in Fig. 3.31. Although there is some degree of scatter the results could be accepted until further data are available from model or prototype measurements. It is noted that some combinations in LWT tests with very short periods and large wave heights, as in Table 3.2, may be unreal, if the results are to be extended to prototype conditions. Hence, for realistic field applications, only the 14 tests utilizing wave periods of 5 seconds or greater were used in deriving the equation:

$$V_{\text{bar}}/H_o L_o = 160(H_o/L_o)\tan\beta + 11,560[(H_o/L_o)\tan\beta]^2 \qquad (3.6)$$

in which $(H_o/L_o)\tan\beta$ is employed. The second term on the RHS of Eq. (3.6) results in a contribution of 7% to 43% of the first term. The range of $(H_o/L_o)\tan\beta$ for most prevailing wave conditions in the field is between 0.0015 to 0.006, where reasonably good fit is evident from the 14 LWT data presented in Fig. 3.31.

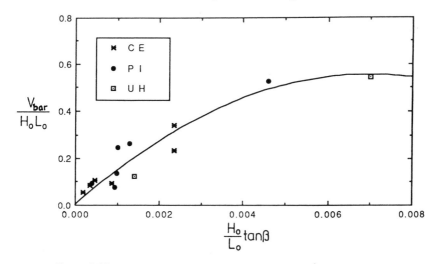

Figure 3.31   Dimensionless bar volume $(V_{bar}/H_o L_o)$ versus $(H_o/L_o)\tan\beta$

The quantity of $V_{\text{bar}}/H_o L_o$ was also favorably related to $(H_o/wT)\tan\beta$. The sediment fall velocity (w) varies with sand diameter, which for a bottom water temperature of 15° C is given in Fig. 3.32. Thus for a given wave height (H_o), initial offshore slope $(\tan\beta)$ and an assumed fall velocity (w), the optimum volume of sand contained within the bar per unit length of beach can also be determined. However, in view of the

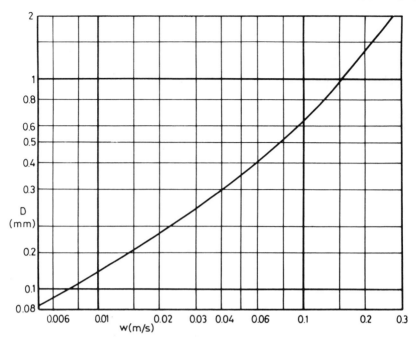

Figure 3.32 Sediment diameter versus fall velocity (w) for a water temperature of 15°C

differential sorting process of sediments during a storm, starting from coarser grain of 1 to 2 mm on the beach face down to some fine grain of 0.2 mm, it becomes difficult to use a constant fall velocity throughout the course of a storm. Thus, it is recommended that the relationship per Eq. (3.6) be used until a better expression can be established for sediment fall velocity during a storm.

Equilibrium point on the profile.   The ideal case of berm erosion extends from an equilibrium point (Dette and Uliczka 1987) in the original beach profile, to the steep face of the receding berm, as can be seen in Figs. 3.20A and 3.26. This point is at distant X_e from the original water line and depth h_e below SWL. It is now desirable to consider an optimum berm erosion model, in which the whole of the bar volume estimated from Eq. (3.6) is assumed to come from the erosion of the berm. Only the five typical LWT cases which comply with this condition are selected from Table 3.2, in addition to the waves with periods greater than 5 seconds, assumed to be realistic for field applications. The dimensionless distance X_e/L_o is plotted against $(H_o/L_o)/\tan \beta$ for the same LWT tests, in Fig. 3.33, with the equation to the curve being:

$$X_e/L_o = 0.96(H_o/L_o)/\tan \beta \qquad (3.7)$$

for which:

$$h_e = X_e \tan \beta \qquad (3.8)$$

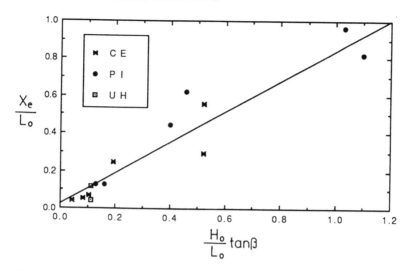

Figure 3.33 Dimensionless equilibrium point distance (X_e/L_o) versus $(H_o/L_o)/\tan\beta$

These dimensions become important when trying to assess the recession of the coast for any given volume of bar.

Equivalence of erosion and bar volumes.

It can be shown in Fig. 3.20 that volume per unit length of beach, or cross-sectional areas of eroded beach and accreted bar are the same. Thus, if the volume within the bar per unit length is known for some storm condition, then the berm recession may be calculated for some known berm height about SWL (h_b) and depth of scouring (h_e). A distinction must be made between low and high tidal ranges as a horizontal berm will only exist with the former.

As per the definition sketch in the inset of Fig. 3.34, it can be shown that the cross-sectional area of berm eroded above the point of equilibrium, E (see Fig. 3.26), or the volume per unit length of beach V_{berm} can be expressed by:

$$V_{\text{berm}} = \frac{h_b(h_b + h_e)}{2\tan\beta} - 0.866 h_b^2 + b\,(h_b + 0.5h_e) \qquad (3.9)$$

where h_b is the berm height, h_e is the equilibrium depth, $\tan\beta$ is the original shore slope, and b is the horizontal erosion distance from the top of the original beach face. The first two terms on the RHS of Eq. (3.9) are grouped as $V_{\text{slope}.b}$, shown in the inset of Fig. 3.34, which denotes the volume of an equivalent monosloped beach, as appears in Fig. 3.35 later, but using h_b rather than h_m in it. Thus, from the transformed relationship of:

$$V_{\text{berm}} = V_{\text{slope}.b} + b\,(h_b + 0.5h_e) \qquad (3.10)$$

a nominal volume $V = V_{\text{berm}} - V_{\text{slope}.b}$ can be established (as the inset of Fig. 3.34), in which the total V_{berm} can be estimated from Eq. (3.9) (alternatively, from its equivalent bar volume of Eq. (3.6) or Fig. 3.31), and $V_{\text{slope}.b}$ from Fig. 3.35. Once the quantity of V

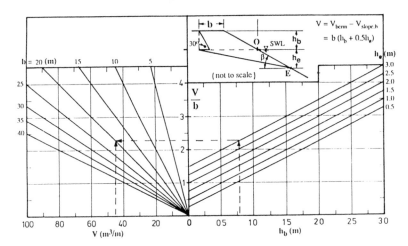

Figure 3.34 Graph to determine beach erosion distance (b) from a given equivalent berm volume (V), berm height (h_b), and equilibrium point depth (h_e)

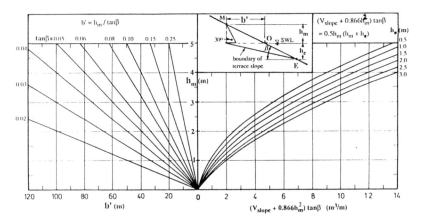

Figure 3.35 Graph to determine erosion distance (b') for a beach with large tidal range, or on a mono-sloped beach

is obtained, b can be conveniently calculated from $V = b(h_b + 0.5h_e)$, or from Fig. 3.34 itself. Thus, for specific values of V, h_b, and h_e, b can be determined.

It is seen in the inset of this figure that the natural slope of an accreted beach (β) is accepted as $8°$ (Larson and Kraus 1989), implying that a representative diameter for coarse sand of 0.5 to 1 mm according to Table 3.1. Also, their "terrace slope" (see Fig. 3.26) from the depth h_e to the SWL varies with the value of b, it being taken to the vertical line from the berm surface at the erosion extremity to SWL (see inset of Fig. 3.34). Since the scarp actually consists of two slopes, an upper steep section (around $50°$ as previously mentioned) and a lower milder slope, this has been averaged at $30°$, for which the volume per unit length is therefore $0.866h_b^2$ as subtracted as in Eq. (3.9).

It should be noted that the SWL is that operating at the peak of the storm and may therefore be applicable to storm surge or high tide level. It is on this basis that berm height (h_b) should be determined. Hence, the case presented is for conditions where either of these is small and a distinct berm is apparent. On the other hand, where a large tidal range exists the beach face is smoothed to a very mild slope and any horizontal berm is well back from the mean waterline. In this case erosion takes place only on a sloping beach (denoted as V_{slope}), as seen in the inset of Fig. 3.35. The volume (V_{slope}) of beach erosion per unit length is then given b:

$$V_{\text{slope}} = \frac{h_m\,(h_m + h_e)}{2 \tan \beta} - 0.866 h_m^2 \qquad (3.11)$$

The same assumptions regarding terrace slope and scarp slopes apply as before. The maximum vertical erosion (h_m) has replaced the beach berm height (h_b) of Eq. (3.9). In Fig. 3.35, Eq. (3.11) is graphed, from which horizontal erosion from the original water line (b') is obtainable by:

$$b' = h_m / \tan \beta \qquad (3.12)$$

Thus, if a volume (V_{bar}) is determined from a given wave condition, as in Fig. 3.31, and then related to Fig. 3.34 or 3.35, the expected horizontal erosion (b or b') can be computed, if other characteristics of the swell-built profile are known (or calculated).

The overall procedures proposed in this section are now assessed using some selective cases in Table 3.2, which meet the requirement of the optimum berm erosion model. Comparison is first made for the calculated and predicted bar volumes with the volume measured, and the predicted beach retreats (b or b') with their corresponding values measured. An overall summary of the comparisons is given in Table 3.3, in which five cases are designated as "monosloped beach," while the other three cases "with (horizontal) berm."

First, the physical dimensions of h_e and h_m (or h_b) and beach slope $\tan \beta$ for the eight selected cases in the original reports (see Table 3.2 and Kraus and Larson 1988, Shimizu et al. 1985, Dette and Uliczka 1987) are measured and their wave conditions recorded. The eroded berm volumes can be *calculated* from Eqs. (3.9) and (3.11) for the cases of "monosloped beach" and "with berm" respectively, which are compared with the bar volumes *measured*, as shown in Fig. 3.36. Good agreement is apparent.

Second, it is also desirable to compare the *predicted* berm volume, obtained directly from substituting a known wave condition (in H_o, L_o, and T) and beach slope $\tan \beta$ into Eq. (3.2) to Eq. (3.11), with the volume *measured*. This is different from the calculated berm volume, which uses the measured h_e and h_m (or h_b) and beach slope $\tan \beta$ directly into Eqs. (3.9) and (3.11). Following the outlines explained previously, the values of h_e and h_m can be predicted from various equations or figures presented in this study, upon substituting relevant wave condition and beach slope. The resultant values of h_e and h_m are listed in Table 3.3, together with various bar and berm volumes predicted. Comparison between the *predicted* berm volumes and the *measured* is depicted in Fig. 3.37, showing also reasonable agreement.

TABLE 3.3 COMPARISON OF MAXIMUM BERM VARIABLES IN SELECTED LWT TESTS

LWT Case No.	Measured h_e (m)	h_m (m)	b'/b (m)	V_{bar} (m³/m)	Calculated V(berm) (m³/m)	h_e (m)	h_m (m)	Predicted V_{bar} (m³/m)	V_{slope} (m³/m)	$V_{slope.b}$ (m³/m)	b'/b (m)
(1)	(2)	(3)	(4)	(5)	(6)	(7)	(8)	(9)	(10)	(11)	(12)
Five selected cases with **mono-sloped** beach			b'								
CE 100	0.71	1.47	21.94	18.5	22.04[c]	1.03	1.34	12.18	22.11		19.96
CE 300	0.92	1.65	24.63	30.1	29.29[c]	1.34	1.63	20.32	33.89		24.36
CE 400	1.06	1.50	22.39	28.6	26.71[c]	1.65	1.37	26.19	29.17		20.43
CE 700	1.20	1.47	21.94	24.5	27.42[c]	1.06	1.61	13.19	29.99		24.09
CE 401	1.82	1.18	17.61	19.6	25.21[c]	1.65	1.37	26.19	29.17		20.43
Three selected cases **with a horizontal berm**			h_b					V_{berm}			b
PI 6-1	2.41	1.63[a]	4.11[b]	36.5	42.28[d]	1.70	1.63[a]	33.88	33.88[e]	24.87	3.63[f]
PI 6-2	1.20	1.16[a]	6.85[b]	25.3	24.58[d]	1.05	1.16[a]	17.48	17.48[e]	11.65	3.46[f]
UH-1	1.64	2.00[a]	13.13[b]	48.7	48.12[d]	1.52	2.00[a]	49.39	49.39[e]	10.62	14.05[f]

[a] Measured berm heights h_b

[b] Measured berm retreats b (see the insert of Fig. 3.34 for definition)

[c] Calculated berm volumes V_{slope} from Eq. (3.11), using the measured h_e and h_m

[d] Calculated berm volumes V_{berm} from Eq. (3.9), using the measured h_e, h_b, and b

[e] Predicted berm volume V_{berm} which is equivalent to V_{bar} predicted, per column (9)

[f] Predicted horizontal distance of berm recession b from Eq. (3.9) or (3.10)

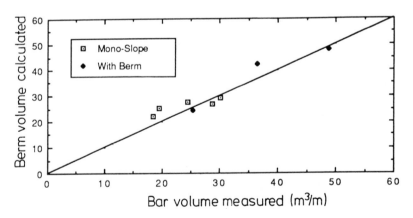

Figure 3.36 Comparison between the calculated berm volume and the bar volume measured, for the 8 selected cases in LWT tests as per Table 3.3

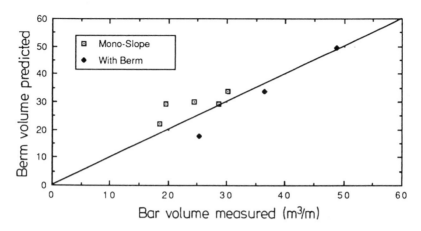

Figure 3.37 Comparison between the predicted berm volume and the bar volume measured, for the 8 selected cases in LWT tests as per Table 3.3

Finally, the horizontal berm retreat b (for the cases "with berm") and beach erosion b' (for cases under "monosloped beach") which can be calculated from Eqs. (3.10) and (3.12) or Fig. 3.34 and Fig. 3.35 respectively, are compared to the measured values, as in Fig. 3.38.

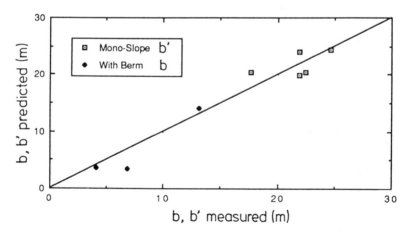

Figure 3.38 Comparison between the predicted beach retreat b and b' with the measured values, for the 8 selected cases in LWT tests as per Table 3.3

Good agreement has been found over the whole procedure of the estimations as demonstrated above, that is, starting from a given combination of wave condition with beach slope and finally ending up with optimum bar and berm recession. For practical applications, the SWL in this study is that operating at the peak of the storm and should therefore apply to storm surge or high tide level. It is on this basis that the actual berm height (h_b) should be determined.

3.4.4 Bar Characteristics from Prototype Measurements

Large wave tank (LWT) test results presented in Section 3.4.3 have shown new relationships can be established for dimensionless parameters involving bar distance, crest depth, equivalent bar volume, and optimum berm retreat. However, there are distinct differences that must be addressed in the prototype. Variables easily measured and varied in flume tests are not so readily recorded in the prototype, so placing different emphases on them in any comparisons. There are also significant differences in the variables measured and analyzed. Nonetheless, tests in LWT should be hailed as near to natural conditions as can possibly be attained and their results regarded as useful, at least for the time being, in the estimation of beach recession during a storm.

Waves in models are generally monochromatic with periods ranging from 3 to 16 seconds and, even if spectral, omit variations in magnitude and the effect of the wind on steepness and run-up. Prototype waves vary in size over time, with peak spectral periods about 10 seconds or greater, but in models they are kept constant. In the LWT tests, some relatively short period waves have combined with high waves, which might be unrealistic in the field. Another problem is the use of breaking wave heights. The breaker depth is normally fixed in a model, generally at a specific location, such as over the bar crest. But, in the prototype breaking could occur far beyond the bar location and the surf zone is much wider than the barred profiles, hence a definition of breaking in spectral conditions is very difficult (Sallenger et al. 1985). Short waves are constantly breaking on longer waves so that the heights and periods of such waves are hard to define. Longshore sediment transport which prevails in prototype conditions can not be produced in any flume test. Bar volumes are measured from differences between the original beach profile and that of the bar. But with the exception of an ideal case, this can involve errors both in the model and prototype situations.

In LWT tests the sediment size can be definitive while in the prototype it will vary throughout the berm and subsequent bar, in which sediment sorting is taking place. From the LWT cases workers can relate bar variables to waves as well as sand grain diameter (D) or sediment fall velocity (w), whereas those observing natural beaches have found a range of sediment sizes is generally present. This implies that both variables D and w are functions of time during a storm. Beach face slope which is initially straight can be varied in a flume, with or without an offshore slope, but is difficult to define in nature. Although such initial beach face slopes can be varied in LWT tests, some of these can be unreal. For comparative purposes similar values of D, w, and $\tan \beta$ should be inserted in any relationships so derived. The groundwater conditions at the beach face which help in causing slumping in the prototype, as discussed in Section 3.4.2, are lacking in the model. The effects of variable winds and storm surge levels are difficult to apply in LWT tests.

The sequence of crest location (X_c) and depth (h_c) as a bar moves offshore is depicted in Fig. 3.39. The average swell-built profile is shown as:

$$d = 0.125X^{0.667} \tag{3.13}$$

as given by Dean (1977), upon which the enlarging bar rests as it proceeds seaward during a storm. The beach recedes but measurements of X_c are made from the initial

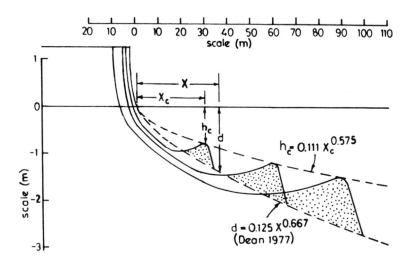

Figure 3.39 Sequential bar profiles as a bar moves offshore during a storm

waterline. Values of X_c and h_c were measured on many profiles provided in the literature, not only of the first but up to the fourth bar, and are plotted in Fig. 3.40, where they fall with very little scatter on a line, giving:

$$h_c = 0.111 X_c^{0.575} \tag{3.14}$$

This relationship is drawn in Fig. 3.39 as a curve through the crests. Thus for any single bar proceeding offshore and for bars in general, including multiple bars, there is a distinct relation between depth and distance to their crests. This fact should be kept in mind by workers researching bar morphology.

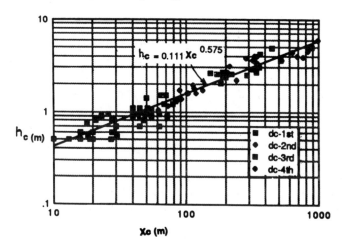

Figure 3.40 Data of crest distance (X_c) and depth (h_c) for single or multiple-bar features

As already mentioned, high and steep waves have been considered of storm character but they are also spectral and multidirectional in nature. This implies that wave steepness is an important parameter, but its specific range is difficult to assess. Sallenger et al. (1985) have defined a storm condition which consists of high energy high frequency storm waves plus high energy low frequency swell. However, wave energy requires further classification both for storm waves and swell, which are considered accretional in building up the berm. In an effort to rationalize the bar movements, it is proposed to use only the spectral peak periods and their normal storm wave heights for quantifying the bar characteristics, such as bar distance offshore, bar crest height, and optimum bar volume.

Although the U.S. Army Corps of Engineers has produced annual reports containing data of waves and nearshore processes at its Field Research Facility (FRF) at Duck, North Carolina since 1982, the first complete record of the formation and movement of an inner bar during a storm can only be found in Sallenger et al. (1985). The nearshore bar system and beach profiles were surveyed periodically and automatically through a storm, using a sled system and amphibious buggy. This enabled the very rapid response of underwater morphology to be monitored during the storm sequence. Beach profiles prior to and after the storm were also measured.

It was found (Sallenger et al. 1985) that the bar became well developed during the increase of wave height in the storm. The rate of offshore migration of the bar was up to 2.2 m/hr, while an overall bar shape was maintained with the ratio of trough to crest depth (h_t/h_c) approximately constant, as seen in Fig. 3.41. Sallenger et al. (1985) also observed that: "The bar crest distance offshore was typically only 10% of the surf-zone width.... . Analysis of the bar distance offshore in terms of a standing wave motion showed that the causative wave period was a minute. Surf-zone wave data showed significant energy in the infragravity band at these periods although no definite link has been made."

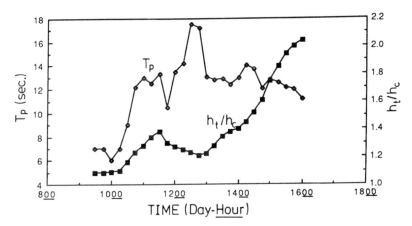

Figure 3.41 Variation of water depth above bar crest and bar trough over time during a storm, as well as peak spectral period (data from Sallenger et al. 1985)

Sallenger et al. (1985) have provided the time history of wind speed, wind direction, significant wave height (H_s), peak wave period (T_p), computed significant breaker height (H_b), breaker zone width (X_b), bar crest distance from shoreline (X_c), bar crest depth (h_c), bar trough depth (h_t), and deepwater steepness (H_o/L_o). From the given data, variables T_p, X_c, and H_b, plus parameters h_t/h_c, $H_b^2 L_b/gT^2$, $\tan\beta/(H_b/L_b)^{1/2}$, and H_b/wT were derived and graphed against time, as in Figs. 3.41 to 3.44, which will be discussed in turn.

In Fig. 3.41 peak period (T_p) and bar depth ratio h_t/h_c are drawn against time, noted in "date-hours." The first digit or two digits represent day in month (i.e., in October 1982) and the last two underlined digits denote the hours of that day. Thus, the $\underline{00}$ mean zero hours or midnight, and the data dots at each 6-hourly intervals can be clearly identified on the figures respectively. The force of the storm was felt from the 9-$\underline{18}$ and ended about 12-$\underline{12}$, when the breaker height started falling. It is seen that T_p increased nonuniformly to 12-$\underline{12}$ (noon on 12th day) and then decreased slowly. The sudden increase from around 14 s to 17.5 s appears rather anomalous and perhaps should be discounted, but the trend as shown above applies. The h_t/h_c curve rises to a maximum at 11-$\underline{12}$ (noon of 11th) and falls to a minimum around 12-$\underline{12}$ just after the time of maximum T_p or peak of the storm.

In Fig. 3.42 A,B,C the crest distance X_c reaches a peak at 13-$\underline{00}$ in A, or about 12 hours after H_b reaches its maximum. This is similar to the optimum of $H_b^2 L_b/gT^2$ in B. Thus, there appears a lag time of half a day between the peak of the storm and the equilibrium distance of the bar crest from the shore. The possible mechanisms responsible for this time lag are worth investigation. The curve for wave steepness (H_o/L_o) reaches a maximum at 10-$\underline{00}$ and then falls swiftly at 10-$\underline{12}$ to very low values. The disparity between wave steepness and H_b indicates that steepness alone is not a good variable to associate with movements of the bar, offshore and onshore. Even when it is combined with bed slope $\tan\beta$ in $\tan\beta/(H_o/L_o)^{1/2}$, as in Fig. 3.43, it does not correlate with the movement of X_c.

An alternative parameter H_b/wT, as used to differentiate bar/berm formation (see Fig. 3.24), is graphed in Fig. 3.44. This curve rises swiftly to a maximum at 10-$\underline{06}$ and then falls more slowly to the end of the storm; it therefore has little correlation with X_c. When it is plotted against X_c, as in Fig. 3.45, it rises from the value of 4.1, defining an initial bar formation (see Fig. 3.24), to a maximum value of 14 within the first 6 hours of the storm attack, when the bar is around 20 m from the original waterline, then decreases slowly while the bar moves offshore to a transitional value around 9.5 when $X_c = 70$ m, which is the equilibrium seaward limit for the storm. As the bar proceeds shoreward due to the wave reduction the relationship between H_b/wT and X_c appears linear to the 4.1 limit as noted previously. The swift rise of H_b/wT at the commencement of the storm means that H_b builds up more swiftly than wave period but that as time passes and perhaps FAS conditions are approached an optimum of around 10 applies.

In Fig. 3.46 bar distance is normalized as X_c/L_o and plotted against $H_b\tan\beta/gT^2$ where bar commencement starts at the top RHS of the figure. As the latter decreases so does X_c/L_o, due to the fact that the increases in wave length associated with peak period

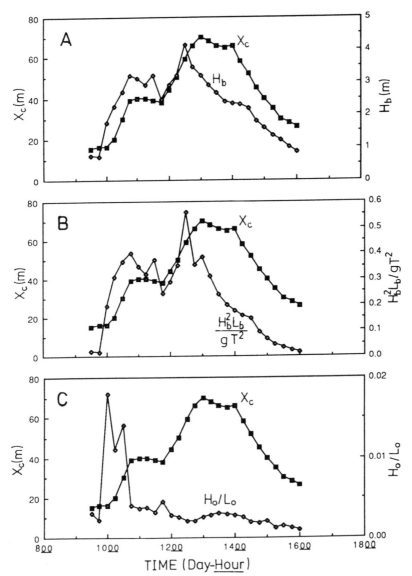

Figure 3.42 Variation of bar crest distance (X_c) as measured throughout a storm, with wave parameters: A. height of wave breaker H_b, B wave energy $H_b^2 L_b / g T^2$, C deep water wave steepness H_o / L_o (data from Sallenger et al. 1985)

are much larger than the bar distance, until a point is reached when X_c / L_o suddenly increases to a peak (due to the falling of peak period at the waning of the storm) before falling almost linearly as the bar moves onshore with a further decrease in $H_b \tan \beta / g T^2$. The curve for LWT tests with the same parameters is drawn in Fig. 3.47 but the difference

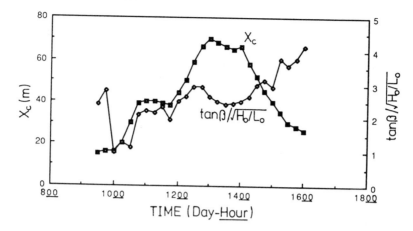

Figure 3.43   Bar crest distance (X_c) and $\tan\beta/\sqrt{H_o/L_o}$ over time

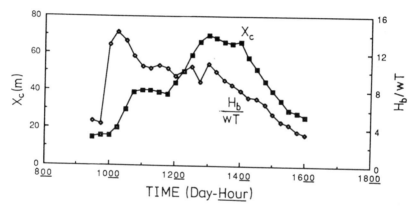

Figure 3.44   Bar crest distance (X_c) with sediment parameter (H_b/wT) over course of a storm, as the bar moved offshore and subsequently returned onshore

in abscissa should be noted, that for the model (Fig. 3.47) are greatly in excess of those for the prototype (Fig. 3.46), which precludes any connection. But, it is seen that the whole figure frame of Fig. 3.46 can be fitted into the lowest left corner of Fig. 3.47 without adjusting their scales respectively, indicating different scales of these phenomena for the prototype and the LWT tests.

In Fig. 3.48, the ratio h_c/X_c is plotted against $\tan\beta/(H_b/L_b)^{1/2}$ for the prototype and in Fig. 3.49 for the LWT tests. Again the discrepancy in abscissa values should be noted even through the ordinates are to the same scale. In the prototype (Fig. 3.48) h_c/X_c decreases to a minimal value of 0.0225 when $\tan\beta/(H_b/L_b)^{1/2} = 0.6$, which is when the bar is at its maximum distance from shore, or in equilibrium during the storm reported by Sallenger et al. (1985). This implies that:

$$h_c = 0.0225X_c \qquad (3.15)$$

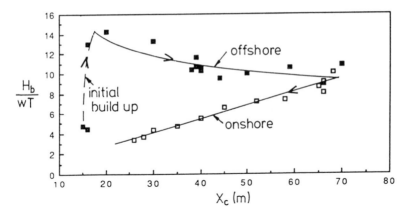

Figure 3.45 Parameter (H_b/wT) and X_c during offshore and then onshore movement of bar

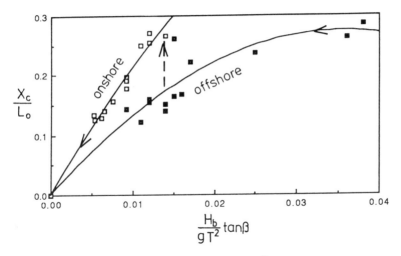

Figure 3.46 Bar crest distance (X_c/L_o) versus $(H_b/gT^2)/\tan\beta$, during a single bar movement offshore and then onshore in prototype conditions (Sallenger et al. 1985)

which differs from the prototype relationship given in Fig. 3.40 for h_c in Eq. (3.14), which match only when $X_c = 43$ m. More research is required both in the field and in modeling with a view to using compatible variables and parameters.

3.4.5 Return of the Bar

Part of the exciting phenomenon of natural coastal defense is the dismantling of the bar or bars when no longer required. This is accomplished by the milder or swell waves following the storm, either immediately or some weeks later. These break over these mounds causing excessive suspension of sand which is carried shoreward to be redeposited on the beach. Although the bar is being moved bodily toward the beach,

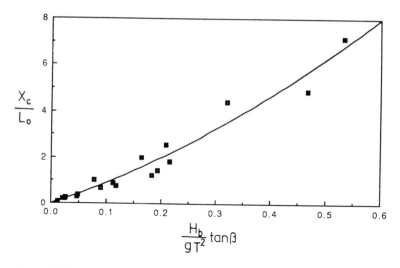

Figure 3.47 Bar crest distance (X_c/L_o) versus a surf parameter $(H_b/gT^2)/\tan\beta$, during a bar movement offshore in LWT tests, compare with Fig. 3.46 but different abscissa and ordinate scales

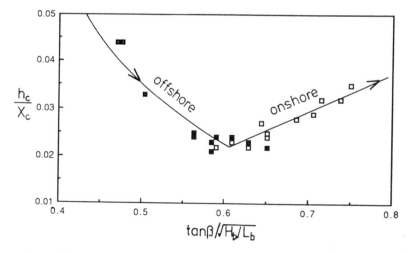

Figure 3.48 Bar crest depth (h_c/X_c) versus a surf parameter $\tan\beta/(H_b/L_b)^{1/2}$, during a bar movement offshore and onshore in prototype conditions (Sallenger et al. 1985)

its landward face is quite steep as material is transmitted over its crest into the trough and ultimately on to the beach face. Resonance occurs for certain components of the wave spectrum as the trough width changes, which tends to maintain the bar as a distinct feature. The newly formed berm face is quite steep and the inner face of the bar likewise, so aiding the resonant action. Whereas berm construction is relatively swift as the bar moves shoreward it increases suddenly as the bar coalesces with the beach.

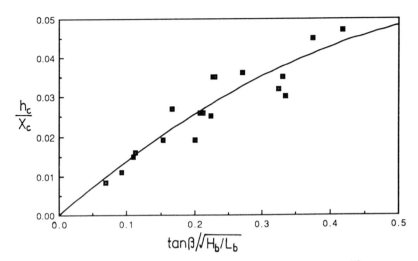

Figure 3.49 Bar crest depth (h_c/X_c) versus a surf parameter $\tan\beta/(H_b/L_b)^{1/2}$, during a bar movement offshore in LWT tests, compare Fig. 3.48, noting different abscissa values

The berm is thus reconstructed to a height equal to the uprush limit of the waves at that particular time. This may be small initially due to the waves being strongly attenuated while traversing the fully developed bar. However, as this mound is depleted and moved bodily shoreward with the surf zone decreasing in width, so the run-up increases and hence the height of the berm. This is why the rear of the beach may be lower than the seaward extremity, with puddles being maintained for a limited period. Larson and Kraus (1989) found a shoreward slope of around 2.5°. The present discussion omits the influence of tides, but sea levels at various stages of a tide during this accretion process can dictate berm levels. More than one bar may form, one at the high and another at the low water stage, but the principles enunciated above still apply.

As noted earlier, not only is suspended sediment carried shoreward by the mass transport in the bores of the surf zone, but also material is rolled from the top of the mound onto the steep landward face, causing the mound to move bodily shoreward. The first and subsequent mounds finally coalesce with the beach. Mason et al. (1984) have recorded a bar movement of 1 m/hr shoreward. They comment: "The rapid response of bar morphology to changing wave conditions indicates the need for more rapid sampling of the surf zone morphology. It appears that previously held concepts of a slowly responding system may be erroneous." This swift natural renourishment has been noted by Kriebel (1987): "A total average sand volume of 7.5 m³/m was returned onshore and deposited above mean sea level over the two month recovery period. Of this total volume, 5.4 m³/m (or 72%) was moved onshore over just the first two days." Also, Katoh et al. (1987) observed an accretion rate of 0.7 m per day, but this could be exceeded with stronger swell.

The seaward face of bars can be as steep as the face of a swell-build beach profile, due to mass transport of the swell straight after the storm. This acts in concert with the offshore current to produce such steep slopes. Subsequent waves traversing this zone will have a higher ratio of breaker height to deepwater height, as seen in Fig. 3.50 derived from Goda (1975). With material filling the trough between bar and beach, these higher than normal breaking waves are traversing a milder inshore slope which retards their attenuation. In Fig. 3.41 the ratio h_t/h_c increases from 1.20 at the peak of the storm to around 2.05 at its end because of the decrease in h_c as the bar nears the beach after 3 days. Thus, the waves in the surf zone are enhanced during this removal of bar to the beach. The effect of this on longshore current generation will be discussed in Section 3.5.

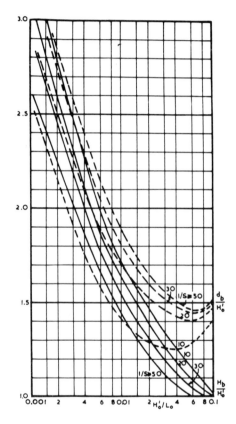

Figure 3.50   Wave characteristics as derived by Goda (1975)

On the margins of enclosed seas, where milder waves after a storm have still a relatively wide spectrum, or are of storm character, the accretionary process is not so effective. It is possible that not all the bar is replaced on the berm. However, as soon as wave energy builds up again, during a subsequent storm, the bar is quickly reformed to a state where it can dissipate these erosive waves. Many such coasts with minimal swell have been studied, with the conclusion reached that all shorelines have continual offshore bars. In fact, the bulk of shorelines around the world have these features for very short periods of the year, approximately for three days after each storm.

3.4.6 Multiple Bar Profiles

The action of the first bar from the beach has been emphasized in the above discussion since its influence in wave dissipation is direct and most influential. Its spacing at the first antinode for the partial standing wave from the surf beat of the storm wave spectrum can also have a second antinode seaward, which forms a second bar parallel to the beach. In fact, on beaches with very mild offshore slopes, where the accretionary waves are almost normal to the coast, a number of bars may be constructed (Exon 1975).

Once the first bar is nearing completion, or is close to breaking all incoming waves being reflected from the steep beach face at the rear of the eroded berm, there is also concomitant reflection from the seaward face of this initial submerged mound. This double reflectivity enhances the seaward movement of the reflected waves and the formation of several antinodes. The longer the duration of the storm the greater the propensity for multiple bar formation. Since the wave lengths of these partially standing waves will be longer in the deeper water offshore, the spacings between bars will increase with distance from shore.

As more bars are formed, so incoming waves are reflected from each bed undulation. Heathershaw and Davies (1985) have derived reflection coefficients for this situation, showing optimum reflection when the bar spacing is half the incident wave length. This is unlikely to be the case since this spacing is dictated more by the surf beat period than these incoming wave trains. Results from their theoretical analysis for ideal conditions are shown in Fig. 3.51, which have been verified by model studies (Heathershaw 1982). It is seen that for ten such mounds reflection is almost complete. Only for very mild slopes offshore could this number be encountered and only then in an accreting situation of almost normal wave approach. Even so, some degree of reflection will occur on beaches with one or two bars, which enhances the ability of such naturally constructed features to stop further erosion of the beach.

Wright et al. (1979) have shown that for the eastern coast of Australia the power spectra of waves landward of the bar were about one tenth of those breaking seaward of it. Transverse currents within this zone also correlated strongly with the outer breaker height. They matched the period of both the incident waves and the surf beat emanating from them. The bar spacings equated to the shallow water celerities of $C = (gd)^{1/2}$ or wave lengths $L = T(gd)^{1/2}$. On oceanic margins the distance of first and subsequent bars, half the above L value, required the insertion of periods around 40 seconds, or four times the normal 10 seconds of the peak waves in the storm spectrum. For beaches on enclosed seas the distance between bars is less, with periods of around 25 to 30 seconds, again four times the local peak period.

3.5 LONGSHORE DRIFT

Should the milder or swell waves still be angled slightly to the beach when breaking, the longshore component of their energy generates a littoral current. If long-term erosion or siltation is being experienced on any section of coast it is conclusive that these waves

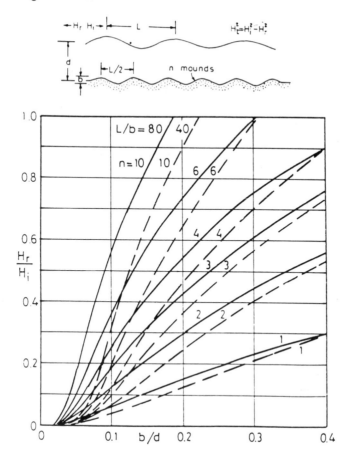

Figure 3.51 Seaward reflection of waves from bed undulations

are arriving obliquely to the shoreline. As noted in the previous section, the higher than normal breaking waves may occur immediately after the storm, while the surf zone width is greatest. This results in excessive suspension while the longshore current is also at its optimum. This results in a pulse of littoral drift which is maximum just after a storm, but reduces in magnitude as the bar is dismantled and moved shoreward. This longshore transport even extends beyond the breaker line, as found by Anderson and Fredsoe (1983).

This excessive littoral drift exists while the wave conditions in the storm-built profile are large. As the normal parabolic shape is approached, with the berm out to its greatest width, waves break close to the shore, in some cases with only one breaker existing at any time. Since the bulk of the sand is now located in the berm, waves have no influence on it, being able only to disturb material on the beach face out to the breaker line close to it. This is so different from the situation where this sand has been placed offshore, to be moved readily by swell waves while coming ashore and simultaneously angled to the coast. It is apparent, therefore, that littoral drift will diminish from a maximum just after a storm to negligible proportions once the swell-built profile again exists.

A temporal distribution of transport rate is hypothesized in Fig. 3.52, which requires verification by field measurements. To carry out such a test would require monitoring of accretion within a sand spit or against some structure at daily intervals over one or two weeks after a storm, or until the bar feature had disappeared. This would demand profiling of the beach to relate the accretion rate with the surf zone width or stage of bar removal. Kraus and Dean (1987) derived an equation which showed that overall transport varied directly with surf zone width that occurs mainly during a storm sequence.

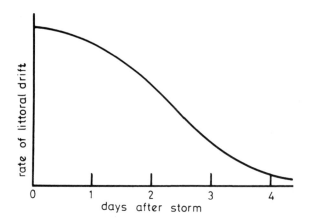

Figure 3.52 Possible littoral drift variation directly after a storm

This impulsive nature of littoral drift is not a new concept, although it appears to have been forgotten since Krumbein and Ohsiek (1950) first observed it. They concluded: "If pulsational transport is the rule along both lake and ocean beaches that are interrupted by inlets, structures, or other obstacles to continuous transportation, consideration of the consequences of such movement becomes significant in terms of periodic increase and decrease of drift as a function of distances downbeach of the obstacle." The authors did not discuss the origin of these fluctuations in drift, only the consequences. It therefore warrants some further discussion as to the reasons for it.

It may be thought that the outgoing current over the developing bar may spread sand more evenly over the seabed, resulting in a mild slope to the bar. But this seaward flow generally takes place at weak spots in this mound, being concentrated at these points in the form of rip currents. These carry material seaward and hence enlarge the depression, which changes locations as longshore drift tends to fill it. It is well known that such dangerous conditions for bathers occur just after storm sequences. Seaward flow is thus minimal over most of the bar length and hence the spreading mechanism is minimized.

As noted previously, this influences the breaker height (see Fig. 3.50) or more particularly the ratio of crest height (a_c) to deepwater wave height (H'_o) for waves arriving normal to the coast. This ratio (a_c/H'_o) is decreased, making the breaker more surging than plunging, which results in greater mass transport shoreward. The wave front becomes steeper (Adeymo 1968) with increased bed slope, again improving littoral current generation.

Another factor worthy of note is the partial reflection from this relatively steep slope of the bar face. Battjes (1974) has shown that when the surf similarity parameter $\tan\alpha/(H_o/L_o)^{1/2}$ has a value of 2.3, it "corresponds to a regime about half way between complete reflection and complete breaking." These partially reflected waves, which must necessarily be angled to the bar, interact with the incident waves to establish a short-crested system (Silvester 1974), whose orbital motions are conducive to sediment suspension, with a large mass transport alongshore. This action could cause excessive drift on the face of the bar, beyond the breaker line, which could exceed values estimated from the littoral current expanding beyond the surf zone.

3.5.1 Littoral Drift Calculations

Mathematical approaches of various workers on littoral current include many variables (Kamphuis 1966). Where sediment motion is involved, empirical force coefficients on sediment particles are employed which are often outside the range of tests. Those of drag and inertia, for example, are subject to great error (Silvester 1974). Correlation coefficients between overall transport rates and some measure of wave energy may be applicable to only one section of coast.

Greer and Madsen (1978) analyzed the studies of three workers on which the U.S. Army Corps of Engineers (SPM 1973) base their formula (equation 1 in the authors' paper). They found all data lacked accuracy and methodology and stated: "The conclusion of our review is therefore that coastal engineers using Eq. (1) for the calculation of longshore sediment transport rates should regard their results as no better than order of magnitude estimates." A later review by some seven authors (Readshaw et al. 1987) reached the following conclusion: "Even if good data are available it is still not certain that reliable estimates of alongshore sand transport can be obtained."

It is normal to associate volumes of accretion against some structure measured on an annual basis, using some average swell condition over the same period. As seen from the previous discussion, swell varies in height and period from hour to hour and hence the selection of any meaningful average is difficult in the extreme. The transport effected on barred profiles differs drastically from that on swell-built profiles, which is not taken into account in these formulae. It becomes obvious that if three storms occur in any year, instead of one the drift could be tripled. This assumes that a reasonable duration of swell occurs between storms or for at least three days before the next sequence.

No account appears to be taken of this interaction between storm waves and swell in littoral drift formulae despite the absurd differences between them, plus noncorrelation with field measurements. Dean et al. (1982) and Dean et al. (1987) measured accretion on the spit leeward of the tip of the breakwater at Santa Barbara harbor over some 380 days. Eight survey periods of approximately equal length were analyzed to compare wave heights (as measured by recorders) with longshore drift. Their results are listed as in Table 3.4, where wave energy was computed by two different methods, giving values of P and S. Immersed transport rate (I) was then divided by wave energy to give a correlation coefficient of K or K^*. The ratio K^*/K was derived to illustrate the sensitivity in the mode of deriving this input energy which gave values ranging from 1.78 to 2.94 indicating a degree of nonconformity.

TABLE 3.4 FIELD RESULTS FROM SANTA BARBARA (DEAN ET AL. 1982)

No. of days	Total vol. change Rate (m³) I (N/s)	Immersed wt. transport Rate I (N/s)	Component wave energy Flux at breaking P (N/s)	$K = I/P$	Component of Mmt S (N/m)	$K^* = I/S$ (m/s)	K^*/K
48	32,820	85.3	52.2	1.63	27.8	3.06	1.90
31	65,070	159.1	101.4	1.57	45.4	3.50	2.13
35	82,810	295.0	352.4	0.84	119.6	2.47	2.94
53	10,290	24.2	76.6	0.32	37.9	0.64	2.00
82	22,220	33.8	31.7	1.07	17.6	1.91	1.78
57	38,760	84.8	63.8	1.33	32.6	2.60	1.95
54	35,640	84.6	64.4	1.31	34.2	2.47	1.89

But accepting the usual constant K, it is seen to vary in the table from 0.32 to 1.63, a range of 400% during this period of just over one year. The authors commented "However the smallest value which exhibits the greatest deviation from the norm is associated with the fourth intersurvey period which is characterized by a very small value of I. If this one point is not included, the ratio of the largest to the smallest of the remaining K values is less than two, which appears reasonable for this type of measurement." This certainly is an admission of some variable not being included in the relationship.

No mention was made in the article of storm sequences during these surveys, even though the extra high storm waves would have been included in the energy calculations, without reference to the changing directions of these locally generated waves or subsequent swell. The values of K were averaged to give a value of 1.23, which was acknowledged to be distinctly different from the previously accepted value of 0.77 (SPM 1973). These variations tend to undermine any confidence in such formulae.

Knowledge of the actual timing of storm sequences over the 1979–80 period was not available. However, inspection of the U.S. Navy Marine Climate (1977) gave, for the North Pacific Ocean, the average number of storms for each month within the 10° latitude longitude square adjacent to the Californian Coast centered on Santa Barbara. These were interpolated for the varying intersurvey periods as designated in Table 3.4 (Silvester 1984). The K values for each period were then plotted against the number of storms in each, as in Fig. 3.53, where it is seen that in spite of some scatter a reasonable

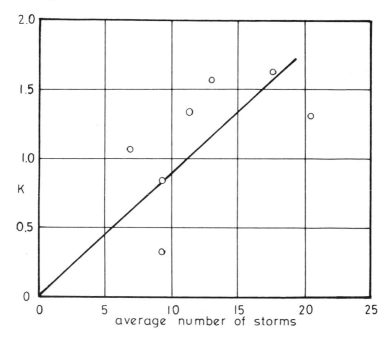

Figure 3.53 Values of K as derived by Dean et al. (1982, 1987) versus number of storms in each measuring period

relationship is indicated of K increasing with the number of storms. It would have been preferable to use the actual number of storms during the drift measurement, but this was not possible from the wave data available. However, it would appear that this variable should be included in any future assessments of littoral drift.

Another example is available of variations in computed drift, model verification, and actual measurements in a sand trap (Pratte et al. 1982). Two computations from different formulae gave 120,000 and 205,000 m³/yr., which is a difference of 170%. A model study indicated an annual drift of only 30,000 m³/yr. Even though waves of the most severe annual storms were used in the model, it is doubtful that a proper bar profile would have formed, with subsequent oblique swell to actually transport material downcoast. The accumulation in the dredged hole to date of publication averaged 110,000 m³/yr., indicating the need for "a permanent dredging plant." These large discrepancies point to some missing link in the chain of events which, it is contended, could be the impulsive nature of littoral drift due to the processes outlined above. Pratte et al. (1982) stated that "it took approximately 12 hours to fill the entrance during storms," which could be interpreted as just after such sequences.

Littoral drift formulae generally assume a uniform slope for the surf zone. Anderson and Fredsoe (1983) have computed littoral drift along a barred beach and compared it to that on an equivalent uniformly sloped profile. Figure 3.54 is derived from this reference where it is seen that the three bars as in (a) cause local increases in H/h as in (b), from which values of q_r have been derived as in (c). The distribution of transport for

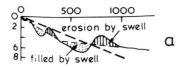

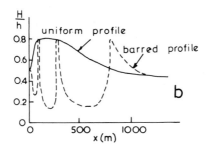

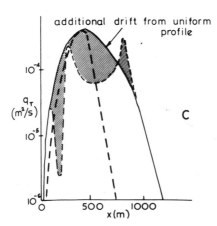

Figure 3.54 Littoral drift on barred beach: (a) shoreline cross section, (b) ratios (H/h), and (c) drift q_r. New analysis full lines, Anderson and Fredsoe (1983) dotted lines

the equivalent uniform slope is shown dotted in (c). Areas under these curves give the total rate of transport, which the authors concluded were essentially the same for both profiles.

An alternative progression is also illustrated in Fig. 3.54, where the swales are filled from the bars seaward of them. This could result in two uniform slopes as indicated in (a), giving the H/h curves as in (b) and hence the modified q_r curve as in (c). The area under the resultant envelope is greater than that for uniform slope as exhibited in Fig. 3.54(c). Despite all the inaccuracies involved in these suppositions, the propensity for greater drift on a barred profile appears worthy of more research.

Verification of littoral currents on fixed bed models could be carried out for the various profiles of barred beach, uniform slope, and the more normal parabolic shape of a swell-built beach, as suggested by Baum and Basco (1986). The distribution of longshore water motion and a final average could determine the expected transport of suspended material and that rolled along the floor.

Another supposition made in littoral drift formulae is uniform size of sediment. However, as shown previously, in Fig. 3.22, a sorting process takes place during the formation of the offshore bar, while it is being brought back to the beach (when the bulk of drift takes place), and selectively while swell is arriving at the final stable profile. Thus, during the pulse of drift the sand sizes being shifted are varying continually, with the finer components moving at the peak of the pulse and the coarsest being transported the least. Such longshore sorting is clearly recognized by researchers who gauge sediment direction by size distribution lengthwise along the coast.

It will be appreciated that the above discussion is heuristic because sampling of beach berms or cores through offshore bars has not proved the mechanisms outlined to exist. However, the principles enunciated are based on physical processes that have been observed in other contexts. They appear a worthwhile line of research for geologists in order to reduce the scatter generally present in sediment samples obtained in the prototype. Normally they are taken in the swash zone, for comparisons of different beach locations, which is not typical for either the coarsest or finest fractions that may exist together on a swell-built beach face.

One important conclusion to be reached from the above discussion on longshore transport formulae is that they discount the lateral movement of sediment in the process. Sand cannot be placed in suspension while it is in the berm, but suffers great disturbance when placed offshore in the form of a bar. This must introduce vastly different rates of transport as this bar is modified and shifted shoreward. As Smith and Gordon (1980) have stated: "The most significant variability in beach response to energy input is clearly demonstrated by the hour to hour and day to day fluctuations in cross-sectional profile. The onshore-offshore movements of beach sediments call into motion much greater volumes of material in the short term, by a factor of many magnitudes, than those that can be measured as alongshore transport. If, as it seems likely, the actual littoral transport represents only the longshore component of onshore-offshore sediment movement under oblique wave energy input, then it would seem essential to study the onshore-offshore processes in their own right before even addressing the ultimate problem of calculating littoral transport."

3.5.2 Longer-Term Variations in Drift

Besides the pulses of littoral drift noted above there are variations in supply of sediment at any section of coast caused by a number of other activities. The most obvious of these is the construction of structures such as breakwaters and groins which intercept littoral drift. Erosion is experienced just downcoast of these impediments almost immediately, and further downcoast at a later date. A less obvious, but more significant, interception is carried out by channels dredged to ports from well offshore. These act like groins of equivalent length, stopping longshore movement both in the surf zone and out to sea. This causes denudation of the bed over large areas downcoast from the channel, when silted material is generally dredged and deposited offshore for convenience. This deepening results in steepening of the beach profile which requires a greater volume of sand to construct the protective bar. Some of this remains offshore to maintain the original slope and hence the renourished berm becomes narrow and beach erosion ensues.

The supply of sediment to the coast comes almost solely from rivers, generally in pulses when flooding occurs. This can vary greatly from year to year and during each year, so resulting in humps of material being available for dispersal along the coast. Since man has had the capability to construct dams across rivers, for purposes of water supply and reduction of flooding potential, so their ability to furnish sedimentary material has diminished. Erosion is experienced just downcoast of the mouth, but over time longer lengths of coast suffer the same fate. Dredging of channels and turning basins within harbors located inside river entrances also interferes with coastal supply.

The wave energy applied to the coast during a storm is concentrated on the edge of a cyclone where the winds are in the same direction as storm movement. This is to the right of the center in the northern hemisphere and to the left in the southern. These extra high waves will demand more beach berm and perhaps frontal dune than further along the coast in each direction. Thus a greater volume of sand is placed into circulation, to be brought back to the beach and moved simultaneously downcoast. Again a hump of sand is transmitted alongshore followed by a dearth of material and a recession in the shoreline.

Many years ago, in the study of sediment movement across an inlet, one of these authors had available aerial photographs taken at about six monthly intervals over about 8 years, with spits forming across the inlet at annual intervals. The changes at monthly intervals were interpolated with silhouettes prepared which were then photographed in sequence on cine-film. The 8-year sequence could then be observed by projecting over 4 minutes. The shoreline was seen to oscillate like a snake as humps of material moved around a headland and into the spit, which elongated one or two kilometers with widths of 50 to 150 m. This was eroded into small islands or disappeared altogether during the winter season, but reformed into various shapes depending on the fresh water discharge during the formative period just after the storms.

This dynamic character of the shoreline is also associated with shoals that do not show above the water surface. They are the continual bane of ships' masters, forming swiftly from a pulse of drift and being denuded more slowly while being moved downcoast. Should a size of dredge be decided upon from some annual rate of drift that discounts fluctuations, it is possible for such a volume to be dumped in a river mouth over a few weeks which could take months to be removed. Knowledge of the sequence and timing of deposition, as per the pulses previously alluded to, should help in dredging programs of harbor mouths and navigation channels.

The coastal engineer should be conversant with what is being planned upcoast that could change supply of sediment to a point under study. Equally, information should be passed on to authorities downcoast, informing them of likely consequences of intended developments, in order that remedial measures may be made immediately at the least cost. The volumes of material involved in erosion or siltation are immense, because they are shifted long distances over very short periods of time. Forewarned is forearmed, and is especially true in marine situations where any activity is costly.

3.6 EFFECTS OF TIDES

Tidal oscillations around the bulk of most oceanic margins are relatively small, being no greater than about 2 m. They are dictated by the width of the continental shelf and it is not until this assumes large proportions that ranges become significant. This is exhibited in the tides experienced around Australia, as in Fig. 3.55. Besides depth changes the influence of converging coasts also concentrates energy which results in larger than normal tides.

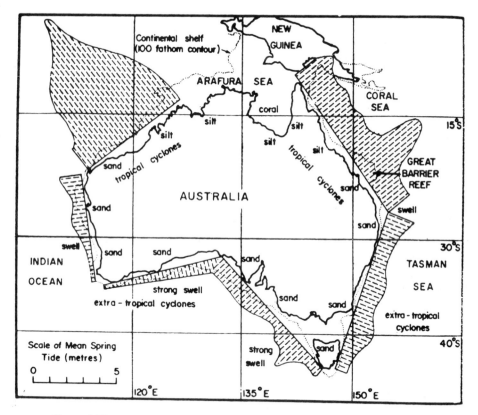

Figure 3.55 Tides around Australia, showing correlation with continental shelf width

The influence of these vertical changes in water level on shoreline processes will be discussed. This will mainly revolve around the construction and dismantling of the off-shore bar during and subsequent to storm sequences. However, the presence of large tidal ranges also alters the general profile of the swell-built beach, causing it to be very mild in slope. Such conditions generally result from a siltation situation, but erosion can occur on certain segments of coast due to an imbalance between supply and demand of sediment.

The horizontal currents produced by tides will then be discussed. These are generally minimal at the bed level whereas wave action, particularly mass transport, is maximum in this segment of the water column. However, currents have an influence on waves and hence their ability to move sediment. For this to happen they must be relatively strong and hence such effects are felt over restricted regions of the coast.

3.6.1 Magnitude of Tides

There are many tomes outlining the sources of tidal oscillations throughout the world's oceans, and a simplified explanation is given in Silvester (1974). These are similar to gravity waves, suffering the same height or range enhancement when they propagate into shoaler water. Because of their wave length they act as shallow-water waves, but are normally reflected completely from a coast. Only in one or two places do tides turn into a bore due to breaking, and only then for spring or extra high tide conditions. In mid-latitudes most oscillations are semi-diurnal with periods of 12 hours 25 minutes or 44,700 seconds. Closer to the poles they are more diurnal, with intermediate regions experiencing a combination of the two.

Engineers are more interested in the tidal range adjacent to the coast where it is increased from the value at the shelf edge, which may be assumed to be 130 m deep (Shepard 1963). Silvester (1974) has presented a figure of range enhancement both for depth variation and width reduction as in triangular-shaped indentations of the coast. Only the former will be discussed here, where the ratio of depth at the shelf edge (d_1) to that near the shoreline (d_2), plus a parameter of shelf width (W) to wave length (T_L) of the tide in depth (d_1), determines the ratio of range (H_2) nearshore to that at the shelf edge (H_1). Curves are shown in Fig. 3.56 for a range of d_1/d_2 giving H_2/H_1 for W/T_L, which should not exceed 5. Substitution of the respective values of $T = 44,700$ sec. and $d_1 = 130$ m in $T_L = T(gd_1)^{1/2}$ gives a value of $T_L = 1,600$ km (994 statute miles, 863 nautical miles). This is applicable to most sandy oceanic margins, but the depth of the shelf edge should be measured from profiles for any specific segment of coast under study.

It is seen in Fig. 3.56 that the tidal range increases swiftly up to $W/T_L = 0.2$, which Dornestein (1961) showed was a resonant length for a wave on a sloped plane. For greater widths enhancement is slower, but it is seen that ratios of 2 to 3 are applicable. To obtain the shoreline range, that beyond the continental rise must be known, which varies throughout the oceans for one component of the possible 62 as in Fig. 3.57 (Pekeris and Accad 1969). This major moon component (M_2) in the deep ocean approximates 0.5 m to which should be added the solar component, about half this value (Ippen 1966). By computing range ratios up an average continental rise and across an average continental shelf (Shepard 1963) results in values of 0.5 m for moon tide in deep ocean, 0.75 m for solar addition, 1.24 m for continental rise, and 2.1 m for continental shelf with $W/T_L = 0.04$. Thus, on average shoreline tides tend to be around 2 m in range.

3.6.2 Influence on Shoreline Processes

With a significant tidal range the attack of storm waves is experienced over a larger swarth of the mildly sloped beach profile. The beach beyond the LWM influence of

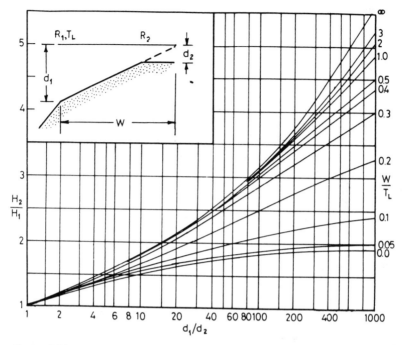

Figure 3.56 Tide enhancement due to shelf width (W) and tide length at edge of shelf (T_L)

the waves remains sensibly the same, but within the tidal area the beach is eroded to a smooth uniform face over a large width. Conditions for bar construction vary during a rising and falling tide, as depicted in Fig. 3.58.

During a storm with a rising tide, as in Fig. 3.58a, the bar commences at LWM and is added to as the water level rises, becoming maximum at HWM. Because of the landward flow of water the seaward face is steeper than normal. During the subsequent falling tide (Fig. 3.58b) the bar is denuded, so placing some material in the previously eroded face and spreading the remainder in almost a horizontal top and a more moderate seaward face. When a storm occurs during a falling tide, as in Fig. 3.58c, the initial bar at HWM is eroded so that ultimately at LWM a horizontal top is provided to the bar with a mild seaward face. During a subsequent rise in sea level a second thickness of beach is demanded to construct a second bar on top of the first, as seen in Fig. 3.58d.

Should the storm continue over a day or more, the above sequences are repeated, so that a double bar system is created. As soon as swell-type waves arrive they will quickly dismantle the bar between LWM and HWM and leave the segment below LWM to be returned to the beach a little later. This is the reason for engineers involved with large tidal range situations believing that substantial bars do not exist after storm sequences.

Since most maritime structures would be situated above the HWM they are not in danger for many stages of the tide. However, care must be taken that storm waves do

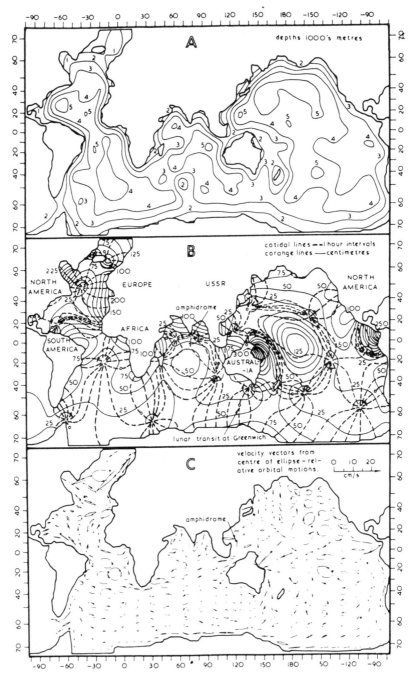

Figure 3.57 World ocean data (Pekeris and Accad 1969): A. Depth contours, B. Cotidal and co-range lines for M_2 tide only, and C. Velocity vectors or relative orbital motions of water particles

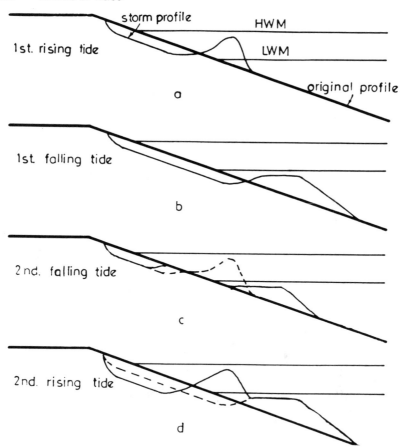

Figure 3.58 Bar conditions during a storm with consecutive tidal cycles

not impact on them, particularly seawalls, at high water levels. The resulting reflection could scour material adjacent to them and put them in jeopardy. In such high tide regions large storm surges are also prevalent, so that designs should take cognizance of this.

Because of the larger range of beach over which longshore drift can occur, deposition in dredged channels from a port extends over a longer length. Depths of channel must be provided for ships to arrive or depart during higher tidal levels, so the extent of siltation must be predictable both during and after storm sequences. Investments in dredging are massive and hence knowledge of deposition is paramount. The proper location of spoil areas is also associated with this problem.

In locations where persistent swell is present, the above siltation problem is magnified. In others, such as 0°–25° latitudes where tropical cyclones are prevalent, calm conditions generally ensue between storms. In this case sediment movement is not so important, but offshore mounds could remain for extensive periods. These may refract waves toward a channel and hence create navigation problems.

In addition to modifying shoreline processes, storms that attack at high tide have caused the most extensive flooding and damage. Pilkey et al. (1983) have reported that: "Both the Great European Storm of 1953 that killed hundreds of Dutch and the Ash Wednesday Storm of 1962 that caused great destruction on the Atlantic coast from New England to Georgia (USA) struck the coast during the highest spring tides." In Japan, the Great Ise Bay Typhoon, which struck Ise Bay and Nagoya region in September 1959 and left 4,500 dead, occurred at a high tide of 3.51 m.

3.7 EFFECTS OF OCEAN CURRENTS

It has been widely believed by geographers, geologists, and some engineers that ocean currents are much more influential than waves in transporting sediment along continental margins. This can be proven incorrect, for the nearshore at least, by comparison of the tidal bed velocities with those of mass transport due to waves. It is also a fact that strong littoral drift occurs on coasts where significant currents are in the opposite direction. In enclosed seas where tides are negligible, longshore transport still takes place at large rates.

Even though surface or mid-depth currents may be substantial, either alongshore or to and from the coast, they generally occur in closed or near-closed loops. Thus, the net movement over a tidal cycle may be very minor. Velocities over this 12-hour period will vary, with some having the capacity to move material suspended by wave action, but the net effect is likely to be small.

Since these velocities reduce as the bed is approached, the circuit produced in the water layers adjacent to the bottom are far smaller than those in the upper reaches of the water column, as seen in Fig. 3.59. The large differential with height above the bed causes eddies to form, both horizontally and vertically, which cause large undulations to

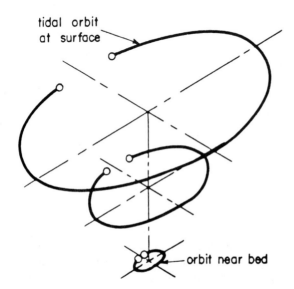

Figure 3.59 Orbital motions of tidal waters at different depths

form. This results in short-term fluctuations in velocity and direction at any point near the bottom. Such swift changes have been observed in tidal current records within 1 m of the seabed in 10 m of water where several 180° changes in direction occurred within 20 seconds of measurement. These eddies could be instrumental in drawing batches of sediment into suspension but net lateral transport could remain negligible.

Most coastal margins suffer very little longshore currents of tidal origin even though they experience significant littoral currents and hence drift by wave action. On the east coast of Florida the strong Gulf current is directed northward while littoral drift is predominantly southward. On the eastern coast of South Africa the Agulhas current is moving southward and the net littoral drift is northward. Such currents may induce eddies close inshore which are opposite in direction to the main stream, but these are sporadic and cannot be accepted as major energy input.

The magnified influence of waves on bed material is understandable when the basis for it is appreciated. As noted in Section 2.3.4, the mass transport of water particles in progressive waves is maximum at the bed, particularly within the boundary layer one or two centimeters from the bed. This optimum net motion thus occurs in a zone where the majority of sediment is suspended or is jumping across the floor. This resulting transport of material is in the direction of wave advance, toward and along the shore. During a storm sequence this mass transport is zero at the bed but counter to the waves slightly higher in the water column, thus during the period when sediment suspension is greater. However, the short duration of storms means that such seaward transport is negligible compared to the almost continuous swell waves.

Currents, depending on their flow direction, can change the characteristics of incident waves and so dictate their capacity to move sediment. For this to happen they must be of reasonable velocity, as caused by concentration at a river or inlet mouth, or between land masses. A contra-current will steepen waves so causing short period components to break while orbital motions in the longer waves will reduce, together with associated velocities (Unna 1942, Yu 1952). A current running with the waves elongates them so increasing their ability to shift sediment (Hunt 1955). Although heights are reduced, they become more peaked so enhancing the mass transport each wave cycle. Thus, a flood tide assists waves in landward transport while the ebb reduces this action.

Variations of sediment movement in the alongshore and cross-shore directions lead to beach profile change, in which strong current may be a contributor, particularly in the early stage of developing a curved planform from an initially straight beach. However, Bird (1984) states that "it was formerly thought that these outlines were shaped by the action of strong currents sweeping sediment along a coast, but is now realised that they are determined largely by wave action." This implies that currents cannot shape the beach, but would mostly help to move sediment to and fro along a beach.

3.8 ESTUARINE CONDITIONS

Estuaries are zones where fresh water debouches from river mouths into the sea causing stratification due to the density difference. These become significant when the mixing waters are contained within a funnel-shaped indentation of the coast. Because of ready wave incidence into such zones, the combined action of waves and currents must be considered.

The lighter fresh water issuing from a river rises to the surface, creating an interface between it and the denser seawater. This slopes downward toward the estuary mouth, so forming a *saltwater wedge*. The shearing effect of the fresh water at the interface causes mixing to take place, with seawater demanded from beneath it. This results in a landward current at the seabed as depicted in Fig. 3.60, which increases as outward fresh water flow builds up (Ippen 1966, Carstens 1970).

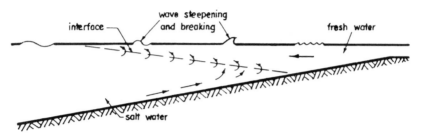

Figure 3.60   Section through an estuary mouth showing the salt-water wedge

This inward current fluctuates over the tidal cycle, being enhanced in the flood tide and perhaps becoming seaward during the ebb (Simmons 1955). Overall it can be accepted that a stronger inward tendency occurs at the bed while at the surface outward motion is preponderant. Longer term fluctuations are due to changes in fresh water flow, high velocities of discharge demand more salt water and thus enhance landward currents at the bed. The wedge itself is translated in and out of the river mouth, so creating changing velocity conditions at any specific plane along it (Schultz and Simmons 1957).

At the toe of the wedge opposite currents occur, an outward flow due to river discharge and an inward current due to the shearing effect on the salt water. Thus, any sediment being carried by either of these will reach a node where horizontal motion is virtually zero. This causes deposition whose location varies with the tide and hence location of the toe (Tully 1953).

As noted in the previous section, currents associated with the outflow of the river or with ebb tide will influence waves entering the estuary. These contra flows, concentrated at the surface, steepen all wave components so breaking the short waves. The longer waves have orbital motions at the bed reduced just when the overall trend is for seaward motion. On the other hand, the flood tide lengthens the waves, so providing more sweeping action on sediment, while the inward current enhances the landward transport as well. This differential in landward currents causes a strong silting tendency in estuaries. Isphording and Imsand (1991) have stated in this regard: "A well established geological axiom states that it is the 'ultimate fate' of an estuary to become filled in and eventually to become part of the sub-aerial environment."

Besides sediment coming from the major river it can emanate from the sea itself and the coastal boundaries of the estuary. As noted in Fig. 3.61 drift may be available

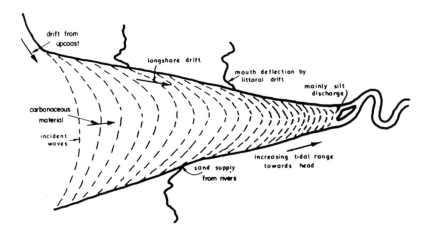

Figure 3.61 Typical estuary showing sources of sediment

from upcoast which is then transported both inside and outside the surf zone into the head of the estuary. Streams or rivers debouching into the feature may supply added material which is equally transported inward by the predominant wave direction. The asymmetric influence of tidal currents in this respect is enhanced in this situation. Shell debris and calcium carbonate precipitation, particularly in warm conditions, can add substantially to the material that can be moved toward the head. In tropical regions the percentage of carbonates in sand can reach 97%, so highlighting the importance of these processes. The bulk of material collected in shoals at the saltwater wedge tip should be analyzed to trace its source; river load, longshore transport, or open sea production. Remedial measures to inhibit further shoaling should take these origins into account. Means for reducing input from alongshore will be discussed in Chapter 6.

Although, as stated earlier, estuaries have a strong tendency to silt up, there are occasions of high storm surge level which create outflowing currents that can scour accumulations of decades over as many hours. A striking example of this action has been described by Isphording and Imsand (1991) at Apalachicola and Mobile bays in the northeast corner of the Gulf of Mexico. Hurricanes in this gulf can generate storm surges of over 7 m and "can cause more submarine erosion and shifting of sediments in a few hours than would be normally experienced in decades, or even centuries."

This has an effect on the ecology of these estuaries. Vogl (1980) has stated: "Most (disturbance) dependent systems follow a tear and build sequence, with death as an essential part of each cycle or rhythm." These catastrophic events, therefore, allow for the generation of future communities. This is another case of Nature tending to maintain the status quo, which is beneficial for man.

Some statistics of hurricane effects on these two bays are worthy of presentation. Their plan forms are shown in Fig. 3.62. Apalachicola Bay suffered hurricane Elena in August 1985 with a storm surge of 3 m which deepened some 50% of the 414 km² bay by 0.5 m. The difference between this erosion and siltation over 50% of this area was

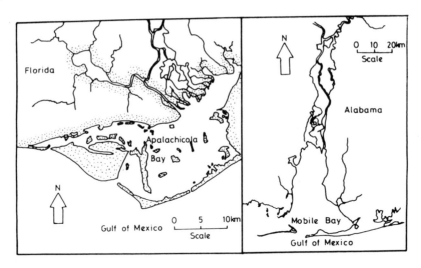

Figure 3.62 Plans of Apalachicola and Mobile Bays on the Gulf of Mexico

92,500,000 m^3 of soil, which added 450 years to the bay's existence from the computed infilling rate of 0.02 m per annum. Mobile Bay, which is 1014 km^2 in area, had hurricane Frederic in August 1979 pass directly over it. A storm surge of 1.92 m together with a high tide caused vast amounts of water to enter, and then leave at about one third of this rate, while the bay sediments were in suspension due to the excessive wave action. The volume of erosion was 463,000,000 m^3 which added some 100 years to the life of this estuary filling at 0.004 m per year.

3.9 COHESIVE SOILS

Clay or silt occurs on shorelines at or downcoast of mouths of rivers which are long and have passed through extensive plains of mild slope. Such examples include the northeast coast of South America (Amazon River), the Louisiana coast of the United States (Mississippi River), the northeast coast of China (Yangstze and Hwangho Rivers), the southwest coast of India (many small rivers sources). Mud can also accumulate at the mouth without being transported alongshore by waves, as in the Ganges-Brahmaputra River system in India/Bangladesh. In this situation many islands form which accrete up to high water level. These are used extensively for farming but become inundated when catastrophic storm surges occur, which are prevalent where the continental shelf is so wide and where tidal ranges are consequently large.

The wave action on coasts consisting of fine cohesive material differs in many respects from that on noncohesive or sandy beaches. One distinct feature of a berm built from sand is the dune formed by particles blown from its surface inland. This not only permits the beach to be clear of the water surface, but the inland can comprise much higher features. This is not so for mud, which dries out during low tide and cakes, thus preventing the wind from blowing it landward. Such shorelines are thus at high tide level and comprise mainly swamps.

Mud can exist in four states, namely, mobile suspension, high concentration suspension, consolidating bed, and settled bed (Mehta et al. 1989a, 1989b). Suspended particles become cohesive where the salinity exceeds 2 to 3 ppt. Cohesion exceeds the forces of gravity for clay ($< 4~\mu$m) much more than for silt (4 to 60 μm, see Table 3.1 for the classification of material on beaches). It is the presence of the clay component in muds that effects the cohesion. In the settling process a balance is present between the weight of the sediment mass and the upward flow of pore water. Moni (1970) has reported increased suspension during monsoons which was triggered by the upward flow of groundwater due to high levels in backwaters produced by precipitation.

Fluidization by waves is due to cyclic loading of the bed. This is confined to a small thickness unless tidal currents exist. "In the absence of current, the result is a high degree of stratification of the water column, with the majority of fluid mud remaining just above the bed" (Mehta 1989). Erosion resistance of the bed increases as consolidation takes place over time, but is less with waves than for current alone, as depicted in Fig. 3.63. As the duration of consolidation increases, the resistance due to these two sources of energy becomes similar.

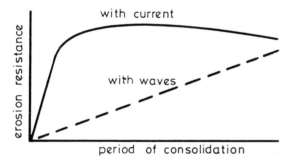

Figure 3.63 Erosion resistance of cohesive bed versus consolidation by waves and currents

The combined influence of currents and waves on a muddy bed is not well understood, but as Mehta (1989) states: "In general, there is sufficient experimental and field evidence to indicate that the fluid mud-generating potential of waves is a critical factor in eroding estuarine beds, particularly in the shallows."

The transport rate of mud by either unidirectional currents or wave action depends on the profiles of sediment concentration, and net water velocities. As seen in Fig. 3.64, the sediment concentration is almost a step function with depth, being uniformly small in the mobile suspension region, down to the surface with the fluid mud known as the *lutocline* (Mehta 1989). Within the fluid mud the concentration increases substantially and then again in the zone termed the deforming bed. The velocity profile of the unidirectional flow (V) decreases from the surface to a small fraction at the lutocline and then further, in an S curve, to zero at top surface of the deforming bed. The horizontal velocities of water-particle orbits due to wave action (U) are seen to decrease in a different fashion with depth

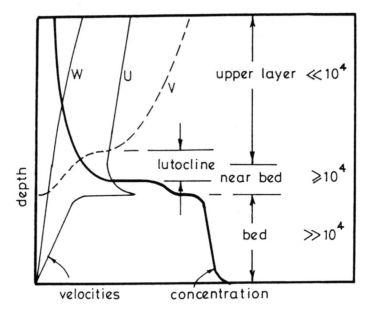

Figure 3.64 Profile of cohesive sediment concentration and velocity V of uniform current; U of horizontal, and W of vertical wave orbital velocity

through the mobile suspension region. It still has magnitude at the lutocline and then reduces through the fluid mud and even into the deforming bed. The vertical velocities of the water particles (W) in wave action are much smaller than the horizontal (U) component and decrease with depth but still exist into the mobile mud and deforming bed.

Mass transport, or net movement forward or backward per wave cycle, is enhanced by a muddy bottom. It occurs when the shear strain in the clay bed exceeds the limit of about 5%, and is proportional to the rate of wave energy dissipation (Yamamoto et al. 1986). Such wave damping increases with wave height (Yamamoto and Schuckman 1984) and hence storm sequences will result in excessive transport of mud. The mudline has been observed to change by several meters during a single storm (Yamamoto et al. 1986). The mass transport is linearly related to the rate of energy dissipation. It is maximum at the lutocline, the surface between water and liquid bed, but decreases and in fact becomes negative as shown in Fig. 3.65 (Shibayama et al. 1990).

Dore (1970) has derived formulae for this distribution and states that: "The mass transport velocity can formally be an order of magnitude larger than obtained by Longuet-Higgins (1953) for a single homogeneous fluid." The parabolic curves, as in Fig. 3.66, require velocities opposite to wave propagation to balance the positive values in order that there is no net drift over a complete vertical plane in both fluid regions. The maximizing of mass transport just above and below the lutocline, where concentrations of mud are large, result in large volumes collecting where wave action ceases. This may cause extra thickness of mobile mud offshore, to act in a similar way to the offshore bar in dissipating waves.

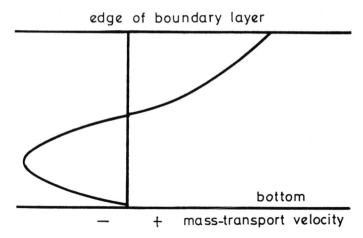

Figure 3.65   Mass-transport velocity within boundary layer for cohesive soil (Shibayama et al. 1990)

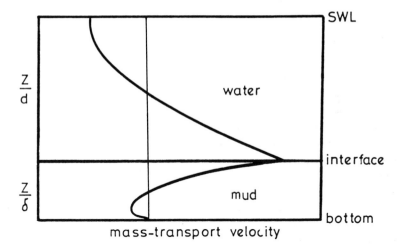

Figure 3.66   Distribution of mass transport with mud bottom (Dore 1970)

In the sea, mud occurs as flocculated material in thickness of one to several meters.This virtual liquid is denser than seawater and has a large influence in attenuating orbital water motions close to the bed. This, in turn, absorbs energy from the waves which are quickly reduced in height. Most swell waves do not reach the shore and many muddy expanses are used throughout the world as refuges for small craft during cyclones.

Besides wave action in transporting mud toward the coast by means of mass transport is an opposing influence of bed currents during a storm surge. These are caused by wave friction on the surface water, which is carried shoreward with a concomitant seaward flow near the bed. As reported by Mehta et al. (1989): "Wind induced currents have a complex structure that includes generally downwind surface currents and upwind bottom currents. These currents may alter sediment transport patterns or, in some cases, particularly where tides are small, be the primary transport mechanism." In the case of estuaries, with fresh water outward flow at the surface (as discussed in Section 3.8), the above tendency may be overshadowed.

The attenuation of the waves in this situation must be kept in mind as it aids in this storing process. Wells and Kemp (1984) have commented: "Surface waves and cohesive sediments interact in these open-coast environments in two ways. First, incoming wave energies are attenuated by cohesive bottom sediments at rates 1–2 orders of magnitude greater than would be expected if waves were propagating over a sandy bottom of equal gradient. Second, these same waves, even though severely attenuated, exert sufficient shear stress to resuspend large volumes of mud on time scales that range from gravity-wave frequency to seasonal frequency (seconds to months)." They provide references on zones around the world where safe anchorages have been used due to mud banks.

Wave damping due to colloidal suspension has been measured in the fluid (Tubman and Suhayda 1976, Wells 1983, Wells and Coleman 1981, Wells and Kemp 1984, Tsuruya et al. 1987) and predicted theoretically (Gade 1958, Dalrymple and Liu 1978, Kraft et al. 1985, MacPherson 1980). Data from these investigations are summarized in Fig. 3.67, where wave energy $(H_{1/3})^2$ reductions are shown against distance traveled. In this presentation reductions in depth with distance would also have been an influence. But it is seen that, compared with the attenuation by bottom friction and percolation for a sandy bottom, the muddy bottom has a profound effect. A ratio of $H/d = 0.20$ is normal for these waves, which tend to assume the form of solitary waves. "Steep symmetrical wave crests were separated by flat troughs that appeared to be at still-water level" (Wells and Kemp 1984).

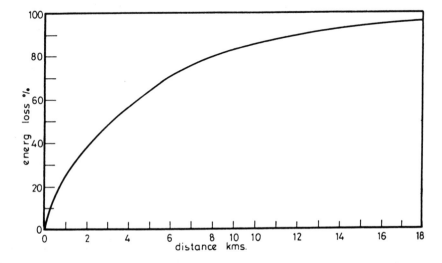

Figure 3.67 Damping of waves traveling over cohesive beds

Moni (1970) has discussed the mud banks on the SW coast of India where he believes that they arise from deposition from rivers, disturbance by waves, and liquefaction due to high backwater levels at the coast. These sources of suspension act together during the monsoon season. These banks are almost semicircular in plan, varying up to 10 km along the coast and to 8 km in width. As Moni (1970) observed: "The general distribution of sediments in the continental shelf of this coast indicate that the inner shelf (up to 20 fathoms) [37 m] consists of greenish black, poorly sorted clays and claying silts and the outer shelf (20 fathoms to 100 fathoms) [37–183 m] consists of well sorted fine and medium sand with abundant shell fragments." This is unlike a normal sandy shoreline where sediments become finer seaward.

Gopinathan and Qasim (1974) have also observed these mud banks over an extended period. These can be located in one position for years but can also be transported downcoast under the influence of oblique waves. As seen in Fig. 3.68, this major displacement may be preceded by a band of mud adjacent to the shoreline being forced downcoast by the littoral current from upcoast, the main body of the bank catching up with it over time.

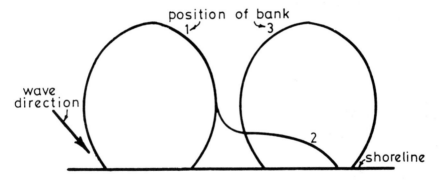

Figure 3.68   Mud transport along shoreline by littoral current, 1. original position of bank, 2. transitional, and 3. final bank location

When a bank is static for a reasonable period, even with oblique storm waves or swell, the waves are refracted toward its center from all boundaries. This creates a null point for littoral drift and therefore accretion of sand from upcoast, as depicted in Fig. 3.69. As noted by Gopinathan and Qasim (1974): "Along the axis of the mudbank, the coastline had a distinct protrusion into the sea." As the suspended mud was carried downcoast as explained above, this "protrusion of the coastline was partly eroded." Downcoast of this littoral-drift sink the sandy coast suffers a deficiency and hence erodes. This will be in a region where oblique waves arrive without diffraction due to the mud bank, as seen in the figure.

Accretion within the mud bank is swift, being in the order of 30 m/year (Moni 1970), whereas the erosion downcoast is spread over longer lengths and is therefore about one-tenth of this. In the intensive study off the coast of Surinam, South America, Wells et al. (1978) have observed: "As a result of alongshore mudbank migration, this erosion/accretion system moves to the northwest at an average rate of 1.5 km/yr,

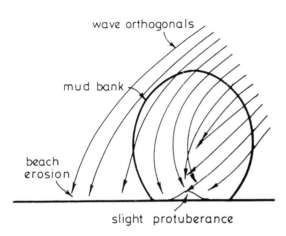

slight protuberance

Figure 3.69   Accretion of sand from upcoast as intercepted by static mud bank

suggesting that any given segment of coast has undergone many cycles of erosion and accretion during the last 5,000 years."

3.10 SHINGLE BEACHES

Pebble or shingle beaches are not widespread but are quite permanent features where they do exist. They have received little attention by coastal engineers as they provide few problems. As noted by Powell (1988): "However, it is only in recent years that their considerable merits as coast protection structures have been fully recognised, and only over the last few years that this belated recognition has been transformed into a more widespread application. This application is still restricted, however, not only by a paucity of information, regarding shingle beach processes under wave action, but also by a general lack of understanding as to how information which is currently available should be applied to a particular problem."

As with any new development researchers devolve their own variables, which differ from each other, that make comparisons difficult. In time these will be standardized and repetitive tests by individual workers will derive worthwhile relationships. This stage has almost been reached with the comprehensive work in the Netherlands (Van Hijim 1974, 1976; Van der Meer and Pilarczyk 1986, 1988). Even among these workers, variables and parameters are still developing.

The first attempt at determining the threshold movement of coarse material was reported by Rance and Warren (1968). Graphs were provided for several parameters but no equations were derived. The tests were conducted in an oscillating water tunnel so that variables were restricted to water orbits and not normal wave characteristics.

In a prototype situation pebbles not only oscillate up and down the face of a relatively steep beach but can also be transported alongshore should the waves arrive obliquely. Such obliquity is unlikely to be great as construction of a beach will assume the equilibrium shape of bay as far as possible, in which case littoral drift is negligible. Carr (1971) inserted sample pebbles into the famous Chesil Beach in England in order to determine rates of longshore movement. The mean value was 0.06 m per minute, although a maximum was recorded of 0.24 m per minute. The pebbles could remain trapped for extensive periods but motion was not necessarily associated with storm conditions. Most samples were recorded at high and low water marks, as also confirmed by Muir Wood (1970).

Some pebbles which were larger than the natural background material were "rejected" from it and hence lay on the surface, there to suffer more from wave action. As noted by Muir Wood (1970): "Moreover, once set in motion by a wave, translational and rotational inertia will tend to cause a large pebble to travel considerably further than a small one." Smaller elements can also become lodged in the interstices of the background material so making further transport almost impossible. Much of the transport is therefore effected by storm type waves when tides or storm surge levels are high. This is the opposite to sandy beaches where these conditions produce an offshore bar and littoral drift is maximum as this is removed back to shore by milder waves.

Nicholls and Wright (1991) conducted tracer tests on pebble beaches in the United Kingdom in order to check the proportionality K as used with sand drift. They used specimens of differing axis lengths which were difficult to compare. However, they state that: "The results from Hurst Castle Spit suggest that K increases with decreasing grain size, as might be expected." In the same paragraph they note that other factors exert importance and so comment: "Therefore, the tracers appear to have been preferentially transported alongshore because of their relatively large size." This project highlights the difficulties of this type of assessment, which is similar to the variable results obtained with sand, as pointed out by Komar (1988).

Field measurements have been carried out by Crickmore et al. (1972), who found that for a 25 mm stone, a maximum orbital velocity of 1.5 m/s was required to move it. This minimal requirement varied with bed conditions, for bed rock it was less and for sand base it was greater. In regard to overall transport they comment: "It appears that mass transport currents associated with heavy wave action are more important than tidal currents in determining the direction of net pebble movement at the nearshore sites. These bring about a landward drift across the tidal stream direction." They found that negligible movement occurred beyond the 18 m depth contour at this particular site. Muir Wood (1970) also confirmed the landward movement of shingle.

The morphology of shingle beaches has been discussed by Carter and Orford (1984), who point out: "Gravel dominated barriers often enclose brackish or freshwater lagoons. Tidal passes or inlets are generally absent.... Absence of tidal inlets is important because without them the barriers are more capable of moving steadily onshore." Because of the lack of concentrated ebb currents material in these beaches is not carried out to sea, as occurs with sandy shorelines. Wave conditions can be so variable that over-topping can take place at certain sites "without invoking the need for change in sea level."

Profiles of two adjacent sections of the Wales coast in the United Kingdom were examined, where limestone ranged in size from shingle to boulders (Caldwell and Williams 1986). The tidal range was extremely high (9.5 m for mean highwater springs) so that the apexes of land-backed beaches were around 7 m above datum. The nearshore slope of these beaches varied little from summer to winter and approximated 1:5. The slopes were quite uniform, pointing to the uniform spreading of this coarse material.

3.11 MARINE CLIFFS

Coastal cliffs, as subaerial scarp or steep slope merging out at the edge of water and having deep submarine bed rock, occupy about 40% of the world's coasts (Coleman and Murray 1976, see also quotation in Walker 1988), as compared to just 20% for sandy shores (The Institution of Civil Engineers 1985). Rocky coasts may be cliffless, in which the rock mass or mountain extends deep into the sea in the form of a steep cliff. Depending on the structure and lithology of the cliff-forming material, waves, tidal level, and weathering, the cliff face and probably its base have always received hydraulic and mechanical action that causes its destruction. The steep-cliffed coasts may have a shore platform or a steep plunging face. The former can include categories of inter-tidal, high tide, and low tide platform.

The adjective of "marine" implies that processes involved in the modification of cliffs as steep features are the actions of the sea. As remarked by Sunamura (1983) who has studied this topic mathematically: "It should be noted, however, that wave action originally induces cliff recession." There are many other factors that help assail these faces, comprising a vast variety of geologic structures, and others that resist their collapse. It is difficult to discuss them on a statistical basis because: "Most of the erosion is the result of local catastrophic events that are often associated with storms" (Bokuniewicz and Tanski 1980). These can be accompanied by storm surge which provides the high water levels above beaches if they exist at the base of the bluff.

Due to the differential strength in resisting abrasion (from sand, pebble, and gravel particles carried by waves), a segment of cliff with nonhomogeneous material may result in many varieties of landform, such as wave-cut notches at the base, smoothed granite mounds, castellated granite cliffs, flaking granite cliffs, columnar basalts, protruding stacks, or even natural stone arches. Zenkovich (1967) and Bird (1984), for example, have given detailed descriptions on the abrasive erosion in natural cliff conditions. On limestone cliffs, coastal lapies, caves, notches, and visors have been developed. Thus, varieties of cliff-forming material at various locations around the world can exhibit variable degrees of erosion. Sunamura (1983) has recorded many different materials and their rates of erosion. These include granitic rocks, limestones, flysh and shale, chalk and tertiary sedimentary rocks, quaternary deposits and unconsolidated volcanic ejects, in descending order of erosion resistance. The most erodible cliffs consist of sand, shale, tertiary clay, or even chalk components. Such a cliff face is subject to recurrent slumping with the slumped material at the base likely to be removed by oblique waves during and subsequent to any storm, so allowing a direct attack on the face during the next onslaught (Lee 1980).

Besides the variety of landforms, the planform of cliffed coasts deserves attention. Bird (1984) has observed that: "Where the predominant wave patterns are refracted by offshore topography or adjacent headlands, the cliffed coasts develop gently-curved outlines in plan, much like those on depositional coasts." These, of course, depend on the contrasts in the geological structure and lithology of the local cliff-forming material. Therefore, it is not surprising to find a curved bay shape, whether produced by a receding cliff or by depositional material, have been developed between two cliffed outcrops or steep coasts, due mainly to the variable resistance against the hydraulic and mechanical actions.

Since the major erosion force is that due to waves, it is worthwhile to discuss conditions that maximize this effect. The waves impacting a near vertical cliff may be standing, breaking, or even broken waves. The worst, as far as cliff erosion is concerned, is that of just breaking (Sunamura 1975). They contain the greatest pressure forces, either as a water body or as compressed air within the tipping crest. This high pressure air enters fissures and then suddenly expands as the water level falls. The requirement of waves breaking at the face implies that the water depth is shallow or is about equal to the wave height at that instant (Sunamura and Horikawa 1969).

There are also shearing forces involved as the crest impacts the face both downward and upward from the crest level, as the horizontal motion of the wave crest suddenly ceases. This occurs mainly for broken waves approaching the face (Sunamura 1975). As noted by Sunamura (1983), the assailing force is proportional to the celerity of the breaking wave squared, which approximates gd where d is the depth at the cliff face. Thus, at higher tide or surge levels the depths are greater and so too the wave forces.

The difference in the excavation hole caused by breaking and broken waves is illustrated in Fig. 3.70, where it is seen that for just breaking (Fig. 3.70a) the indentation (X) is large and extends below SWL. The cavity for broken waves (Fig. 3.70b) commences at SWL and reaches higher than the former. A typical time related increase in X is shown in Fig. 3.71, where the standing wave is shown with zero effects (Sunamura 1975).

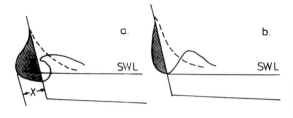

Figure 3.70 Profiles of caves excavated: (a) by breaking waves, and (b) by broken waves

A cliff face with little or no debris at its foot and relatively deep water will have only standing waves, which have little influence in eroding it. But, if a cliff has collapsed rocks of all sizes plus sand present at the base forming a beach, this may preclude waves from reaching the face, in which case erosion ceases. However, with storm waves and higher than normal water levels, reflection will produce high orbital velocities which so modify the beach profile to bring it below SWL. At the same time large boulders will be smashed into small components. A typical profile (Sunamura and Horikawa 1971)

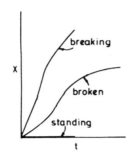

Figure 3.71   Indentation distance X, as in Fig. 3.70, measured over time t for various types of wave (Sunamura 1975)

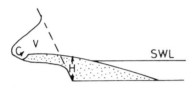

Figure 3.72   A schematic profile of cliff erosion with debris collected as a beach

is depicted in Fig. 3.72 where a beach is seen to protrude into an excavated cave. As noted by Robinson (1977) from prototype studies: "The level of maximum erosion for the sand and pebble beach was at 14.5 cm below the beach surface." The presence of this sediment aids erosion by it being used as projectiles in the vortex motion of the breaking wave. The importance of the beach in this process was reported by Robinson in his long-term field observations: "Erosion of 0.087 cm was recorded here during a year when there was no beach in front of it. In only the next six months when a beach was intermittently present erosion was 0.636 cm, a rate which was fifteen times higher than formerly." However, these values should only be regarded as local characteristics, and should not be applied universally.

Despite the plethora of geologic conditions, the erosion mechanism has been studied in models (Sanders 1968, Horikawa and Sunamura 1968, Sunamura 1977, Sunamura 1983). A weak cement of plaster mortar was cast into a steep cliff against which waves impacted at varying levels and intensities. A typical result is depicted in Fig. 3.73, where it is seen that over time the scour apex rises. The debris collects in front of the cliff face to form a beach, up which the wave propagates prior to its collision with the developing cave boundary.

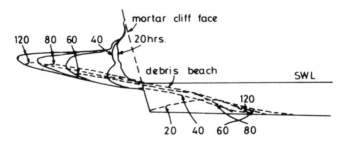

Figure 3.73   A typical erosion profile over time obtained from models (Sunamura and Horikawa 1969)

Sunamura and Horikawa (1971) in their model studies have indicated the volume (*V*) removed over time (*t*) as in Fig. 3.74, which results in an *S* curve, with a slow rate prior to debris collecting to form a beach. There is then a swift increase as quarrying is enhanced by suspension of sediment, which has been termed "wedging" (Robinson 1977). Later the erosion decreases as friction at the top surface of the cave, plus return water from the previous wave, reduces impact velocity. The rate of erosion (*dV/dt*) over time (*t*) is shown in Fig. 3.75, together with the beach built up (*H*) from the foot of the cliff.

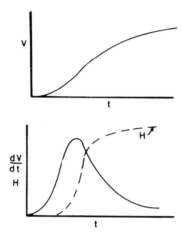

Figure 3.74   Volume (*V*) removed over time (*t*) from a cliff (Sunamura and Horikawa 1971)

Figure 3.75   Rate of erosion (*dV/dt*) and buildup of beach (*H*) over time (*t*) (Sunamura and Horikawa 1971)

In prototype situations beaches will comprise sand and pebbles, but the finer component is readily removed to form a bar during a storm and moved downcoast if subsequent swell is angled to the coast. The pebble beach can remain in place for much longer. Williams and Davies (1987) have observed: "Fringing beaches are invariably composed of coarse clastic rock fragments, typical beach gradients being 1:4 or 5." In other words, these represent a slope between 14° and 11°, which coincide with the values tabulated in Table 3.1 and shown Fig. 3.6. for particle size between granules and pebbles. This sudden shoaling produces plunging breakers that impact with great force against the bluff face.

Cliff erosion is episodic in that certain conditions of tide and wave intensity must be met before erosion is effected. It is not until removal has reached a critical stage that the cave collapses and the face subsides. This is likely to occur during a storm when shaking of the whole structure takes place due to massive wave forces. As seen in Fig. 3.76,

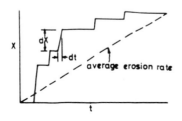

Figure 3.76   Cliff erosion rates as measured over short- and long-terms (Sunamura 1983)

the erosion rate as measured over small time intervals is likely to vary significantly from an average calculated over many years, which serves very little purpose even in comparisons of similar types of shoreline (Sunamura 1983). In this respect Kuhn and Shepard (1983) have noted: "In examining the history of beaches and sea cliff erosion it is important to include available information concerning the history of storms in the area, in that the present weather conditions may not be particularly representative." In referring to the San Diego County they state: "Prior to 1900 the sandy beach served as the most heavily traveled 'highway' along the Californian coast. During the 1880s the coastal sand beaches varied in width from 30 to 80 m." Of course, this would have been with horse-drawn vehicles. They also believe that the lack of floods in the past 40 years has caused the present dearth of sediment on this coast.

Besides waves other factors are instrumental in causing cliff collapse. These include overland runoff, groundwater seepage, wind, rain, and ground freezing. Where runoff is concentrated in natural or man-made channels erosion can become severe. Groundwater adds to the weight of soil, lubricates slide surfaces, and increases solution of rocks such as limestone. Excessive irrigation of residential subdivisions on cliffs can add up to 1,500 mm to the annual precipitation which becomes groundwater (Kuhn and Shepard 1980, 1983). The mechanical forces of raindrops and wind can disintegrate sedimentary rocks, while freezing of water in fissures can cause major fractures in large boulders.

Lumbsden and Kirk (1991) report the installation of a cheap gabion-type wall 6–10 m out from a 30–40 m high cliff to reduce wave energy. However, the real culprit in this cliff collapse appears to be water penetration at the top. They state: "When the site was visited in 1987 by the authors it was apparent that undercutting of the toe of the cliff was not a significant cause of collapse of the cliff face. Complex and obvious (up to 300 mm wide) cracking was noted along the cliff top at distances up to 14 meters from the cliff edge." These permitted water to reach rock structures some 20 m down the face, which upon freezing caused bulging and final failure. This excessive water input was probably caused by residential development which was reported near the top of this high cliff. Thus the cause of collapse is unlikely to be solved by foot protection, but more by prevention of sprinkling lawns and gardens and providing controlled runoff.

Mass movements of rock or mud occur from sliding, either in rotational or planar mechanisms. As noted by Sunamura (1983): "The occurrence of mass movement is clearly influenced by the lithology, geologic structure, and geotechnical properties of the cliff-forming material." Caves which subsequently collapse are sculptured by waves in rocks such as limestone. This differs in abrasive strength both vertically and horizontally due to the cementation processes from previous high sand dunes from which the limestone was formed. Columns can exist where limewater has percolated through holes in a roof structure and hence may retard cave subsidence (Williams and Davies 1987). But, as noted by Sunamura (1977): "There is no relationship between the long term erosion rate of coastal cliff and their height."

Because of the many variables involved in cliff recession, Jones and Williams (1991) attempted a statistical analysis to find the most influential. In order of importance these were found to be:

1. Beach face volume (V_b)
6. Beach terrace height (T_h)
2. Tidal coverage
7. Height of beach (B_h)
3. Total beach volume (V_t)
8. Orientation of coastline
4. Beach face width (B_w)
9. Beach terrace slope (α_t)
5. Beach face slope (α_b)
10. Strength of cliff material

These variables are shown in Fig. 3.77, where it is seen that total beach volume (V_t) is taken down to zero ordinance datum. In respect to coast orientation, peaks of erosion were indicated in another of their figures which coincided with positions of crenulate shaped bays on the coast. In positions of greatest indentation, where recession would be the highest for bays in dynamic equilibrium, the removal of voluminous beaches in front of cliffs would also be the greatest. Similar differential erosion has been reported: "Glacial deposits receded more rapidly on the downdrift side of artificial structures such as groins and jetties, where the beach is poorly developed at the cliff base, as compared with the updrift side having a well developed beach." Even the upcoast accreted zone can disappear during a strong storm sequence but these are likely to be less frequent than beach removal on downcoast zones, leaving only a submerged slope up to the cave opening.

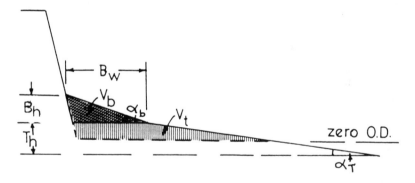

Figure 3.77 Definition sketch for variables as in analysis by Jones and Williams (1991)

The conclusion of Jones and Williams (1991) appears important: "Regression analysis suggested that the volume of beach face material was the dominant explanatory variable in short-term cliff erosion." This matches with the observation of Giese and Aubrey (1987): "As a result, we conclude that change in beach form, rather than storm frequency, is the major factor determining the degree of erosion at a single site over a single year in such cases." It must be remembered that storms will always reduce the beach deposit to the minimal, which is a sloping face, probably of cobble size, so that frequency of storms must be an over-riding factor in cliff erosion.

Sunamura (1983) has discussed the topic of shore platform formation. While providing a comprehensive reference list on this and cliff erosion in general, he observes: "However, an explicit explanation has not yet been provided to the following basic question: at what level, by what kind of waves, and how fast are platforms originated at a place with given topographical, geological, and tidal conditions?" He warns against using agents "visibly at work on them at present," since tectonics may have played a role. As noted by Carter and Oxford (1984) with respect to pebble beaches, there is no need to invoke eustatic changes in sea level to explain their features. The deposition of rock at the base of bluffs are but one example of these coarse clastic accumulations.

In concluding, it is fitting to quote (Bird 1984) that: "The morphogenic system leading to the evolution of cliffed coasts can be analyzed in terms of the effects of geological structure and lithology, the action of marine and subaerial processes, tidal conditions, and the inheritance of changing levels of land and sea and changing climatic conditions. The tempo of change on a cliffed coast shows marked variations from place to place, even on adjacent flanks of an embayment or headland, according to rock resistance and degree of exposure to marine attack." Neither a uniform finding for the phenomenon of cliff erosion, nor a universal conclusion can be derived for field applications, as compared to the depositional sandy beaches to be discussed in Chapter 4.

REFERENCES

ADEYMO, M. D. 1968. Effect of beach slope and shoaling on wave asymmetry. *Proc. 11th Inter. Conf. Coastal Eng., ASCE 1*: 145–72.

ANDERSON O. H., and FREDSOE, J. 1983. Transport of suspended sediment along the coast. *Inst. Hydro. and Hyd. Eng., Tech. Univ. of Denmark,* Prog. Rep. 59, 33–46.

Atlas of Sea and Swell Charts: South Atlantic, Publ. 799B, 1948; Western Pacific Publ. 799CE, 1963; North Eastern Pacific Publ. 2990, 1963. US Hydrographic Office, Washington, D.C.

BATTJES, J. A. 1974. Computation of set-up, longshore currents, run-up and overtopping due to wind-generated waves. *Delft Univ. of Tech.,* Rep. 74-2.

BAUM, S. K., and BASCO, D. R. 1986. A numerical investigation of the longshore current profile for multiple bar/trough beaches. *Proc. 20th. Inter. Conf. Coastal Eng., ASCE 2*: 971–85.

BIRD, E. C. F. 1984. *Coasts: An Introduction to Coastal Geomorphology,* 3rd ed. Canberra: Aust. National Univ. Press.

BOKUNIEWICZ, H., and TANSKI, J. 1980. Managing localized erosion of coastal bluffs. *Proc. Coastal Zone '80, ASCE 3*: 1883–98.

CALDWELL, N. E., and WILLIAMS, A. T. 1986. Spatial and seasonal pebble beach profile characteristics. *J. Geology, 21*: 127–38.

CARR, A. P. 1971. Experiments on longshore transport and sorting of pebbles: Chesil Beach, England. *J. Sed. Petrology 41*: 1084–1104.

CARSTENS, T. 1970. Turbulent diffusion and entrainment in two-layer flow. *J. Waterways and Harbors Div., ASCE 96*(WW1): 97–104.

CARTER, R. W. G., and ORFORD, J. D. 1984. Coarse clastic barrier beaches: a discussion of the distinctive dynamic and morpho-sedimentary characteristics. *Marine Geology 60*: 377–89.

CHAPMAN, D. M., and SMITH, A. W. 1980. The dynamic swept prism. *Proc. 17th. Inter. Conf. Coastal Eng., ASCE 1*: 1036–50.

COLEMAN, J. M., and MURRAY, S. P. 1976. Coastal sciences: recent advances and future outlook. *Science, Technology, and the Modern Navy,* Office of Naval Research, Arlington, 346–70.

CRICKMORE, M. J., WATERS, C. B., and PRICE, W. A. 1972. The measurement of offshore shingle movement. *Proc. 13th Inter. Conf. Coastal Eng., ASCE 2:* 1005–25.

DALLY, W. R. 1987. Longshore bar formation—surf beat or undertow? *Proc. Coastal Sediments '87, ASCE 1:* 71–86.

DALRYMPLE, R. A., and LIU, P. L.-F. 1978. Waves over soft muds : a two-layer fluid model. *J. Physical Ocean. 8:* 1121–31.

DAVIES, J. L. 1964. A morphogenic approach to world shorelines. *Annals of Geomorphology 8:* 127–42.

DEAN, R. G. 1977. Equilibrium beach profiles: U.S. Atlantic and Gulf coasts. *University of Delaware,* Newark, Ocean Eng. Report No. 12.

DEAN, R. G., BEREK, E., GABLE, C. G., and SEYMOUR, R. J. 1982. Longshore transport determined by an efficient trap. *Proc. 18th. Inter. Conf. Coastal Eng., ASCE 2:* 954–88.

DEAN, R. G., BEREK, E. P., BODGE, K. R., and GABLE, C. G. 1987. NSTS measurements of total longshore transport. *Proc. Coastal Sediments '87,* ASCE *1:* 652–67.

DETTE, H. H., and ULICZKA, K. 1987. Prototype investigation on time-dependent dune recession and beach erosion. *Proc. Coastal Sediments '87, ASCE 2:* 1430–44.

DOLAN, T. J. 1983. Wave mechanisms for the formation of multiple longshore bars with emphasis on the Chesapeake Bay. *MCE thesis,* Dept. of Civil Eng., University of Delaware.

DORE, B. D. 1970. Mass transport in layered fluid systems. *J. Fluid Mech 40:* 113–26.

DORNESTEIN, R. 1961. Amplification of long waves in bays. *Florida Exp. Stn.,* Tech. Pap. 213.

EXON, N. F. 1975. An extensive offshore sand bar field in the western Baltic Sea. *Marine Geology 18:* 197–212.

GADE, H. G. 1958. Effects of a non-rigid, impermeable bottom on plane surface waves in shallow water. *J. Marine Res. 16*(2): 61–82.

GIESE, G. S., and AUBREY, D. G. 1987. Bluff erosion on outer Cape Cod. *Proc. Coastal Sediments '87, ASCE 2:* 1871–76.

GODA, Y. 1975. Irregular wave deformation in the surf zone. *Coastal Eng. in Japan 18:* 13–26.

GOPINATHAN, C. K., and QASIM, S. Z. 1974. Mud banks of Kerala—their formation and characteristics. *Indian. J. Marine Sc. 3:* 105–14.

GREER, M. N., and MADSEN, O. S. 1978. Longshore sediment transport data: a review. *Proc. 16th Inter. Conf. Coastal Eng., ASCE 2:* 1563–76.

HEATHERSHAW, A. D. 1982. Seabed-wave resonance and sand bar growth. *Nature 296:* 343–45.

HEATHERSHAW, A. D., and DAVIES, A. G. 1985. Resonant wave reflection by transverse bedforms and its relation to beaches and offshore bars. *Marine Geology 62:* 321–28.

HOGBEN, N. 1988. Experience from compilation of global wave statistics. *Ocean Eng. 15*(1): 1–31.

HOGBEN, N., and LUMB, F. E. 1967. *Ocean Wave Statistics,* Natl. Phys. Lab., U.K., HMSO.

HOGBEN, N., DACUNHA, N. M. C., and OLLIVER, G. F. 1986. *Global Wave Statistics.* London: Unwin Bros.

HORIKAWA, K., and SUNAMURA, T. 1968. An experimental study on erosion of coastal cliffs due to wave action. *Coastal Eng. in Japan 11:* 131–47.

HUNT, J. N. 1955. Gravity waves in flowing water. *Proc. Roy. Soc., Ser. A 231:* 496–504.

IPPEN, A. T., ed. 1966. *Estuary and Coastline Hydrodynamics.* New York: McGraw Hill.

ISPHORDING, W. C., and IMSAND, F. D. 1991. Cyclonic events and sedimentation in the Gulf of Mexico. *Proc. Coastal Seds. '91, ASCE 1*:1122–36.

JONES, D. G., and WILLIAMS, A. T. 1991. Statistical analysis of factors influencing cliff erosion along a section of the west Wales coast, U.K., *Earth Surface Processes and Landforms,* Vol. 16.

JOHNSON, J. W. 1956. Dynamics of nearshore sediment movement. *Bull. Am. Ass. Petrol. Geol. 40*: 2211–32.

KAJIMA, R., SHIMIZU, T., MARUYAMA, K., and SAITO, S. 1983a. Experiments of beach profile change with a large wave flume. *Proc. 18th Inter. Conf. Coastal Eng., ASCE 2*: 1385–1404.

KAJIMA, R., SAITO, S., SHIMIZU, T., MARUYAMA, K., HASEGAWA, H., and SAKAKIYAMA, T. 1983b. Sand transport experiments performed by using a large water wave tank. *Central Res. Inst. for Electric Power Industry*, Japan, Data Rep. No. 4-1. (In Japanese)

KAMPHUIS, J. W. 1966. Mathematical simulation of bottom sediment motion by waves. *Proc. 10th. Inter. Conf. Coastal Eng., ASCE 1*: 766–92.

KATOH, K., YANAGISHIMA, S., MURAKAMI, H., and SUETSUGU, K. 1987. Daily changes of shoreline position and its tentative predictive model. *Rep., Port and Harbour Res. Inst.* Japan 26(2): 64–96. (In Japanese)

KOMAR, P. D. 1988. Environmental controls on littoral sand transport. *Proc. 21st Inter. Conf. Coastal Eng., ASCE 2*: 1238–52.

KRAFT, L. M., HELFVICH, S. C., SUHAYDA, J. N., and MARIN, J. 1985. Soil response to ocean waves. *Marine Geotec. 6*(2): 173–204.

KRAUS, N. C., and DEAN, J. L. 1987. Longshore sediment transport rate distributions. *Proc. Coastal Sediments '87, ASCE 1*: 881–96.

KRAUS, N. C., and LARSON, M. 1988. Beach profile change measured in the tank for large waves, 1956–1957 and 1962. U.S. Army Corps of Engrs., *Coastal Eng. Res. Center*, Waterways Exp. Stn., Vicksburg, Miss., Tech. Rep. CERC-88–6.

KRIEBEL, D. L. 1987. Beach recovery following Hurrican Elena. *Proc. Coastal Sediments '87, ASCE 1*: 990–1005.

KRUMBEIN, W. C., and OHSIEK, L. E. 1950. Pulsational transport of sand by shore agents. *Trans. Amer. Geophysic. Un. 31*: 216–20.

KUHN, G. G., and SHEPARD, F. P. 1980. Coastal erosion in San Diego County, California. *Proc. Coastal Zone '80,* ASCE 3: 1899–1918.

———. 1983. Beach processes and sea cliff erosion in San Diego County, California. In *Handbook of Coastal Processes,* ed., P. D. Komar. Florida: CRC Press, 267–84.

LARSON, M. 1988. *Quantification of Beach Profile Change.* Dept. Water Resources Eng., *Lund University,* Sweden, Report No. 1008.

LARSON, M., and KRAUS, N. C. 1989. SBEACH: Numerical model for simulating storm-induced beach change. Rep. 1. Empirical foundation and model development. U.S. Army Corps of Engrs., *Coastal Eng. Res. Center*, Waterways Exp. Stn., Vicksburg, Miss., Tech. Rep. CERC-89-9.

LEE, L. J. 1980. Sea cliff erosion in southern California. *Proc. Coastal Zone '80, ASCE 3*: 1919–38.

LONGUET-HIGGINS, M. S. 1953. Mass-transport in water waves. *Phil. Trans. Roy. Soc., Ser. A 245*: 535–81.

———. 1983. Wave set-up, percolation and underflow in the surf zone. *Proc. Roy. Soc., Ser. A 390*: 283–91.

LUMSDEN, J. L., and KIRK, R. M. 1991. Cliff protection at Motunau—a low cost approach. *Proc. 10th Aust. Conf. Coastal and Ocean Eng.* 277–82.

MACPHERSON, H. 1980. The attenuation of water waves over a non-rigid bed. *J. Fluid Mech.* 97(4): 721–42.

MASON, C., SALLENGER, A. H., HOLMAN, R. A., and BIRKEMEIER, W. A. 1984. DUCK '82 - a coastal storm process experiment. *Proc. 19th. Inter. Conf. Coastal Eng., ASCE* 2: 1913–28.

MCARTHUR, W. M., and BARTLE, G. A. 1980. Land resources management. Series 5, 6 & 7, Commonwealth Scientific & Industrial Res. Organisation, Australia.

MCGILL, J. T. 1958. Map of coastal landforms of the world. *Geogr. Rev. 48*: 402–05.

MCINTYRE, A. D. 1977. Sandy foreshores. Chapter 2 in *The Coastline,* ed. R.S.K. Barnes. London: John Wiley & Sons.

MEHTA, A. J. 1986. Characterisation of cohesive sediment properties and transport processes in estuaries. In *Estuarine Cohesive Dynamics,* ed. A. J. Mehta. Lecture Notes Coastal Estuarine Studies, Vol. 14, Springer-Verlag, 290–325.

———. 1989. On estuarine cohesive sediment suspension behaviour. *J. Geophys. Res. 94*(C10): 14303–314.

MEHTA, A. J., HAYTER, E. J., PARKER, W. R., KRONE, R. B., and TEETER, A. M. 1989a. Cohesive sediment transport. I: process description. *J. Hyd. Eng., ASCE 115*(8): 1076–93.

MEHTA, A. J., MCARALLY Jr., W. H., HAYTER, E. J., TEETER, A. M., SCHOELLHAMER, D., HELTZEL, S. B., and COREY, W. P. 1989b. Cohesive sediment transport. II: application. *J. Hyd. Eng., ASCE 115*(8): 1094–1112.

MENARD, H. W. 1961. Some rates of regional erosion. *J. Geology 69*: 154–61.

MONI, N. S. 1970. Study of mud banks along the southwest coast of India. *Proc. 12th Inter. Conf. Coastal Eng., ASCE* 2: 739–50.

Monthly Meteorological Charts: Atlantic Ocean, 1959; Indian Ocean Mo. 519, 1949; Western Pacific Mo. 484, 1956; HMSO, London.

MUIR WOOD, A. M. 1970. Characteristics of shingle beaches: the solution to some practical problems. *Proc. 12th Inter. Conf. Coastal Eng., ASCE* 2: 1059–75.

NICHOLLS, R. J., and WRIGHT, P. 1991. Longshore transport of pebbles: experimental estimates of K. *Proc. Coastal Sediments '91, ASCE 1*: 920–33.

PEKERIS, C. L., and ACCAD, Y. 1969. Solution of Laplace's equations for the M2 tide in the world oceans. *Phil. Trans. Roy. Soc., Ser. A 265*: 413–36.

PILKEY, O. H., SR., PILKEY, W. D., PILKEY, O. H., JR., and NEAL, W. J. 1983. *Coastal Design: A Guide for Builders, Planners, and Home Owners.* New York: Van Nostrand Reinhold.

POLOUS, H. G. 1988. *Marine Geotechnics.* London: Unwin Hyman.

POWELL, K. A. 1988. The dynamic response of shingle beaches to random waves. *Proc. 21st Inter. Conf., Coastal Eng., ASCE* 2: 1763–73.

PRATTE, B. D., WILLIS, D. H., and PLOEG, J. 1982. Harbour sedimentation—comparison with model. *Proc. 18th. Inter. Conf. Coastal Eng., ASCE* 2: 1119–26.

RANCE, P. J., and WARREN, N. F. 1968. The threshold of movement of coarse material in oscillatory flow. *Proc. 11th Inter. Conf. Coastal Eng., ASCE 1*: 487–91.

READSHAW, J. S., GLODOWSKI, C. W., CHARTRAND, D. M., WILLIS, D. H., BOWEN, A. J., PIPER, D., and THIBAULT, J. 1987. A review of procedures to predict alongshore sand transport. *Proc. Coastal Sediments '87, ASCE 1*: 738–55.

RICHMOND, B. M., and SALLENGER, A. H., JR. 1984. Cross-shore transport of bimodal sands. *Proc. 19th. Inter. Conf. Coastal Eng., ASCE 2:* 1997–2008.

ROBINSON, L. A. 1977. Marine erosive processes at the cliff foot. *Mar. Geol. 23:* 257–71.

SALLENGER, A. H., and HOLMAN, R. A. 1984. On predicting infragravity energy in the surf zone. *Proc 19th. Inter. Conf. Coastal Eng., ASCE 2:* 1940–51.

SALLENGER, A. H., HOLMAN, R. A., and BIRKEMEIER, W. A. 1985. Storm-induced response of a nearshore-bar system. *Marine Geology 64:* 237–57.

SALLENGER, A. H., and HOWD, P. A. 1989. Nearshore bars and the break-point hypothesis. *Coastal Eng. 12:* 301–13.

SANDERS, N. K. 1968. Wave tank experiments on the erosion of rocky coasts. *Proc. Roy. Soc. Tasmania, 102:* 11–16.

SAWARAGI, T., and DEGUCHI, I. 1980. On-offshore sediment transport rate in the surf zone. *Proc. 17th Inter. Conf. Coastal Eng., ASCE 2:* 1195–1214.

SCHULTZ, E. A., and SIMMONS, H. B. 1957. Fresh-water-salt-water density currents, a major cause of siltation in estuaries. *Proc. PIANC 19th Inter. Nav. Congr.,* Sect. II, Comm. 3, 43–64.

SHEPARD, F. P. 1963. *Submarine Geology.* 2nd ed. New York: Harper Row.

SHIBAYAMA, T., OKUNO, M., and SATO, S. 1990. Mud transport rate in mud layer due to wave action. *Proc. 22nd Inter. Conf. Coastal Eng., ASCE 3:* 3037–49.

SHIMIZU, T., SAITO, S., MARUYAMA, K., KASEGAWA, H., and KAJIMA, R. 1985. Modeling of onshore-offshore sand transport rate of distribution based on the large wave flume experiments. *Civ. Eng. Lab.* Rep. No. 384028, Abiko City, Japan. (In Japanese)

Shore Protection Manual. 1973. U. S. Army of Corps of Engrs., Coastal Eng. Res. Center, U.S. Govt. Printing Office, Washington, D.C.

SILVESTER, R. 1974. *Coastal Engineering Vols. 1& 2.* Amsterdam: Elsevier.

———. 1984. Fluctuations in littoral drift. *Proc. 19th Inter. Conf. Coastal Eng., ASCE 2:* 1291–1305.

SILVESTER, R., and VONGVISESSOMJAI, S. 1978. Spectral growth of waves to the fully arisen sea. *Proc. Inter. Conf. on Water Resources Eng.,* Asian Inst. Tech., Bangkok *1:* 375–94.

SIMMONS, H. B. 1955. Some effects of upland discharge on estuarine hydraulics. *J. Waterways and harbors Div., ASCE 81* (Sept.), No. 792.

SMITH, A. W., and GORDON, A. D. 1980. Secondary sand transport mechanisms. *Proc. 17th Inter. Conf. Coastal Eng., ASCE 2,* 1122–39.

SNODGRASS, F. E., GROVES, G. W., HASSELMEN, K. F., MILLER, G. R., MUNK, W. A., and POWERS, W. H. 1966. Propagation of ocean swell across the Pacific. *Phil. Trans. Roy. Soc., Ser. A 259* (1103): 431–97.

———. 1984. U.S. Corps of Engrs., Coastal Eng. Reg. Center, U.S. Govt. Printing Office, Washington, D.C.

STODDART, D. R. 1969. World erosion and sedimentation. In *Water Earth and Man,* ed. R. J. Chorley. London: Methuen.

SUNAMURA, T. 1975. A laboratory study of wave-cut platform formation. *J. Geology 83:* 389–97.

———. 1977. A relationship between wave-induced cliff erosion and erosive force of waves. *J. Geol. 85:* 613–18.

———. 1983. Processes of sea cliff and platform erosion. *Handbook of Coastal Processes,* ed., P. D. Komar. Florida: CRC Press, 233–65.

SUNAMURA, T., and HORIKAWA, K. 1969. A study on erosion of coastal cliffs by using aerial photographs, report no. 2. *Coastal Eng. in Japan 12*: 99–120.

———. 1971. A quantitative study on the effect of beach deposits upon cliff erosion. *Coastal Eng. in Japan 14*: 97–106.

SUNAMURA, T., and MARUYAMA, K. 1987. Wave geomorphic response of eroding beaches—with special reference to seaward migrating bars. *Proc. Coastal Seds. '87, ASCE 1*: 788–801.

The Institution of Civil Engineers. 1985. *Coastal Engineering Research*. Maritime Eng. Group, UK, London: Thomas Telford Ltd.

TSURUYA, H., NAKANO, S., and TAKAHAMA, J. 1987. Interactions between surface waves and a multi-layered mud bed. *Rep., Port and Harbour Res. Inst.*, Japan 26(5): 137–73.

TUBMAN, M. W., and SUHAYDA, J. N. 1976. Wave action and bottom movements in fine sediments. *Proc. 15th Inter. Conf. Coastal Eng., ASCE 2*, 1168–83.

TULLY, J. P. 1953. On structure, entrainment and transport in estuary embayments. *J. Marine Res. 17*: 523–35

UNNA, P. J. H. 1942. Waves and tidal streams. *Nature 149*, 219–20.

U.S. Navy Marine Climate. 1977. *Atlas of the World Vol. 2, North Pacific Ocean*. Naval Weather Service, Washington, D.C., 50-16-529.

VAN DER MEER, J. W., and PILARCZYK, K. W. 1986. Dynamic stability of rock slopes and gravel beaches. *Proc. 20th Inter. Conf. Coastal Eng., ASCE 2*: 1713–26.

———. 1988. Large verification tests on rock slope stability. *Proc. 21st Inter. Conf. Coastal Eng., ASCE 3*: 2116–28.

VAN HIJUM, E. 1974. Equilibrium profiles of coarse material under wave attack. *Proc. 14th Inter. Conf. Coastal Eng., ASCE 2*: 939–57.

VAN HIJUM, E. 1976. Equilibrium profiles and longshore transport of coarse material under oblique wave attack. *Proc. 15th Inter. Conf. Coastal Eng., ASCE 2*: 1258–76.

VOGL, R. J. 1980. The ecological factors that produce perturbation-dependent ecosystems. *Damaged Ecosystems*, ed. J. Cairns, Ann Arbour Sc., 63–94.

VONGVISESSOMJAI, S., MUNASINGHE, L. C. J , and GUNARATNA, P. P. 1986. Transient ripple formation and sediment transport. *Proc. 20th Inter. Conf. Coastal Eng., ASCE 2*: 1638–52.

WALKER, H. J. 1988. Artificial structures and shorelines: an introduction. *Artificial Structures and Shorelines*, ed. H. J. Walker, Dordercht: Kluwer Academic Publishers, 1–8.

WELLS, J. T. 1983. Fluid-mud dynamics in low-, moderate-, and high-tide-range environments. *Canadian J. Fish. and Aquatic Sc. 40* (Sup 1): 130–42.

WELLS, J. T., and COLEMAN, J. M. 1981. Physical processes and fine-grained sediment dynamics, coast of Surinam, South America. *J. Sed. Petrology 51*: 1053–68.

WELLS, J. T., and KEMP, G. P. 1984. Interaction of surface waves and cohesive sediments field observations and geologic significance. In *Estuarine Cohesive Dynamics*, ed. A. J. Mehta. New York: Springer-Verlag, 43–65.

WELLS, J. T., COLEMAN, J. M., and WISEMAN JR., W. J. 1978. Suspension and transportation of fluid mud by solitary-like waves. *Proc. 16th Inter. Conf. Coastal Eng., ASCE 2*: 1932–52.

WILLIAMS, A. T., and DAVIES, P. 1987. Rates and mechanisms of coastal cliff erosion in lower lias rocks. *Proc. Coastal Sediments '87, ASCE 2*: 1855–70.

WRIGHT, L. D., CHAPPELL, J., THOM, B. G., BRADSHAW, M. P., and COWELL, P. 1979. Morphodynamics of reflective and dissipative beach and inshore systems: south eastern Australia. *Marine Geology 32*: 105–40.

WU, J. 1975. Wind-induced drift currents. *J. Fluid Mech. 68* (Part 1): 49–70.

YAMAMOTO, T., and SCHUCKMAN, B. 1984. Experiments and theory of wave-soil interactions. *J. Eng. Mechs., ASCE 110*(1): 95–112.

YAMAMOTO, T., NAGAI, T., and FIGUERO, J. L. 1986. Experiments on wave-soil interaction and wave-driven soil transport in clay beds. *Continental Shelf Res. 5*: 521–40.

YU, Y. 1952. Breaking of waves by an opposing current. *Trans. Amer. Geophys. Un. 33*: 39–41.

ZENKOVICH, V. P. 1967. *Processes of Coastal Development.* ed. J. A. Steers, trans. C. G. Fry. Edinburgh: Oliver and Boyd.

Engineering Aspects
of Coastal Geomorphology

Coastal geomorphology is the study of coastal landforms, their evolution, the processes on them, and the changes now taking place (Bird 1984). Traditionally, geologists have given many different classifications for coasts, notably from a genetic point of view, which evolve around the dichotomy of submergence and emergence (Johnson 1919), as well as the influence of sea level changes. Shepard (1937, 1973) considered the effects of wave action and the nature of the processes that form the coastal character. A more recent approach (Davies 1958, 1964) is based on the nature of wave climate (storm and swell), their geographical location (east and west coasts), and orientation to the influencing waves. Consequently, landforms in each of these environments respond to the local wave characteristics, which interplay with various geological structures of the coast, such as clays, muds, sands, shingles, and rocks.

Studies in coastal sedimentation by coastal engineers have been conventionally based on the mechanisms of sediment movement by waves and wind, and from model studies of two-dimensional beach profile changes, as noted in Chapter 3. Although this approach has promoted some fundamental insight into sediment movement along coasts, it has not produced desirable results for prototype conditions. The lack of understanding in coastal processes and Nature's capacity to modify beaches has resulted in the failure of various solutions for coastal defense, as will be noted in Chapter 5. Since coastal evolution is the reflection of Nature's work, whether in cyclic action of storm and swell or in a pulsative manner, it is required that coastal geomorphologists and engineers observe Nature and emulate Her as much as possible (Silvester 1979).

4.1 GEOMORPHOLOGY OF COASTS

The narrow belt directly landward from the coastal zone, where about two thirds of the world's population live (Soucie 1973), has diverse fauna, flora, and aquatic organisms. Within this zone a variety of physiographically complex features co-exist, including rocky shores, sandy beaches, estuaries, deltas, salt marshes, and mangrove swamps. These coastal environments have been modified dynamically and continuously by waves, tides, and currents. It is unfortunate for man that these transitions have not often been peaceful but very volatile and sometimes catastrophic. Along limited lengths of coast, erosion and deposition occur either continually or in sequences, due mainly to the action of waves. Some accretional landforms are depicted in Fig. 4.1, of which bays and barrier beaches require special attention because of their close link with human activities.

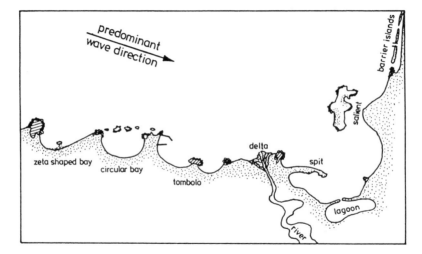

Figure 4.1   Various depositional landforms along a hypothetical coast

4.1.1 Configuration of Beaches

A coastline is rarely straight; some segments may curve gently in plan, while others are more indented, or even of more intricate configuration with various shapes and sizes. These planforms have various names such as bay, gulf, and sea, depending on their indentation and magnitude of water area enclosed. Shalowitz (1964) has discussed bays, of which those containing sand are the most dynamic with regard to beach processes and are of paramount importance. Those formed between headlands on the coast can be observed on maps and aerial photographs, and are universal. It is essential to know why and how they are formed and to determine their stability. In this chapter their evolution and constancy will be examined in detail, because they are a pointer to how man can stabilize a shoreline. There is no better teacher than Nature.

Under the influence of the global wind systems most coasts receive persistent swell waves from some oblique direction, as discussed in Chapter 3. As Lewis (1938)

has observed: "Beaches tend to be built up transverse to the direction of approach of the most important beach constructing waves." This early recognition of bay orientation took some decades to sink in, but is applicable on a global scale. Such features are observable on nautical charts and aerial photographs and can be correlated with the sea and swell charts available for the various oceans. The consistency of orientation along large lengths of coastline was noted by Halligan (1906): "Along the east coast of Australia from Sydney to Fraser Island extends an almost continuous series of asymmetrical curved bays joining one headland to the next." In Fig. 4.2 are shown a series of bays on the southern coast of Western Australia due to the incessant SW swell.

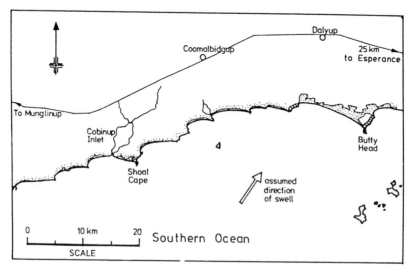

Figure 4.2   A chain of crenulate shaped bays to the west of Esperance, Western Australia, receiving persistent swell from the Southern Ocean

The orientation of these bays is an excellent indicator of the direction of net sediment movement along the coast. By observing these shapes along large reaches of shoreline it can be shown that littoral drift is in the same direction for many kilometers or even hundreds of kilometers. This knowledge can help man predict future trends in shoreline movement over large lengths of time.

The shapes of these bays have been called a variety of names in the literature, such as zeta bays (Halligan 1906, Silvester et al. 1980), half-heart bay (Silvester 1960), crenulate shaped bays (Ho 1971, Silvester and Ho 1972), spiral beaches (Krumbein 1944, Le Blond 1972), curved or hooked beaches (Rea and Komar 1975), headland bay beaches (Le Blond 1979, Wong 1981, Phillips 1985), and pocket beaches (Silvester et al. 1980). Half-Moon Bay in California, as shown in Fig. 4.3, is one of the best known examples of this kind, in which the curved waterline comprises three main segments, a downcoast tangential section, a curved element approximating a logarithmic spiral, and an almost circular beach in the lee of the upcoast headland. Carter (1988) has cited 28 zeta bays around the British Isles, under the influence of Atlantic swell waves penetrating into

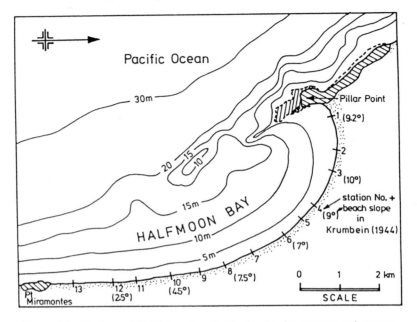

Figure 4.3 Planform of Half-Moon Bay, California, showing contours and average beach slopes along various stations numbered (Krumbein 1944)

the North Sea and Irish Sea. Sandy bays can easily be identified on large scale maps of Gower Peninsular bounding the Bristol Channel, United Kingdom, as depicted in Fig. 4.4.

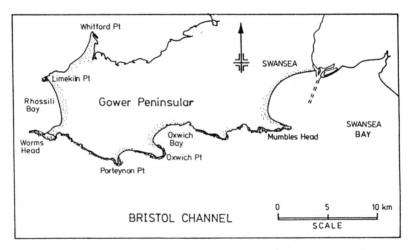

Figure 4.4 Bays formed downcoast of headlands on Gower Peninsular bounding Bristol Channel, United Kingdom

These crenulate shaped bays are ubiquitous, not only on oceanic margins but also along coasts of enclosed seas, lakes, and river shorelines. They indicate Nature's method of balancing wave energy and load of sediment transport. In this manner coasts have been maintained in position for thousands, or even millions of years. If it were not for headlands being located spasmodically along the coast, vast bays would have formed during the removal of material downcoast to some region where persistent swell arrives normal to the beach.

Geologists and geographers were the first to be interested in the existence of such bay shapes (Halligan 1906), but the shape as a stable physiographic feature was first recognized by Jennings (1955), without full knowledge of the waves involved. Davies (1958) realized the importance of wave refraction. The sculpturing process was later included in a textbook (Greswell 1957). The unique zeta-shaped beaches along the New South Wales coastline in Australia were noted by Langford-Smith and Thom (1969) without scientific analysis of their shapes. Yasso (1965) measured the planforms of a number of prototype bays in the United States of America and showed that they were equivalent to a logarithmic spiral. This empirical relationship has been accepted for almost 25 years until Hsu and Evans (1989) developed a more universal relationship.

Although the dynamic processes of this geomorphic feature have been studied by many researchers (Bird 1984, 1985; Carter 1988; Davies 1973; Davis 1985; King 1972; Shepard 1973; Yasso 1965; and Zenkovich 1967), their state of stability, that is, "how permanent they are," has only been examined by coastal engineers (Silvester 1960, for example). It is believed that they have been formed over some thousands of years, at least, but have been eroded, or more greatly indented, in the past hundred years or so due to reductions in sediment supply to the coast for many reasons. In this relatively short period in geologic time man has been able to harness rivers by damming, for water supply or flood mitigation, which impedes the transport of sand and stone to their mouths, and from whence it is spread alongshore by wave action.

It will be shown that the complete periphery of each bay can be better defined by a polynomial equation utilizing arcs from the point of diffraction at the upcoast headland to points on the coast. These radii are angled to the wave-crest alignment at this point, with a specific value for the bay extremity whose angle to this alignment defines the obliquity of arrival. A ratio of arc lengths to various points around the periphery is related to their respective angles in a manner that predicts the complete bayed shoreline for stable conditions, when no further littoral drift takes place. Even where a seawall is exposed within a developing bay the beach segments on either side will assume the crenulate shape as though it did not exist. This is displayed in the progressive recession of the coast at Sandy Hook, New Jersey in Fig. 4.5, where a building has fallen into the sea, but a highway was maintained by a rock embankment.

Where prototype bays are located on islands or promontories of mainland coasts (Reader's Digest 1983), it can be concluded that the availability of sediment is negligible or almost zero, due to no river system feeding material to the coast. Their shapes could thus be accepted as stable or in static equilibrium. Their predictable peripheries can be compared to existing bays for the same wave obliquity, to see whether they are stable, or

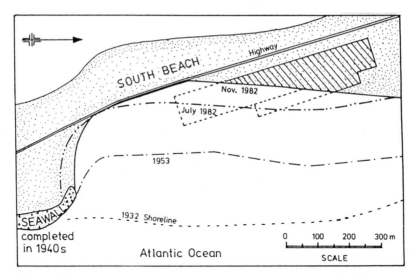

Figure 4.5   Beach flanking downcoast of a shore-based stone seawall on Sandy Hook, New Jersey, showing bay development (Phillips 1985)

whether they are likely to suffer greater indentation as littoral drift is reduced or ceases altogether.

4.1.2 Barrier Beaches

Another geomorphological feature which occurs around all continental margins is the barrier beach. This comprises a sand or pebble spit springing from a sharp change in direction of the mainland or running almost parallel to the shoreline, with an open or enclosed body of water between it and the mainland. Because of their importance as tourist attractions they have received much attention from geologists, geographers, ecologists, and engineers. Even so, their origin is the subject of much argument. Any explanation must account for present environmental conditions as they are observed to be still forming, or being modified by existing input forces.

Barrier beaches (as shown in Fig. 4.1), either as islands or long spits that partially or fully enclose bodies of water, are an important geomorphological feature of significance around the coastlines of the world. It is contended by these authors that the major energy input is wave action, particularly pulsational littoral drift. The more recent age of most of these features indicates a constructional process that discounts major changes in sea level.

Since these barrier spits are narrow they can be overtopped by severe storm waves and broken through to form inlets or island features, hence, the more general term *barrier islands*. Like the bays discussed above this physiographic feature is ubiquitous, being found along margins of oceans or enclosed seas. The main requirement for their formation and maintenance is persistent oblique swell or locally generated waves and abundant sediment supply.

4.1.3 Engineering Significance

Although workers in several distinct professions have studied the beaches and coastal processes, Komar (1976) has commented: "Often two or more groups attack the same problems without recognizing the contributions being made by their counterparts, and generally they have different approaches and sets of prejudices." Geographers and geo-morphologists have interest in the evolution of various coastal landforms, comparing the existing with the old, and more recently in coastal planning and management. Geologists and sedimentologists have contributed to the understanding of the nature and source of sediments in geological terms, and the development of barrier islands and spits. Oceanographers have promoted the physical and chemical properties of sea water and circulation of oceanic currents. On the other hand, coastal engineers have the task of protecting coasts from erosion by designing and constructing various marine structures, through understanding waves and the resulting sediment movement. The coastal engineering profession has frequently been asked to model the wave motions and coastal processes in the nearshore zone. Interdisciplinary collaboration should benefit all workers involved in this common environment. Guilcher (1974) suggested that coastal engineers should have learned the experience gained by geologists, geographers, and geomorphologists, and vice versa.

Silvester (1960) initiated the research on crenulate shaped bays from a coastal engineer's point of view. He first reported the results of a model study in which waves were directed at 45° to an initially straight beach and allowed it to erode without sediment replenishment from upcoast. The resultant embayment was referred to as a *half-heart bay*. The only variables dictating the curvature of this bay were the obliquity of the waves, which were persistent in their direction, and the distance between the artificial headlands upcoast and downcoast. He supervised research on the formation of such bays at the Asian Institute of Technology, Bangkok (Vichetpan 1969, Ho 1971), under a new name called *crenulate shaped bays*. From the model tests conducted, relationships between the bay periphery and wave obliquity were established, to fit the logarithmic spiral relationship available at that time (Yasso 1965).

When such a bay is in *static equilibrium* (i.e., it is stable), the tangential section downcoast (shown in Fig. 4.6) is found experimentally to be parallel to the wave crests approaching the coast from offshore. It is in this condition that the incoming waves will refract and diffract into the bay and will break simultaneously around the whole periphery. This, in turn, means no longshore component of breaking wave energy and hence no littoral drift within the embayment. At present, bays in *dynamic equilibrium* can not be predicted, mainly because the littoral drift that is still occurring is difficult to assess. Thus, only static equilibrium shapes can be related to the wave input.

The stabilization of mainland coasts by some economical and practical means can equally be applied to barrier islands, where in many cases it is more urgent. They are in greater jeopardy due to their narrowness and current lack of sediment supply. Since any new impediment to littoral drift can affect a downcoast region, this becomes a predicament when the shoreline has too little width to withstand erosion. Practical methods to overcome this problem are discussed in Chapter 6.

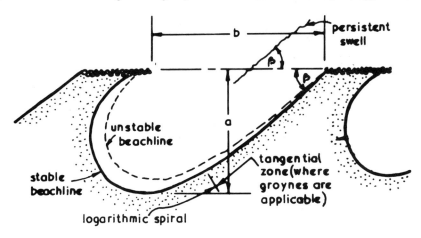

Figure 4.6 Definition sketch of zeta shaped bays in static equilibrium

4.2 CRENULATE SHAPED BAYS

This section contains many subsections due to its importance as a means of stabilizing coastlines. The reasons for the specific orientation of bays to the oblique swell are discussed, followed by earlier and later mathematical equations for predicting the shapes of these features. Verification of these both in models and prototype situations are presented, as well as practical problems in applying the principles. A simplified method of judging the stability of an existing bay without drawing the complete shape is then shown, involving what is termed the *indentation ratio*. There are difficulties at times in identifying bay extremities due to islands or reefs causing extraneous diffraction patterns, which are discussed in detail. The erosion of bayed beaches is not uniform around their peripheries and varies with the angle of attack of the storm waves. Finally, the case is presented of bays enclosed by small gaps, when diffraction with little refraction can produce different peripheral planform from the normal crenulate shape.

4.2.1 Bay Orientations

When any reasonable length of coast is considered, the direction of the persistent swell is sensibly constant, even when refracted across the continental shelf to the beach. Should a series of headlands be almost aligned along it, the wave orthogonals should be parallel and hence the downcoast tangents of successive bays are also similarly oriented. Such a case is illustrated in Fig. 4.7, where a section of the Japanese coast exhibits bays of this form. Many coastlines of the world have bays sculptured between headlands of similar orientation. To obtain the incident orthogonal for each, the one normal to the downcoast tangent in any bay can be refracted into deep water and then transposed farther along the coast to be refracted back into other bays, in order to obtain the approach direction for them. From these the correct values of wave obliquity can be obtained to check for stability. Even though rocky coasts may separate bays, with no beaches present, it does

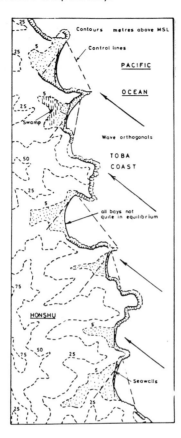

Figure 4.7 Similar bay orientation along
a section of the Japanese coast

not mean that sand is not being transported along that section of shoreline. Reflection
from these near vertical faces creates short-crested waves which can transmit material at
greater depths.

Where headlands are not aligned, as may occur at a sharp change in mainland
orientation (see Fig. 4.8), the offshore contours are likely to follow this general curvature.
As seen in this figure, the deepwater swell orthogonals have the same direction but are
refracted by differing amounts before reaching each successive headland. The angle of
their crests to each *control line*, joining the upcoast diffraction point and the downcoast
extremity of the bay, varies as illustrated. This implies that the indentation or shape of
adjacent bays varies, with the downcoast bay in this case being more indented than its
upcoast compatriots. As seen in Fig. 4.9, the bays either side of Pt. Reyes in California
are distinctly different in shape even though their orientations are in the same general
direction. The probable wave orthogonals have the same direction, from the northwest in
this location, but are refracted by the underwater contours by vastly differing amounts.

A series of bays along the New South Wales and Queensland coasts, Australia,
which are of vastly different dimensions and indentations, are illustrated in Fig. 4.10.
They have similar orientations due to the preponderance of waves from the SE quadrant

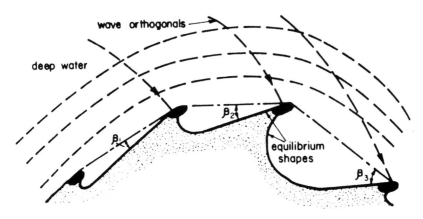

Figure 4.8   Bay shapes around a mainland protuberance with curved offshore contours

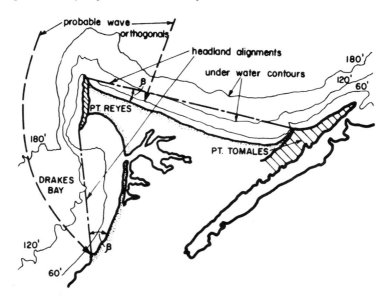

Figure 4.9   Bays either side of Point Reyes, California

(Patterson and Patterson 1983). In one instance, as seen in the figure, bays have partially formed in the lee of an island offshore. Such salients and cuspate forelands will be discussed in Chapter 6.

As seen in Fig. 4.6, a bay can be in an unstable condition as defined by material moving through it from upcoast or being supplied from a river within it. A stable shape or static equilibrium limit is reached when no further supply is available and the shoreline has eroded to a condition where littoral drift is reduced to zero. Even before this stage is reached the bay assumes a crenulate or zeta shape, but it is only for the final stable shape that its characteristics are known.

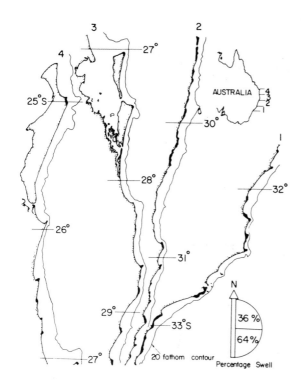

Figure 4.10 Portion of coastline of New South Wales and Queensland, Australia

In this static equilibrium condition the sole input variable is the angle β, as in Fig. 4.6, between the crests of the persistent waves and the *control line*, joining successive headlands. This is the same as the angle between this same line and the tangent to the relatively straight beach at the downcoast extremity. This condition of no littoral drift implies that breaking waves are arriving normal to the beach. This, in turn, infers that waves are breaking simultaneously around the whole periphery of the bay. In fact, such an observation of persistent swell doing so, which is beautiful to behold, is a clear field indicator of stability. Kana et al. (1986) have concluded: "Orientation of artificial beaches should be aligned into the predominant wave approach to minimize sand transport." Tsuchiya (1982) has expressed a similar view: "Decrease the incident angle of breaking waves by changing either the bottom topography or the inclination of the shoreline to the incidence of predominant waves." Berek and Dean (1982) have also stated: "Following a change in wave direction, the active contours in an idealized pocket beach respond by rotating such that they approach a perpendicular orientation relative to the incoming wave rays." A picture of a wave breaking simultaneously around the periphery of a bay in static equilibrium is shown in Fig. 4.11.

Model tests are generally necessary to obtain characteristics of bays for static equilibrium, when upcoast supply is absent. Some prototype situations can be studied

Figure 4.11 Simultaneous breaking of wave around periphery of a stable bay on the coast of South Australia

for this condition, as when bays are located on islands or promontories, in which case continued supply of material can be considered negligible. Scaling factors do not arise so long as the waves used in models have sufficient height and length to move sediment across the complete bed between headlands. It is preferable to commence with a straight beach between these fixed points and introduce monochromatic waves oblique to this shoreline. Sand will be transported toward and past the downcoast headland, which can then be removed. A rig as used by Ho (1971) is illustrated in Fig. 4.12, where it is seen that the original straight beach is shaped into a bay. The wave guide serves as

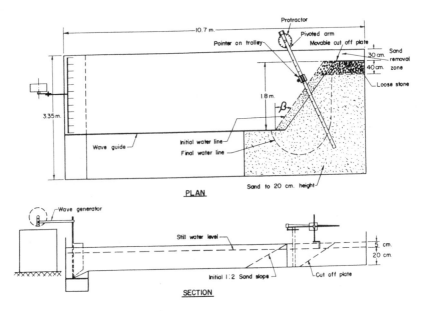

Figure 4.12 Test rig as used by Ho (1971) to derive stable bays

a headland since diffraction occurs at its tip, an essential component in the sculpturing process. Various tests could be applied to prove the stability of the final shoreline as follows:

1. No further sand is deposited in the trap provided at the downcoast end.
2. The beach is not receding any further.
3. Waves are breaking simultaneously around the periphery.
4. Dye inserted in the surf zone does not move along the beach.

In terms of stability, these bays may be in dynamic equilibrium with continual sediment supply or in static equilibrium when no further littoral drift is taking place, while many existing bays, as measured from maps or hydrographic charts, may not be in static equilibrium but can remain in dynamic equilibrium for long periods. This is when constant supply of material from upcoast or within the embayment is passing through the bay, so maintaining a shoreline that is not so indented as that of the static equilibrium version.

However, if the upcoast supply of material is cut off, an ever present problem these days, the bay will become more indented until littoral drift ceases. The plan shape is then in static equilibrium and it is for this condition that it can be related to the wave obliquity. At this stage all waves arrive normal to the beach. Obliquity is measured by the angle of wave crests at the upcoast headland to a control line joining the point of diffraction with the downcoast limit of the bay, as seen in Fig. 4.6. This is the same angle as between the control line and the downcoast tangent to the bay, as already mentioned.

Silvester and Ho (1972) have derived a typical bay with offshore contours derived from a number of bays in static equilibrium, by averaging offshore profiles around their peripheries. With β determined they then took several diffracted orthogonals and refracted them to the shore where they were shown to be within $1°$–$2°$ of normal to the beach at each point. This is illustrated in Fig. 4.13. Dyer (1986) has also sketched wave

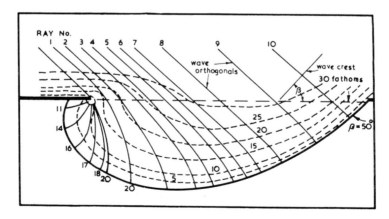

Figure 4.13   Wave refraction patterns in static equilibrium bay (Silvester 1972)

refraction patterns and transformations in a stable bay, showing normal approach around the complete periphery.

4.2.2 Logarithmic Bay Shape

Two empirical equations have been proposed in deriving bay shaped beaches in the past 30 years, these being referred to as *parabolic* and *logarithmic spiral*. The former had received very little attention until a recent re-analysis by Hsu et al. (1987), while the latter has been applied extensively by geographers and coastal engineers alike since its introduction. Krumbein (1944) examined beach processes of Half-Moon Bay near San Francisco, California, and suggested the approximation of a logarithmic spiral. Yasso (1965) measured planforms of four natural headland bay beaches on the east and west coasts of the United States, and concluded that they approximated the same form. A definition sketch of such spirals is given in Fig. 4.14 of which the equation is:

$$R_2/R_1 = \exp(\theta \cot \alpha) \tag{4.1}$$

where θ is the angle between radii R_2 and R_1 (where $R_2 > R_1$) and α is the constant angle between either radius and its tangent to the curve. According to Yasso (1965), the centers of the logarithmic spirals where the radii were measured could not be fitted to the ideal spot at the upcoast headland where wave diffraction took place. The shifting of these centers ranged from 0.3 to 2000 m respectively for the four cases presented

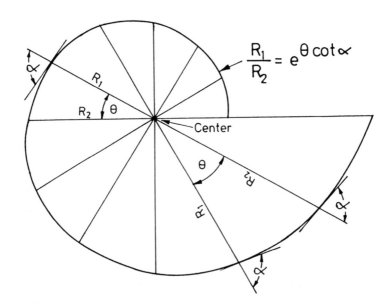

Figure 4.14   Definition sketch of logarithmic spiral

in his original paper. Wave characteristics and wave obliquity were not included in the logarithmic expression.

Employing the concept of logarithmic spiral, Silvester (1970) showed from model tests (Vichetpan 1969) that the ratio of R_2/R_1 or α changed as the bay was eroded from a straight alignment to its final equilibrium shape. There were specific values for these variables for the stable condition that varied with the wave obliquity β only, as depicted in Fig. 4.15. In the static equilibrium condition, the value of β was the same as that between the downcoast tangent and the control line, as seen in Fig. 4.6.

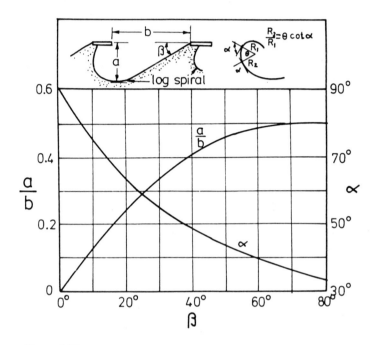

Figure 4.15 Relationship between α versus wave obliquity β for Eq. (4.1)

Thus there was only one value of R_2/R_1 or α of the logarithmic spiral relationship for a stable bay for any given wave obliquity β. Transparent templates of curves with varying R_2/R_1 or α unique to each spiral were employed on the receding shorelines, as exemplified in Fig. 4.16 (Ho 1971). In this figure $\beta = 45°$ but several obliquities were tested. The log-spiral constant was plotted against wave duration in hours, which resulted in a steep curve initially, but became asymptotic to the horizontal as static equilibrium for the bay was reached. The value of α at this limit was then plotted against β, giving a smooth curve, not only for the model tests, but also for some prototype bays accepted as stable because of the sediment supply conditions. Silvester (1970, 1974) presented this as a design curve for defining static equilibrium. The similarity of bay peripheries to logarithmic spirals has also been shown by LeBlond (1972, 1979), Rea and Komar (1975), Parker and Quigley (1980), and Finkelstein (1982).

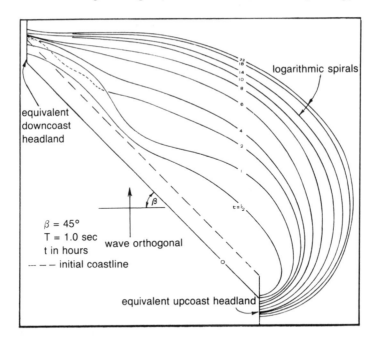

Figure 4.16 Example of test by Ho (1971) for $\beta = 45°$

Although the scale of reproduction for maps or aerial photographs does not matter, since any part of a complete spiral can be used, the fitting of templates is not precise. The spiral applies only to the curved section of beach in the shadow zone of the upcoast headland, and its center does not match the point at which diffraction takes place. It should be emphasized that bays assume this crenulate shape even before stability is reached and hence are sometimes assumed to be in equilibrium. For this reason, and maybe others, it was difficult for engineers to apply this criterion of stability. It was found later (Hsu et al. 1987) that the spirals did not apply to the downcoast periphery of the bays if their centers are fixed at the upcoast control point, at which wave diffraction takes place, as seen in Fig. 4.17. This difference is accentuated more for smaller values of β, which usually apply.

Despite the log-spiral lacking physical justification in describing a coast in equilibrium, Walton (1977) has applied them to certain features along the Florida coast. It is unlikely that any of these would be in static equilibrium but even so they appeared to follow this curvature for portions of their shoreline. No attempt was made to relate the constant in the equation to wave obliquity. The author thought the fit "may be fortuitous in view of the ease of fitting log-spiral smooth curves to many phenomena found in nature." Parker and Quigley (1980) also applied the log-spiral to a curved shoreline downcoast of Port Stanley on the north shore of Lake Erie, Canada, using an arbitrary point A (see Fig. 4.18), which clearly does not fit the complete 1973 waterline farther downcoast of the port, compared with the parabolic bay shape to be presented in Section 4.2.3.

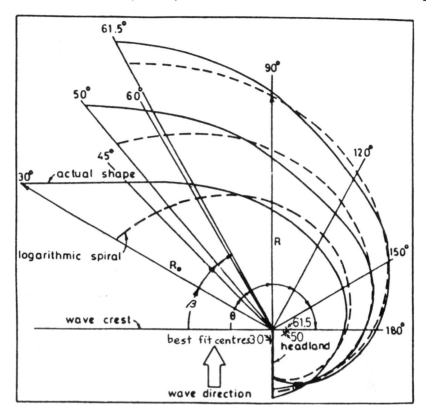

Figure 4.17 Comparison of log spirals with actual bay shapes for various β

4.2.3 Parabolic Bay Shape

The term parabolic was first mentioned by Mashima (1961) to describe the geometric configuration of the bow-shaped stable coasts found on the coastal margins of Tokyo Bay, Sagami Bay, and Boso Peninsula of Japan, as seen in Fig. 4.19 (see also Fig. 7.31). Mashima (1961) derived the annual wind and wave energy roses which he found to be semi-ellipses in shape. A close relationship existed between the wave energy ellipse and the bay shape. He derived a parabola

$$y = px^2 - b \qquad (4.2)$$

as seen in Fig. 4.20. There were complications such as centering the parabola for any particular bay, not taking diffraction into account, or the wave obliquity. Only the curved waterlines were shown as in Fig. 4.19 and not the headlands, nor their points of diffraction upcoast.

Reanalysis of model data (Vichetpan 1969, Ho 1971), together with those from prototype bays known to be in static equilibrium from sediment source conditions, suggested a new approach (Hsu et al. 1987). This is illustrated in Fig. 4.21, where radii (R)

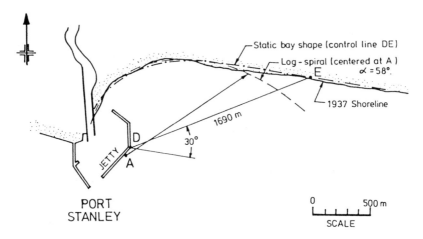

PORT
STANLEY

Figure 4.18 Application of the log-spiral shape to a bay at Port Stanley, Lake Erie, Canada, by Parker and Quigley (1980), compared with the parabolic prediction of Eq. (4.4)

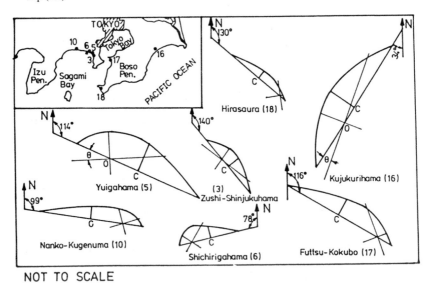

NOT TO SCALE

Figure 4.19 Examples of bow-shaped bays in Japan (Mashima 1961)

are drawn from the point of diffraction to the beach at angle θ to the wave-crest line. One such radius is R_o or the *control line* at angle β to the same wave-crest line. For a bay in static equilibrium this angle (β) is the same between R_o and the tangent to the downcoast beach line. Even though the bay may not be completely stable, this tangent alignment is likely to be reached prior to the bay eroding back to its limiting shape. The subsequent erosion takes place at the deepest indentation zone (Everts 1983).

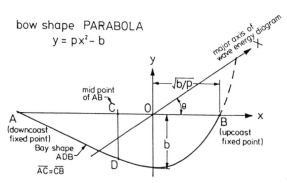

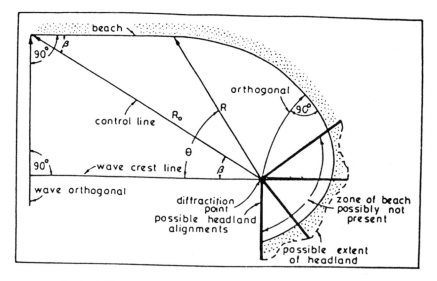

Figure 4.20 Coordinates of a parabola defined by Mashima (1961)

Figure 4.21 Definition sketch of a new parabolic approach to bay shape

For engineering applications, nondimensional parameters are preferred. The value of R at any angle θ has to be normalized by taking its ratio to R_o, which is the length of the control line. It is desirable that arc ratio R/R_o be accurately predicted for any value of θ for a given bay with known wave obliquity (β). The ratio R/R_o can then be related in some way to β and θ, separately or in combination (Hsu et al. 1989a, 1989b). A relationship was then tested between the dimensionless ratios of R/R_o and θ/β but, as seen in Fig. 4.22, curves for varying β could not provide a unique solution. For specific angles θ, $\log(R/R_o)$ was then plotted against $\log \beta$ which resulted in a series of nearly parallel straight lines. This enabled θ to be plotted against R/R_o for constant values of β and the values smoothed. This then gave an equation

$$R/R_o = 0.81\beta^{0.83}/\theta^{0.77} \qquad (4.3)$$

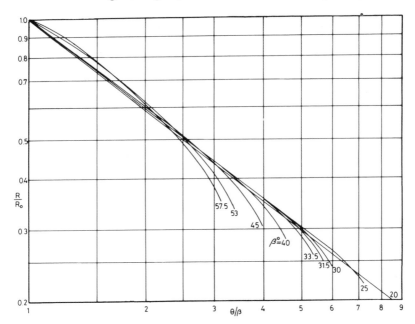

Figure 4.22   Radii ratio (R/R_o) versus angle ratio (θ/β) obtained from tests (Ho 1971)

which has been drawn in Fig. 4.23, with data from specific model bays and for θ as provided in the legend. It is seen that for $\theta = 45°$ to $90°$ the data points follow the curve closely but for $\theta = 120°$ to $180°$ they are more scattered. It was found that for large β and θ values the exponents for the best computer fit differed from those in Eq. (4.3), together with the constant 0.81. Hence this relationship was not considered universal.

Values of R/R_o versus β for a range of θ were then plotted on a linear scale, not only for model tests by Ho (1971) but also for prototype bays around Australia located on islands and peninsulars (Reader's Digest 1983), where sediment supply was minimal or zero and hence the bays could be considered in static equilibrium. Their locations and specific values of β are listed in Table 4.1. The resulting curves are presented in Fig. 4.24, where it is seen that values were available for $\beta = 22.5°$ to $72°$ from which extrapolation was carried out for $\beta = 0°$ to $80°$ (Hsu et al. 1989a). These curves can be used directly to determine the final shape for a specific value of β, which is readily measured between the control line and the downcoast tangent. If the predicted shoreline is landward of that being tested it must be accepted that the bay is unstable, implying that with any reduction of sediment supply the beach will erode or become more indented, back to the limit as specified by the specific β on the curve.

It is seen in Fig. 4.24 that lines up to $\theta = 150°$ are rising for β up to $80°$. For values of $\theta > 180°$ the curves have a negative slope for $\beta > 55°$ approximately. The lines for $\theta = 240°$ to $270°$ are very closely aligned, implying similar values of R/R_o, or almost a circular arc in this shadow zone. Although most prototype cases will not

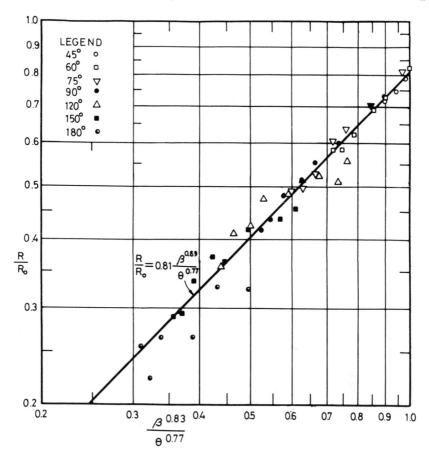

Figure 4.23 Relationship of R/R_o versus parameter $\beta^{0.83}/\theta^{0.77}$ from tests by Ho (1971)

have a beach in areas of $\theta > 150°$, curves should be provided for them. It is likely that the prototype bays in this zone may differ slightly from the predicted stable condition (i.e., be more seaward than them) because the removal of material in these areas is very slow due to the small height of waves in this greatly diffracted condition. Storm waves oblique to the bayed shoreline may transport material into the leeward area which may not spread evenly to the correct stable shape readily.

The curves in Fig. 4.24 have been computerized (Hsu and Evans 1989) to derive a polynomial of the form

$$R/R_o = C_o + C_1(\beta/\theta) + C_2(\beta/\theta)^2 \tag{4.4}$$

It was found that an equation to this order was sufficient to obtain bay outlines which matched actual bays extremely well for the complete periphery. For any specific β and R_o the ratio of β/θ, together with coefficients, provide ratios of R/R_o from which arc R

TABLE 4.1 LOCATION OF DATA IN FIG. 4.24
FOR AUSTRALIAN BAYS

Location	β^o
Magnetic Island—Queensland	22.5
Mission Island—Queensland	35
Phillip Island—Victoria	38
Magnetic Island—Queensland	39
Wilson Promon.—Victoria (upper)	43
Keppel Island—Queensland (lower)	43
Wilson Promon.—Victoria	44
Dongara—Western Australia	52
Freycinet Penin.—Tasmania	55.5
Phillip Island—Victoria	57
Jurien Bay—Western Australia	61
Kangaroo Island—South Australia	68
Freycinet Penin.—Tasmania	71.5
Port Arthur Penin.—Tasmania	72

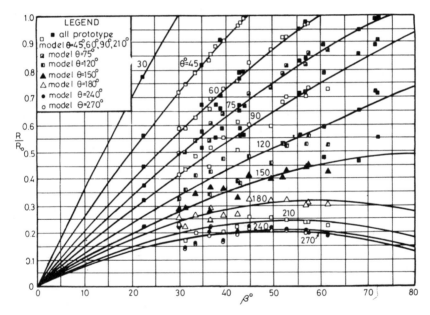

Figure 4.24 Radii ratio (R/R_o) versus β for a range of θ from model tests (Ho 1971) and prototype bays (Table 4.1)

is derived. The values of C_o, C_1, and C_2 vary uniformly with β as in Fig. 4.25. Values of these coefficients and R/R_o are listed in Table 4.2 for convenience. Most bays will fall between values of β from 20° to 80°.

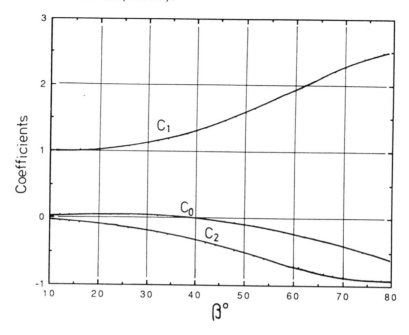

Figure 4.25   The three C coefficients as in Eq. (4.4) over a range of wave obliquity β

4.2.4 Verification of Parabolic Shapes

The parabolic shape as derived from Fig. 4.24 or Eq. (4.4) is compared to a model bay by Ho (1971) for $\beta = 30°$ in Fig. 4.26. It is seen that this curve predicts the shape extremely well around the complete periphery, whereas the logarithmic spiral (centered on the diffraction point) deviates considerably, especially at the downcoast section, as also exhibited in Fig. 4.17. Another comparison is presented in Fig. 4.27 for a model bay by Ho (1971) for $\beta = 57.5°$. Here again the deviation from the actual bay shape is quite modest.

  The shapes of prototype bays are also well predicted by the new relationships. For example, in Fig. 4.28 the natural and parabolic curves are very close. The deviation at the large θ values (adjacent to the upcoast headland) is due, as noted earlier, to the slow removal of sediment by the very small waves in this region. Not only is diffraction taking its toll of these heights, but also dissipation along the rugged headland reduces the effect of waves in sand transportation. However, for the bulk of the bay the parabolic relationship is applicable.

  On Magnetic Island off the coast of Queensland, Australia, supply of sediment can be considered as negligible. On the SE portion of this island are three bays, as depicted in Fig. 4.29, which should be stable. The predicted curvatures of these are shown seaward of the beaches themselves for clarity. The values of β vary from 22.5° to 51.5° with incident wave orthogonals varying slightly, as indicated by the orientation of the downcoast tangents of each beach. Although the beach lines near the upcoast

TABLE 4.2 MEANS FOR DETERMINING RADII RATIOS (R/R_O)

β^o	Coefficients in Eq. (4.4)			Values of R/R_o for $\theta^o =$							
	C_o	C_1	C_2	30	45	60	75	90	120	150	180
20	0.054	1.040	-0.094	0.705	0.497	0.39	0.324	0.280	0.225	0.191	0.168
22	0.054	1.053	-0.109	0.768	0.543	0.426	0.354	0.305	0.244	0.206	0.181
24	0.054	1.069	-0.125	0.829	0.588	0.461	0.383	0.330	0.263	0.222	0.194
26	0.052	1.088	-0.144	0.887	0.633	0.497	0.412	0.355	0.281	0.237	0.207
28	0.050	1.110	-0.164	0.944	0.677	0.532	0.442	0.379	0.300	0.251	0.219
30	0.046	1.136	-0.186	1.000	0.721	0.568	0.471	0.404	0.319	0.266	0.230
32	0.041	1.166	-0.210		0.763	0.603	0.500	0.429	0.337	0.280	0.242
34	0.034	1.199	-0.237		0.805	0.638	0.529	0.453	0.355	0.294	0.252
36	0.026	1.236	-0.265		0.845	0.672	0.558	0.478	0.373	0.307	0.262
38	0.015	1.277	-0.296		0.883	0.706	0.586	0.502	0.390	0.320	0.272
40	0.003	1.322	-0.328		0.919	0.739	0.615	0.526	0.407	0.332	0.281
42	-0.011	1.370	-0.362		0.953	0.771	0.643	0.550	0.424	0.344	0.289
44	-0.027	1.422	-0.398		0.983	0.802	0.670	0.573	0.441	0.356	0.297
46	-0.045	1.478	-0.435			0.832	0.698	0.596	0.457	0.367	0.304
48	-0.066	1.537	-0.473			0.861	0.724	0.619	0.473	0.378	0.311
50	-0.088	1.598	-0.512			0.888	0.750	0.642	0.489	0.388	0.317
52	-0.112	1.662	-0.552			0.914	0.775	0.664	0.505	0.398	0.322
54	-0.138	1.729	-0.592			0.938	0.800	0.686	0.520	0.408	0.327
56	-0.166	1.797	-0.632			0.960	0.823	0.707	0.535	0.417	0.332
58	-0.196	1.866	-0.671			0.981	0.846	0.728	0.549	0.425	0.336
60	-0.227	1.936	-0.710			1.000	0.867	0.748	0.563	0.434	0.339
62	-0.260	2.006	-0.746				0.888	0.768	0.577	0.441	0.342
64	-0.295	2.076	-0.781				0.908	0.787	0.590	0.449	0.345
66	-0.331	2.145	-0.813				0.927	0.805	0.603	0.456	0.346
68	-0.368	2.212	-0.842				0.945	0.823	0.615	0.462	0.348
70	-0.405	2.276	-0.867				0.963	0.840	0.627	0.468	0.349
72	-0.444	2.336	-0.888				0.981	0.857	0.638	0.473	0.349
74	-0.483	2.393	-0.903				1.000	0.874	0.649	0.478	0.348
76	-0.522	2.444	-0.912					0.891	0.660	0.482	0.347
78	-0.561	2.489	-0.915					0.909	0.670	0.486	0.346
80	-0.600	2.526	-0.910					0.927	0.680	0.489	0.343

headlands are not clearly defined in the photograph, the curvatures of the actual and predicted bays for the bulk of bay peripheries are extremely close. The central bay in this figure is that reproduced in Fig. 4.28.

Some variables associated with natural bays, such as their beach profiles and the wave characteristics other than wave obliquity, are not included in the present method of prediction. It is felt, however, that the primary parameter in shaping the curved waterlines is the value of β, whereas beach profile, wave height, period (or steepness) are of secondary importance. The best-fit relationship and parabolic form are derived empirically from a blend of model and prototype bays, which would have represented various wave conditions. Therefore, the present prediction method appears versatile and suitable for field applications.

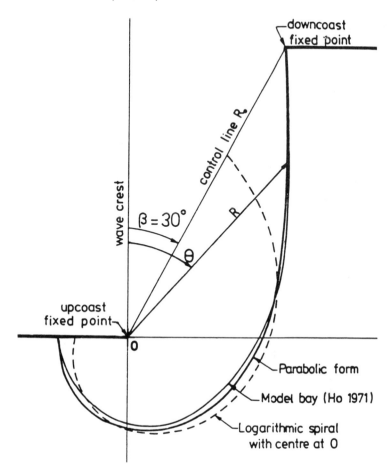

Figure 4.26 Comparison of parabolic and log-spiral shape with actual test (Ho 1971) for $\beta = 30°$

It appears that the term *parabolic* here may be a misnomer, since Eq. (4.4) contains a linear term of (β/θ), which makes it quadratic. Although model tests appeared to show that the C constants varied with wave steepness, this variable is difficult to define and varies continually in prototype conditions. Swell waves are of very low steepness anyway and therefore its omission seems logical and consistent. Thus, the simpler relationship in terms of wave obliquity (β) only is recommended for engineering purposes.

4.2.5 Practical Applications

In applying this new parabolic form to natural bays, one difficulty is in defining the downcoast control point. This affects the value of *control line* length (R_o) and the value of wave obliquity (β). The latter is generally measured between the control line and

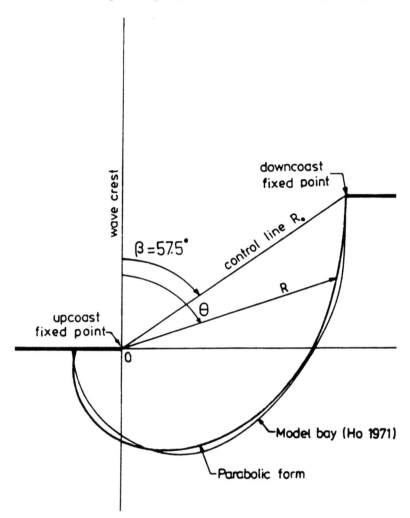

Figure 4.27   Comparison of parabolic shape with model test (Ho 1971) for $\beta = 57.5°$

the tangent at the extremity of the bay. For a bay still in dynamic equilibrium this tangent may not be oriented to that for static equilibrium, although it will not change much as further erosion takes place, as might be gathered from Fig. 4.16. However, the scale of the bay being used from maps or aerial photographs could introduce a slight error. Taking a β value slightly greater than that measured by protractor could offset this difference.

Some aerial photographs may contain crests of breaking waves which may be angled to the shore at the downcoast extremity of the bay. Using a tangent to these breaking crests for the determination of β could be erroneous, because the waves on the particular day may not be indicative of the swell just after a recent storm, which dictates

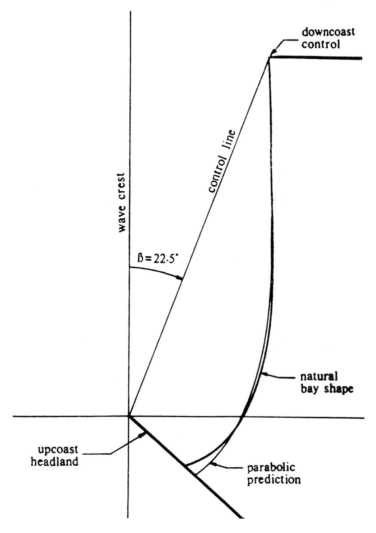

Figure 4.28   Predicted and actual bay shape on Magnetic Island, Queensland, Australia
(Table 4.1)

the more persistent waterline. Thus, use of visual wave crests is not recommended in determining the tangent for β evaluation.

Generally, there is no difficulty in locating the upcoast control point where wave diffraction takes place. It may not be a point on the mainland but a small island or even reef offshore. An aerial photograph showing wave crests is an obvious advantage, but maps of suitable scale can exhibit points of diffraction. Once the wave orthogonal normal to the downcoast tangent is transferred across to the upcoast headland, it is the intercept with any of the above features that determines this control point.

Figure 4.29   Predicted shapes for bays on Magnetic Island, Queensland, Australia, (Reader's Digest 1983)

Sensitivity of the location of the downcoast control point can be gauged by varying it slightly to increase R_o with a commensurate decrease in β, or vice versa. This has been done for a bay on Magnetic Island as in Fig. 4.29, where $\beta = 22.5°$, which was also drawn in Fig. 4.28. The value of β was varied $\pm 5°$ and the bay curves derived as in Fig. 4.30, where it is seen that little difference occurred in the predicted waterlines.

The procedure to be followed is exhibited in Fig. 4.21, showing the transferred orthogonal intercepting the upcoast point of diffraction, with the normal to it at this point representing the incident wave-crest alignment. It is from this line that angle θ is measured. The lowest limit of θ is that of β, so that radii are drawn at various $\theta (> \beta)$. Suggested intervals are $15°$ up to $90°$ and $30°$ for $\theta > 90°$. The lengths of these arcs (R) to the predicted shoreline are in proportion to R/R_o values obtained from Fig. 4.24 or Eq. (4.4). In this way the stable shape is obtained for the case of zero littoral drift.

The wave-crest alignment is accepted as straight, which may not be the case if a shoal or depression is present in this region. Also it is assumed that the offshore area within the bay consists solely of sand with no rock outcrops or substantial seaweed

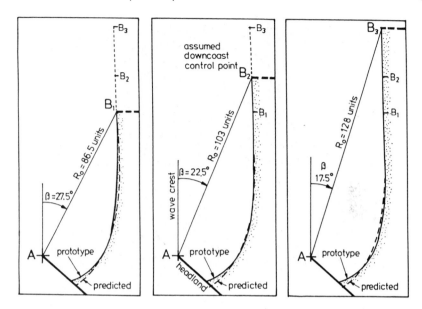

Figure 4.30   Effect of varying β by 5° either side of 22.5° as in Fig. 4.28 showing little change in predicted bay periphery

growth. Contours should be uniformly curved within the bay for waves to sculpture the predicted outline. Since for static equilibrium the waves arrive simultaneously around the bay periphery (Le Blond 1979), any change in wave celerity due to local depth variations will cause a deviation in the waterline. For example, if a rocky shoal of reasonable dimensions exists within the bay, waves will take longer to arrive and so cause a protuberance on the coast, as shown in Fig. 4.31. Equally, if depths are greater, a shorter time is taken by the waves and so a landward deviation will occur. All such variations should be smoothed when comparing existing to predicted shorelines.

As noted earlier, an actual bayed shoreline appearing seaward of the predicted version implies that its beaches could erode back to this limit. As seen in Fig. 4.6, such denudation would be greater in the zone of largest indentation. At the tangential downcoast limit of the bay, recession would be small, as also in the zone adjacent to the upcoast headland. Thus, *set-back* limits decided by authorities to cater for future unknown erosion should not be uniform, as is generally accepted. There is a specific limit to erosion should all sediment supply cease, as predicted by the relationships provided above. Encroachment on the beach can be greater when stability is reached, but should be limited to landward of the beach berm that can be removed by storms at least once every 2 years. In fact, major structures should be kept slightly landward of this limit to account for the more severe storm that might usurp portions of the frontal dune less frequently. Since the persistent swell will be arriving normal to the bayed periphery, the offshore profile will be milder in slope than for more obliquely arriving waves. This will demand less volume for the offshore bar before it reaches its protective height and hence less width of berm is likely to disappear each storm sequence.

Figure 4.31 Island offshore causing small salient in a bay north of Tandy Point, Queensland, Australia (Reader's Digest 1983)

The question constantly arises: "What happens if swell waves arrive from a different direction from that assumed in deriving the obliquity β?" As noted in Chapter 3, swell waves have a persistent direction of approach. This does not mean that they cannot arrive at angles differing from this, but their duration compared to the annual input of energy must be taken into account. However, as already noted, it is the direction of milder waves for a few days after a storm that dictates the net movement of sediment alongshore. These occur when sand is offshore, or *in circulation*, when bar material is being returned shoreward. Subsequent swell has much smaller effect as the bulk of the sand is stored within the berm causing only the beach face (to the proximity of the breaker line) to be disturbed. In a stable condition the swell for a few days after each storm will arrive normal to the periphery and hence move sand directly back to a position in the berm from whence it was removed to form the bar. Dean (1978) concludes: "It appears that planforms in nature may be smoothed by a range of effective wave periods and directions."

Dean and Maurmeyer (1977) have attempted to predict the shape of a small bay formed at the extremity of a jetty and extending to a revetment placed to anchor the beach at Shinnecock inlet on Long Island, New York, as in Fig. 4.32. The control line (R_o) in this case was only 90 m in length, which is the order of swell wave length from the Atlantic Ocean. Diffraction and refraction was conducted on an interactive basis which employed wave height as a variable. Method I, as indicated in the figure, assumed the jetty was thin, while Method II "attempted to account for the fact that the jetty was not 'thin' by an intuitive diffraction procedure." The authors stated that "the predicted and actual plan forms presented are in reasonable qualitative agreement."

If Fig. 4.24 or Eq. (4.4) are used to predict this shape the major input variables are angle $\beta = 38°$ and $R_o = 90$ m from which R is derived from R/R_o for specific values of

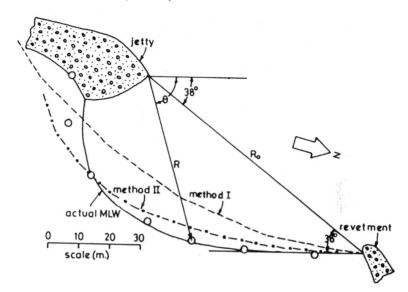

Figure 4.32   Comparison of parabolic prediction (open circles) with two methods used by Dean and Maurmeyer (1977) for a small bay at Shinnecock Inlet, Long Island, New York

θ. Open circles give the result from this method in Fig. 4.32 and are seen to follow the actual MLW curve exceedingly well. The only deviation is at the more curved section of the beach, or large θ, where the actual shoreline is seaward of the predicted one or accretion has taken place. Dean and Maurmeyer (1977) noted that waves generated across Shinnecock Bay came mainly from the NW where the fetch was only 3 km. Even so, these could force material toward the southern extremity of the bay, where greatly diffracted swell waves from the ocean may not remove it during the time available.

4.2.6 Indentation Ratio

Sometimes it is not essential to derive the actual static equilibrium shape to prove instability of an existing bayed coast. An alternative method that is swifter to apply could be what is termed the *indentation ratio*. As seen in the inset of Fig. 4.33, the greatest indentation (a) is measured normal from the control line to the point of largest retreat of the shoreline. This is obtained by drawing a tangent parallel to the control line which is asymptotic to the beach. The ratio a/R_o has been correlated with β as in this figure (Hsu et al. 1989a, 1989b) for both model tests (Ho 1971) and prototype bays (Table 4.1). The equation of this curve is

$$a/R_o = 0.0014\beta - 0.000094\beta^2 \tag{4.5}$$

This indentation ratio can be measured expeditiously. Should the point fall below the line in Fig. 4.33, the bay is unstable and can erode back to a point on the line, but not above it.

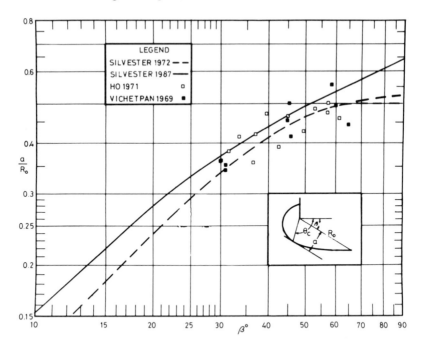

Figure 4.33 Indentation ratio (a/R_o) versus β, showing value of θ_c at greatest indentation

As also seen in Fig. 4.33, the angle θ_c to the radius at the point of contact with the greatest indentation can also be related to β, as in Fig. 4.34, using data from the same sources as in Fig. 4.33. The equation so derived is:

$$\theta_c = 63° + 1.04\beta \tag{4.6}$$

This can also be expressed as:

$$\theta_c - \beta = 63° + 0.04\beta \tag{4.7}$$

Values of a/R_o, θ_c, and $\theta_c - \beta$ are listed for $\beta = 10°$ to $90°$ in Table 4.3, or values can be taken directly from Fig. 4.34.

When sketching the static equilibrium bay from the indentation ratio (a/R_o), it is commenced at the downcoast end at angle β to the control line and curved to become asymptotic to the tangent distant (a) from it. The beach line is then extended with greater curvature into the shadow zone until the upcoast headland is reached. Later, the more specific waterline can be drawn with the aid of Fig. 4.24 or Eq. (4.4). Everts (1983) has employed indentation ratio rather than shoreline shape to derive denudation stages over time once littoral drift has been intercepted. He attempted to define it from net and gross drift, but acknowledged that more field data were required to refine the procedure.

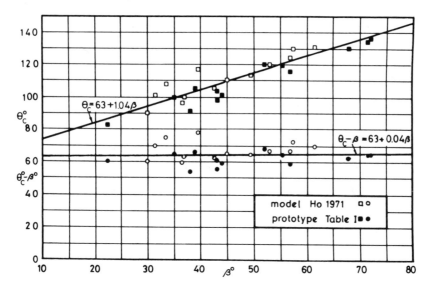

Figure 4.34   Values of θ_c or $(\theta_c - \beta)$ versus β

TABLE 4.3 INDENTATION RATIO
a/R_O, CRITICAL ANGLE $\theta_C - \beta$
OVER A RANGE OF β

β^o	a/R_o	θ_c^o	$\theta_c - \beta^o$
10	0.15	73.4	63.4
15	0.22	78.6	63.6
20	0.28	83.8	63.8
25	0.33	89.0	64.0
30	0.37	94.2	64.2
35	0.41	99.4	64.4
40	0.44	104.6	64.6
45	0.47	109.8	64.8
50	0.49	115.0	65.0
55	0.51	120.2	65.2
60	0.54	125.4	65.4
65	0.56	130.6	65.6
70	0.57	135.8	65.8
75	0.60	141.0	66.0
80	0.61	146.2	66.2
85	0.63	151.4	66.4
90	0.65	156.6	66.6

4.2.7 Bay Extremities

As indicated by Silvester et al. (1980) there are sometimes difficulties in defining the upcoast point of diffraction and the downcoast limit of a bay. Also the uniform parabolic shape may not be present due to some secondary effect of reflection from natural or man-made structures, or diffraction behind protruding features.

Where a downcoast headland runs some distance into the sea there may be insufficient material to build a beach to its upcoast extremity. These have been termed *pocket beaches* and should be in static equilibrium unless a river is still supplying sand to some point within them. This condition is depicted in Fig. 4.35, where it is seen that the control line joins the diffraction point to the downcoast limit of the bayed beach. As more sediment is added, the tangential section of coast proceeds seaward, normal to the wave orthogonal dictating shape. This accretion could take place until a straight beach exists from the downcoast tip of the upcoast headland normal to the wave orthogonal. Any further addition of sand would result in a new bay situation, extending to a headland further upcoast than the existing one shown. In each case the bay should be stable if sufficient time has elapsed for any sediment input from upcoast or from a river to have spread evenly throughout the bay.

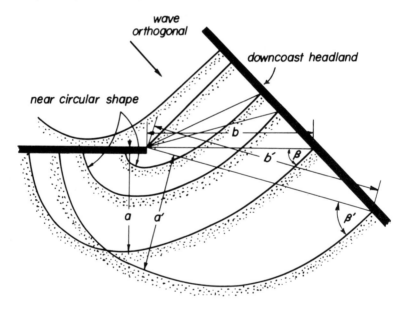

Figure 4.35   A case of protruding downcoast headland

An example of this situation is contained in Fig. 4.36, which is a bay on the Honshu coast facing the Sea of Japan. The wave orthogonals are normal to the tangential downcoast section of the bay. That on the left determines the control point of the upcoast headland, while the control line intercepts the meeting of the beach with the protruding downcoast headland. The value of $\beta = 34°$ predicts the static equilibrium bay as shown dotted. It is seen that at the upcoast or western end extra accretion has occurred due to wave diffraction around the jetty constructed at the river outlet. Otherwise the existing coast is in static equilibrium or nearly so, which calls into question the need for the

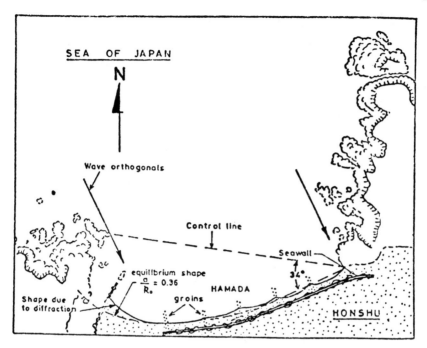

Figure 4.36 A bay exhibiting the control line for protruding downcoast headland

seawall and groins constructed along the tangential section of the beach. This is a case where sand renourishment might have solved the transient erosion problem.

Another complication can arise with the downcoast headland if it protrudes into the bay, as depicted in Fig. 4.37. Here diffraction takes place behind it, causing a deviation in the mild curvature of the downcoast end of the bay. A transition point between these two curves is determined to which the accepted orthogonal is normal. The limit of the bay then becomes this transition point, to give R_o, and the angle β is found between the transition tangent and the control line from which R/R_o ratios provide the predicted form. The indentation (a) can also be determined as an extra test of stability by comparison with a/R_o in Fig. 4.33.

An example of this problem is given in Fig. 4.38, which is a bay on the Shikoku coast of Japan facing the Pacific Ocean. The increased curvature at the western or downcoast section of this pocket beach indicates diffraction by the protruding downcoast headland. The wave orthogonal, as shown, must be normal to the transition tangent (dotted), which when transferred across to the upcoast headland gives the control point in that region. The control line is thus drawn between it and the transition point, from which R_o and angle β are found. In this case it is seen that the actual bay requires extension into the mountainous structure if the greatest indentation (a) is to be measured for calculation of a/R_o for the secondary test of stability. Another example of the need to determine a transition point is exhibited in Fig. 4.39, where a tombolo has formed

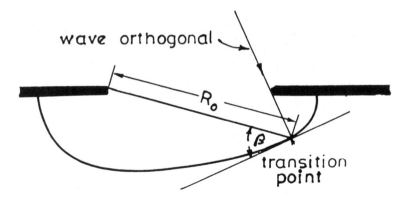

Figure 4.37   Diffraction due to protruding downcoast headland

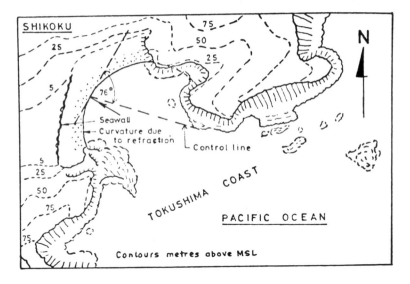

Figure 4.38   Example of a bay with double curvature

behind an offshore island. Diffraction around its southern extremity has produced extra curvature at the downcoast extremity of the bay.

In most cases the point of diffraction on the upcoast headland is clearly evident but sometimes another rock outcrop or reef may occur within the bay. This is illustrated in Fig. 4.40, where headlands A and C may be present. The downcoast tangential line is angled β' to the control line AD which will give the bay shape, up to the point where the orthogonal from A to C is diffracted. From there to the lee of C further diffraction and perhaps refraction will occur, so causing greater curvature of the shoreline. If headland C extends to the point B, the original incident orthogonal intersects both A and B so making a new control line BD, with angle β to the downcoast tangent. This creates a fresh static equilibrium shape which can be continued to its contact with headland CB.

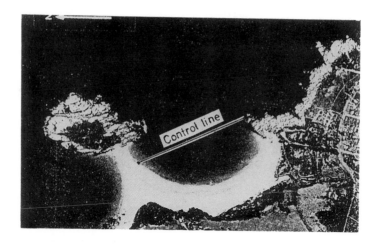

Figure 4.39 Tombolo formation behind island just north of Bicheno, on the east coast of Tasmania, Australia (Reader's Digest 1983)

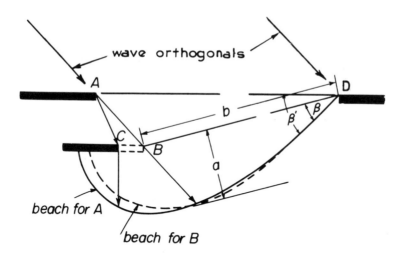

Figure 4.40 Bays formed by different upcoast headland control

These bays can be drawn using relationships from Fig. 4.24 or Eq. (4.4). The alternative criterion using a/R_o could also be used for a rough estimate of the bay shape.

Figure 4.41 shows a bay on the Shikoku coast of Japan facing the Pacific Ocean, where several islands exist in proximity to the upcoast headland. At the downcoast limit a deviation from the parabolic shape also occurs due to diffraction and dissipation from the many islands in the eastern area. The transition point has been ascertained and the wave orthogonal transposed to the upcoast headland from which the control line gives

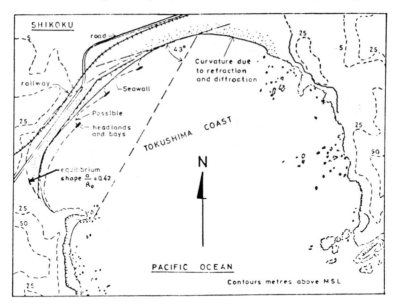

Figure 4.41 Example of defining control at upcoast and downcoast headlands, also showing possible stabilization by headlands

$\beta = 43°$. The static equilibrium bay has thus been drawn which indicates instability. This is confirmed by the extensive seawalls shown, which have provided protection for the road and rail links, a necessary feature of most Japanese coastlines. A new method of stabilizing such a bay is indicated by headlands, to be discussed in Chapter 6.

The parabolic form of bays, either in dynamic or static equilibrium, provides a uniformly changing curvature, any deviations from which should be investigated. For example, if a rocky shoal occurs offshore, waves will be refracted around it to cause a shoal in its lee with perhaps a protuberance in the waterline on its orthogonal centerline and erosion either side. These aberrations should be smoothed out when determining β or the indentation ratio (a/R_o).

Another source of shoreline deviation could be reflection from headlands, two types of which are depicted in Fig. 4.42. At the downcoast end orthogonals to the smoothed shoreline may reflect from the downcoast headland protruding into the sea. The reflection commences at the tip, for which the limiting reflected orthogonal is drawn. Within this almost triangular area of water a short-crested wave system is established which transports material parallel to the headland. This results in a new shoreline, to where a protuberance occurs close to the intersection of the limiting reflected orthogonal with the parabolic shoreline. As seen in the figure, use of the incorrect downcoast control point could give a wrong value of β. At the upcoast end of the bay reflection from a headland, as shaped in Fig. 4.42, could produce a zone of short-crested waves which will reshape the shoreline as shown. This may result in a wrong evaluation of indentation (a) being used in stability analysis. Examples of either type of shoreline deviation are not available, but should be smoothed out if they appear on any actual bay shapes.

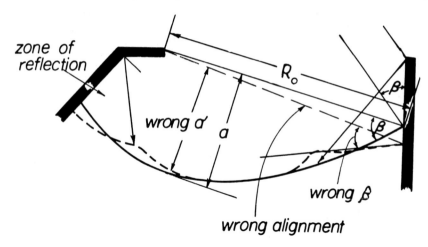

Figure 4.42   Influence of wave reflection on smooth bayed beach line

There is the possibility of control points changing over time, either induced by man or by natural processes. These are depicted in Fig. 4.43, where alterations in bay shape in *A, C,* and *D* are due to man while that in *B* is produced naturally. It is a common practice to extend headlands in the form of a breakwater, as in Fig. 4.43A, to provide calmer water for a port in the lee of a natural headland. A new control line length R_o results which causes accretion in the zone of the harbor because of the new point of diffraction. This has caused siltation probably before facilities have been constructed (Laurent 1955). In the case of a bay already in static equilibrium, this transfer of material into the lee zone will cause erosion further around the bay (see also Section 6.9).

Where an extensive spit forms on the leeside of the tip of a breakwater the diffraction point shifts to the inner extremity of this feature. As seen in Fig. 4.43B this causes a new shape of bay, which is unlikely to be in static equilibrium because some sediment is transmitted across the bay to the inner shoreline. But the main deviation to be expected is accretion in the lee of the breakwater-spit combination.

In Fig. 4.43C the reshaping of a bay due to the intrusion of a large structure, such as a massive port complex or airport runway, is illustrated. Between *A* and *B* reflected waves could reorient the beach as shown dotted. Just downcoast of the new structure accretion will occur which could cause erosion adjacent to it if the bay is already near equilibrium.

In the case of the downcoast headland being extended seaward, as shown in Fig. 4.43D, a new control line and hence β results, which creates substantial accretion upcoast of the new structure. This is one location where a groin, running normal to the beach, can be effective in retaining sand where it is desired. An excellent example of this type of development is shown in Fig. 4.44 where, at Kirra Point on the coast of New South Wales, Australia, the headland was extended by a groin type structure and a beautiful beach provided for Coolangatta. Prior to this, a stone embankment was built for protection which caused deepening of the bay due to oblique wave reflection.

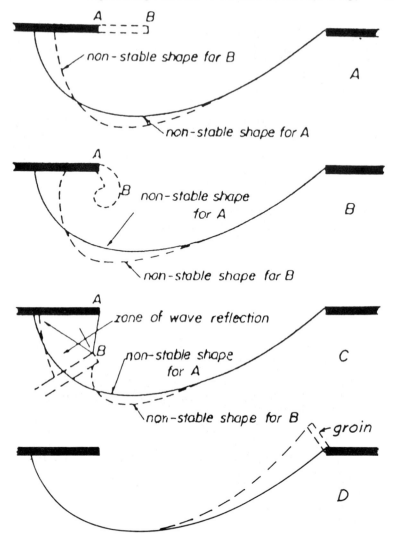

Figure 4.43 Examples of natural or man-made features that influence bay shape

It was proposed to renourish this beach after completion of the new headland but it was saturated with sand while it was being built.

With respect to downcoast conditions it is not necessary to have a headland, as instanced by the salient shown in Fig. 4.45 formed behind an offshore reef. This apex will not be very stable as it could be eroded badly during storms, but should be reformed very quickly as swell waves arrive. To stabilize this downcoast point a synthetic headland could be installed to maintain the beaches either side in position.

Figure 4.44 Headland extension at Kirra Point, New South Wales, Australia, which produced a stable beach at Coolangatta

Figure 4.45 Downcoast bay limit established by a salient found leeward of an offshore reef, just east of Peterborough, Victoria, Australia (Reader's Digest 1983)

4.2.8 Bay Under Storm Attack

Storm waves can arrive from a variety of directions as a storm center traverses the point of interest. It is necessary to investigate the behavior of beach states which respond to major storm waves from a specific approach angle. This is illustrated in Fig. 4.46, where it is seen that the width of beach berm eroded varies around the periphery. The

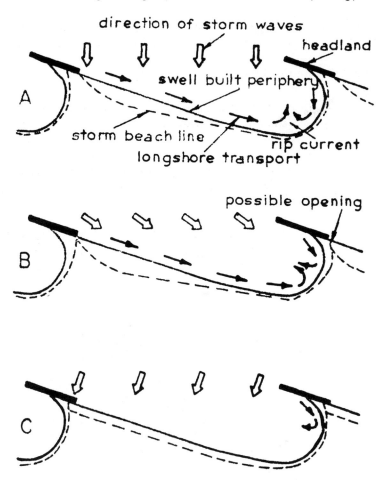

Figure 4.46 Influence of storm wave direction on a bayed beach

most severe denudation occurs when the wind arrives obliquely from the downcoast quadrant. Material will be transported from the downcoast headland into the bay and from the leeside of the upcoast headland. This could result in the upcoast headland being disconnected from the mainland. This will only be temporary as swell waves will, within a matter of days, reform the beaches and reattach the headland to shore, as discussed in Chapter 3. However, if this experience of detachment is not desired, the headland could be lengthened to prevent it.

The action of the storm waves in generating this longshore drift places more material at the zone of greatest indentation and hence provides optimum protection where the original shoreline is closest to the new waterline. It is expected that sediment removed from the beach berm in the bay will remain within the system, unlike that for a straight reach of shoreline or in a system of groins or offshore breakwaters, in which sediment will eventually be lost offshore and alongshore. The transient beach erosion and bar distance offshore can be estimated using the model presented in Section 3.4.3.

4.2.9 Bays with Small Gaps

A special case arises when the incident waves traverse a gap in a breakwater or reef before diffracting and refracting to the shoreline. Diffraction takes place in the lee of both segments (see Fig. 4.47), with the incident waves proceeding between the limiting orthogonals unabated except for height reduction due to the energy demand of the diffraction process. Since in most cases the protected coast is accreting outward, either naturally or by man, the depths offshore could be of reasonably uniform depth so that diffraction takes precedence over refraction. In this event the planforms will differ from

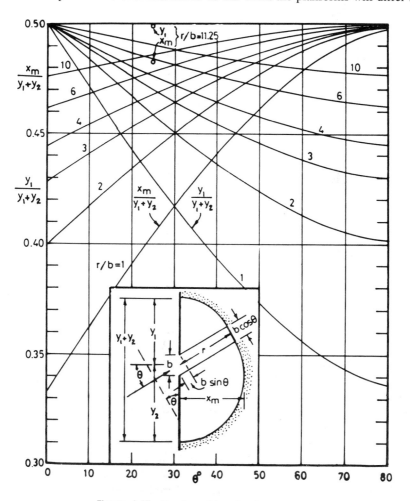

Figure 4.47 Bay formed by a breakwater gap

the parabolic shape previously derived and will be mainly arcs of circles. The beach segment at the end of the ribbon of waves not diffracting will tend to be straight. From there to the breakwater the waterlines will be arcuate shape with a constant radius. Should depths throughout the embayment not be uniform, refraction will cause a deviation in these areas.

Consider the situation as depicted in the inset of Fig. 4.47, where a gap of width b permits waves angled θ to the breakwater to enter across a band width of $b\cos\theta$. These proceed for a distance r to form a beach sensibly normal to these orthogonals. As seen in the figure the radius from one gap extremity is $r - (b\sin\theta)/2$ and the other $r+(b\sin\theta)/2$. Where the beaches reach the leeward side of the breakwater, the distances as measured from the center of the gap are

$$y_1 = r - (b\sin\theta)/2 + b/2 \text{ and } y_2 = r + (b\sin\theta)/2 + b/2. \qquad (4.8)$$

The total distance $y_1 + y_2 = 2r + b$ results with no reference to wave obliquity θ. However, the ratio of $y_1/(y_1 + y_2)$ varies with θ as seen in Figure 4.47. Also the maximum indentation ratio $x_m/(y_1 + y_2)$ varies with θ as also shown. For the case of normal wave approach $y_1 + y_2 = 2r + b$, which is the same as proposed by Dean (1978).

Inspection of Fig. 4.47 shows that as r/b increases (i.e., the gap becoming smaller) so the beach planform gets closer to a semicircle in which $y_1/(y_1+y_2) = 0.5$, particularly for θ approaching zero. For such normal approach of waves the same ratio applies for smaller r/b values, although the central part of the beach is relatively straight. The maximum indentation ratio $x_m/(y_1 + y_2)$ also approaches 0.5 as θ goes to 80° since the shoreline is again semicircular with $b\cos\theta$ very small. For $r/b = x_m/b = 1$ and normal wave approach ($\theta = 0°$) this ratio becomes 0.333. At $\theta = 30°$ the ratios $y_1/(y_1+y_2)$ and $x_m/(y_1+y_2)$ are identical, with the maximum at $r/b = 10$ of 0.488 and the minimum at $r/b = 1$ of 0.417.

An example of such a quadrant shaped beach is that at Wanboro Sound in Western Australia, depicted in Fig. 4.48. It is seen that this section of coast is fronted by limestone reefs rising to within a meter or two of the sea surface, where the tidal range is around 0.5 m. These cause wave breaking and hence virtually total dissipation, except where gaps occur as seen in the figure. The predominant swell waves arrive in deep water from the SW, which when refracted to the 15 m contour near the gaps arrive 25° S of W. The passages for penetration to the Sound are indicated by the orthogonal lines drawn, the largest opening being 450 m wide.

The bulk of the sound has an almost uniform depth of 18 m but, as the bed contours indicate, shoaling has occurred at the N and S limits adjacent to the reefs. Waves will therefore suffer diffraction alone before reaching the beach except at the N and S edges, where refraction causes a new orientation in the shorelines. Values of $y_1/(y_1 + y_2)$ and $x_m/(y_1 + y_2)$ have been computed by measuring orthogonal radii to the shore from the two appropriate reefs which are slightly offset from the N-S direction. These have been plotted in Fig. 4.47, which are applicable to $r/b = 11.25$. Whereas $y_1/(y_1+y_2)$ is close to the derived value, that for $x_m/(y_1+y_2)$ is slightly reduced, due probably to the refraction just inside the entrance.

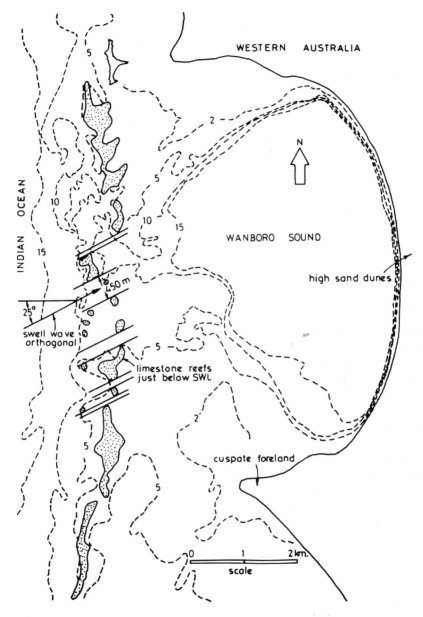

Figure 4.48   Almost circular shape of beach in Wanboro Sound, Western Australia

The fronting sand dunes rise some 30 m from the berm, due to the beaches not suffering much degradation during storms, nor any long-term erosion. The cuspate foreland has formed in the lee of the major reef at the southern extremity of the bay. Both swell and some storm waves can arrive at its apex from the WSW causing sand to be transported into the Sound, resulting in the limited accretion shown for about 1.8 km. This shoreline orientation is dictated by the diffracted and refracted waves arriving normal to it. Radii of equal length normal to the uniformly curved beach meet at a location about 1 km east of the southern extremity of the major gap in the reef. This is because of the influence of the small gaps south of this.

It will be shown in Section 6.2 that with offshore breakwaters closely spaced sufficient transverse motion of sediment takes place for the depths within the embayment not to be constant. This causes refraction to play an important role so that parabolic relationships as for crenulate shaped bays must be utilized. Thus, for the applicability of the circular beach form the bed contours throughout the system must be known. If they follow the beachline uniformly, as in Fig. 4.48, with constant depths beyond, diffraction is paramount, and hence arcs of circles result. Even with open bays it has been noted that in the very sheltered region the shoreline assumes this shape, due to offshore contours being circular.

4.3 PHYSIOGRAPHIC UNITS

As seen in Section 4.2.1, the orientation of bays is a clear indicator of the direction of net sediment movement along the coast. Another excellent source of such information is the sea and swell chart for individual ocean areas (*Atlas of Sea and Swell Charts*). It is the persistent swell, particularly that arriving soon after each storm, that dictates the direction that material placed into circulation in the offshore bar is moved while being brought onshore. Silvester (1962, 1968) examined hydrographic charts of the continental margins of the world to determine net movements from the orientation of crenulate shaped bays plus data from swell charts.

Silvester (1968) introduced a term *physiographic unit* for sections of coast where net movement was in the same direction along its length or there was no clear indication of such net transport. In either case there could be material moving in both directions at different times of the year, but in the first case a resultant was apparent while in the second this movement was virtually in balance. Adjacent physiographic units could either have a change in direction, particularly where the mainland changed direction significantly, or a change from net movement to no resultant.

Such vectors have existed over decades, centuries, or even millions of years due to the repetitive positions of storm zones in the oceans as presented in Fig. 2.9. From these cyclonic centers swell emanates and traverses either one or two oceans to reach a distant shore with practically no loss of energy, even though it is spread thinly in fans from the generating areas. The adjective *persistent* is most applicable to this type of wave. As noted in Chapter 3, the west coasts of continents receive swell incessantly from the continual cyclones traveling eastward in the 40° to 60° latitudes. Davies (1964) has shown that waves generated from SSW to NW in such Southern Hemisphere centers can arrive from the SW quadrant in latitudes from 20° N to 40° S through great circle propagation. Swell on east coasts is more variable, dependent as it is on the paths followed by storm centers as they cross the coast and travel out to sea.

One major advantage of using this bay direction indicator is that it is an integrator of wave energy over geologic time. Should tracer or other tests, which are spot measurements in time, indicate a longshore drift contrary to that indicated by the bay shape, a check should be made of the uniqueness or otherwise of the weather conditions at the time of the experiment. In this context it should be realized that littoral drift reaching an upcoast headland will be transmitted across the bay, with very little transported around the shadow zone, as shown in Fig. 4.49. This will, of course, depend on the protrusion of the headland.

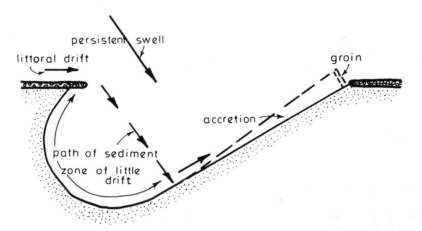

Figure 4.49   Transmission of sediment through a bay in dynamic equilibrium

Hughs (1957) and Baker (1961) have reported prototype tests from which Fig. 4.50 has been derived. By dumping large volumes of sand similar to but distinct in character from local material, they were able to trace its motion over many weeks. It was moved across large depths by the mass transport of the longer period swell that just did not suffer diffraction. It landed on the less curved segment of the parabolic periphery and then mainly downcoast to the tip of the bay. It could be accepted that some of this will be spread into the shadow zone, in order to maintain the uniform curvature of the parabolic shape. Thus, the caution noted above with respect to difference in direction from the overall net movement must take account of position in a bay periphery. There is also the possibility of littoral current being generated into the shadow zone by differential breaker height as waves are diffracted (O'Rourke and LeBlond 1972; Gourlay 1974, 1981).

The net movement previously discussed does not infer similarities in wave climate, geology, or geography throughout the physiographic unit, not even rates of transport and type of sediment. If sediment is available from rivers it will be moved along the coast, both in the surf zone and offshore, in the direction shown, or be accumulated in zones of no net movement. In some areas where a rocky shoreline exists over long distances, or lengthy expanses of sandy coast having no headlands for bay formation, other criteria may have to be used to judge direction of net movement. These include (a) isolated bays, (b) contour spreading, (c) shoals, and (d) end conditions.

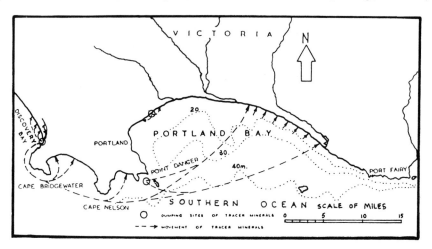

Figure 4.50 Portland harbor in Victoria, Australia, where tracer tests showed sediment bypassing of bay

4.3.1 Isolated Bays

In the event of a coast consisting of cliffs over a long length of coastal margin, bays may exist at both ends which should be oriented in the same direction. Since net movement is the same at these extremities it can be accepted it is similar along the rugged coast, even though no beaches exist. Transport is effected in deeper water due to the reflection of the oblique swell, by the generation of a short-crested wave system as will be discussed in Chapters 5 and 7. The orientation of such bays could not be opposite in direction on a relatively straight section of coast because the wave climate over the length of coast could be considered the same. For example, the bays at each end could not be directed toward the center of the cliffed coast because a large shoal should exist there with the possibility of a sedimentary shoreline. In this context it should be remembered that many thousands of years are implied for this net movement and hence large volumes of accretion would have taken place.

In the case of a rocky shoreline changing direction, as illustrated in Fig. 4.51A, the bays at either extremity could be oriented in opposite directions. The persistent swell could be moving material away from the rocky coast. Such a promontory will have its cliffs run into very large depths, as illustrated in Fig. 4.51B, because any sediment that may be fed to this coast is swiftly removed elsewhere. On the other hand, if such a promontory has sediment transported toward it, as exhibited in Fig. 4.51C, a large submerged shoal will exist, probably accompanied by a wide expanse of sedimentary shoreline.

4.3.2 Contour Spreading

Successive surveys of offshore areas made at long intervals of time may exhibit long-term stability, indicating that sediment is passing along it at a steady rate. Material

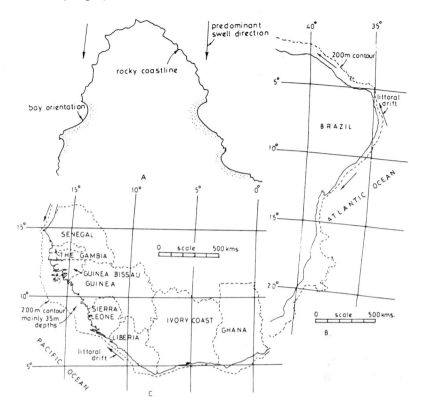

Figure 4.51   Long-term features (A) for a protruding coast, (B) for the eastern tip of South America, and (C) for the western coast of Africa

passing through or across one bay will do likewise in the adjacent downcoast bay. If the overall rate differs, then accretion or denudation is occurring somewhere along the physiographic unit. Although the overall rate is constant this does not mean that rate per unit width normal to the coast is constant, because active zones differ across the profile with the energy supply. The changing distances of the 20 fathom contour in Fig. 4.10 suggest variable needs of waves in transporting sediment at a constant rate. This does not imply that all wave activity is confined to this depth limit.

As sediment is fed to the coast at intervals along the length of a coastline, where net movement is in the same direction, more material must be transported by a given wave energy input if no accretion is to occur (see Fig. 4.52A). Thus, for a uniform longshore distribution of wave energy the active zone must widen, or shoaling occur, in order to cope with this enlarged load. Similarly, for a reduction in wave energy from a storm source, as in Fig. 4.52B, contours must widen downcoast. This general spreading of contours, even to the edge of the continental shelf, should not be confused with the more local changes described above, which are more dependent on wave energy variations closer inshore.

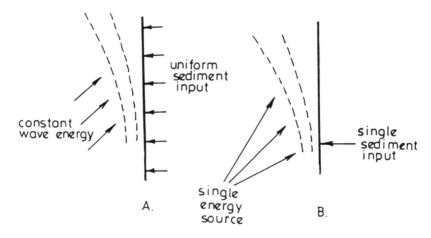

Figure 4.52 Contour spreading due to (A) increasing sediment input with constant wave energy, and (B) decreasing wave energy with constant sediment input

Another factor in this milder slope of seabed downcoast is the change in sediment size alongshore. Finer material travels faster and is located farther downcoast from a river source. The continuous abrasion of particles in the transporting process also aids in this size distribution variation. Wave energy decreases at greater distances from the storm source and, hence, for a single load of sediment, contours will spread downcoast, giving waves a better transfer of energy to the seabed.

4.3.3 Shoals

Just downcoast of a headland, where reflected waves can expedite transport, sediment comes under the influence of incident waves only. Not until the depths are shoaled can these waves build up sufficient oscillatory movement at the bed and transport it by mass transport. Finer particles are moved first, so leaving the coarser fraction just downcoast from the headland or rocky expanse. However, large volumes are transmitted across reasonable depths if given sufficient geologic time to do so, as exemplified in Fig. 4.50.

A rocky coastline is likely to have similarly rugged islands offshore from it. If these are within depths of disturbance of swell waves, a shoal will occur in their lee. The orientation of these bed contours will indicate the direction of this persistent swell and hence of the general net movement along the coast. Should waves be sensibly normal to the coast, accretion could form a salient or cuspate foreland from the adjacent mainland, as will be discussed in Chapter 6.

Where there has been or still is a surfeit of sand to be transported alongshore, spits are formed known as barrier beaches. These are of vastly differing dimensions, springing from swift changes in mainland direction or across river mouths. As has been discussed, this type of accretion is aided greatly by the pulsative nature of littoral drift. This particular physiographic feature will be discussed in Section 4.4. However, the direction in which such spits point is that of net sediment movement.

At inlet or river mouths, where ebb tides and freshwater discharges occur, littoral drift arriving at them is deflected seaward in an arcuate form. These intercept drift which decreases in magnitude offshore, so causing a decrease in their width seaward. Recognition of such shoals is a clear indication of net sediment movement. These features are also detailed further in Section 4.4 on barrier beaches.

4.3.4 End Conditions

Another general indicator of net sediment movement along a physiographic unit is that the downcoast region must contain an abundance of sediment, while the upcoast zone suffers a dearth of it. This accumulation will be reflected in a large expanse of sedimentary shoreline or a wide continental shelf, or both. In the upcoast region the lack of marine material results in a narrow shelf and bays indented to their full static equilibrium shape.

The accretion being considered here has taken place over thousands or even millions of years. If longshore drift has continued at the same rate as it is today, and there are indications it has been much exceeded in the geologic past, vast volumes of material have been accumulated in a horizontal manner as accreting shorelines. Figure 4.53 shows the areas and thickness of sediment that could result from a modest littoral drift of 100,000 m^3 per annum. For a duration of 1 m years, which is small geologically speaking, an area of 500 km^2 of 200 m thickness could be accumulated (shown dotted in the figure). It is little wonder, therefore, that the bulk of land masses are sedimentary and most of this deposited under marine conditions. This also implies that much of this area has been constructed at sea level or just above (Silvester 1974). Where this sediment is sand

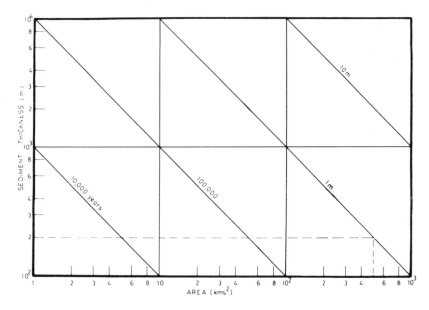

Figure 4.53 Area and thickness of sediment accreted by a constant littoral drift of 100,000 m^3 per annum

it can be blown into dunes which can stand many tens of meters above MSL. It can also be blown onto rocky rises and reach higher than normal levels due to large tidal ranges, some of which could have exceeded, by an order of magnitude, the largest being experienced today, as large embayments filled out to present shorelines.

The weathering of rock and its transport to the sea by gravity via the river systems obeys all the laws of Nature. The final distribution of this load to some trap or sump by waves is also a natural process, since the energy required is provided by the wind systems of the world. These forces have been available from when the oceans were first formed and weathering commenced. This duration has been stated by Stoddart (1969) to be in the order of 2000 m years. Initially boulders and gravel would have been accreted on the coasts, but as rivers became longer and weathering continued, sand was the order of the day. With many rivers now being extremely long and flatter in slope, only silt material is delivered by them.

The same process that provided the original sea floor when the earth's crust cooled would have also provided the zones that remained above sea level. The former have been protected from erosive forces by the oceans themselves, while the latter have suffered all the weathering forces of cold, heat, rain, and run-off. The precipitous valleys still exhibited on some coastal margins, termed *submarine canyons*, indicate the probable ruggedness of the initial land areas. But most of these steep submarine margins have now been buried by sediment, in the form of continental shelves. Where filling is incomplete, due to a lack of sediment, the precipitous marine valleys still exist, even though the outer trenches have been partially filled with vast volumes of material that have slid down the upcoast edges of these features.

This implies that the majority of continental shelves are sedimentary, which is confirmed in the statistics of Table 4.4 provided by Hayes (1967), who subdivided shelves into various climatic categories, but in all these the percentage of sedimentary material never fell below 74%. Since coral reefs cannot exist unless supported by pile action within a sedimentary base (Silvester 1965), the 6.4% in Table 4.4 should be added to the left-hand column, making an average of 86.8% sedimentary.

TABLE 4.4 BOTTOM SEDIMENT TYPE OF THE INNER CONTINENTAL SHELF (HAYES 1967)

sand	43.5	rocky	13.2
mud	28.0	coral	6.4
shell	4.8		
gravel	4.1		
Total:	80.4%	Total:	19.6%

While gravel, sand, and mud are swept along the coast by wave action they are sorted, with the finer sand fractions being moved to the outer edge of the shelf. This limit is now recognized as at 130 m depth or the limit of disturbance of the most predominant swell waves in any ocean basin. As has been noted in Chapter 2, higher wind velocities with commensurate fetches and durations generate waves in the peak of the spectrum with periods of 13 seconds (as discussed in Section 2.2.3). These can produce orbital motions at the bed to half their deepwater wave length, which for these waves is $L_o/2 = gT^2/4\pi = 132$ m. This correlates with the edge of the oceanic continental shelves, beyond which is the continental rise, where sediment collapses or falls below the reach of most ocean waves. Such limits of disturbance and hence accretion are smaller for enclosed seas, where wave periods are less due to fetch limitations.

4.3.5 Continental Patterns

By observing crenulate shaped bays and swell charts, Silvester (1962, 1966, 1968) examined coastlines of major continents and islands, in order to determine net sediment movement along their shores. Although the hydrographic charts used could distinguish between rocky and sedimentary segments, or inferred from their smoothness of outline, a check of coastal landforms was available from McGill (1958). This provided general geographical information about coastal plains as a whole, whereas interest in this case was concentrated on rock outcrops at the waterline, or mobile material being transported along the coasts.

Information is becoming available on the nature of continental shelves in all areas. However, a ready assessment of sediment accretion to form or fill out the shelf is provided by the smoothness of the bed contours. If these are indented then an undulatory surface is indicated, which is unlikely to be produced if sand or silt is spread by wave action. This criterion is more applicable to deeper zones since regions nearer shore are more affected by rivers and islands in the deposition of material.

The maps so derived are presented in Figs. 4.54 to 4.59. These show arrows where net drift is evident and plus (+) signs where no clear signs of such drift is apparent. Also indicated are continental shelves to a depth of 90 m, where the majority of wave action takes place. The picture so painted is simplified drastically and hence some restrictions must be observed when reading these figures:

1. Rate of sediment movement is not specified. The arrows indicate only that if material is available it will move in the direction shown.
2. Net-longshore transport is inferred. Transient reversals of drift, due to storms or secondary swells, may occur spasmodically or periodically.
3. Where drift is substantially retarded, or equal in both directions, plus (+) signs are used. This does not necessarily presume a progradation of beaches as sediment may not be available for this.
4. The general coastline only is considered. Sediment movement within or behind minor physiographic features, such as islands or promontories, is excluded. However, on all coasts facing a major ocean the drift is as shown.
5. The stippled zones indicating sediment do not differentiate between shingle, sand, or silt, for which the reader is referred to McGill (1958).
6. The cross-hatched zones indicating rocky shoreline apply strictly to the waterline and not to the coastal plain in general, to which McGill (1958) is again directed.

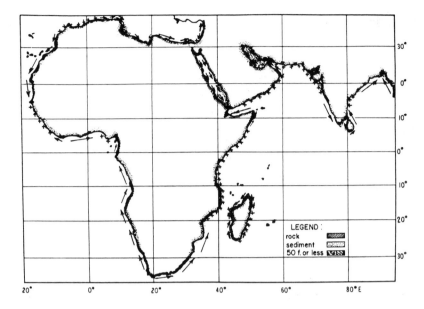

Figure 4.54 Net sediment movement around the African continent

It does not seem worthwhile to discuss these continents in detail as has been done by Silvester (1966, 1974). However, it is noteworthy that net sediment movement is indicated in the Arctic regions as in Figs. 4.57 and 4.58. These water areas will be covered with ice for large proportions of the year with no wave action. But there will be periods when large waves are generated by strong winds over limited fetches.

The senior author in his teaching program of coastal engineering has set assignments where students are required to study the geology, climate, and vegetation to derive sediment sources, and then derive net movement of sediment along the coast from knowledge of storm waves, swell, tides, and possibilities of storm surge. Such islands useful for this educational process are: Sri Lanka, Sumatra, Madagascar, Java, Tasmania, New Guinea, New Zealand (North or South Island), Luzon (Philippines), Taiwan, and Borneo.

There are many others with reasonable river systems, so long as data are available on wave climate. Silvester (1966, 1974) has presented a study of the islands of Japan in this regard. Such exercises could aid in the assessment of long-term accretion or erosion which helps coastal engineers take a macroscopic view of the problems encountered on the coast. Specific features such as bays, barrier beaches, coastal lagoons, deltas, and other features can be detailed.

The net drift around continents as in Figs. 4.54 to 4.59 can be combined into a world pattern presented in Fig. 4.60. Similar criteria apply to this figure as for others of specific continents. Continental shelves to 183 m are hatched as well as wind systems for January and July, representative of winter and summer conditions in either hemisphere. These should be read in conjunction with Figs. 2.7 and 2.9 showing wind systems and storm patterns for the various oceans.

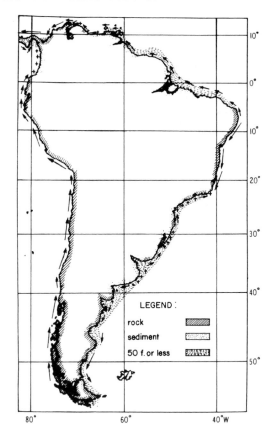

LEGEND :

rock

sediment

50 f. or less

Figure 4.55 Net sediment movement around the South American continent

From Fig. 4.60 it can be seen that on continental margins where winds blow predominantly in the same direction throughout the year, a net drift extends for complete lengths of the landmasses. For example, on the west coast of South America (Fig. 4.55) the drift is toward the equator, as is so on the southwest coast of Africa (Fig. 4.54), and also the western coast of Australia (Fig. 4.59). This equator-ward transport has been confirmed by Davies (1964) as due to the persistent swell from cyclones passing from west to east in the 40° to 60° latitudes. On many of these western margins few large rivers debouch so that continental shelves are narrow or nonexistent. On eastern coasts, where the major sediment load is discharged, the longshore drift is dictated more by local storms, either as tropical cyclones traveling from the equator poleward with a western tendency, or extra-tropical cyclones traveling eastward away from the coastal margins. Major indentations of the coastline also influence the local net movement (see Fig. 4.60).

4.4 BARRIER BEACHES

Barrier beaches are accumulations of sand of limited width, generally comprising sand dunes, which stretch lengthwise along the coast, either to partially or fully enclose bodies

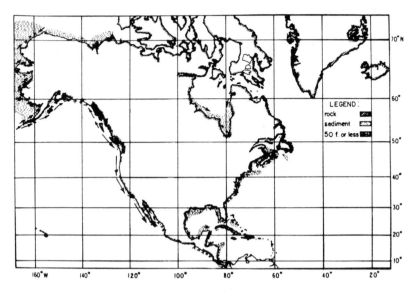

Figure 4.56 Net sediment movement around the North American continent

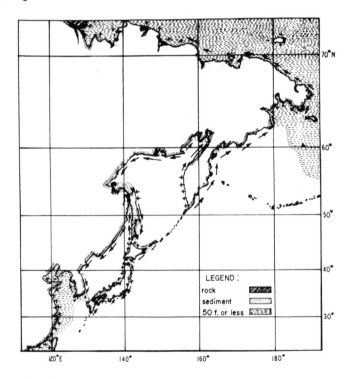

Figure 4.57 Net sediment movement around the North West Pacific coasts

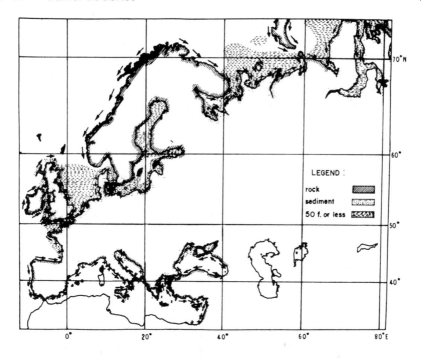

Figure 4.58 Net sediment movement around the European and Arctic coasts

of water between them and the mainland. They may exist as elongated spits springing from a sharp change in shoreline direction, as exemplified in Figs. 4.61 to 4.63. They may also extend parallel to the coast and in places reattach to the mainland, thus enclosing lagoons, as depicted in Figs. 4.64 to 4.66. These physiographic features are found along most shorelines, be they along margins of oceans, or enclosed seas of any size. The main requirement for their formation is plentiful supply of sediment which is transported by obliquely arriving swell or predominant locally generated waves.

The age of these spits is generally less than 6,000 years and hence they have been mainly constructed during a still stand of sea level (Fairbridge 1961). For example Roy and Stephens (1980) in referring to barrier formation comment: "Considerable radiocarbon age data on shelly beach and nearshore facies in prograded beach ridge systems ... show that most Holocene Barriers accreted between 6000 and 3000 years ago, with a few continuing to grow up to 1000 years ago. Over this time span, sea level on the tectonically stable southeast Australian margin has remained fairly constant (+1 m)." Suggestions for their origin need not therefore invoke eustatic changes. Since locations and orientations of spits are associated with specific wave climates, it is this energy source that must seriously be considered as the dominant constructional mechanism. This implies that wave directions have not changed along continental margins over many thousands of years.

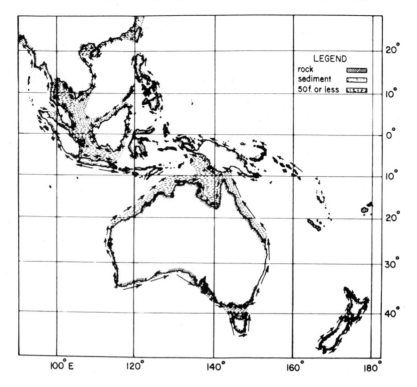

Figure 4.59   Net sediment movement around the Australasian coasts

The global temperature distribution from the sun could not have changed much from the time that water initially spread over 70% of the earth to form the oceans. Certainly, land masses have enlarged as weathered material from high mountains has dispersed over river plains and finally along the coast. However, the repetitive wind systems have generated storm waves in specific regions which have then arrived as swell to distant shores, with practically no loss of energy.

Even within enclosed seas, such as the Mediterranean, and the Baltic, the annual storm centers follow distinctive paths, to generate waves with persistent obliquity to these coasts. This constancy of conditions is supported by Meyerhoff (1970) through other criteria: "Axisymmetry of cool and evaporate belts about the present rotation axis, the coincidence through time of the earth's two horse latitude belts, and the occurrence of 95 percent of the world's evaporates—by volume and by area in areas now receiving less than 100 cm of rainfall—show that the planetary wind and ocean current pattern has been essentially the same for 800–1000 million years." If these winds have been so repetitive, so have been the waves generated therefrom.

The sediment supply to the coasts by means of rivers has changed naturally over geologic time. As barrier beaches have formed they enclose portions of coast so that rivers are now flowing into lagoons, thus cutting off their supply of material to the

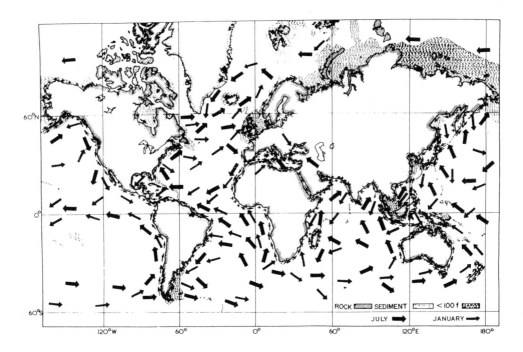

Figure 4.60   Net sediment movement around the coastlines of the world showing the global wind patterns

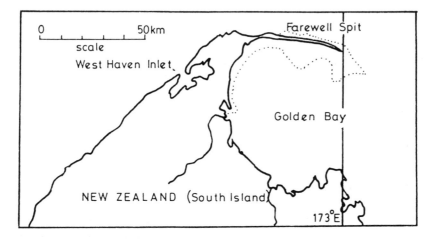

Figure 4.61   Barrier beach on south island of New Zealand

outer shoreline. Another element that has entered the picture in the last 200 years is man's ability to change flow conditions in waterways. By harnessing rivers, with the construction of dams for various reasons, he has intercepted large volumes of sediment

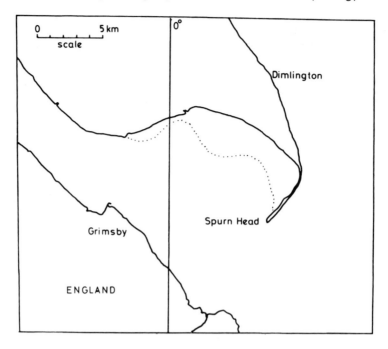

Figure 4.62   Barrier beach on east coast of England

that would otherwise be distributed along the coast. Tsuchiya (1982) has observed: "Beach erosion has advanced actively in many countries of the world, especially in Japan (Tsuchiya 1980) due to decrease in sediment input from rivers and the construction of coastal structures."

Due to the above effects, natural and man-induced, barrier beaches that have depended for their existence on a constant or increasing supply of sand are now suffering denudation. An additional influence on littoral drift is the construction of coastal barriers that intercept its path. As noted previously, dredged access channels across the shelf are probably more effective in this regard. All these changes take time to exert their effects, but the general coastal erosion being suffered today is a fact of life which must be engineered.

4.4.1 Historical Review

Barrier beaches have been the purview of geologists in the past because they are a disproportionate part of the geomorphology of the coast. A plethora of authors has discussed this feature, but even today there is no consensus on their origin. Coates (1973), as Proceedings editor of a symposium on the subject, has commented: "The reader may wonder why one type of coastal landform has been singled out for special treatment, since an entire section of the book is devoted to the topic of barrier islands. The reason is threefold: (l) they are world wide in extent and constitute one of the

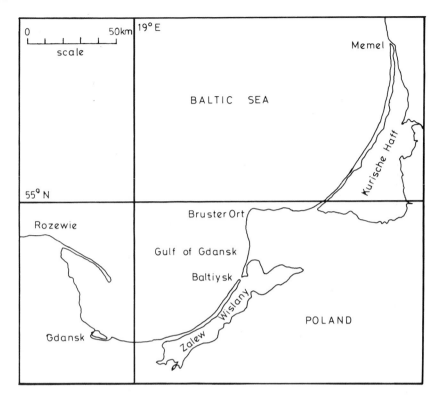

Figure 4.63   Barrier beaches on the coast of Poland

major landforms of today's coastlines, (2) their origin and developmental history has been argued more than 125 yr., with increasing amount of research during the past 12 yr., (3) man's utilization of these areas is greatly expanding, as they are favored sites for buildings and recreation."

Theories on the formation of barrier beaches have revolved mainly around the source of sediment. De Beaumont (1845) suggested upward building of offshore bars, which idea was strongly supported by Johnson (1919). However, Gilbert (1885) proposed that material was transported alongshore rather than from offshore. This was criticized by Johnson (1919) due to his criterion of the offshore slope having to match that of the water mass enclosed by the barrier beach. This is not necessarily so if subsequent accretion takes place within the partially or fully enclosed bay or lagoon. However, the eminence of Johnson as a geologist thwarted critical review for decades until Fisher (1968).

Price (1963) has noted small barriers forming offshore during periods of high water associated with storms; these were probably exposed as the storm surge receded. Zenkovich (1962) promoted the concepts of a sinking wave-built terrace or submergence of an alluvial plane. Leontyev and Nikiforov (1966) have suggested that barriers form offshore bars during lowering of sea level. They also thought that local uplift has pushed

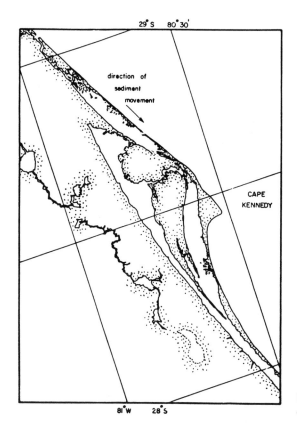

29° S   80° 30'

direction of
sediment
movement

CAPE
KENNEDY

81° W   28° S

Figure 4.64 Barrier beaches on the east coast of the United States

bars above the surface. Neither of these seem tenable because offshore bars are very temporary features (Silvester 1974).

Hoyt (1967) was critical of previous explanations: "Formation of barrier islands from emergent bars is also rejected, because evidence from many areas of the world does not support a sea level higher than present during the Holocene. Also unacceptable is the hypothesis of continuous barrier development throughout the Holocene submergence because it does not explain the original formation. Barrier islands which form from barrier spits or, in some instances, from bars are accepted, but these methods are not regarded as the general mechanism of barrier island formation." He then proposed his own hypothesis of submergence subsequent to ridges' being constructed by water and aeolian sources. This would have to be proved by dates of such features and data on sealevel rise over the period.

Fisher (1968) disputed Hoyt's (1967) theory while proposing that complex spits were first formed which then became barrier beaches by submergence. He acknowledged that inlets between islands are of more recent origin and not areas of low elevation, as suggested by Hoyt's submergence theory. He supported the contention by Gilbert (1885)

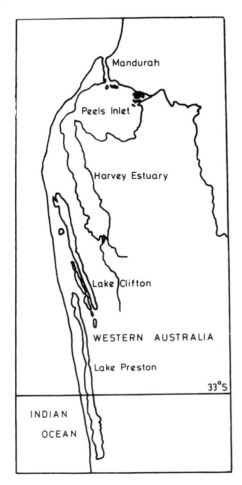

Figure 4.65   Barrier beaches and lagoons on the Western Australian coast

that spit formation by horizontal transport is an essential element in barrier formation. In conclusion he stated: "Thus from the author's point of view, three problems are yet to remain in the study of barrier chain shorelines. First, are they definitely features of submergence? Second, are they primary features (complex spits) or second-cycle features (drowned subaerial dunes or emerged offshore bars)? Third, is there genetically, as well as descriptively, more than one type of barrier island shoreline?".

Kwon (1969), in a detailed study of the Gulf of Mexico, concluded that: "In all cases, littoral drift is the primary factor in transporting barrier-forming sands from the source and in initiating or sustaining barrier island development along the northern Gulf coast. No genetic distinction can be made between spit and barrier island growth." Otvos (1970) also examined this location extensively and stated: "During the past period of stabilized sea level, shore and barrier island progradation started at several locations." He also commented as follows: "A bay-mouth bar or spit may develop fast enough so as not to allow time for the deposition of noticeable volumes of open marine sediments before

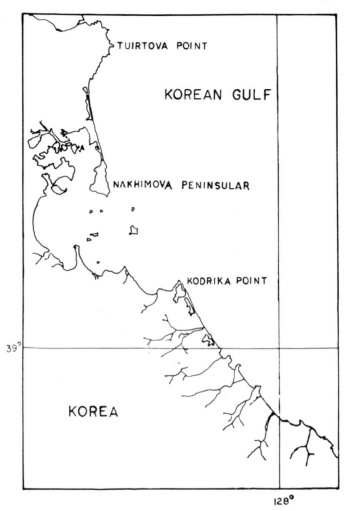

Figure 4.66   Barrier beaches on the coast of Korea

sedimentation becomes lagoonal." He gives figures for the growth of islands downcoast which range from 20 to 400 m/year. Perhaps if measurements had been made at more frequent intervals, such rates may have occurred over much less time than 12 months.

Melville (1984), from many borings made over a number of years on the New South Wales coast of Australia, has constructed a detailed sequence of barrier beach buildup over the past 7,000 years. His cross section is presented in Fig. 4.67. He stressed the importance of headlands and offshore islands in locating such beaches along the coast, together with bay formation. This is the major difference between the eastern coasts of Australia and the United States where, along the latter, barrier chains run for hundreds of kilometers without the presence of fixed features.

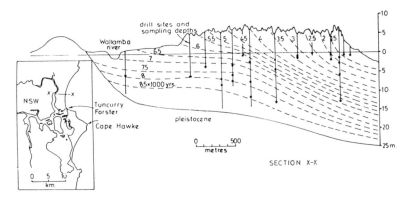

Figure 4.67   Beach buildup on a section of the New South Wales coast, Australia

4.4.2 Mode of Spit Construction

It has already been noted that the maintenance of a barrier beach is dependent on the continual supply of sediment from upcoast. This applies to any shoreline where oblique waves are transporting material along it. Any fluctuation in supply is likely to cause either siltation or erosion. However, the added factor in the formation and elongation of barrier beaches in the form of sand spits is the pulsational nature of littoral drift, as stressed previously. It has been accepted in the past, by geologists and engineers alike, that this drift is uniform throughout the year, virtually "a river of sand" (Encyclopaedia Britannica Educational Films 1965) considered even more uniform than transport down a river bed. It is, in fact, the reverse of this, being a significantly fluctuating variable. Tanner (1987) concludes: "The 'river of sand' idea may look like a good teaching model, for reasons of simplicity, but it does not represent very much of reality, and it does not advance our understanding of coastal sediment transport processes."

As discussed in Chapter 3 a pulse of littoral drift occurs after each storm sequence, while the offshore bar material is being returned to the beach downcoast. Refraction at the tip of the spit can create almost a reversal of the littoral current, as seen in Fig. 4.68. This causes deposition in this region and extension of the spit while a surfeit of sand is being supplied, or during the bar removal process upcoast from the spit extremity. Hine (1979) has concluded: "Berm-ridges develop along the strongly recurved portion of an active spit and represent the most rapid form of beach progradation."

During a long period of swell action, when littoral drift is minimal (as noted in Chapter 3), the end of the spit may be eroded or deflected to the lee side of the tip. This sequence of stagnation and elongation is depicted in Fig. 4.69. If some years are experienced with little or no storm action, a spit may be so deflected landward, resulting in it making contact with the mainland, so enclosing a lagoon. At the next storm sequence a new spit or barrier beach may form enclosing in time another lagoon parallel to the coast, as illustrated in Figs. 4.64 to 4.66.

It can be surmised that the above mode of barrier beach construction does not need to invoke eustatic or other changes in sea level. Such processes are taking place today

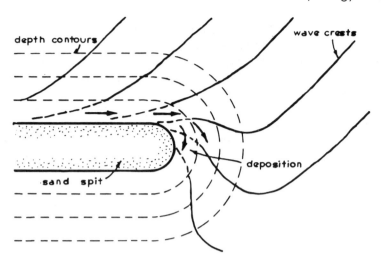

Figure 4.68 Refraction taking place at the end of a spit

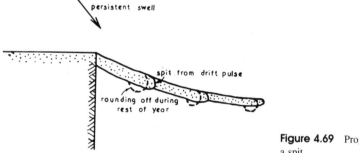

Figure 4.69 Progressive development of a spit

in seas with zero tidal range. As commented by Roy and Stephens (1980): "Thus it is reasonable to assume that barrier building occurred under marine conditions broadly similar to those operating today." Spits have been filling very deep zones as well as shallow segments of coast, despite wave climates varying from open ocean to enclosed seas. The only requirements are that there be an adequate supply of sand and that intermittent storms place it into circulation in the form of a bar, and then be taken onshore and alongshore by the milder or swell type waves. Future data gathering and analyses of these features could well take this suggested construction mechanism into account.

4.4.3 Erosive Conditions

Barrier beaches vary from a few meters to some kilometers in width, even within the length of a single feature. These dimensions are dictated by the speed of sediment addition and the duration of swell waves with little or no littoral drift, as already discussed. But the present paucity of sand being fed to the coast storm waves, perhaps accompanied by high water levels due to storm surge, can overtop the narrow sections and thus break through an opening. This forms a waterway from the ocean to the lagoon or body of water landward of the barrier beach.

Such inlets, or more correctly outlets, are very unstable in position since sand is deposited on the upcoast side in pulses, while the downcoast region is temporally devoid of material and hence erodes. Because of the longer ebb flow through these openings, added to by runoff from the hinterland, the accretion of littoral drift is generally in the form of an arc running some distance out to sea, as depicted in Fig. 4.70. Depending on the strength and duration of the swell, these arcuate shaped submerged bars can accumulate large volumes of material on their updrift side (Smith and Laracy 1985).

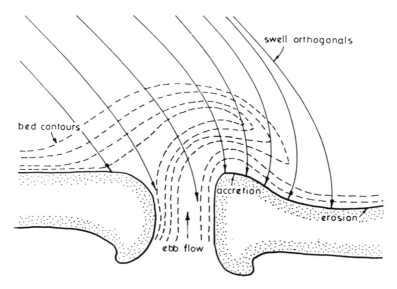

Figure 4.70   Arcuate shaped shoal formed at coastal inlet

Erosion downcoast of an inlet may be a natural process when an inlet is formed during a storm accompanied by storm surge, but it may also be caused by man when dredging such an opening. Such a case was reported by Vallianos (1970) where erosion was caused some 6 km downcoast of it. Hydrographic surveys showed accretion in the offshore shoal of 2,500,000 m^3 over 17 years. This requires some expensive remedial measures.

This feature also creates a new refraction pattern for waves, as indicated in Fig. 4.70 where accretion may occur near the downcoast side of the inlet but erosion takes place further along the shoreline (Douglas 1987). It has also been concluded: "The refraction pattern of waves in the immediate neighbourhood of the inlet as well as the effect of jetties on that pattern should carefully be studied. Refracted waves may transport sand locally in a direction opposite to that of littoral transport along a nearby straight beach"

(Partheniades and Purpura 1972). Confirmation of this has also been reported by Hayes et al. (1970) and Morton (1977). When a storm arrives, producing waves from a large fan of directions (Silvester 1974), the shoal is disturbed greatly and may be transmitted downcoast or upcoast with respect to the persistent swell direction. The arcuate bar is then reformed by the following littoral pulse and the strong outward current.

Sometimes these arcuate shaped shoals may accrete above sea level, to form elongated spits. This causes a swift change in direction of outlet flow which continually erodes the landward edge at this point. As large changes are accompanied by storms, the seaward face is also attacked resulting in a new breakthrough. The island so formed then proceeds to be worn away and transmitted downcoast. This has been observed by Sexton and Hayes (1982): "Channel formation and abandonment facilitate bar-bypassing at the inlet, as opposed to continuous channel migration."

Barrier beaches, with their surfeit of sand and dune vegetation, are ideal sites for holiday resorts and public recreation. However, they have been formed during the recent geologic age when sediment was abundantly available from rivers. Today, or since the last 100 years or so, an extremely short time span in geologic terms, this sediment source has diminished drastically from man's entry to the coast. Once supply has been reduced, or perhaps cut off completely, erosion commences immediately on the shoreline just downcoast of the source. This condition progresses along the physiographic unit where net littoral drift is in one direction. As noted in Section 4.3, such units extend for hundreds or even thousands of kilometers, which is just as applicable to barrier beaches as to mainland margins. Means for stabilizing all these eroding beaches will be discussed in Chapter 6.

REFERENCES

Atlas of Sea and Swell Charts: South Atlantic, Publ. 799B, 1948; Western Pacific, Publ.799CE, 1963; North Eastern Pacific, Publ. 2990, 1963; U. S. Hydrographic Office, Washington, D.C.

BAKER, G. 1961. Portland's all weather harbour. *The Aust. Scientist 1*: 107–13.

BEREK, E. P., and DEAN, R. G. 1982. Field investigation of longshore transport distribution. *Proc. 18th Inter. Conf. Coastal Eng., ASCE 2*: 1620–39.

BIRD, E. C. F. 1984. *Coasts: An Introduction to Coastal Geomorphology.* 3rd ed. Canberra: Australian National Univ. Press.

———. 1985. *Coastal Changes: A Global Review.* Chichester: John Wiley & Sons.

CARTER, R. W. G. 1988. *Coastal Environments: An Introduction to the Physical, Ecological and Cultural Systems of Coastlines.* London: Academic Press.

COATES, D. R., ed. 1973. *Coastal Geomorphology.* London: George Allen and Unwin.

DAVIES, J. L. 1958. Wave refraction and the evolution of shoreline curves. *Geogr. Stud. 5*: 1–14.

———. 1964. A morphogenetic approach to world shorelines. *Zeits. für Geomorph. 8*, Sup. No. 127*–142*.

———. 1973. *Geographical Variation in Coastal Development.* New York: Hafner.

DAVIS, R. A., ed. 1985. *Coastal Sedimentary Environments*. 2nd rev. expanded ed. New York: Springer-Verlag.

DEAN, R. G. 1978. Diffraction calculation of shoreline planforms. *Proc. 16th Inter. Conf. Coastal Eng., ASCE 2*: 1903–17.

DEAN, R. G., and MAURMEYER, E. M. 1977. Predictability of characteristics of two embayments. *Proc. Coastal Sediments '77, ASCE* 848–66.

DE BEAUMONT, E. 1845. *Lesons de geologic practique*. ed. P. Bertrand. Paris, 223–52.

DOUGLAS, S. L. 1987. Coastal response to jetties at Murrels Inlet, South Carolina. *Shore and Beach* 55(4): 21–32.

DYER, K. R. 1986. *Coastal and Estuarine Sediment Dynamics*. Chichester: John Wiley & Sons.

Encyclopaedia Britannica Educational Films. 1965. *Beach: A River of Sand*, color, sound, 20 min.

EVERTS, C. H. 1983. Shoreline changes downdrift of a littoral barrier. *Proc. Coastal Structures '83, ASCE* 673–89.

FAIRBRIDGE, R. W. 1961. Eustatic changes in sea level. *Physics and Chemistry of the Earth 4*: 99–185.

FINKELSTEIN, K. 1982. Morphological variations and sediment transport in crenulate-bay beaches, Kodiak Island, Alaska. *Marine Geology 47*: 261–81.

FISHER, J. J. 1968. Barrier island formation: discussion. *Geol. Soc. of Amer., Bull.* 79: 1412–26.

GILBERT, G. K. 1885. The topographic features of lake shores. *U.S. Geol. Surveys*, 5th Ann. Rep., 69–123.

GOURLAY, M. R. 1974. Wave set-up and wave generated currents in the lee of a breakwater or headland. *Proc. 14th Inter. Conf. Coastal Eng., ASCE 3*: 1976–87.

———. 1981. Beach processes in the vicinity of offshore breakwaters. *Proc. 5th Aust. Conf. Coastal and Ocean Eng.*, 129–34.

GRESWELL, R. K. 1957. *The Physical Geography of Beaches and Coastlines*. London: Hulton Ed. Publ.

GUILCHER, A. 1974. Studies in coastal geomorpholoy contributing to coastal engineering. *Proc. 14th Inter. Conf. Coastal Eng., ASCE 1*: 1–19.

HAILS, J.R., ed. 1977. *Applied Geomorphology*. Amsterdam: Elsevier.

HALLIGAN, G. H. 1906. Sand movement on the New South Wales coast. *Proc. Limn. Soc. N.S.W. 31*: 619–40.

HAYES, M. D. 1967. Relationships between coastal climate and bottom sediment type on the inner continental shelf. *Mar. Geology 5*: 111–32.

HAYES, M. D., GOLDSMITH, V., and HOBBS, C. H. 1970. Offset coastal inlets. *Proc. 12th Inter. Conf. Coastal Eng., ASCE 2*: 1187–1200.

HINE, A. C. 1979. Mechanisms of berm development and resulting beach growth along a barrier spit complex. *Sedimentology 26*: 333–51.

HO, S. K. 1971. Crenulate shaped bays. Asian Inst. Tech., *Master Eng. thesis*, No. 346.

HOYT, J. H. 1967. Barrier island formation. *Geol. Soc. Ann. Bull.* 78: 1125–36.

HSU, J. R. C., and EVANS, C. 1989. Parabolic bay shapes and applications. *Proc. Instn. Civil Engrs.* 87: 557–70.

HSU, J. R. C., SILVESTER, R., and XIA, Y. M. 1987. New characteristics of equilibrium shaped bays. *Proc. 8th Aust. Conf. Coastal and Ocean Eng.*, 140–44.

————. 1989a. Generalities on static equilibrium bays. *Coastal Eng. 12*: 353–69.

————. 1989b. Static equilibrium bays: new relationships. *J. Waterway, Port, Coastal and Ocean Eng., ASCE 115*(3): 285–98.

HUGHS, E. P. C. 1957. The investigation and design for Portland Harbour, Victoria. *J. Instn. Engrs. Aust. 29*: 55–68.

Institution of Civil Engineers. 1985. *Coastal Engineering Research.* Maritime Group, UK, Thomas Telford, London.

JENNINGS, J. N. 1955. The influence of wave action on coastal outline in plan. *Aust. Geogr. 6*: 36–44.

JOHNSON, D. W. 1919. *Shore Processes and Shoreline Development.* New York: John Wiley and Sons.

KANA, T. W., AL-SARAWI, M., and HOLLAND, M. 1986. Design and performance of artificial beaches for Kuwait waterfront project. *Proc. 20th Inter. Conf. Coastal Eng., ASCE 3*: 2545–58.

KING, C. A. M. 1972. *Beaches and Coasts.* 2nd ed. London: Edward Arnold.

KOMAR, P. D. 1976. *Beach Processes and Sedimentation.* Englewood Cliffs, N.J.: Prentice Hall.

KRUMBEIN, W. C. 1944. Shore processes and beach characteristics. U.S. Corps of Engrs., *Beach Erosion Board,* Tech. Mem. No. 3.

KWON, H. J. 1969. Barrier island of the northern gulf of Mexico coast: sediment source and development. Coastal Studies Inst., *Louisiana State Univ.,* Tech. Rep. No.75.

LANGFORD-SMITH, T., and THOM, B. G. 1969. New South Wales coastal geomorphology. *J. Geol. Soc. Aust. 16*: 572–80.

LAURENT, J. 1955. Le regime de la rade de Tanger. *Proc. 5th Inter. Conf. Coastal Eng., ASCE* 364–78.

LEBLOND, P. H. 1972. On the formation of spiral beaches. *Proc. 13th Inter. Conf. Coastal Eng., ASCE 2*: 1331–45.

————. An explanation of the logarithmic spiral plan shape of headland bay beaches. *J. Sed. Petrology 49*: 1093–1100.

LEONTYEV, O. K., and NIKIFOROV, L. G. 1966. An approach to the problem of the origin of barrier bars. *Proc. 2nd Inter. Oceanographic Congr.,* Abstracts of Papers, 221–222.

LEWIS, W. V. 1938. The evolution of shoreline curves. *Proc. Geol. Ass. 49*: 107–26.

MASHIMA, Y. 1961. Stable configuration of coastline. *Coastal Eng. in Japan 4*: 47–59.

McGILL, J. T. 1958. Map of coastal land forms of the world. *Geogr. Rev. 48*: 402–05.

MELVILLE, G. 1984. Headlands and offshore islands as dominant controlling factors during late quarternary barrier formation in the Forster-Tuncurry area, New South Wales. *Aust. Sed. Petrology 39*: 243–71.

MEYERHOFF, A. A. 1970. Continental drift: implication of paleomagnetic studies, meteorology, physical oceanography, and climatology. *J. Geol. 78*: 1–51.

Monthly Meteorological Charts: Indian Ocean, Mo. 519, 1949; Western Pacific, Mo. 484, 1956. H.M.S.O., London.

MORTAN, R. A. 1977. Nearshore changes at jettied inlets, Texas coast. *Proc. Coastal Sediments '77, ASCE* 267–86.

O'ROURKE, J. G., and LEBLOND, P. H. 1972. Longshore currents in a semicircular bay. *J. Geophys. Res. 77*: 444-52.

OTVOS, E. G. 1970. Development and migration of barrier islands, northern Gulf of Mexico. *Geol. Soc. of Amer., Bull. 81*: 241–46.

PARKER. G. F., and QUIGLEY, R. M. 1980. Shoreline embayment growth between two headlands at Port Stanley, Ontario. *Proc. Canadian Coastal Conf.*, 380–93.

PARTHENIADES, E., and PURPURA, J. A. 1972. Coastline changes near a tidal inlet. *Proc. 13th Inter. Conf. Coastal Eng., ASCE 2*: 843–63.

PATTERSON, C. C., and PATTERSON, D. C. 1983. Gold coast longshore transport. *Proc. 6th Aust. Conf. Coastal and Ocean Eng.* 251–56.

PHILLIPS, D. A. 1985. Headland-bay beaches revisited: an example from Sandy Hook, New Jersey. *Marine Geology 65*: 21–31.

PRICE, W. A. 1963. Origin of barrier chain and beach ridge. *Geol. Soc. of Amer.,* Abstracts for 1962, Special Paper 73, 219, 1963.

REA, C. C., and KOMAR, P. D. 1975. Computer simulation models of a hooked beaches shoreline configuration. *J. Sed. Petrology 45*: 866–72.

Readers' Digest. 1983. *Guide to the Australian Coast.* ed. R. Pullan. Sydney, Australia: Reader's Digest Publ. Co.

ROY, P. S., and STEPHENS, A. W. 1980. Geological controls on process-response S.E. Australia. *Proc. 17th Inter. Conf. Coastal Eng., ASCE 1*: 913–33.

SEXTON, W. J., and HAYES, M. O. 1982. Natural by-passing of sand of tidal inlets. *Proc. 8th Inter. Conf. Coastal Eng., ASCE 2*: 1479–95.

SHALOWITZ, A. L. 1964. *Shore and Sea Boundaries: with Special Reference to the Interpretation and Use of Coast and Geodetic Survey Data.* Volumes 1 and 2, U. S. Department of Commerce, Publication 10–1.

SHEPARD, F. P. 1937. Revised classification of marine shoreline. *J. Geology 45*: 602–24.

———. 1973. *Submarine Geology.* 3rd ed. New York: Harper & Row.

SILVESTER, R. 1960. Stabilization of sedimentary coastlines. *Nature 188* , Paper 4749, 467–69.

———. 1962. Sediment movement around the coastlines of the world. *Proc. Instn. Civil Engrs.* 289–315.

———. 1965. Coral reefs, atolls and guyots. *Nature 207*(4998): 681–88.

———. 1966. Sediment transport and accretion around the coastlines of Japan. *Proc. 10th Inter. Conf. Coastal Eng., ASCE 1*: 469–88.

———. 1968. Sediment transport, long-term net movement. In *Encyclopaedia of Geomorphology,* ed. R. W. Fairbridge. London: Reinhold: 985–88.

———. 1970. Development of crenulate shaped bays to equilibrium. *J. Waterways and Harbors Divn., ASCE 96*(WW2): 275–87.

———. 1974. *Coastal Engineering,* 2. Amsterdam: Elsevier.

———. 1979. A new look at beach erosion control. *Annual, Disaster Prevention Res. Inst.,* Kyoto University, Japan, 22(A): 19–31.

SILVESTER, R., and HO, S. K. 1972. Use of crenulate shaped bays to stabilize coasts. *Proc. 13th Inter. Conf. Coastal Eng., ASCE 2*: 1347–65.

SILVESTER, R., TSUCHIYA, Y., and SHIBANO, Y. 1980. Zeta bays, pocket beaches and headland control. *Proc. 17th Inter. Conf. Coastal Eng., ASCE 2*: 1306–19.

Sᴍɪᴛʜ, A. W. S., and Lᴀʀᴀᴄʏ, S. W. 1985. The dynamics of training walls at barrier island estuarine outlets. *Proc. 7th Aust. Conf. Coastal and Ocean Eng. 1*: 547–56.

Souᴄɪᴇ, G. 1973. Where beaches have been going: into the ocean. *Smithsonian 4*(3): 55–61.

Sᴛᴏᴅᴅᴀʀᴛ, D. R. 1969. World erosion and sedimentation. *Water, Earth and Man*, ed. R. J. Chorley. London: Methuen.

Tᴀɴɴᴇʀ, W. F. 1987. The beach: where is the river of sand. *J. Coastal Res. 3*: 377–86.

Tsuᴄʜɪʏᴀ, Y. 1980. Principles and technical measures in shore protection. *J. Japanese Soc. Civil Engrs., 65*: 2–8. (In Japanese)

———. 1982. The rate of longshore sediment transport and beach erosion control. *Proc. 18th Inter. Conf. Coastal Eng., ASCE 2*: 1326–34.

Vᴀʟʟɪᴀɴᴏs, L. 1970. Recent history of erosion at Carolina beach N.C. *Proc. 12th Inter. Conf. Coastal Eng., ASCE 2*: 1223–42.

Vɪᴄʜᴇᴛᴘᴀɴ, N. 1969. Equilibrium shapes of coastline in plan. Asian Inst. of Technology, Bangkok, *Master Eng. Thesis*, No. 280.

Wᴀʟᴛᴏɴ, T. L. 1977. Equilibrium shores and coastal design. *Proc. Coastal Sediments '77, ASCE*, 1–16.

Wᴏɴɢ, P. P. 1981. Beach evolution between headland breakwaters. *Shore and Beach 49*(3): 3–12.

Yᴀssᴏ, W. E. 1965. Plan geometry of headland-bay beaches. *J. Geology 73*: 702–14.

Zᴇɴᴋᴏᴠɪᴄʜ, V. P. 1962. Some new exploration results about sand shores development during the sea transgression. De Ingenieur No. 17, Bouwen Waterbouwkunde *9*: 113–21.

———. 1967. *Processes of Coastal Development*. ed. J. A. Steers, trans. D. G. Fry. Edinburgh: Oliver & Boyd.

Coastal Defense

Beaches are unique features on the coast, being relatively flat ribbons of sand between the sea and the hinterland generally comprised of dunes rising many meters above the berm level. This is the case where sand is predominant which can be blown landward from the beach. In the case of a surfeit of this material, vegetation attempting to grow may be buried, resulting in what are termed *blowouts*, as noted in Section 3.1.3. Such features have existed for thousands of years, long before man had been involved in these areas. Means of overcoming such burial events have been addressed, but investment in terms of water supply appears to be the main factor.

The obliquity of waves when breaking is the main cause of longshore drift, which occurs in pulses during storms rather than uniform transport by swell waves. Besides such changes in supply to any point on the coast, longer-term fluctuations occur due to natural conditions and the influence of man in intercepting this littoral drift. Transient interruption of flow can cause humps of material to move along the coast, so causing intermittent erosion and siltation. Thus, beaches are a very dynamic section of the coastal margin, the understanding of which requires many measurements at frequent and longer-term intervals. Although there had not been any standard quotation as to what would be considered as severe beach erosion, Bird (1984) suggested that a loss of more than 10 meters per year can be regarded as exceptionally rapid.

For stability the waterline needs to be retained in a relatively constant position. This must necessarily fluctuate as storm waves demand part or all of a beach berm, even more during fiercer sequences, but is restored in a matter of days by the milder waves. It is the swell-built profile, which exists for most of the year on oceanic margins, for which the waterline is required to be static. Even where some structure is installed to accrete a new beach line it will not be stable until the offshore profile has reached an equilibrium with respect to the general wave climate.

A plethora of papers has been written on shoreline processes and the influence of marine structures purporting to stabilize the shoreline. These have been distinctly unsuccessful in coming to grips with the situation, as instanced *"Effective Use of the Sea"* (1966), by a Panel on Oceanography, where it was stated: "The nation needs to improve the technology for constructing coastal zone structures, which will make the national expenditure on breakwaters, harbors, beach erosion, docks, etc., more effective. The panel was distressed to find a high failure rate of construction projects in the surf zone and on beaches; the destruction of beaches by breakwaters designed to extend the beaches; the silting of harbors and marinas as a result of construction designed to provide shelter; and the enhancement of wave accretion by building jetties supposed to lessen wave erosion are but a few examples of the inadequacy of our knowledge and practice in coastal engineering."

This indictment of the profession by an erudite group of scientists in 1965 is equally applicable today. The conservatism of coastal engineers, who wish to copy passed procedures rather than try novel approaches, has duplicated mistakes continually. The above publication instigated the preparation of *"Our Nation and the Sea"* (1968) by a Commission of Marine Science, Engineering and Resources, established by the U.S. Congress in 1966. Its report comprised three complementary volumes entitled: I. Science and the environment, II. Industry and technology, and III. Marine resources and legal-political arrangements for their development. More recently many millions of dollars have been spent on research into coastal processes, with piers constructed solely for monitoring beach profiles.

It is salutary to sit back and take stock of where research has been going, to see whether slight changes in direction might effect great savings in investments on the coast. Applications of new theory, model data, and field measurements take many years to implement. Papers that synthesize various inputs are discouraged because they contain no new data, which appears to be the major criterion for acceptance. All variables need to be given equal prominence in any investigation and not just those fitting into some hypothesis. Although some may be difficult to measure, theory can relate them to others that can.

Seawalls and revetments have been constructed for many decades. They have served as transient protectors of certain coastal sections, generally at the expense of the adjacent downcoast region, pointing to the need of extension along this eroded section. The walls themselves have often subsided as scouring occurs in front of them. They have been replaced with larger units without reasons for their previous failure. The desire of having frontal beaches has not been met as they do nothing about the original problem of decreased supply of sand from upcoast.

Groins have been installed across the path of littoral drift in order to retain beaches where they have been needed. However, the less frequent storm waves remove this material and disperse it farther offshore, to be moved farther downcoast than if they did not exist. Like the seawall, they cause erosion downcoast of the groin field because they must continually be refilled with sediment arriving from upcoast. This prompts the extension of this ineffective solution.

Offshore breakwaters, built parallel to the coast with reasonably small gaps between them, have received great attention in the past two decades. Salients or even tombolos have formed in their lee, thus providing beaches where seawalls had previously caused them to disappear. Somewhat like groins, storm waves can cause this material to be swept out to sea by rip currents through the gaps, there to be transported downcoast by the oblique reflection of swell waves from these structures. It is this action that scours material from in front of them, which enters these openings to provide the beaches. With a dearth of upcoast supply, only transient reprieve is provided.

Since beaches defend themselves by constructing an offshore bar during storms, some engineers believe that a permanent submerged structure may achieve the same result. These purport to provide a "perched beach" with a wider berm. Since the angled swell, which created the problem in the first place, will scour in front of and behind the offshore structure they will require much maintenance. Any interception of littoral drift will cause erosion downcoast, even if only temporally.

In recognition of beaches providing their own defense, bulk renourishment of coastal sections has been undertaken. However, monitoring of waterlines has shown these filled areas to erode more swiftly than normal shorelines. This has required additions of this expensive material biennially in order to maintain a status quo. The swift addition of large volumes of mobile sand on the coast can cause siltation problems downcoast.

Since the only way to retain sand where it is wanted is to reorient the beach normal to the incoming waves, a new concept has been derived where headlands achieve this. It is a combination of the offshore breakwater concept and renourishment, but structures are spaced well apart, in order that reasonably sized beaches are formed. Bays are sculptured between these headlands of crenulate shape, such that diffraction and refraction causes accretionary waves to arrive normal to the coast. Thus, material removed from the beach to form the offshore bar during storms is replaced directly back on the berm from whence it came, within the embayment. This results in a stable shoreline, even if sediment is still passing through a system of headlands. However, if supply should cease altogether the beach recession reaches a limit which is predictable. Downcoast conditions will be no worse than what they would have been without the headlands.

The above process is known as *headland control* and emulates Nature, which has maintained stable beaches by such natural structures over thousands of years. By proper shaping and orientation, these headlands can be protected by a beach in front of them so their height and size of armor units can be modest. Examples will be furnished where this concept has worked extremely well. It is economical and so should be considered in any beach stabilization program.

Thus, this chapter will discuss all the above topics in order that the reader can obtain an overall view of the status of physical coastal management. Millions of dollars have been lost over many years in repeating past mistakes. This has been caused by many devices being described soon after construction, before their full effects have been experienced and deficiencies exhibited. It is not easy to publicize failures, but more is learned from them than from successes, where overdesign may achieve this at great cost.

5.1 CAUSE OF EROSION

The causes of beach erosion can best be summarized by Bruun (1972), who started with a time-honored Dutch quotation: "Water shall not be compelled by any 'fortse', or it will return that fortse unto you." He divided the main causes of erosion into those by Nature and by Man. It must be accepted that wave obliquity, storm attack, imbalance of sediment transport, and perhaps sealevel rise, are the primary factors. However, there are many ways to cause sand loss from a beach, including removal alongshore, blown inland, swept into inlets or lagoons, lost to the deep sea, and extraction from the beach for constructional purposes.

Bruun (1972) pointed out that "Nature has not only demonstrated how to erode but also how to protect. It may safely be said that there is no protection initiated by man which has not beforehand been invented by nature, and nature obtained all the good results as well as all the bad results before man did. Consequently we can learn from nature if we will only make the effort of opening our eyes and looking. It must be admitted that nature has been more imaginative and has had more success than man." Apart from the causes of discontinuity in sediment supply from upcoast, the effects of storms, though infrequent, are often devastating. Man would like to ensure a full or near full recovery of beach berm after any storm. It is important to understand and utilize the principle of dynamic beach response to the cyclic action of persistent swell and storm sequences.

5.1.1 Main Causes

A distinction must be made between the transient erosion of a beach berm during a storm and the longer-term denudation due to lack of supply of sand from upcoast. The repetitive removal of beach material by storm waves has to be accepted as a fact of life. This excessive input of energy, generally over a short period of a day or two, requires Nature to deal with it, which she does admirably by constructing a bar to dissipate the incoming high waves. This is followed naturally, over a few days normally, by dismantling of the bar and renewal of the beach berm ready for the next onslaught. Even if the bar is not completely removed due to a lack of swell-type waves, it is there to be quickly reformed to a stage where wave breaking is accomplished. As noted by Carey (1907), no better defense can be provided for these transitory erosion events than the beach itself.

However, in the process of the bar material coming ashore it is also moved along-shore by the obliquity of the accretionary waves. If such waves arrive normal to shore, there can be no long-term erosion on that section of beach but probably a problem of accumulation. Thus, it is the littoral drift, plus offshore net movement, that is associated with long-term erosion of a segment of coast. It is the fluctuations in supply of this medium that causes the waterline to oscillate either landward or seaward. With a continual dearth of sediment from upcoast, erosion will worsen, until the shoreline is reoriented to be parallel to the crests of the most persistent waves, as noted in Section 4.1.1.

Alterations in longshore drift can be due to natural or man-induced causes. The former may be related to rivers or the wave climate. The latter are structures built on the coast either to directly impede movement, or do so as a consequence of other goals. The initial effect is for the offshore area to be scoured or deepened and hence the nearshore bed slope to be steepened. This subsequently requires more material in the bar for it to achieve its protective role of dissipating storm waves. Less of the offshore volume finds its way back to the beach as it stays in place in an attempt to maintain the previous profile. The waterline thus recedes and erosion is experienced.

A change in supply conditions can occur at the upcoast end of a section of coast where net movement is taking place. Whereas sediment may still be available from beaches still proceeding landward, perhaps to some equilibrium bay configuration, the amount is decreasing each year. It behooves the coastal engineer to look upcoast to see the source of any sediment available. If there are no rivers debouching onto the said coast, then it must be concluded that the only sediment supply is that due to slow erosion of this upcoast zone. In time this will mean no more material is available, which may be a blessing if siltation is a problem, or a curse if local beaches are likely to be degraded.

The rivers themselves are contrary in discharging sediment to the coast. The bulk of it is provided during floods which occur sporadically, either as bed or suspended load. The longer the river the more attenuated are the peaks of water flow. Also the source of water, through plain run-off or melting of snow, will influence the fluctuations in discharge. The degree of vegetation and slopes of catchments can significantly affect the nature of floods. Their geology, whether sand or clay, will also influence the volumes of material that reach the mouth on each storm or winter occasion.

The bypassing of littoral drift across an inlet or river mouth is pulsative in character. As noted in Chapter 4, shoals or even bars may be formed which retain sediment to various lengths of time. While this transient siltation takes place, downcoast beaches cannot receive material, so resulting in erosion for relatively short periods of time. Any dredging that may be carried out for navigation purposes, with spoil disposed of out to sea, can cause a more permanent reduction in supply with dire consequences.

As has been explained in Chapter 3, only swell waves returning the offshore bar to the beach have any significant effect on littoral drift. Subsequent swell, for the bulk of the year, can only influence the beach face and bed out to the breaker line, which is close to the waterline. For this reason, longshore transport within this narrow surf zone is minimal. Thus, the supply from upcoast can vary considerably, dependent as it is on the frequency of storms on that section of coast.

The volume of material placed into circulation for transport along a segment of shoreline is dictated by the size of the bar formed by any particular storm event. Cyclonic centers and the resulting capacity to generate storm waves varies around their peripheries. Where shoreward winds coincide with storm center movement the waves will be the largest, so resulting in a larger bar in that zone. There will be another section of coast where winds are seaward as a storm crosses a coast and hence no bar is formed. It is readily seen that batches of material from these larger bars are moved alongshore, virtually as humps. This can result in oversupply of sand with consequent accretion for short periods of time. Downcoast of the offshore wind zone little or no material is available for some period. It is little wonder, therefore, that the shoreline oscillates almost in a snakelike action.

As far as man's influence on longshore drift is concerned, his attempts at retaining sand on certain sections of shoreline by way of groins can cause erosion downcoast. This implies that one beach's salvation can be another's destruction. This may be transitory or more permanent depending on the length and number of these structures. Since they are installed only where an erosive situation already exists, there was already a dearth of sediment available on that section of coast. This lack of supply has not been dealt with and is exacerbated by the retention of part of it on any segment.

Similar effects are experienced when breakwaters are constructed on a mobile shoreline to serve as harbors. These generally protrude further seaward than groins and therefore exert a greater influence just downcoast of such complexes. This is purportedly overcome by seawalls or groins in an attempt to protect facilities associated with the port itself. As will be noted in Chapter 6, these solutions are not as effective as they could be and an alternative is suggested. Where a port is located in the lee of a headland the longshore drift may be minimal, since the bulk of sand is transmitted downcoast across the bay so formed (Hughs 1957). This situation is to be discussed in Chapter 6 in detail.

Associated with any port in the context above is generally a deep channel excavated across the inner continental shelf to provide access for deep-draft vessels. Longshore drift, both within the surf zone and beyond to the limit of disturbance of the persistent waves, will be deposited in such a depression. When this is dredged sporadically the spoil is normally dumped out at sea for convenience. This channel therefore becomes a complete interceptor of longshore drift, as effective as a groin-type structure of the same length. Thus, the seabed out to its seaward extremity is devoid of material, which results in the deepening and steepening previously alluded to. The consequence is severe erosion of the downcoast area. The magnitude of the problem is dependent on the wave climate, especially the obliquity of the persistent swell.

5.1.2 Basic Requirements

As can be seen from the above, the basic requirement in coast stabilization is to provide a beach which retains its waterline from one year to another or one decade to another. Even though short-term storm erosion must be accepted as a fact of life, the limit of such action should be predictable (see Section 3.4). In this case the coastal margin can be used for any activity up to the limit of the landward limit of the beach berm. Structures should be kept even a little shoreward of this, or of the foredune, to account for the more severe storms every five years or so.

Where a section of coast is suffering erosion it is due to the constant obliquity of the waves incessantly arriving on it. While this condition continues, so will fluctuations in sand supply occur. Unless something is done about this longshore drift on a permanent basis so will the problem continue. As will be shown in Section 5.3, seawalls, groins, beach nourishment, perched beaches, and offshore breakwaters do nothing to continue supply from upcoast. In most cases they expedite the passage of material downcoast.

Although the essence of the solution is to retain sand where it is required, the volume of material needed to achieve this is small in relation to the annual throughput on any section of coast. A downcoast region may suffer erosion temporarily but ultimately it is no worse than it was previously. If a new solution is adopted with concurrent reclamation then the transient erosion downcoast can be kept to a minimum.

As distinct from erosion, which takes a long time to exert its influence, accretion adjacent to some man-made obstacle can be very swift, since much of the action takes place within the surf zone in pulses. It is this speedy reaction that gives coastal engineers the impression of immediate success with erosion-abatement structures. But the new waterline is unlikely to be stable until the offshore profile is accreted to a stable profile commensurate with the wave climate of the area. Unless the replenishment problem is solved "in depth," long-term stabilization has not been achieved.

5.2 DUNE STABILIZATION

Engineers involved with coastal management want a solution to these large bodies of moving sand. To deal with the root of the problem, the supplying beach could be narrowed by inserting walls on the berm. The obliquely reflecting waves would expedite removal of sand downcoast and thus reduce the width of beach from which the blown sand is derived. This mechanism was discussed in the above section and will be discussed further in Chapter 9. Although reducing sand input, the existing area could receive an injection of capital in terms of water by opening it up for housing, recreation, or industry. Specific sand dunes or blowouts could be retained for aesthetic or cultural reasons, but the bulk of them could be vegetated or covered by roads and permanent buildings.

Dunes have been accepted as Nature's safeguard against future unknown coastal erosion. As noted in Chapter 3, Nature has not been very efficient in such a provision, for where most material is needed the supply is the least. This concept of a safety factor might be better termed a "factor of ignorance." In fact, when a dune is attacked by a larger than normal storm sequence, the tendency is to dump stone or build a revetment to protect it. This negates the idea of the dune coping with an erosion problem.

There is also a school of thought which proposes to retain dunes by vegetation in order to prevent erosion by storm waves. This is unlikely to work, since the root systems incorporated in the sand mass will hold the slope vertically, resulting in more efficient reflection of storm waves. This increases wave heights at the face and orbital motions of water particles, which aids removal to the offshore bar. Thus, in an erosive situation vegetation will exacerbate the problem of dune removal.

Accepting that the width of the first major dune with scant vegetation, for most of its length at least, is no longer required to act as a safeguard for beach erosion, its volume could be reduced to serve other purposes. As its surface is reduced in height, any vegetation introduced in the form of grasses, shrubs, and even appropriately selected trees can reach the water table more readily. This prized section of shoreline could be turned into playing fields and parking spots for the enjoyment of beach goers. A green expanse is provided adjacent to this beautiful strip of sand, dividing it from the sea. The problem of wind-blown sand could be dealt with by suitable plants to prevent landward movement of this vexatious sand which silts up roads and drainage systems.

The material of the dune could be utilized to reclaim extra area from the sea in conjunction with headlands to stabilize the beach against successive erosive events. This could be accompanied economically through pumping by means of water jet pumps (Silvester and Vongvisessomjai 1971) using seawater as the driving force. The lowering of this land area permits residents in houses located on the second dune to have a better view of the sea which at present is impaired by the high sandy first dune. The above views may raise the blood pressure of some ecologists and others, but they are presented with a view to making better use of the coastal margins and have aesthetic appeal.

5.3 PREVIOUS SOLUTIONS

Bruun (1972) gives an interesting comparison between nature's coastal protections and man's counterparts. For examples, a seawall resembles shore rock in the natural condition, a submerged bulkhead or mound is identical to rock reef, offshore breakwaters are similar to rock islands, large breakwaters perpendicular or angled to shore are comparable with natural headlands, groins simulate large rock outcrops perpendicular to shore, and artificial nourishment from land sources corresponds to various sediment transport mechanisms.

It has been accepted that the natural beach can defend itself extremely well, but is generally accompanied by littoral drift by pulses. However, if the shoreline is reoriented to become parallel to the breaking waves, this longshore drift can be reduced or even made zero. This is a natural process where headlands cause crenulate shaped bays to be sculptured between them, indicating the direction in which design should go. Reorienting the beachline more normal to the orthogonals of the persistent waves provides greater stability since the longshore energy of breaking waves is reduced substantially or completely. The stabilization of sections of coastline can be carried out by reclamation with the simultaneous installation of means to keep it in place, so precluding downcoast erosion.

It is salutary to observe the results of various solutions to coastal defense, to see where, when, and why they have succeeded or unfortunately failed. It is not known to the authors which came first, the seawall or the groin, but it is easy to conceive that the more obvious method of protecting property was to construct a walled barrier against the waves. Later, the longshore travel of sand, as perceived by the littoral current in the surf zone, may have suggested some impediment in the form of groins. More recently the concept of offshore breakwaters, constructed parallel to the shore, echelon fashion, has been promoted strongly. At the same time, where material is available, renourishment of the beach has been implemented. All of these apparent solutions have deficiencies, sufficient to state that they do not achieve their desired goals. Many structures that are designed to protect shorelines in the first place have often ended up needing to be protected against damage to themselves. It is time to stop and take stock of the paths previously followed to see if some deviations might lead to a more rewarding destination.

5.3.1 Seawalls and Revetments

The seawall family consists of walled structures such as bulkheads, revetments, and the normal seawalls. They are constructed parallel to the beach, to halt shoreline erosion by receiving the impact of waves, or to prevent coastal flooding due to storm surge. Despite their various locations of construction in relation to local beach and surf, they are designed for blocking wave energy physically. The face fronting the wave action may be vertical or sloping, with or without armor protection. Bulkheads are normally vertical walls and built on the upper reach of a beach, perhaps fronting or replacing the first dune, while revetments have sloping surfaces armored with various kinds of larger blocks. These two structures are supposed to receive wave action only during storms. Normal seawalls are often constructed with limited width of fronting beach or even without, and so may receive direct wave action on the face of the structure, when energy is increased.

Some decades ago an eminent oceanographer said to the senior author: "There is no doubt about the effectiveness of seawalls, they reflect the waves back to sea, never to be seen again." He, as many engineers likewise, did not realize that while propagating seaward these reflected waves applied a second batch of energy to the sea floor, with resultant sediment transport. When incident and hence reflected waves are oblique to the wall they do not act independently of each other, but set up a short-crested wave system as discussed in Chapter 2. The water particle orbits in these are quite complex and their influence on a mobile bed worthy of discussion.

As noted in Chapter 2, the simple case of two waves angled at 90° to each other with equal heights and periods (and hence wave length) in a constant depth, could be considered (Hsu 1979). These are given once again in Fig. 5.1 for ease of discussion. A distinction must be made between the wave length (L) of the short-crested system, between crests of *island-type* waves propagating parallel to the wall, and the wave length (λ) of the incident and reflected components. The new variable of crest length (L'), between island crests propagating in tandem and measured normal to the wall, can then be used for identifying orbital motions. These crests at the intersection of the two components and the troughs between produce a wave height which is double that of the incident wave. Heights are additive when heights of these components differ. Because of this, and the smaller d/L ratio due to the increased length, the result is larger oscillations at the bed, with commensurately greater suspension of sand grains.

Thus, along alignments parallel to the wall at distance ratios $Z/L' = 0, 1/2, 1, \ldots$ (see Fig. 5.1), the orbital motions are similar to incident waves progressing along the wall, but with double their height and increased wave length. As noted in Figs. 2.60 and 2.61, the mass transport of the island waves for $\theta = 45°$ is about double that of incident waves and hence sediment placed in suspension due to the oscillating motion at the bed is swiftly moved parallel to the wall, in the direction of island crest propagation. Bed dunes formed in this action are normal to water motion and thus normal to the wall.

Halfway between these crest alignments, or $Z/L' = 1/4, 3/4, 5/4, \ldots$ the orbits are horizontal rectilinear oscillations, as seen in Fig. 5.1 on Section $X/L = 0$. Even though

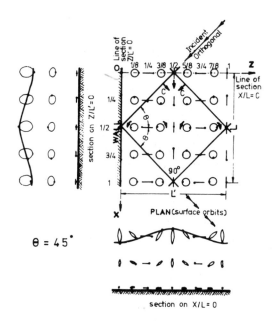

Figure 5.1 Short-crested wave system adjacent to a reflecting wall, showing water-particle motions

these motions are normal to the wall, they have a net movement parallel to it, as seen in Fig. 2.60, which is about 50% of the mass transport of incident waves moving parallel to the wall. The orbital motions and hence ability to suspend sediment can be less or greater than along $Z/L' = 0, 1/2, 1, \ldots$ alignments, depending on the wave obliquity. For $\theta < 45°$ they are greater and for $\theta > 45°$ they are smaller. These locations of optimum motion determine which alignment scouring will take place first. The bed dunes close to the alignment of $Z/L' = 1/4, 3/4, 5/4, \ldots$ will be parallel to the wall.

At intermediate alignments ($Z/L' = 1/8, 3/8, 5/8, \ldots$) orbits are elliptical, and angled to the horizontal throughout the water column (see Fig. 5.1), but become horizontal at the bed. As noted in Chapter 2, these are virtual vortices, even though for $\tan\theta \neq 45°$ the motions are elliptical and for $\theta = 45°$ they are circular. The action on a sedimentary bed is for a suction to be applied at the centers of these vortices and for sand grains to be thrown almost radially into circular mounds. Bed undulations take on a snakelike appearance. These vortex motions generate others, so increasing macroturbulence, which keeps sediment suspended for longer and hence influenced more by the mass transport. This net motion, parallel to the wall, at this alignment ($Z/L' = 1/8, 3/8, 5/8, \ldots$) is some 40% greater than for the incident wave.

The overall pattern of dune formation on a sedimentary bed experiencing the above simplified short-crested system of two wave trains angled 90° to each other is shown in Fig. 5.2. The orientation of dunes is as discussed above, extending over more than one crest length from the wall. Contours shown on the bed for this short duration of waves indicate that material is in transit toward the downcoast end, the accretion of which can cause refraction of incident and reflected waves and hence the slight deviation in the alignments parallel to the wall. Dunes are both normal to or parallel to the wall, or at

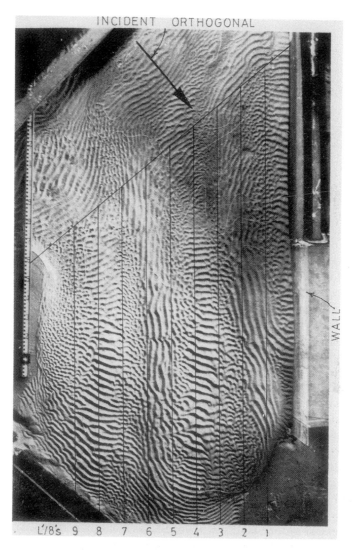

Figure 5.2 Sand ripple patterns produced by a simple short-crested wave system

$Z/L' = 1/8, 3/8, 5/8, \ldots$ they are circular. The depression at the downcoast tip of the wall, as in Fig. 5.2, will be discussed in detail in Section 7.3.

It is pertinent to now examine what happens to a section of coast where a seawall is installed along it, as depicted in Fig. 5.3. The incident waves refract as they approach the wall but are still angled as they strike it. The reflected waves also refract in section B as they proceed seaward. The orthogonals for these through the extremities of the wall are shown, which are seen to intersect some distance offshore, so forming what is known as a *wave caustic*. This implies a concentration of wave energy which results in wave

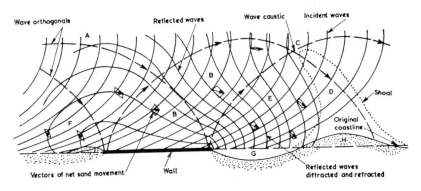

Figure 5.3 Effect of shore-based seawall on a section of beach downcoast

breaking, not only of the reflected component but also of the incident waves themselves. Thus the region (D) in the shadow of this caustic will have substantially reduced energy.

In front of the wall (section B) the short-crested system so formed causes a net sediment movement to the right which results in bed scour. Even beyond the limits of B, in section E for example, the reflected waves diffract and also refract toward the shore. This maintains the short-crested system, with net movements shown diagrammatically as vector resultants of incident and reflected components. This causes sediment movement both toward and alongshore, with scouring continuing in this zone.

Further along the coast (section D) the reduced wave energy cannot maintain long-shore drift until the area is shoaled. This may be accompanied by a protuberance on the coast, as illustrated in Fig. 5.3. The immediate response of an unsuspecting engineer, observing erosion between the wall and the silted condition downcoast, is to extend the wall. This, of course, magnifies the reflection whilst shifting the problem downcoast, perhaps into a region under someone else's jurisdiction. With the knowledge now available this appears an obvious mistake, but the authors have seen such errors still being perpetrated around the world's shorelines at great cost.

One example of such seawall extensions has been reported from the island of Sylt, West Germany (Dette and Gärtner 1987), where the authors conclude: "The seawall in front of Westerland and the necessity of continuous elongation of the coastal structures due to unavoidable lee-erosion and the helplessness against the beach excavation in front of the structures which necessitates the construction of 'generations' of toe protection may be considered as a typical and classical example of creating an 'armoring' ('Verfelsung') of the coastline at the expenses of beach loss."

Very little research has been done on the dimensions of scour adjacent to seawalls, despite some laboratory experiments and field monitoring (Kraus 1987). Although it has been stated that: "Failure by toe-scour is probably one of the most common causes of failure for vertical or near vertical walls." (Institution of Civil Engineers 1985). The only attempt, in the knowledge of the writers, is that by McDougal et al. (1987), who introduced normal waves to a wall with the coast set back on the downcoast side. Certain characteristics of the crenulate shaped bay formed adjacent to the wall, but the normal wave approach does not warrant their inclusion here. Oblique waves would magnify the resulting bay so that experiments are required for this condition.

Seawalls are still being installed and extended, with "the bigger the better" being the main criterion of design. They are reported in the technical literature soon after construction, before their full impact has been exhibited. It would be more instructive for papers to be presented some years later so that readers could judge more accurately the success or otherwise of these developments. If they had failed in their expectations it may be difficult to publicize this fact, but failures are far better learning experiences than successes. Most advances in engineering have emanated from observations and analyses of catastrophies.

Although many workers believe seawalls have little influence on the formation of beaches fronting them, the statement by Magoon and Edge (1978) on this subject is worthy of note: "But, they do not really stabilize the shoreline in any long-term sense. These structures are more related to storm protection than stabilizing the shoreline over the daily processes which are so important in shaping the beach."

There is a need for seawalls and breakwaters to be tested in three-dimensional models with movable beds (see Chapter 7), in order that probable scour can be predicted. This does not overcome the need for monitoring of prototype situations by hydrographic surveys for several months after seawall installation. In modeling there are certain scale effects as, for example, the use of waves 1/50th the size in nature whilst the sand is probably a little finer than in nature. Rates of scouring for a given wave duration cannot be predicted with confidence. The limit of removal for given wave characteristics with zero input of sand cannot yet be given.

Seawalls may be vertical, sloping, stepped, or concave seaward and may be smooth or have projections on their faces. Such elements are small in character, causing waves to treat them as planar. When structures are placed well back from a beach, considered as stable, they provide protection against storm surges and tsunamis. But it is essential to prove that the fronting beach is unlikely to erode. Baba and Thomas (1987) have suggested that: "A combination of a seawall and a frontal beach is a better coastal protective measure in the case of an equilibrium beach with cyclic erosional-accretional processes," because: "the longshore bar formed during the peak erosional phase (June-July) acts as a submerged breakwater. Higher energy waves break at the longshore bar permitting only low energy waves to reach the seawall directly."

As noted in Chapter 3, the complete beach berm is likely to be demanded by storms at least each year and on occasions even the foredune or first dune may be attacked at less frequent intervals. A wall located somewhere on this expanse of sand reduces the volume of material available for the construction of the offshore bar. Thus, it is not built sufficiently high to prevent waves from reaching the seawall and hence reflecting from it, running up it and perhaps overtopping it. Subsequent reflection of swell waves obliquely helps transmission of sand downcoast and in the case of reduced upcoast supply, which was probably the reason for the wall in the first place, it becomes difficult for a suitably wide beach to be reformed. Wave attack will thus be experienced each storm season.

A section of coast has been monitored over 2 years where part was plain beach and another contained seawalls. Sayre (1987) concludes: "A undeveloped island's beach recovered quickly after winter-time and hurricane caused erosion. ... The two other sites, on highly developed barriers and backed by seawalls, have suffered greatly. One narrow beach was completely destroyed by a hurricane and only partially recovered." Thus, a fronting beach should be wide enough to provide all the material required for a fully fledged protective bar.

Van de Graaff and Bijker (1988) have commented: "All over the world, shore-parallel structures suffer from damage. It is argued that these structures are in fact frequently built at places where they shouldn't." Also, Thieler et al. (1989) have observed: "Seawalls, buildings and storm rubble cause beach narrowing. On the north coast, the post-storm beach in front of seawalls and storm rubble is not as wide as areas backed by dunes."

The above quotations from references appear to conflict with the findings of Kraus (1987), who reviewed some 100 papers and concluded that walls had no effect on beach oscillations. "Sediment volumes eroded by storm at beaches with and without seawalls are comparable, as are post-storm recovery rates." In this context it should be remembered that when a severe storm occurs a larger than normal bar is constructed, which results in a larger pulse of sediment moved along the coast. This can be deposited in front of a seawall as readily as along the adjacent "unprotected" sections. A shoreline will always end up a straight line or a smooth curve, so making the wall appear to have no influence.

Tait and Griggs (1990) have reviewed 40 papers on beach response to seawalls, "few of which were focussed extensively on the effects of seawalls and beaches; they contained a minimum of relevant information." With respect to field studies they expressed the need to include a variety of wall types, coastal environments, and to observe over sufficient seasons. They conclude: "More studies are needed, especially in light of the fact that the processes and controls involved are largely speculative, and have not been measured in the field."

There would appear to the present authors nothing wrong with heuristic discussion of any phenomenon if it suggests variables to be included in any research project. Field observations should indicate all aspects of the problem both long and short term. It is to be hoped that the discussion in this section has clarified some of the problems.

5.3.2 Groins

These were probably man's second attempt at coast stabilization. Excessive turbulence in the surf zone, plus a longshore current, indicated a zone of large transport downcoast. If structures could be placed across this path then material would be retained in place. These generally ran normal to the coast, out to the limit of the breaking waves, and were termed *groins*. There are several types of groins, these being short groins (with length not exceeding the surf zone), long groins (extending through the surf zone), high groins (with crown height above the breaker), low groins, permeable groins, and impermeable groins. Various kinds of construction materials have been utilized.

Groins have been indicated on charts in England since the sixteenth century, and general rules governing the construction of these structures can be found from documents available since the very early nineteenth century. However, "The passage of time does not seem to have added much to the knowledge of the behaviour of the groynes" (Institution of Civil Engineers 1985). By this it refers to the most suitable type of groin to be constructed at any given section of coast, its geometric dimensions (i.e., length, height, and distance apart) and orientation to beach and waves for achieving the best performance. Considerable difference of opinion still exists in spite of the experiences over many centuries.

Groins do nothing for the lack of sediment supply which is applicable not only in surf zone but also in the offshore region. Thus, beyond these structures erosion continues and the bed deepens and steepens, as already noted. This problem was recognized by Lehnfelt and Svendsen (1958): "The groynes can, under favourable conditions, protect the bed of the sea from erosion as far as their extremities but not much further. Beyond the end of the groyne, erosion continues and it is evident that after a certain time this erosion will attack the groyne and later the land itself." This message, like many others, has not been heeded by the profession at large. This deepening and concomitant steepening offshore permits the waves to approach more obliquely and hence have a greater longshore component of mass transport and so increase sediment transport.

Following the construction of a shore-normal groin, as shown in Fig. 5.4, sand accumulation can be found at the upcoast side and erosion downcoast, thus proving the prevailing direction of longshore transport under the angled waves. But even the material retained between groins is in jeopardy because the storm waves, arriving from a large fan of directions, will be more angled to the new beachline than to the original coast. The resulting longshore current is intercepted by each groin, as in Fig. 5.4, and is deflected seaward as a rip current. This carries a large volume of material offshore, much farther than for an offshore bar formed on the original shoreline. Some of it will stay there to make up for previous offshore erosion, while the remainder will move farther downcoast before arriving back on the berm. It is contended, therefore, that a field of groins will expedite longshore drift, which it was meant to reduce.

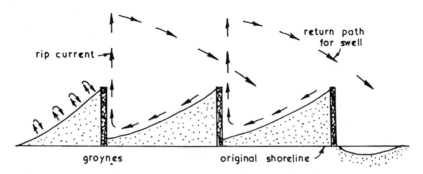

Figure 5.4 Shore-normal groins, showing upcoast accretion and erosion downcoast of the last, with rip current formation during storms

With respect to a field of groins when each compartment is filled to capacity, sand is deflected seaward downcoast of each groin. As noted by Everts (1979): "The seaward deflection to deep water in the zone where the sand is moving reduces the volume of sand reaching shore at the downdrift end, and downdrift of the groins." The greater the number of groins the greater the erosion, much like the influence of extending a seawall. When subsequent swell arrives, sand is accumulated against the upcoast groin and then over time proceeds to fill other compartments along the field. During this filling process sand is not available downcoast of the last groin so that erosion increases. Again, the obvious conclusion is to extend the system, which naturally exacerbates the problem since more cells need to be filled before sediment is available at the downcoast end.

Balsillie and Berg (1972) examined the effectiveness of groins worldwide and concluded they have a 50% chance of fulfilling their role. This is, of course, a subjective assessment, but by any stretch of the imagination cannot be considered good engineering practice. There are locations where groins can accrete a beach without too many adverse affects, but most of these are not long-term erosive situations in any case. But on sections of coast where the shoreline is receding, groins can accomplish very little for the reasons given above. The same authors conclude: "By the very nature of beaches, that no two are the same, it is erroneous to assume that one specific groyne design will provide the answer to all shore erosion problems."

Riedel and Fidge (1977) in a historical review of a coastline in Australia have stated: "Consequently the groyne fields have either been 'drowned' by sand when there has been an oversupply from the ocean beaches, or the groynes have been 'flanked' by water when there was an undersupply," and also: " ... on beaches with a very irregular littoral drift supply, groynes are unlikely to be effective." Lillevang (1965) confirmed this with the statement: "Under unusual but possible site conditions beach erosion could result because of a groin installation, even though the groin had not caused any beneficial effect as intended."

Balsillie and Bruno (1972) presented a bibliography on groins up to 1971 which contained a multitude of subheadings that included construction methods, groin types and dimensions, beach dynamics, materials, permeability, and location. It also contained items on wave climate, groin orientation, and spacing, downcoast erosion and scour at the toe. Most articles describe some new field with glowing reports of effectiveness, initially at least, with few discussing long-term trends.

In spite of the plethora of papers on this topic, Berg and Watts (1965) made a statement which is just as applicable today: "Of the numerous type of structures developed for shore protection purposes, the groin is probably the most widely utilized although its functional behaviour is perhaps the least understood." This is probably due to the difficulty of replicating storm conditions in models and hence reproducing the rip current that is a significant feature of its mal-operation. Monochromatic waves are generally used which represent the swell period of beach accretion. When viewed in toto, structures running normal to the shoreline are unlikely to prevent attack by storm waves. The resultant combination of these with the swell is seen to expedite longshore movement rather than reduce it.

As noted by Balsillie and Berg (1972): "There are numerous examples where groins have fulfilled their purposes and as many others which have not, indeed some have actually intensified the problems they were intended to solve." The ubiquity of the crenulate shaped bay has been presented in Chapter 4 and it is in this context that groin installation should be considered. As already noted, sand reaching an upcoast headland is transmitted across the bay to the slightly curved downcoast region (Hughs 1957). Thus, in the more indented region the longshore drift will be only from material being removed as the bay is reaching its static equilibrium limit. Hence, any groins inserted in this section will not prevent the shoreline receding to its natural shape, aligned with the crests of the diffracting and refracting waves. But as noted in Chapter 4 groins placed on the tangential downcoast segment of a bay will accrete long segments of beach as waves are angled only slightly to the beach, which is almost stable in any case.

There is precious little theory that can be applied to groins, prompting Wiegel (1964) to observe: "There is an extensive literature on the construction and operation of groin systems, much of which is controversial." Since then Bakker (1968) and Bakker et al. (1970) have presented theory. The first was based on a paper by Pelnard-Considère (1954) who assumed incorrect shapes of shoreline either side of the groin. The second paper corrects this by discussing diffraction on the downcoast site. Even so, no practical suggestions emanate from the study.

The graphs and tables contained in Chapter 4 for determining shapes of static equilibrium bays can be used to predict the shoreline between groins. For this purpose the deepwater direction of persistent swell should be known and this orthogonal refracted to depths between the groin tips. This decides the obliquity β, which is also the angle between the tangential downcoast extremity of the partial bay so sculptured and the control line. Since the indentation ratio (a/R_o) is more relevant, the length of groin (B in Fig. 5.5) from the tip to the waterline of the original shoreline becomes (a), assuming that no existing beach is to be lost. Once β is known and the spacing of the groins (D) decided, their lengths (B) can be found from Fig. 5.5. It can be shown that for $\alpha_o = 25°$ to $50°$ and depths from 3 to 10 m and waves of 8 to 14 second periods the angle of crests to the shoreline or β ranges from $15°$ to $30°$. This gives groin spacing to length ratios (D/B) of 5 to 3.

Nagai (1956) and Horikawa and Sonu (1958) have suggested 2 to 3 times groin length as suitable. Thom and Roberts (1981) have recommended 1.5 to 2 for this ratio. But as seen from Fig. 5.5, the approach angle of the waves in water depths between the groin tips should determine the spacing. For waves more normal to the coast, groins can be placed farther apart, which is rather obvious. This is where wave climate assumes importance, as depicted in Fig. 5.6.

In order to protect the heel of the groin from scour, Barcelo (1970) has suggested inclining groins downcoast by $20°$ when the waves are angled by a similar amount to the coast. Anderson et al. (1983) have utilized a spur normal to the groin on the downcoast side to achieve this, particularly where the coast there is set back from the updrift side. The longer the groin the more likely is the heel to be exposed due to erosion.

The above presentation is not sympathetic to groin installation. It is not that some marine structures are not as effective as desired, but that adjacent shorelines may be affected. Lillevang (1965) expressed similar misgivings: "There are more individuals who are certain a groin will be beneficial than there are engineers whose experience tells

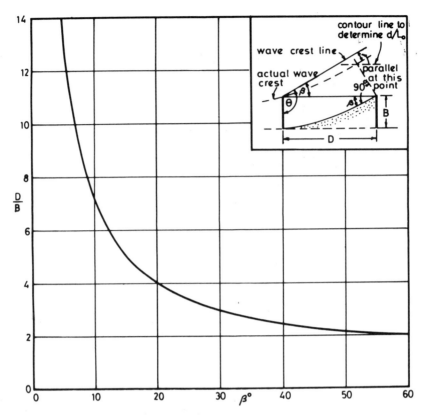

Figure 5.5 Spacing between groins in relation to approaching wave angle

them that one does not achieve unmixed blessings when a groin is placed on a coastline."
Magoon and Edge (1978) made a statement which can be considered applicable today:
"Although thousands of groins have been built, it is unfortunately true that the technology
of groin design has not been investigated adequately either in the laboratory or in the
field. With this in mind, it is easy to understand why groins have so often not only
failed but have created worse conditions than they were designed to prevent." The lack
of support for groins is echoed by Berg and Watts (1965): "Of the numerous types of
structures developed for shore protection purposes, the groin is probably the most widely
utilized although its functional behaviour is perhaps least understood. Examples may be
found when a single groin or groin system achieved its intended purpose, while similar
structures in different locations have provided little or no protection. In addition, there
are a seemingly endless number of variations in groin designs, some of which have
been purported to be the ultimate answer to shore erosion." Unlike the predictability for
a bay shape in static equilibrium sculptured between headlands, which is presented in
Chapter 4, prediction of waterlines between groins is still difficult. Perhaps the approach
in Fig. 5.5 may aid future research efforts.

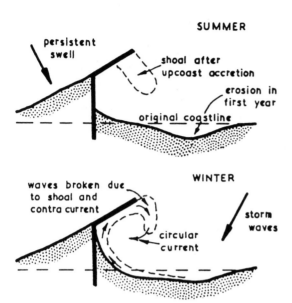

Figure 5.6 Wave action during summer and winter with angled tip of groin

5.3.3 Offshore Breakwaters

These structures are constructed offshore in 3 to 5 m depth parallel to the coast, with spacings varying from 1/2 to 5 times their individual lengths. Swell waves entering the gaps diffract and refract and so sculpture salients parallel to their curved crests, as will be discussed in Section 6.2. The greatly reduced height of these diffracted waves implies little run-up in the formation of beach berm so that a greater width must be taken during a storm to form an adequate bar in limiting erosion.

This mode of coast stabilization has been equated to headland control (Loveless 1986, Dally and Pope 1986), but the action is significantly different in that the beach length in proportion to structural dimensions is much smaller. Magoon and Edge (1978) have also not distinguished between these alternatives: "Offshore breakwaters or artificial headlands are the most promising means of stabilizing a shoreline and also developing a recreational beach (Silvester 1976). Along the Pacific coastline are numerous examples of the success of natural headlands in providing protection to the point that tombolos are actually formed." It is worth noting that headlands are spaced much farther apart than offshore breakwaters. They are spaced seaward for continuous tombolo contact, in order to form lengthy beaches which are self protecting.

These structures have been installed in Europe and the United States since the mid-1960s but have proliferated in Japan from the 1970s, with some 2,500 units by 1981 (Seiji et al. 1987) and around 4,800 by 1989 (Uda, personal communication). A typical example of this system can be seen on the Kaike coast of Honshu facing the Sea of Japan, as can be seen in Fig. 5.7. In many cases the motivation has been the

Figure 5.7   Offshore breakwater field on Kaike coast, Japan (see Fig. 7.31 for location)

rejuvenation of beaches that have disappeared due to the prior construction of seawalls to which have been added fields of groins. They have not achieved the desired result as extra breakwaters have been inserted across the existing gaps, to protect facilities facing them.

The sediment movement both inside and seaward of these structures will be examined, to see how effective they are in the long term. In this context it should be remembered that they have been installed due to a lack of sand supply from upcoast. This infers that the swell waves are arriving obliquely to the coast and therefore will be reflected obliquely from these breakwaters in reasonable water depths. As seen in Fig. 5.8, this establishes a short-crested system, either partial or full, because swell is well reflected, even from precast concrete blocks that are purported to be permeable.

An interesting question to ask is: "Where did the sediment for the salient come from?" The answer must be that it came from the seaward region of the breakwaters,

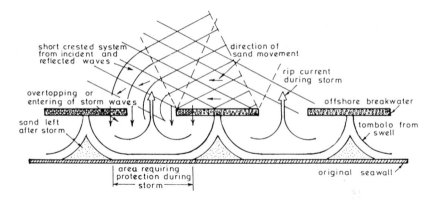

Figure 5.8   Short-crested wave system formed in front of offshore breakwaters, showing rip current between them during storm

especially if there was little sand between them and the seawall previously constructed. The efficient scouring capacity of the short-crested waves, noted previously, moves material along to the gap where it is carried by diffracting waves into the central region of the shadow zone. There may be sufficient material for the beach protuberance to reach the breakwater, in which case a *tombolo* is formed, as indicated in Fig. 5.8. This accretion takes place while milder or swell waves are arriving, which are dissipated or reflected at the structure.

When a storm occurs, large waves enter the gap and probably overtop the breakwaters. An excessive volume of water is poured into the space between them and the beach (or wall) and must return to the sea. This it does in the form of a strong rip current (see Fig. 5.8) through the gaps, carrying with it the material previously salvaged from the sea. During these cyclonic sequences, the shoreline fronting each gap has any berm available removed offshore to form a bar in close proximity to this opening, as applies also to the salient apexes. It is thus in a good position to be removed by the rip current, to be deposited as an arcuate shaped shoal seaward of the breakwater alignment. As seen in Fig. 5.7 it has been found necessary to place blocks in front of the original seawall to protect it and reduce overtopping. This is because insufficient berm volume was available to construct the protective bar.

With regard to storm events, Toyoshima (1974) conceded: "It is impossible to prevent completely a fatal coast erosion even by detached breakwater works." The reason given was: "As the shoreline just beyond the opening between the neighbouring two breakwaters is used to be scoured, a counter plan for shore protection should be considered beforehand." Another practical aspect, for the Japanese at least with their large fishing fleet, is the difficulty of bringing craft into the relatively sheltered water leeward of the breakwaters when a strong rip current exists in the entrance. The storm waves reflected from adjacent structures spread toward the gap to form a very complex sea of steep waves. In other countries where this solution may be adopted, similar navigation hazards could arise for small craft caught out at sea at the outset of a storm.

The material deposited out at sea then comes under the influence of the incident and reflected waves, to be transported downcoast, with some entering the next space during the following swell season. In an erosive situation, which it must be for these structures to have been installed in the first place, the retention rate will decrease as offshore areas deepen due to scouring. The proportion of sediment moving downcoast to that retained temporally is hard to gauge, but ultimately decreasing amounts are available for beach formation. It is not correct to assume that longshore drift has not been prevented by off-shore breakwaters; in fact, it could be increased by the action of reflected waves on them.

This solution is virtually equivalent to a revetment wall for two thirds of the coastline. Since it is constructed offshore, it is an expensive version of protection. Because breakwaters consist of armor units resting on sand, the inevitable scouring causes them to subside and hence reduce their height, thus more overtopping. This deficiency has been recognized by Toyoshima (1976), who has concluded: "The most important problem is subsidence of the breakwaters. Up to now, the subsidence of the breakwaters on the Kaike coast is considered to be moderate, but it is urgently needed to develop methods by which the speed of subsidence can be minimized." The sea bottom slope fronting these structures has been steepened drastically (Toyoshima 1982). Walker et al. (1980) have shown bed profiles through an offshore breakwater where 2.6 m of scour has occurred 35 m from the face within one year of construction. In time the collapsed elements serve as a foundation for new blocks to be added, but the whole exercise is extremely expensive while not serving its purpose in the long run. Thus, offshore breakwaters cannot be accepted by these authors as a worthwhile solution for coastal defense.

5.3.4 Perched Beaches

As noted in Chapter 3, storm waves form a protective bar which limits erosion. However, some workers have suggested copying this mechanism by inserting submerged structures some distance from the beach. This concept purports to replace the offshore bar formed naturally during storms with a permanent rubble-mound structure, which may also be linked to the beach at each end (Douglass and Weggel 1986). These are generally associated with reclamation in order that nearshore levels are built up in what is termed a *perched beach*. It can be accepted that the milder offshore slope would demand less material from the berm to construct a bar of sufficient height to effect wave breaking. However, there is more to this problem than just the onshore and offshore movement of sediment.

The oblique swell waves will be reflected from the seaward face of this submerged mound and so cause scour on its seaward edge, with consequent subsidence. But even shoreward of the structure it has been found that sand is eroded by the increased wave height in the shoaled area, so that some protection for a considerable width is advocated (Inman and Frautschy 1965). Any free edges of any reclaimed material will be worn away readily so that large losses will accrue, with a hump of material traversing the coast. In the meantime, littoral drift may be impeded at the perched beach, which will result in erosion downcoast.

Chatham (1972) has noted that: "Given the beach profile in a given area is governed by the wave climate in that area, in a beach-fill project the existing profile must be reproduced for a considerable distance seaward if the fill is to be stable." In fact, the toe structure should be located at a position where the bar is likely to form, which is one quarter wave length of the surf beat with periods many times that of the incident storm waves or T_{max} of the spectrum. This is 50 m or more from the beach. Chatham (1972) conducted tests for the equivalent prototype distances of 107 to 335 m, while Sorensen and Beil (1988) have recommended a distance of 40 to 50 times the significant wave height, presumably measured at the site of the structure.

Chatham (1972) stated: "Since turbulence induced by the oncoming waves as they travel over the submerged toe structure could transport quantities of beach material seaward, a stone rip-rap apron might be needed along the shoreward edge of the toe structure crest to reduce seaward migration of sand." It is understandable that if the bar were to be located in the region of the structure, some of its contents must necessarily spill over to the region seaward of it. It is felt by the authors that this purported protection could not prevent such material from being lost to the perched beach. In the event of the bar being formed either landward or seaward of the structure, the breaking of waves over it could erode sand adjacent to it, which is apparent from some of the cross sections shown by Chatham (1972).

Since the milder or swell-type waves must be angled to the coast for an erosive situation to exist, any model studies of the concept should be conducted under these conditions. Very few studies have been made but all have used flumes which imply that all waves are normal to the coast. Because of this they have not examined a real prototype situation and hence any conclusions drawn must be viewed with some skepticism.

Although very little change was observed for milder waves in the flume situation, they would have a drastic effect if angled to the shoreline since the decreased depths could expedite the downcoast removal of the perched beach. Until such three-dimensional tests are carried out, any design data so presented should be used with great caution. Sorensen and Beil (1988) observed that as the height of the structure was increased (to SWL in their case), "the horizontal distance from the submerged toe structure of the beach face is decreased." Such a situation is closer to the offshore breakwater discussed in the previous section and in more detail in Chapter 6. A salient forms due to the diffraction occurring in the leeward region. The need for three-dimensional studies was recognized by Chatham (1972), who stated: "It is well known that quantitative three-dimensional movable-bed model investigations are difficult to conduct and each area where such an investigation is contemplated must be carefully analyzed." He suggested the following requirements:

1. Littoral drift assessment using wave structures.
2. Sand size distribution over the site.
3. Simultaneous measurement of incident waves, bottom profiles, alongshore plus on/offshore transport of sand.

In spite of the difficulties in analysis and data gathering, such comprehensive tests are necessary. But before the investment is made, a qualitative study is required of the phenomenon to see whether the procedure is worth the expense. As emphasized in this tome, some sort of headland structures of modest height, even of reef character, are to be preferred. Some prototype perched beaches have been reported. Douglass and Weggel (1986) conducted surveys on an installation at Delaware Bay where the sand deposit disappeared within 4 years. They observed that one shortcoming of the sill structure is that sand removed landward of the structures was deposited seaward of it. This material would quickly be moved downcoast due to the oblique reflection of swell waves from the submerged breakwater.

Deguchi and Sawaragi (1986) described a submerged breakwater 80 m in length and 85 m offshore where 5,000 m^3 of sand was placed in its lee. By the end of two months, 50% of this volume had been removed, and 10% in the following two months. No discussion was available on wave climate during this period, except that during the winter-spring 35% of waves were higher than 1 m and arrived from directions both sides of the normal to the coast. There is little doubt that these structures do not retain material as intended and are likely to subside due to oblique reflection of persistent waves.

5.3.5 Beach Renourishment

To summarize the effects of seawalls and groins, what better statement can be presented than that of Bijker and Van de Graaff (1983): "There is a tendency to avoid the construction of groynes, seawalls, etc. as long as possible. Most types of erosion problems can be solved by proper sand suppletion." By this is meant beach renourishment, of which Bijker is a strong advocate. At least this confirms the view of the authors that the provision of beaches is the best defense against the onslaught of the sea. However, the concept of renourishment needs re-examination to see if it fulfills stabilization requirements economically.

The placement of new material on the beach is one of the latest panaceas for coastal protection, promoted vigorously by organizations selling dredging services. It should be remembered at the outset that the same wave climate exists as before, which is oblique persistent waves in an erosive situation. It is pertinent to examine whether this fill material is removed downcoast as quickly, if not more so, than that of the original shoreline. Although surprisingly few studies of this effect have been carried out effectively, Pilkey et al. (1983) believe that a good minimum figure for an artificial beach is an erosion rate 10 times that of the natural system. The scales of beach renourishment vary tremendously. The same authors have commented: "At one end of the spectrum, it may consist of a home owner hiring a dump trunk or two to unload sand in front of a home, or a bulldozer pushing sand from the lower beach to the upper beach. At the other end of the spectrum is the Miami Beach replenishment project: $64 million for fifteen miles of new beach. Between these two extremities is the city of Virginia Beach,Virginia, which each summer hauls thousands of dump-trunk loads of sand to her beaches."

Spoil, when dumped on the beach by trucks, or offshore from it by dredgers, will rest at almost the angle of repose for sediment in this saturated condition. This can be either steeper or flatter than the natural profile that has been under continual wave action. Nature does not countenance such an imbalance of energy and load, so quickly smooths this material out to greater depths, in order to approach the original bed shape, as depicted in Fig. 5.9 for a steep spoil profile. Everts et al. (1974) have acknowledged that 30% to 50% of any initial renourishment is lost in this process. This, of course, is accompanied by downcoast movement as the waves are oblique as before.

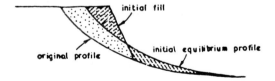

Figure 5.9   Steep beach profile formed by dumping spoil on a renourished site

As seen in Fig. 5.10, the downcoast extremity of this widened beach is more angled to the breaking waves and hence will erode faster. But even during the complete removal of this protrusion, the waterline is more oblique to the waves than the original shoreline. It could be expected, therefore, that this filled zone will be denuded faster than the natural beach. This has been verified by waterline monitoring over a long shoreline where reclamation had been carried out over certain sections at various times (Winton et al. 1981, Winton 1983, Chou et al. 1983). For these particular widened beaches, Winton (1983) has stated: "The initial response was found to be an exponential loss rate with up to 80% to 90% of the beach width eroded after 15 to 24 months. Subsequently the beach face retreated at a fairly constant rate with a magnitude similar to the pre-fill long-term erosion rate."

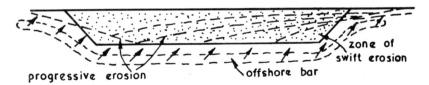

Figure 5.10   Downcoast extremity of an artificially renourished beach, showing widened beach angle to breaking waves

The renourishment carried out at the Wrightsville beach in 1970 (Winton et al. 1981) was not uniform along the 3 km of coast, as seen in Fig. 5.11, which is different from the situation depicted in Fig. 5.10. The profile lines WB13 to WB29 are marked which straddle this filled area. The excursion distances after the initial exponential losses are also shown, which indicates around 80% loss in relatively quick time.

The model for waterline movements as devised by Winton et al. (1981) is depicted in Fig. 5.12, where a long-term erosion rate is given by the slope α. At some time $t = 0$ the beach is widened by amount Y, which is then swiftly eroded back in time t_1 to the long-term erosion rate line, which reaches the original denudation limit at time t_e. Extrapolation of this long-term line back to $t = 0$ gives the effective initial excursion Y'. The authors found that semi-log plots of excursion distance (Y_t) versus time on any profile normal to the beach gave an exponential decrease such that

$$Y_t/Y' = 1/10^{kt} \tag{5.1}$$

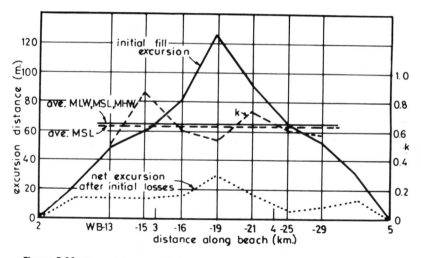

Figure 5.11   Renourishment carried out at Wrightsville, showing nonuniform fill along beach

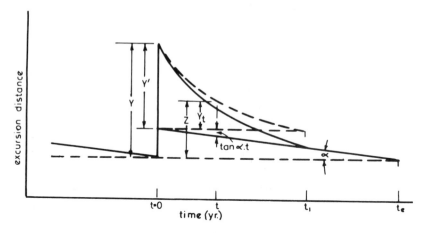

Figure 5.12   Model of long-term beach erosion on a renourished beach, showing variables involved (Winton et al. 1981)

with variables as depicted in Fig. 5.12. As k increases, Y_t is a smaller proportion of Y' at any given time t. Plots of Y_t/Y' against time t for any specific profile were found to be a straight line on this semi-log plot, as seen in Fig. 5.13, but the value of k varied at each location for MSL excursions. These are plotted in Fig. 5.11, together with their average of 0.64, which corresponds well with the average for LW, MSL, and MHW by Winton et al. (1981) of 0.66, also plotted.

Of more particular interest than percentages of Y' lost is that of fractions of Y from the original shoreline. This is the excursion width that will be observed on the beach. It can be seen from Fig. 5.12 that

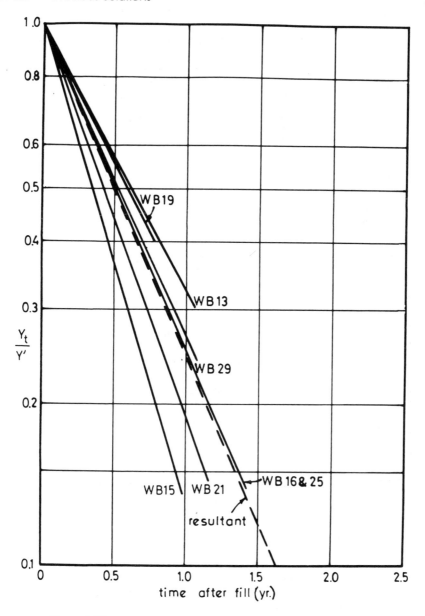

Figure 5.13 Excursion distance of beach fill material versus time elapsed

$$Z = Y - Y' + Y'/10^{kt} - t \tan \alpha \tag{5.2}$$

Nondimensionalizing by dividing through by Y gives

$$Z/Y = 1 - Y'/Y + (Y'/Y)/10^{kt} - (t \tan \alpha)/Y \tag{5.3}$$

Winton et al. (1981) have obtained average values of $Y'/Y = 0.8$, where $\tan\alpha$ is the average long-term rate of loss per unit time or 1 year. Using $k = 0.64$ as derived above, and wishing to determine the time for $Z = 0$, Eq. (5.3) can be rewritten as

$$Y = t\tan\alpha / [(0.8/10^{kt}) + 0.2] \qquad (5.4)$$

As t increases so $0.8/10^{kt}$ reduces such that the denominator in the square bracket becomes 0.2 when $t = 3$ years, after which Eq. (5.4) becomes

$$t = 0.2Y / \tan\alpha \qquad (5.5)$$

For the average excursion value of 76.6 m at Wrightsville beach and the long-term erosion rate $\tan\alpha = 3.8$ m, then $t = 4.03$ years. Actually the denominator in Eq. (5.4) is then 0.202, which gives from Eq. (5.5) $t = 4.07$ years, a negligible difference. For convenience in design, Eq. (5.4) has been plotted in Fig. 5.14, for various values of $\tan\alpha$ = long-term erosion rate in meters/annum. Thus, for some initial excursion (Y) the time (t_e) for its complete disappearance is given for some value of $\tan\alpha$. For example, as shown dotted in the figure, a value of $Y = 64.5$ m with $\tan\alpha = 4$ the value of t_e is 3.3 years. Another set of curves can be derived for a net excursion of 10% the initial value rather than zero. This would be reached more quickly as given by the graph to the right of Fig. 5.14. As seen by the dotted curve this would be at 65% of the 3.3 years for complete disappearance, or 2.1 years after the fill was carried out.

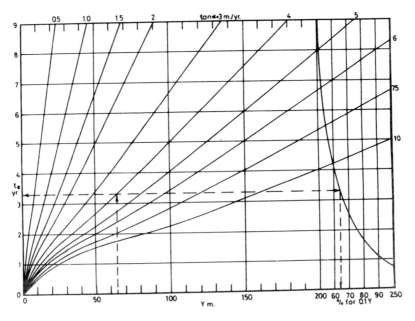

Figure 5.14 Reduction of beach width on a filled site versus long-term erosion rate and time, per Eq. (5.4)

When the original renourishment has been reduced to one tenth of its width some alarm would be raised, either by the coastal engineers concerned or by the rate payers who have financed it. At this receded beach alignment a larger than normal storm could obliterate the berm and attack facilities or dunes landward of it. Thus, at this stage of denudation authorities would be considering an addition of sand. These authors have observed in the literature such repetition at about 2 yearly intervals.

Some examples of renourishment are worth reporting. Fisher and Felder (1976) reported: "The 1973 beach nourishment at Cape Hatteras placed approximately 465,000 m^3 on the subaerial beach. Eighteen months later, about 51 percent of this material remained on the beach. During this period there were relatively few storms, and this mild wave climate is largely responsible for this high retention. Our analysis at Cape Hatteras suggests that in attempting to develop a relationship between storms and volume changes, that the post-storm wave conditions may be important." This fact has been highlighted in Chapter 3.

Everts et al. (1974) in discussing beach fill at Atlantic City, New Jersey, observed: "Survey results show that following replenishment, losses of the fill material above MSL were between nine and twelve times the losses measured in adjacent non-fill areas." The island of Sylt, West Germany, is protected in part by a seawall, revetment, and tetrapods placed in front of these and portions of the plain beach. Progressive erosion instigated a renourishment program by means of constructing a sand groin normal to the coast at the center of the affected area (Fohrböter 1974, Wenzell 1978). A continuous bar rising to -2.5 m some 525 m from the fixed boundary contained a trough to 5.2 m between these features. Wenzell (1978) concluded that: "The beach nourishment extended only to the beach zone down to a depth of about -5 m. Particularly the bar of the central 1.5 km shoreline remained unaffected by the fill." A further replenishment was to be carried out in 1978 by filling 900 m alongshore.

Beach nourishment was carried out at Virginia Beach, Virginia, during 1966, 1967, and 1968 with costs rising from \$1.41 to \$2.91 per cubic yard during this period. It was considered that 141,000 cubic yards would be necessary each year "to maintain present beach dimensions." This is likely to cost \$0.5 m annually, a very expensive alternative to headland control of this coast. On the cost of renourishment projects, Pilkey et al. (1983) suggested that: "In general, it is reasonable to assume a minimum cost of replenishment of an open ocean beach of \$1 million per mile (1.6 km)."

Grove et al. (1987) studied the migration of a stockpile of 200,000 cubic yards over two years at San Onofore, California. They found this hump of material moved downcoast at 0.73 km/annum. However, its crest decayed to 41% in one year, to 9.1% in 3 years, and 1% in 5 years. Thus, when the hump had moved 2.2 km it had been reduced to 9.1% and when at 1% it had only moved 3.65 km. But a significant observation was: "On the other hand, the erosion wave preceding the sand hump elongated its range much farther downcoast, to at least 9 km in less than 2 years of existence of the hump." Thus, the renourishment has the effect of preventing littoral drift at its upcoast edge.

Beach renourishment was carried out on 3.5 km of shoreline just south of Barnegat Inlet at Long Beach Island, New Jersey, where converging jetties force littoral drift seaward to form a large shoal (Ashley et al. 1987). Material was removed both northward due to refraction over this shoal and southward due to normal processes. After monitoring over 7 years the authors reported: "Residence time of the fill at the emplacement site was approximately 3 years. Given a total budget of $4.6 million the fill cost $1.5 million/year. The computed volume per annum was 330,000 m^3. Thus, in hindsight, the previously projected beach nourishment residence time of 8 years and a volume of 100,000 m^3/yr was a serious underestimate."

A similar situation was reported by Vallianos (1970): "Immediately following the construction of the Carolina Beach project, rapid erosion was manifest along the entire length of the fill structure. Though initial adjustments were expected, the actual changes, particularly those evidenced along the onshore section of the project, were much greater than anticipated during the planning and design phases of the project." In this region the net sediment movement is southward but much of the original fill was moved northward in the evening out of the shoreline. This prompted the addition of 360,000 cubic yards of material plus the construction of a 123 m groin at the northern terminal of the project. "In the year following the implementation of the emergency measures, approximately 203,000 cubic yards of emergency fill were lost to erosion and the major portion of the shoreline returned to about the same position it had prior to the emergency work."

Pilarczyk et al. (1986) outlined the contents of the *Manual on Artificial Beach Nourishment* (1986) giving background knowledge to renourishment. They distinguished between short- and long-term erosion as noted in Chapter 2. The active zone of wave action to where bar material cannot be returned to shore depends on the fineness of the fill material and the wave climate. The authors comment: "It should be stated beforehand that the seaward limit of the profile that is shaped by the action of the waves cannot be ascertained." In their concluding comments is the following statement: "Still, the more is learned about coastal processes, the more it is realized how little is known yet." This certainly appears applicable to the subject of renourishment as gauged by the empirical approach evident in the references just alluded to.

Not only is the problem of erosion not being solved effectively or economically, but at the same time it could also be creating problems downcoast. This sudden influx of sand placed into circulation travels downcoast as a hump and quickly silts up harbor and river mouths. Also, the prehump erosion noted above (Grove et al. 1987) can create problems. Such adverse trends can occur outside the boundaries of the authority carrying out the renourishment and may even extend beyond national borders. Perhaps when litigation can have such tendencies proven, the extra costs of these lawsuits may instigate a re-evaluation of this approach to beach stabilization.

Papers have been published which discuss the benefit of adding coarser material than exists on the coast (James 1974, 1975; Hobson 1977). The basis of such arguments is that it will not be moved so readily by waves. But in the sorting process that takes place during the formation and return of the offshore bar during and after storm sequences (Silvester 1984), the coarse sediment will end up in the beach face at the site and downcoast of it. These slopes will be steeper due to this larger sized material and will tend to reflect the incoming waves. Partial short-crested waves will be established with greater wave heights and greater propensity to move material alongshore. Thus, Nature alters its mechanism to maintain its littoral drift rate so that variation in size distribution will have little effect. No matter what material is used, it will be sorted by waves with the fine particles moving offshore and the coarser components resting in the beach berm above MSL. Material moved offshore provides a milder offshore profile which requires less material for the construction of the protective bar.

The ideas expressed above agree with those reported by Chapman (1980): "Whilst coarser non-cohesive sediment is less likely to be moved by any given wave than is fine sediment, it does not necessarily follow that coarser sediment will ultimately produce a more stable beach or more desirable beach state than native material. ... A beach will be built, by waves, of whatever suitable material is available. ... Sediment size is but one of a matrix of factors responsible for beach behaviour at a site under any given wave climate; varying sand grain size may simply cause a shift into a different stage in the morphodynamic continuum rather than producing a beach of different erosion resistance."

From the various examples discussed above, it can be realized that: "A beach nourishment scheme rarely provides a permanent solution" (Institution of Civil Engineers 1985). Although initial capital is relatively low for such a scheme, maintenance costs are high, so producing the nomenclature *renourishment* for repetitive action. Some engineers consider that beach renourishment works best at beaches where longshore transport is low, but it is arguable whether beach erosion at such a place might have become serious in the first place.

Renourishment, to the authors, is symbolic of the pill used by the medical profession to ease the pain without treating the real problem, in this case the lack of sand supply from upcoast. It is a palliative which reduces the incentive to deal with the cause of the illness. These engineering doctors recognize the transitory nature of the remedy and so recommend an annual consumption of painkiller to the extent of 20% of the initial dose. The actual need, of course, will depend on the wave climate, or more specifically the number of storms experienced each year. As time goes on these pills become more expensive as sediment in the locality becomes scarce, necessitating longer hauls. The alternative of dredging from offshore is now being contemplated, even though it may have some drastic side effects.

The reader might consider that too much space has been devoted to this topic. Perhaps there has, but it is because the authors feel that the two papers of Winton et al. (1981) and Winton (1983) should have received much wider publicity. Many eminent coastal engineers are still promoting this pill for ailing shorelines without recognizing the ongoing costs of such a prescription. Perhaps this presentation may help in a re-assessment. Magoon and Edge (1978) have observed: "The cost of replacing the sand lost annually from a beach stabilized by replenishment may be enough to justify the cost of structures to reduce or eliminate this loss."

The latest treatise on this topic (*Artificial Beach Nourishment* 1991) includes 8 papers by various authors on the following aspects:

1. philosophy and coastal protection policy
2. execution methods
3. monitoring program

4. particle size effect
5. numerical modeling
6. socio-economic aspects
7. environmental aspects
8. sea level rise

These invited papers were necessarily in favor of the procedure as they omitted 15 out of 21 references included in this section. Particular omissions were those of Winton et. al. (1981) and Winton (1983), which the authors have recommended above should have received wider publicity because of their erudite treatment of fill losses over time. There are many papers by Pilkey and his associates at Duke University that have questioned the economics of renourishment. The editors of this contribution in the initial paper compare it to painting cost on bridges and houses: "Repeated painting is never considered a waste of money, but as conservation of valuable investments; why should we not consider repeated nourishing of beaches in the same way?" No thought appears to be given to the frequency of repetition and percentage of overall cost in these two instances. While these papers are recommended reading some alternatives to keeping material in place should concurrently be considered conscientiously.

5.4 NEW CONCEPT

The best mode of coastal defense is to produce a situation where the shoreline is reoriented parallel to the crests of the incoming waves, thus minimizing or eliminating totally the sediment transport alongshore. This condition is seen to be provided by Nature in her ability to sculpture crenulate or zeta-shaped bays between headlands. Man's aim should be to cooperate with Nature and not confront her, for she holds the upper hand.

5.4.1 Headland Control

As presented in Chapter 4, crenulate shaped bays are ubiquitous, with a large spectrum of sizes on margins of oceans, enclosed seas, lakes, and rivers. They exist where waves arrive persistently from an oblique direction to the general coast. They therefore indicate to man Nature's way of maintaining beaches that are in dynamic or static equilibrium. This natural phenomenon can be utilized to retain beaches in locations where previously erosion was being experienced. Such a process has been termed *headland control* and is applicable to all shapes of coastline, be they straight, bayed, or convex in plan. The headlands themselves can be designed to minimize the effect of storm waves on them. The annual or more frequent removal of beach berm offshore during storms may not be uniform around a bayed periphery, but the naturally renourished waterline will be retained at the predicted shape.

The concept of providing Nature with beaches for her to defend in her own inimitable fashion has been discussed fully in Chapter 3. She has carried out this task over millions of years and will continue to do so, in spite of human interference. The wave energy applied to the coast will continue to fluctuate tremendously, with the formation of protective bars and their return to shore, accompanied by a pulse of littoral drift during this process. The only change required by man is to have the swell waves returning the bar to the beach arrive normal to it, so that longshore movement is reduced or brought to zero. In this event material moved offshore is returned exactly from whence it came.

This involves a simple reorientation of the beachline, as depicted in Fig. 5.15, in which it is seen that refraction causes wave crests to curve seaward. If a structure is located offshore, but close enough to create a tombolo to it, a shoreline will be constructed parallel to the incoming waves. On the downcoast side of the structure, waves will diffract and refract to form a curved beach parallel to these curving crests. With successive headlands a bay is formed through which continuing littoral drift will pass while maintaining its crenulate shape. In such an event the waves will not be breaking parallel to these waterlines because sediment must still be transmitted through the system. However, if this supply should cease the bay will erode back until static equilibrium is reached, which shape is predictable by the parabolic relationships given in Chapter 4. The availability of the equations, graphs, and tables in this tome should obviate the views as expressed by Magoon and Edge (1978) that: "However, no one has, as yet, demonstrated the ability to accurately predict, with either a mathematical or a physical model, the result on a shoreline of an artificially created headland. ... Only when one can adequately simulate the natural headland (or enclosed beach) in the laboratory or with the computer will one be able to predict the success of an artificial headland structure." Most progress has been made since 1978, although the use of the logarithmic spiral delayed application by some decades.

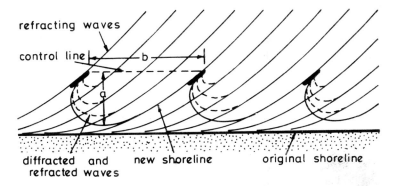

refracting waves

control line

b

a

diffracted   and   new shoreline   original shoreline
refracted waves

Figure 5.15   Reorientation of beachline, showing refraction over curved contours

Bays sculptured between headlands can be employed to stabilize straight, concave, or convex coasts (Hsu et al. 1989). This requires headlands at different spacings or distances offshore to produce stable accretion along the coast. Filling a bayed system from natural littoral drift requires the downcoast headland to be constructed first and others in turn after deposition has occurred. This interception of material could create substantial erosion downcoast of the last headland which may be of concern, or an added benefit in the case of a river outlet or harbor entrance where shoaling is to be prevented. The cost of headlands, spread over many years, has to be balanced by the land accreted from the sea which can be sold or leased, in the knowledge that it is no longer suffering erosion as did the original shoreline. Bishop (1983) has also suggested headlands as an alternative for beach stabilization.

5.4.2 Possible Applications

This concept has been employed successfully (Silvester and Ho 1972) along the southern coast of Singapore, where land has been reclaimed for the construction of high rise apartments and commercial activities. As seen in Fig. 5.16, the major wave input is swell from the South China Sea arriving from an easterly direction. This is quite oblique to the new shoreline and would have eroded it, moving material westward into the channel of the inner harbor. The original scheme was to provide a revetment wall along this 30 km length of coast. However, this started to subside due to the obliquity of the waves, and at a short course in coastal engineering, conducted in 1968 by the senior author at the then SEATO Graduate School of Engineering in Bangkok (now the Asian Institute of Technology), the problem was presented. It was suggested, and later accepted by the Singapore Housing and Development Board, to install two pilot headlands at the eastern end of the reclamation. These formed a bay immediately which gave a value of β that could be applied to other headlands of which there are now 48 along this segment of coast.

The swell waves from the east are mild in character, being at the most 6 seconds in period and 1 m high. They refract to become about 20° to the control lines joining the headlands, thus giving $a/R_o = 0.25$ (Fig. 4.33). In this region the tidal range is around 3 m and gabion-type structures (see Fig. 5.17) were initially placed at LWM some 70 m from HWM. This required a spacing of 280 m to maintain the greatest indentation at the original shoreline. The headlands themselves were 30 m in length and had to rise more than 3 m above the bed. During the tide waves diffracted and refracted to produce tombolos and finally a shoreline, as seen in the figure.

Due to people cutting the wire frames around the stones to reach the larger mussels inside the gabions, the Board adopted the idea of overfilling, digging a mound and associated trenches either side, and placing rip-rap stone to form the headlands. The surplus sediment was quickly removed from in front of these fixtures by the waves and moved westward to form new tombolos and establish bays in static equilibrium. In fact, there was a surfeit of material, so causing some headlands to be located landward of the shoreline. The Parks and Gardens Division of the Board wanted to plant trees and construct shelters on this protruding land but were advised to keep such amenities behind the predicted bay shapes which would ultimately be operable.

It is pertinent to note that the sediment obtained by flattening nearby hills, and transporting it by endless belt to the site, consisted of both clay and sand. However, the waves sorted the finer fraction offshore, leaving beautiful white sandy beaches. These were the first experienced in Singapore and hence a park established along 30 km length of coast has provided recreation for its populace. The Singapore Housing and Development

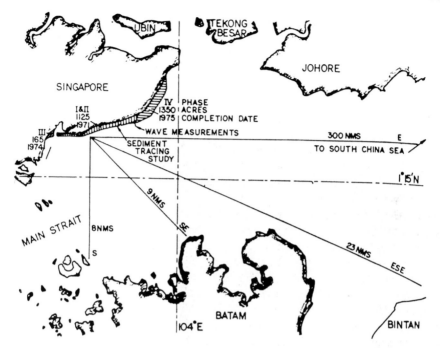

Figure 5.16 Headland control at Singapore, showing oblique wave input from South China Sea

Figure 5.17 Gabion-type structures served as headlands for the beach reclamation at Singapore

Board was proud of this achievement and hence printed the photo as in Fig. 5.18 on the cover of its 1973–74 annual report. Other aerial photographs of the result taken at different times are exhibited in Figs. 5.19 to 5.21.

Figure 5.18 Front cover of an annual report from The Singapore Housing and Development Board, showing the results of headland control

The advantages of this headland control approach over the seawall solution are as follows:

1. The coastline was stabilized with no sediment being transported by reflected and incident waves westward (see Fig. 5.16).
2. The cost savings, compared to the alternative sea revetment plus capping stones, has been estimated at 50% for the gabion headlands and 25% for the rip-rap version. This did not take into account the maintenance cost that would have ensued had the revetment been installed.
3. The beaches are safe for swimming whereas the revetment would have had deep water adjacent to it and would have been dangerous for children.

Figure 5.19 Aerial photograph of Singapore south coast looking east. The large groin and the seawall in the foreground were previously installed

4. The mild offshore slopes break waves prior to reaching the beach and hence the spray produced does not affect vegetation of the park. With the revetment the short-crested wave system so generated would have been accompanied by much spray onto the adjoining land.

This was the first major application of headland control (Silvester and Ho 1972) which prompted an invited paper (Silvester and Ho 1974). Further descriptions have been provided by Chew et al. (1974) and Wong (1981). Most of these bays have now reached full stability since no further sediment is available from upcoast.

Another application of this concept was to prevent sand incursion into a cooling water channel for a refinery (Silvester and Searle 1981). The location of a BP refinery was on a cuspate foreland within Cockburn Sound on the coast of Western Australia, as seen in Fig. 5.22. This salient had been formed by SW swell penetrating the gap between the mainland (Point Peron) and Garden Island and diffracting around the northern tip of this island. When a causeway and bridge were constructed to provide better access to the island, the wave energy through the restricted gap was reduced substantially. The major wave input was then these diffracted swell waves plus storm waves from the NW.

As seen by the location of the water intake in Fig. 5.22, the apex of James Point commenced to erode and would have formed a beach to the tip of the water intake

Figure 5.20 Aerial photograph of
Singapore looking west. The pool was
being constructed for leisure purposes

so silting the entrance. Three headlands were inserted north of the intake as indicated
in Fig. 5.23 in order to hold the beach in static equilibrium. As is evident from the
new circular shape of the northern half of the foreland, no further littoral drift was
expected from that quarter. Thus, the bays so formed could be considered to be in static
equilibrium. From the aerial photograph subsequently taken, as in Fig. 5.24, it is seen
that their shape is not of parabolic form as might be expected. This is due to incessant
sea breezes from the south during summer which tend to move sand northward. In fact,
the central bay has a slight protuberance due to this bi-directional transport. Should
excessive sediment be collected between the southernmost headland and the intake, all
that is required is for some of the berm to be removed and placed landward out of the
reach of waves.

The construction of these headlands in the shallow water offshore was effected
by providing sand spits by which limestone was dumped to the cross section shown in
Fig. 5.25. Their modest height of 1.8 m above LWM (the tidal range being only about
0.5 m) equates to that of the beach berm itself. With sand shifting and limestone being
available nearby, the cost for the three headlands, each some 50 m long, was $62,000 in
Australian currency in 1980. This was very much less than a seawall considered as an
alternative, which would have exacerbated the problem in any case. Although contact

Figure 5.21   Aerial photograph showing result of headland control in progress in Singapore

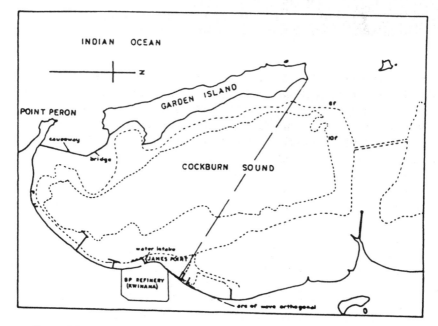

Figure 5.22   Location of a BP refinery at Cockburn Sound, Western Australia, showing cooling water intake

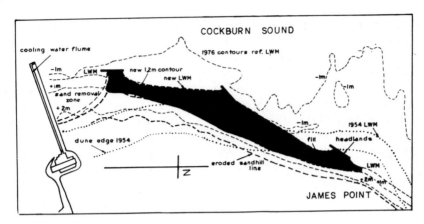

Figure 5.23 Three headlands inserted on Cockburn Sound to prevent sand incursion into cooling water intake of BP refinery

Figure 5.24 Aerial photograph showing the resultant waterline of the bays formed by headland control at Cockburn Sound

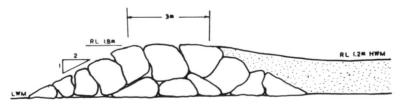

Figure 5.25 Cross section of the limestone headland at Cockburn Sound, Western Australia

with the mainland is lost in the lee of the headlands during storms, it is quickly regained when milder waves return.

Another example in Japan is the Kaike Coast of Honshu from which the photograph of Fig. 5.7 was taken. The formation of this type of barrier beach is indicated in Fig. 5.26 where the supply from a river causes sediment to be transported back into the lee of the prominent upcoast headland. The progressive lengthening of this spit by pulsative littoral drift is indicated in the figure. While this is still proceeding, the actual bay accretion has not reached the static equilibrium shape as predicted from the β value from the control line joining the bay extremities. Only after this has been reached can the β value be reduced to cover the case for the beach line being seaward of this stable shape shown dotted in Fig. 5.26.

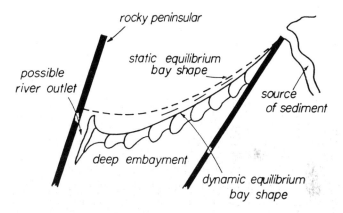

Figure 5.26   Formation of barrier beach in the lee of an upcoast headland with sand supply from a river

The case of the Kaike coast, which is one section of such a barrier beach, is shown in Fig. 5.27, where the Hino River had in the past provided material for its development. The widening of the spit adjacent to the Shimane Peninsula has been caused by tidal flow to and from Lake Nakaumi. Now that sediment supply from the Hino River has been drastically reduced due to dam construction upstream, the Kaike coast has suffered erosion. As noted already, the solution has been the construction of many offshore breakwaters (Toyoshima 1976) which have temporally retained some sand where it is required.

As seen in Fig. 5.27, some headlands spaced along this barrier beach at relatively large spacing could achieve a much more positive result. It is noteworthy that the orientation of these bays is opposite to those depicted in Fig. 4.41, due to the fact that filling is taking place in the main bay rather than removal. The most easterly headland in Fig. 5.27 provides shelter for the mouth of the Hino River in its lee, which is the more normal location for such outlets (Bascom 1954). This is due to the waves in the more curved section of the bay being diffracted strongly and hence are of smaller height with less run-up. The berm so constructed is therefore lower in elevation, resulting in

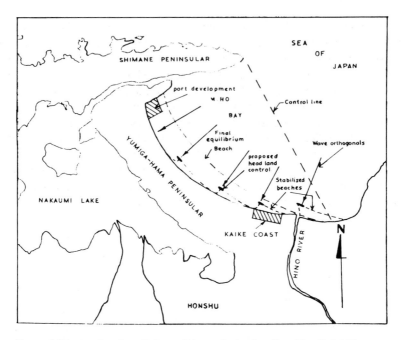

Figure 5.27   Beaches, from Kaike to Shimane Peninsula, affected by diminishing supply from Hino River, western Honshu, Japan (see Fig. 7.31 for location)

less material to be removed by flood flow or storm waves before the river can discharge normally. This section of bay has little longshore drift and hence shoals are less likely to form across the river mouth.

5.5 SHINGLE BEACHES

In many ways these beaches have been treated the same as sandy beaches, even though the actions of transient erosion during storms differ somewhat, as discussed in Section 3.10. The influence of a seawall on a sandy beach is to expedite longshore drift and as noted by Muir Wood (1970) for pebbles: "When a seawall is present, extreme flattening of the beach occurs when storm waves come in contact with the wall, and yet higher rates of littoral drift may therefore be associated with the consequent change of profile with each tide." He also discusses Seaford on the SE coast of England, which is a shingle beach: "For many years sea walls and conventional groynes have been constructed at Seaford during which period the sea continued to encroach, causing considerable damage and the collapse of the seawalls."

    As with a sandy shoreline, the removal of material downcoast can be prevented by either reducing the wave energy reaching the beach or by reorienting the beach more normal to the breaking wave orthogonals. This is attempted with groins, but as has been seen in Section 5.3.2 these exacerbate the problem of retaining sediment in place by causing rip currents during storms, which carry large volumes offshore. In the case of shingle this is not the case so that this is one important occasion where groins may serve a useful purpose. Unlike the straight version normal to the coast as is usual, a curved groin can be used as depicted in Fig. 5.28a. This runs from the heel, which is normal to the coast, to become parallel to the wave orthogonal normal to the developing beach. This is, in fact, shorter than the former for any angled waterline. The volume of sediment retention is slightly smaller but the persistent swell waves impact only the end of the groin. Depending on the wave climate some storm waves may arrive obliquely to this curved structure, so causing scouring adjacent to it, particularly if the offshore consists of sand. Only experience will confirm this action, but bed protection can be swiftly carried out, as outlined in Chapter 8. Similar types of action may be necessary on straight groins, should oblique storm waves be frequent enough.

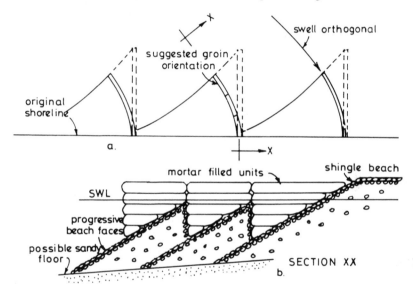

Figure 5.28 Groins for shingle beaches: (a) suggested orientation, (b) progressive building after beach accretion

Since the beach face slope of shingle beaches is steep, the nearshore depths are likely to be deep, which makes structures expensive. To overcome this problem groins may be taken partway out to the toe of the beach face only (see Fig. 5.28b) after which longshore drift will take the waterline to the extremity of the structure. By this time the steep beach face at the groin will have progressed seaward, so allowing for extension of the groin, as seen in the figure. The structure proceeds seaward in a curve following the specific wave orthogonal.

Thought could be given to making these structures from geotextile containers filled with cement mortar. This makes them easier to construct since little infrastructure is necessary to pump the mortar from the beach to its location in the beach face. Handling large stones or precast concrete blocks on steep pebble slopes is costly and dangerous.

The mortar filled units assume the shape of the pebbles or boulders beneath them, or the shape of the cast units already in position. Even if the supporting pebble base subsides due to subsequent wave action, these massive units will still have large volume in their broken sections and can readily receive new units which will be stable.

Shingle beaches have been suggested as a form of shore revetment or breakwater (Van Hijum 1974). For this purpose the natural shaping of the beach is required to find the volume required to protect a given facility. An extensive series of tests has been carried out in the Netherlands in both small and large scale flumes, in order to cover a large range of sediment size and wave characteristics. The major work has been reported by Van der Meer ad Pilarczyk (1986, 1988) and Van der Meer (1988).

The above authors derived a parameter $H_s/\Delta D_{n50}$, where H_s = significant wave height, Δ = relative mass density of stone, and D_{n50} = nominal diameter of average stone mass. Gravel or shingle beaches have this parameter varying from 15 to 500, which is quite an extensive range over which sedimentary material acts in a similar manner. Van der Meer and Pilarczyk (1986) found that : "For small diameters, however, it was concluded that some parts of the profile, for example the crest height, were not influenced by the diameter." Besides H_s and D_{n50} it was found that wave period T_z and number of waves N had an influence on the dynamic profile. The wave spectral shape and grading of material had little effect. Beyond a value of $H_s/\Delta D_{n50}$ of 15, the initial shape of the beach had no influence.

The definition sketch in Fig. 5.29 shows the variables that are here considered pertinent. Measurements are taken from the intersection of the SWL with the final equilibrium beach face. From these the height of the crest (h_c), its horizontal displacement (l_c), the depth to the change in underwater profile (h_s), and its displacement seaward (l_s), provide most of the input to the dynamic profile. The actual profile from the point A at the crest through origin (O) to point C can be defined by two curves, one above SWL and the other below. Landward of point A the accretion of pebbles can be considered to be the angle of repose or even a little steeper. This angle either side of A could be accepted as 45° as well as the deposited slope seaward of point C.

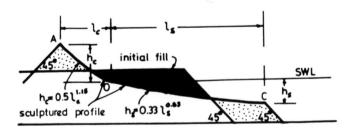

Figure 5.29 Definition sketch of shingle beach

The above variables can be combined with either D_{n50} or H_s together with a wave duration factor (N) to some power, as derived by Van der Meer and Pilarczyk (1986). Another parameter they derived included equal influence of the wave height (H_s) and period T_z in the form of:

$$H_o T_o = (H_s/\Delta D_{n50})(g/D_{n50})^{1/2} T_z \qquad (5.6)$$

Another function utilized was that of wave steepness (H_s/L_m) where H_s is the wave height at the toe of the structure and L_m is the deepwater wave length based on the mean zero crossing period $T_z = T_m$. This becomes:

$$S_m = H_s/L_m = H_s/(gT_m^2/2\pi) \qquad (5.7)$$

and may be termed a "fictitious" steepness by combining a local wave height with a deepwater length, as is done in the surf similarity parameter (Battjes 1974) or Irribarren Number (Irribarren 1938).

Van der Meer and Pilarczyk (1986) obtained the following relationship:

$$H_o T_o = 21(l_c/D_{n50}N^{0.12})^{1.2} \qquad (5.8)$$

In a plot of results with a curve Van der Meer and Pilarczyk (1988) gave no equation, but the following equation was obtained by these authors:

$$H_o T_o = 42.25(h_c/D_{n50}N^{0.15})^{1.325} \qquad (5.9)$$

A mean exponent for the RHS term of 1.2625 could be assumed, which from both equations gives:

$$1 = 2(h_c/l_c N^{0.03})^{1.262} \qquad (5.10)$$

If $N = 3,000$ waves of 10-second periods the duration of the storm is 8.3 hours which appears reasonable for a high tide duration ($N^{0.03} = 1.27$). This results in the ratio:

$$h_c/l_c = 0.735 \qquad (5.11)$$

In a similar way, the two following equations were obtained:

$$H_o T_o = 3.8(l_s/D_{n50}N^{0.07})^{1.3} \qquad (5.12)$$

and

$$H_o T_o = 38.37(h_s/D_{n50}N^{0.07})^{1.246} \qquad (5.13)$$

from which a mean exponent of 1.273 is used to give:

$$1 = 10(h_s/l_s)^{1.273} \qquad (5.14)$$

from which:

$$h_s/l_s = 0.164 \qquad (5.15)$$

Thus, if h_c and h_s can be predicted from some given wave condition, so l_c and l_s are readily computed. These vertical measurements were related to wave steepness by Van der Meer and Pilarczyk (1986) as follows:

$$h_c/H_s N^{0.15} = 0.089(L_z/H_s)^{0.5} \qquad (5.16)$$

which for $N = 3,000$ waves and $T_p/T_z = 1.15$ (for a Pierson-Moskowitz spectrum, where T_p is the period of the spectral peak) gives:

$$h_c = 0.102T_p(gH_s)^{0.5} \tag{5.17}$$

The second vertical dimension h_s was represented by:

$$h_s/H_sN^{0.07} = 0.22(L_z/H_s)^{0.3} \tag{5.18}$$

which, for similar conditions to the above, results in:

$$h_s = 0.205T_p^{0.6}g^{0.3}H_s^{0.7} \tag{5.19}$$

The equation given by Van der Meer and Pilarczyk (1986) for the beach face above the SWL was:

$$h_c = al_c^{1.15} \tag{5.20}$$

but applying Eq. (5.11) this results in:

$$a = 0.71l_c^{-0.15} \tag{5.21}$$

For the face below the SWL the equation was:

$$h_s = a'l_s^{0.83} \tag{5.22}$$

but when Eq. (5.15) is applied gives:

$$a' = 0.223h_s^{0.17} \tag{5.23}$$

Calculating values of $l_c^{-0.15}$ and $h_s^{0.17}$ for Eqs. (5.21) and (5.23) over the range for either from 5 to 13 m gives an average value of $a = 0.5$ and of $a' = 0.32$. Thus, the full equation for the beach profile is known.

Equation (5.17) has been plotted in Fig. 5.30 where T_p and H_s provide a value of h_c in the ordinate from which l_c is obtainable from Eq. (5.11) or the line in the figure. The $H_s - T_p$ relationship gives straight lines at varying slope getting steeper as H_s increases. However, not all these wave heights are available since at a limiting deepwater steepness of around 0.1 the heights are restricted, as per the line indicating "deep-water breaking limit."

Since the maximum impact or shaping function will be by storm waves, it is the condition for a FAS that should be examined. Silvester (1974) has shown that for this state $T_p = 4.4(H_s)^{1/2}$ for H_s in meters. The points on the $H_s - T_p$ curves have been joined to give the curve titled FAS. Silvester (1974) has also shown that, for a fetch of 5% of that for the FAS, $T_p = 3.8(H_s)^{1/2}$ for which points are joined to give the curve titled $F/F_{FAS} = 5\%$. It is the $H_s - T_p$ curves between these two limiting sea states that the h_c should be derived. He has also shown that for FAS $T_p = U_{19.5}/3$, where $U_{19.5}$ is the wind velocity measured 19.5 meters above the sea surface in knots. Thus, wind velocities in knots can be marked along the FAS curve which is producing optimum wave conditions.

As wind velocities exceed 30 knots, the fetches required for FAS become extremely long and are not likely to exist. Walden (1963) has observed: "The fetch required for

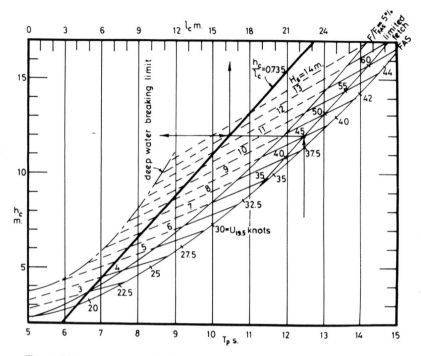

Figure 5.30 Values of h_c and l_c from wave characteristics of storm condition as per Eq. (5.17)

higher winds (up to about 30 knots) to produce fully developed seas occurs frequently in most areas. Wind speeds greater than 30 knots are rarely associated with fetches great enough to produce fully developed seas." Silvester (1974) therefore studied the effect of reducing fetches by larger percentages as wind speeds increased. This was interesting as it limited all $H_{1/3}$ values to around 10 m and T_p to 13 s, which agreed with data supplied by Scott (1968) and Moskowitz et al. (1962) covering the Atlantic Ocean and the Irish Sea. Silvester (1974) obtained a relationship of:

$$H_{1/3} = 0.2U_{19.5} \tag{5.24}$$

where $H_{1/3}$ is in meters when $U_{19.5}$ is in knots. The ratio $T/(H_{1/3})^{1/2}$ was reduced commensurately with the percentage of F_{FAS} employed so that T_p was derived as per the curve in Fig. 5.30 denoted "limited fetch." This virtually ceases at $T_p = 13$ s and $H_s = 10$ m as noted already, but has been extended to allow for greater wind velocities in higher latitudes (Scott 1968). It is thus the limited fetch curve that should be used to derive the maximum h_c and l_c for any expected wind velocity in knots that might have an extended duration at the site. An example is shown in Fig. 5.30 of 45 knot winds with limited fetch which produce 9 m high waves with 12.4 s period. These result in $h_c = 12$ m and $l_c = 16.3$ m.

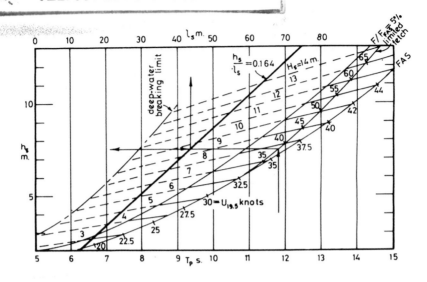

Figure 5.31 Values of h_s and l_s from wave characteristics of storm condition as per Eq. (5.19)

In a like manner Fig. 5.31 is presented for h_s and l_s from Eq. (5.19), with similar curves for FAS, $F/F_{FAS} = 5\%$ and limited fetches for higher wind velocities. The example shown in this figure is for $U_{19.5} = 39$ knots producing with limited fetch waves of 7.8 m and 11.8 s. These result in $h_s = 7.5$ m and $l_s = 43.5$ m. Van der Meer and Pilarczyk (1988) report that a computer software package has been produced to predict these values and those above.

It may be thought that h_s represents the threshold condition for shingle movement but active spreading from this depth must be taking place even though particles are only being oscillated backward and forward. The Hydraulic Research Station at Wallingford in UK has presented a graph of threshold depth for shingle of different sizes for various wave heights and periods obtained from their pulsating water tunnel. Table 5.1 lists wind speeds producing either a FAS or FAS with limited fetch (as previously discussed) which give certain values for H_s and T_p. These will be slightly different from the monochromatic waves implied in the water tunnel conditions but for threshold purposes could be accepted as equivalent. Accepting a condition of $H_s/\Delta D_{n50} \geq 200$, the D_{n50} for each H_s is listed in Table 5.1 from which the threshold depth h_{th} could be obtained. The value of h_s is given in Eq. (5.19) or Fig. 5.31. It is seen that the ratio of h_{th}/h_s is around 5, indicating that movement of shingle at h_s is quite pronounced.

Ahrens (1990) and Ward and Ahrens (1991) report similar tests to those by Delft Laboratory in the Netherlands but conducted for shallower conditions. Van der Meer and Pilarczyk (1988) did conduct a few tests with similar toe depths, but these data could not be separated from the remainder because no beach dimensions were listed in the table of results. As seen in Fig. 5.32 the ratio of h_c/L_o graphed against H_o/L_o to give:

$$h_c/L_o = 0.23(H_o/L_o)^{0.5} \qquad (5.25)$$

TABLE 5.1 VALUES OF h_s AND h_{th} FOR
$H_s/\Delta D_{n50} = 200$

	FAS	Limited Fetch		
$U_{19.5}$ (knots)	30	35	40	50
H_s (m)	5	7	8	10
T_p (s)	10	11.5	12	13
h_s (m)	5	7	7.8	9.5
D_{50} (mm)	15	21	24	30
h_{th} (m)	25	30	40	50
h_{th}/h_s	5	4.3	5.1	5.5

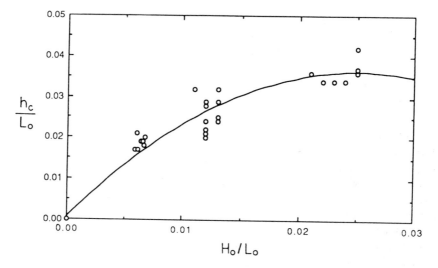

Figure 5.32 Normalized crest height h_c/L_o versus H_o/L_o for shingle beach

which can revert to:

$$h_c = 0.29T(H_o)^{1/2} \tag{5.26}$$

and compared to Eq. (5.17) which reduces to:

$$h_c = 0.32T(H_o)^{1/2} \tag{5.27}$$

It is thus seen that the ratios of h_c for Ahrens (1990) to that for Van der Meer and Pilarczyk (1988) is 0.9, accepting that H_o for the former equals H_s for the latter reference, whose measurement refers to just in front of the structure. For 50% of these results the d/L_o values were such that $H/H_o' = 0.92$ which implies Eq. (5.27) becomes:

$$h_c = 0.302T(H_o)^{1/2} \tag{5.28}$$

which is close to those of Ahrens (1990) in Eq. (5.26).

The distance l_c in Ahrens (1990) is plotted in Fig. 5.33 to give:

$$l_c/L_o = 0.418(H_o/L_o)^{0.533} \tag{5.29}$$

which could be equated to:

$$l_c/L_o = 0.36(H_o/L_o)^{0.5} \tag{5.30}$$

which from Eq. (5.25) gives:

$$h_c/l_c = 0.64 \tag{5.31}$$

which is 87% of the value given by Eq. (5.11).

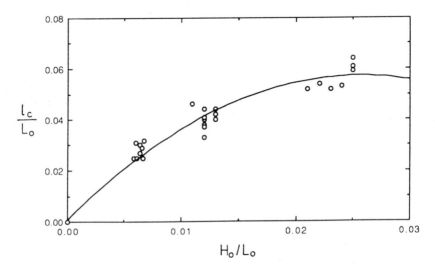

Figure 5.33 Normalized crest distance l_c/L_o versus H_o/L_o for shingle beach

It is believed that in the tests of Ahrens (1990) and Ward and Ahrens (1991), with its shallower conditions there was insufficient material to be shaped into the crest and deposited offshore. This is shown in Fig. 5.34, where curve S indicates sufficient, I inadequate, and F failure, in defense of the walled structure. The cross sections show that the underwater profile, with a steep depositional slope as at C in Fig. 5.29, was not formed, so preventing an assessment of l_s for analysis. These profiles would have caused more overtopping, which is supported by the milder slope shoreward of the beach crest. The subaerial beach slope from SWL was also steeper than the h_s/l_s obtained by the Delft studies of 0.164 (Eq. 5.15), these approximated 0.3 (varying very little with wave steepness). Since this portion of the profile is a hyperbola, (see Eq. (5.22)), its steepness will be greatest close to the origin at SWL, tending to be 0.3 as per this equation.

The main purpose for deriving the above relationships is possibly to ascertain the volume of material required to be stable in its own right (as a breakwater for example), or to protect some facility backing on to it. It is envisaged the shingle would be dumped outward to form a berm with a horizontal top and beach face with a slope of almost

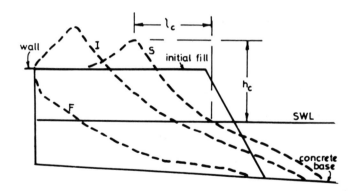

Figure 5.34   Model profiles of shingle beach as obtained by Ahrens (1990)

45°. As seen in Fig. 5.29, the volume to be eroded from this addition must equal that deposited in the crest section plus that removed seaward beyond the original beach face slope. It is extremely difficult to calculate these areas in terms of profile variables (as was done for sandy beaches in Section 3.4.3), so a transparent template with squares marked on it might be placed over the predicted profile in order to balance the eroded and accreted areas by eye. The original bed slope will be known from which volumes per unit length can be computed.

When determining the SWL for the calculation of the equilibrium profile, high water spring tide should be used. This lasts long enough (approximately 8 hours) for the shaping by storm waves. It is believed that a storm surge duration (usually about 2 hours) is insufficient for profiling of shingle to be effected. This may not be the case for sandy beaches which are readily moved. The size of the shingle does not appear to be an important variable as it was canceled out in deriving profile characteristics. Perhaps $H_s/\Delta D_{n50}$ greater than 200 is applicable for the above relationships.

This topic of dynamic revetments is again discussed in Section 8.1, where berm breakwaters are considered. There is little doubt in the minds of these authors that permitting Nature to shape its own profiles on the coast, bars for sandy beaches and S-shaped profiles for larger sized material, will effect economies in design.

5.6 COHESIVE SOILS

The topic of muddy shorelines was discussed in Section 3.9, where it was shown that very little research has been conducted on it. As Midun and Lee (1989) have stated: "There has not been a well developed theory regarding the erosion and suspension of clayey materials under wave and current actions. This discrepancy in attention results in the preponderance of empiricism and speculations that attend attempts to explain the phenomenon. Various researchers have tendered differing explanations on the erosion process in mud coasts." There is nothing wrong in using empirical relationships as long as all relevant variables are included; theory can follow later.

The attenuation of waves by mud is well known and is a fact that must be taken into account when this type of beach erosion is being considered. The relative importance of wave and current input to this action must be assessed. Muddy shoreline profiles differ from those for sandy coasts, but are rather similar from place to place. Recession of the waterline on the former can be an order of magnitude greater than on the latter. Reasons for it, and its variability, have to be understood if suitable remedial measures are to be undertaken. Solutions need to take into account the poor foundation conditions, so that some novel approaches are indicated. Solutions already suggested and some new ones will be presented.

5.6.1 Waves and Currents

Mud may exist on a section of coast as a layer from HWM to a depth of around 10 m or more. The only waves likely to reach the near vertical bluff, which defines the beach as such, are those from a storm. Even these will be modified greatly during their transit across the bed of denser fluid. Yu and Bao (1987) have noted: "Wave is the main power to erode beach." Their prototype measurements showed that mud concentration varied with the square of the maximum orbital velocity due to the waves. They add: "On the other hand, if no wave, the sediment on these erosion beaches is very difficult to be started by the tide current. This has been proved by many field observations." The situation changes drastically when waves and currents combine, as noted by the same authors: "In general, the tide current can't start the sediment in bed, but, as soon as wind wave exists, the sediment started by wave can keep suspending for a long time under the relative strong turbulence of waves. These sediments can be carried away for a long distance by the tide current."

The macro-scale orbits of tidal currents need to be known, to see whether these are essentially normal to the coast or along it. Measurements should be made over the complete tidal cycle during all stages. Since the offshore region is at a very modest slope, the waterline can range from some kilometers from high to low water mark. Wave breaking can take place over this same region which provides the turbulence for suspending material. If this is carried out of the area by tidal action or the seaward bed current generated by strong waves and not replaced by upcoast supply, then the beach recedes. The waves, in the absence of other currents, have their own mass transport net movement in line with their orthogonals, and therefore accumulate mud at the bluff.

5.6.2 Shoreline Profiles and Mode of Erosion

Midun and Lee (1989) have addressed the problem of stabilizing muddy coasts, of which a general profile is presented in Fig. 5.35, which is typical of tropical coastlines where mangroves flourish. Those in higher latitudes may not have this vegetation. Protection is afforded by mangroves as noted by Ibe et al. (1989): "From the main, prop-rooted stem, fibrous roots spread laterally through the muddy soil forming a broad based anchor approximately equal in extent the foliage above." Mangroves can deteriorate naturally if fresh water declines, but as Midun and Lee (1989) comment: "When fresh water streams are diverted and discharged to the sea at discrete points through tidal control gates, the

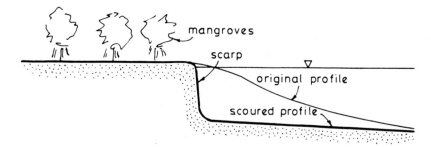

Figure 5.35   General profile on a muddy coast of Malaysia (Midun and Lee 1989)

affected stretch of mangrove belt suffers from a deficit of detritus supply." As recorded by Ibe et al. (1989), man's intervention has also caused this natural protective device to be decimated: "The mangrove trees are felled for erecting buildings, ... is also used for a variety of purposes—in staking out fish traps, tying up boats, cooking, constructing dry racks, etc."

As noted, erosion commences in the offshore area by deepening. Midun and Lee (1989) state: "Erosion normally starts with the lowering of the mudflats in front of the foreshore scarp, which may occur due to changes in the offshore environment. This drawdown permits higher depth limited waves to break directly on the scarp. The eroding scarp then strips off the mangroves." Pinchin and Nairn (1986) have also addressed this question: "A cohesive coastal profile recedes primarily from downcutting of the nearshore bottom part of the profile, not from erosion of the toe of the bluff portion of the profile. If this were not so, waves could not continue to reach the toe of the receding bluff and eventually bluff recession could cease." The action is similar to the liquefaction occurring at the scarp on a sandy beach, where groundwater is moving upward at the toe in its return to the sea, as discussed in Chapter 3.

Erosion of a muddy shoreline can be caused by a large mud shoal shifting location over a matter of months or years (Midun and Lee 1989). Where a river has changed its course and hence its outlet, the downdrift coast will suffer degradation (Yu and Bao 1987) to the tune of 20 km over the past 100 years. Recession amounts of 12 to 35 m/yr have been reported by Ibe et al. (1989) on the Nigerian coast over a 17-year period. This leaves little doubt that the subject is of paramount importance.

5.6.3 Previous Solutions

In respect to the Malaysian coast, Midun and Lee (1989) have reported on attempts to solve this problem in the 1960s: "The earlier protection works are mainly done on an *ad hoc* basis using temporary low cost technology, rock dump and sand bags. They were erected as an emergency response to a crisis situation without proper understanding of the underlying problem and thus, proper planning." The need, as is the case with sandy shorelines, is to reduce the wave energy on the affected coast, as noted by Yu and Bao (1987): "Hence, in order to prevent these beaches from erosion, we have made wave break farther away from beach and to reduce the wave action on the beach."

The first solution to be considered is that of accumulating a large batch of mud in front of the eroding shoreline. As Moni (1970) has suggested: "Within the mudbank, littoral material accumulates, thereby accreting the shore within it. Absence of waves preclude littoral movement through the mudbank and hence material once trapped within it is prevented from moving out. The mudbank acts as a storehouse of littoral motion." He records shoreline advances 25 to 30 m/yr. This added material is sand from upcoast of the mudbank, which is subject to movement downcoast unless means can be found to retain it in position to serve as a protector.

The second alternative is to actually dump sand on top of the mud for a reasonable distance offshore, as suggested by Pinchin and Nairn (1986). They recommended headlands between which sand nourishment is carried out, making due allowance for a storm profile. With the presence of the mud bottom offshore the wave intensity is not likely to be the same as on a normal beach and hence, the offshore bar should be smaller and closer to shore. Sand and mud will mix during this process, but, during subsequent return of the bar to shore, sorting of material will take place, so that the coarser sand grains will form the berm and the mud forced offshore. Midun and Lee (1989) have recognized the shortage of sand on muddy coastal areas.

The third possibility is to utilize rock structures, even though the shoreline to be protected provides very poor foundations for them. To protect their earth bunds, built some 400 m landward of the mangroves, Midun and Lee (1989) finalized on rock revetment with a mild seaward face, using small stones since: "those that are too big will result in foundation failure." Yu and Bao (1987) report that on the Chinese mud coasts rock is in short supply, even though they suggested T-shaped groins to prevent soil removal alongshore by tidal and other currents. Another structure they researched was the offshore breakwater, for which they found accretion occurred adjacent to the structure and not in the form of a salient from shore, as is the case for sand (see Section 6.2). Another difference was the fact that: "accumulation thickness is greater behind a detached breakwater during storming days." The opposite is true for sandy salients in the lee of such structures.

The fourth alternative is a thin mattress-like structure laid on the mud surface which should be able to support it. This suggestion has been put forward by Midun and Lee (1989), while Yu and Bao (1987) have recommended fabric sacks filled with beach soil to form as a core of their offshore breakwaters. These were then covered with stone armor, but the overall structure was lighter than with full rock employment.

The final solution so far tried involves floating structures which are an obvious choice with poor foundations. Ibe et al. (1989) suggested two goals to be achieved from this approach: "So, such floating breakwaters should be anchored to the seabed in such a way that they not only serve to dampen wave energy but that they also serve as sediment fences to trap and retain any sediment that would otherwise be lost 'forever' to deeper waters." They mention the tyre breakwater that is well known, but also another novel feature of "disused oil filled hoses linked together to form an underwater 'forest'."

Midum and Lee (1989) have termed the research effort as rather academic, stating: "Taking into account the typical long lead time to field application, which is easily in terms of years, a better understanding of the mechanics of mud coast erosion is unlikely to emerge in the immediate future. This is especially so in developing countries such as Malaysia." The phenomenon of this type of erosion is quite complex, with both model and field work difficult to carry out. However, it is a case of lateral thinking, as also applies to sandy coasts. It is hoped that with the resume above and the discussion in Section 3.9, some new concepts can be tried, at least on a pilot model basis, since full-scale testing is required in this difficult situation.

5.6.4 New Solutions

As discussed in Section 5.6.2, the erosion of a muddy shoreline is due to the offshore area becoming denuded, so that waves can then attack the toe of the bluff. If some economical means can be provided of covering this zone for a reasonable width with a lightweight covering, then the bed will not be lowered. This is where a mattress-like structure may be appropriate.

Such bed protection will be suggested in Section 8.3, where suitable fabrics are sewn to form containers for weak cement slurry mixes, which flow to all segments without the use of vibrators. These can take the form of a complete flat surface, top and bottom, or have longitudinal intermittent seams to provide sausage-like undulations at the top surface. For the purpose under study, the average thickness of these protection covers need not exceed 10 cm. Those recommended for use on sandy beds and adjacent to large breakwaters may need to be 0.5 to 1.0 m thick in order not to be lifted by wave action, as noted in Section 8.3. However, in the mud situation they will suffer less uplift due to the milder wave climate and the mud floor will more readily preclude water flow beneath them, so inhibiting upward motion. The mattress will virtually be "stuck in the mud" and so require a tremendous force to lift it.

These units are cast in place, but the mixing and pumping can be accomplished onshore with only lightweight pipes operating in the water. If there is a reasonable tidal range, much of the work can be carried out in the dry. The admixtures discussed in Section 8.3.3 still apply, so that costs can be minimized. It would not appear to be necessary to protect the full length of a coastline, but segments in echelon fashion with spaces of equal length to those covered. This method is worthy of some prototype trials as a pilot study, to check out costs as well as physical survival.

Midun and Lee (1989) compared the costs of rock structures with those of pre-fabricated articulated mattresses. They state: "Technical aspects apart, the cost of these systems is normally less competitive compared to rock revetment." Where such slabs are prefabricated they must be reinforced in order to withstand forces in handling, which is also rather difficult in mud environments. The unreinforced cast in situ version discussed above will have no forces applied to it even in service, so that its physical strength need be very moderate, with certainly no reinforcement.

The second novel approach involves a floating offshore platform as depicted in Fig. 5.36, taking the form of a submerged plate, to be discussed in Section 8.4. It is seen that pipes, possibly of plastic nature, are bolted in place between two timber beams to form a platform. These pipes are filled with fluid mud with a density between that at the surface and that close to the bed. The structure is then hung from 2 or more pipes

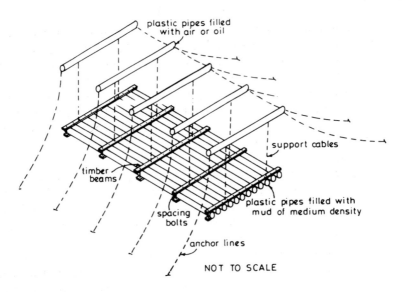

Figure 5.36   Floating offshore platform, consisting of pipes filled with mud

floating at the surface with cord lengths to give a submergence for most of the tidal cycle as designed in Section 8.4.1. The floating pipes could also be of plastic material to obviate corrosion. They are unlikely to bend due to wave action if linked at intervals to the platform below.

Such a mobile structure could serve as an offshore breakwater, which could accumulate mud on its leeward side, particularly during storms and so protect the shore where it is located. It could be expected that as the tide subsides the platform would rest on muddy bottom and perhaps the floating pipes on top of it. As the tide rose again these in turn would rise from the bed and continue their action of reflecting or attenuating waves. The anchors would be such that they could dig deep into the mud bottom. These floating offshore breakwaters could be spaced a short distance apart along the coast to be protected, and if found too close or too far apart can readily be moved to suitable locations. This is another remedial measure deserving of research on a prototype scale.

5.7 CLIFF RECESSION

In Section 3.11 the factors involved in cliff subsidence have been discussed, where the main one was seen to be wave action, of which beach formation at the foot was either the greatest inhibitor or greatest enhancer of scouring. There is no doubt that if a beach can be maintained in front of a cliff face then waves cannot reach it and so erode its base. Thus, any coastal defense measures, as with any other segment of shoreline, should have this as its major aim. Sunamura (1973) has observed: "On rocky coasts, however, the section once eroded away can never be restored to an original state even by artificial ways. Therefore, perfect preventive measures should be taken."

Except at places where roads, services, building and community infrastructures are located, cliff erosion has not been regarded as a threat. In many cases, the land value of an inhabited cliff top at risk is not sufficient to make any protection cost effective (The Institution of Civil Engineers 1985). In view of the complexity in the resistance to abrasion among various cliff-forming materials, uniform cliff protection is difficult. For an eroding cliff, which provides longshore drift, protection at the cliff base by various means will only disturb the existing sediment supply, hence causing further erosion downcoast. As will be explained below, an integrated view has to be devised when cliff protection is planned upcoast.

While bluff erosion has provided material for longshore supply of sediment to beaches downcoast, the dependence on this supply is at great cost to valuable property and is also haphazard. As noted in Section 3.11 the collapse of cliffs is episodic, as will be the sand or cobbles delivered downcoast. With regard to utilizing this source, Giese and Aubrey (1987) have commented: "On the other hand, since bluff erosion supplies the sediment which maintains beaches and dunes, efforts to retard their erosion are often prohibited." A similar approach has been taken by some U.S. governmental authorities to normal beaches, where all man-designed structures are now banned. There is little doubt that some attempts to stabilize coasts have met with little success, and as Kuhn and Shepard (1983) have observed: "The erosive effects of improperly located artificial protection and other structures have not been fully understood, especially during periods of large storm and high tides. It has been observed that where such protective measures project or extend seaward beyond adjacent unprotected lots, there is immediate erosion and notching of the latter."

Sunamura and Horikawa (1972) have discussed seawalls made of concrete or steel sheet-pile placed near or adjacent to a cliff foot. They also describe sections of a coast where concrete blocks of various shapes have formed trapezoidal mounds with side slopes around 1:1, that were displaced from the cliff face. Measurements made from aerial photographs over 10 years permitted a comparison between unprotected and protected regions. Although the erosion rates over this 9 km length of coast were extremely scattered, those over the protected region were substantially smaller. This solution is very expensive and a further detraction is that the shoreline is made unattractive to people who wish to utilize the water margin. If such blocks are placed on sand or small cobbles, they are subject to subsidence due to scouring of the bed along their seaward face, as discovered with offshore breakwaters built in a similar manner (Toyoshima 1982).

What is suggested here to stabilize a cliffed coast is the same as for a sandy shoreline, normally the insertion of headlands to form crenulate shaped bays between them. This places the full reliance on Nature to maintain beaches in position, either in a dynamic or static equilibrium shape. Since material is bound to be arriving from upcoast for many centuries to come, most bays so formed will generally be in dynamic equilibrium. However, since the supply of material, in this case from episodic collapses of cliffs, can become zero for lengthy periods static equilibrium bay shapes should be used in design.

As with sandy shores, the waterlines being predicted by the formulae provided as in Chapter 4 apply to the summer or swell-built profile. To safeguard facilities on the coastal margin, the cliffs in this case against wave attack, allowance must be made for the beach berm to disappear, especially in the case of a sandy shoreline. Estimation of berm retreat can be made by the optimum beach erosion model as outlined in Section 3.4.3. Where cobbles are the major component, bars are not so likely to form and hence the waterline is not likely to retreat far. As noted in Section 5.5, the major effect of storm waves is to push the whole beach face landward slightly and build up its crest. Relationships presented there should permit profiles to be predicted.

A basic premise in coastal engineering design is that the coast must be looked at on a macro-scale, or in terms of kilometers of coast if not tens or hundreds of kilometers. Hence, if a cliffed section which is supplying sediment to downcoast beaches is to be stabilized, the effect of this new condition on adjacent beaches must be planned for, prior to implementing this action. This again may utilize headland control, but if a large bay occurs downcoast of a rocky segment it may already be near static equilibrium and hence no deleterious effect may be felt. If natural headlands existed all along coastal margins, then littoral drift would be minimal and most coastal engineers dealing with sediment problems would be out of work.

It may appear somewhat peculiar that this coastal defense measure is being discussed without some figure to complement it. But since the same principles apply as for a normal beach, except perhaps that cobbles are present instead of sand, the bay shapes for design purposes will be similar. In fact, this coarser material will reach static equilibrium in plan much more readily than sand. If these new embayments are to reach their equilibrium condition from natural supply of sediment from upcoast, it is obvious that treatment of a segment of coast should commence in the downcoast region. Should an intermediate section require some remedial measure urgently, headlands could be installed with concurrent reclamation, in which case additional material from upcoast can bypass the controlled section to feed further alongshore.

5.8 SEA LEVEL RISE

Although it has been stated in Chapter 3 that tides of 2 m or less will not greatly affect the processes outlined there, the subject of sea level rise has assumed such importance in recent time that it is felt it should be addressed. Certain hypotheses on its influence have been presented which need close scrutiny, as well as proposals for obviating the trend, if it actually exists. The changes in sea level can be broadly divided into the Holocene (between 15,000 and 6,500 years before the present BP) and from 6,500 years BP to this day (Gordon 1988). The evidence for the purported 120 m rise from the first to the current period might be questioned, based as it is on the average profile of the continental shelf with its edge at about this depth. But the more important issue is the relatively small change during the past 6,500 years. Is it likely to remain static or will the heating of the atmosphere by man's industrialization and effect on the stratosphere by chemical emissions cause an acceleration in sea level rise?

The accuracy of measurement of these long-term fluctuations must be addressed since changes need to be recorded over decades. Were the tide gages used back that far sophisticated enough to compute averages over some years down to parts of a millimeter? Modern day electronic instruments with frequent spot records of tide level permit computer analysis of such data, but older pen records are dubious. Physiographic features on the coast, plus levels of faunal and floral relics, are used to determine ancient sea levels, but are not extremely accurate. Some observers have not had the knowledge of coastal processes, which are now only coming to light from the accelerated research effort in this field. Tidal records appear to show places only 100 km apart having rises, falls, and static sea levels, which indicates some variables not being taken into account.

The major cause in future sea level rise is atmospheric temperature, which will heat the oceans and melt the ice masses. The change in water volumes from these two sources is not known accurately and so requires continued attention. The addition of sediment to the ocean from rivers does not seem to be considered very influential. The changes in their shape and proportions over geologic time, as land masses have accreted horizontally, must have changed the tidal oscillations in them. The impacts of elevations in mean sea level on the coast in general and on marine structures constructed thereon requires much research before this new variable is included in design and development. Differences need to be taken into account for sandy and muddy coasts. The hypotheses put forward by some eminent protagonists of the inevitability of sea level rise need to be questioned. In this regard, the lifetime of structures being planned has to be kept in mind, since designing them for a century ahead is not economical or effective. As far as the beaches themselves are concerned, the predicted erosion may or may not occur, and any modifications would be so mammoth that national economies could not stand the cost. It is a matter of deciding to accept the changing conditions or to act just in the most affected areas, if these can be determined.

This section is not meant to be a comprehensive survey of all aspects of the greenhouse effect and its concomitant rise in sea level, nor even a review of papers dealing with the latter subject. It will put forward some new views on tidal measurements, the criteria used in fixing old dating sea levels, some causes of change, relevant impacts on the coast, and decisions to be made. These new hypotheses have not been proven but they are food for thought for researchers in the field.

5.8.1 Tide Measurements

Sea level fluctuations are due to many causes besides the normal tidal oscillation. They are affected in the short term by storms due to air pressure variations, storm surge due to wind stress on the ocean, and flooding from adjacent rivers. In the long term, glaciological, seismological, and tectonic events can influence water levels. To these must be added the errors that may arise in the recording, on which Bird (1988) comments: "but it should be remembered that most tidal gages were established for port operation and navigation purposes rather than for scientific research. They have low resolution, and their datum levels have often been poorly maintained."

Lengthy records are available from very few locations, the earliest being at Amsterdam in 1682, around the Baltic Sea from 1839, and in Cornwall, UK since 1919 (Embleton 1982). They consist of the addition of some 62 sinusoidal components of different frequencies, which have to be averaged over many years before a MSL can be established. In fact, it should be assumed over one lunar nodal year of 18.6 years. The highs and lows in tides vary continually and tables given of spring and neap to distinguish highest and lowest variations throughout a day. What is being sought are variations in the mean value of 1 or 2 mm per annum with fluctuations of one to several meters. This level of accuracy demands excellent input, for which commercial operation of tide gages cannot generally meet. This applies particularly to older installations where floats and pens on drums was the order of the day.

5.8.2 Sea Level Change over Time

Some authors believe that mean sea levels can change differently at various continental margins due to the lack of correlation experienced world wide. Lisitzin (1974) has stated: "However, it must always be remembered that, since all oceans and seas are interconnected, sea-level changes in one part of the Earth's globe must respond to related fluctuations in others." This has been reiterated by Titus (1990). Thus, rates of rise should be apparent in all tide records around the world if the evidence is to be accepted at all.

When the sea level data during the Holocene is viewed, as in Fig. 5.37 (Carter 1988), and the curves of different workers put together in an envelope, the differences at any one time can run into several meters. This calls into doubt the suppositions on which the levels are based. It is generally accepted that some fossils, which would have been deposited near sea level at the time and whose date of origin can be fixed, give the required level at that date. This does not take into account that the shoreline on which it was placed on its demise may not have been subsequently eroded and this material slumped or transported to some depression further along the coast. They thus could be found over a range of depths varying by several meters. As Carter (1988) states: "These difficulties, added to the common problem of making widespread stratigraphical correlations, normally hampers accurate elucidation of sea-level trends less than one million years long. ... the 'search' for a global eustatic sea-level curve has now been all but abandoned."

Van de Plassche (1986) has listed the many marine fauna and flora plus sedimentological evidences used in determining ancient sea levels. Even the dating of these can be called into question. Bird (1988) suggested that other forces may have played a part: "It is then necessary to decide whether observed changes are simply the results of sea-level rise, or whether they have been caused (at least in part) by other geomorphological and ecological processes."

Although coral reefs grow and die at certain levels connected with the existing sea level at the time, the dating of core sections at different depths need not indicate changed sea levels in that time period. Silvester (1965) has shown that these massive structures must be vibrated down into a sedimentary bed, like a tooth in a gum, in order to survive structurally. Thus, sections of coral that died just below sea level are moved downward by tens of meters over centuries, so giving erroneous points on a tide-time graph.

It is disconcerting to read of places only hundreds of kilometers apart that show varying rates of sea level rise, or even some falling or remaining static. This makes

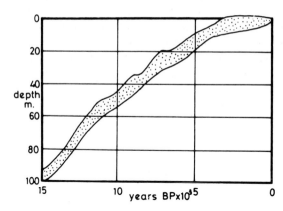

Figure 5.37 Sea level data during the Holocene, as suggested by different workers

correlation worldwide very difficult (Pirazzoli 1986). Applying regression analysis to tidal records from 23 selected cities around the world, Rossiter (1967) and Hicks (1973) have shown MSL trends ranging from −4.1 to +5.7 mm/year. All this conflicting evidence points to some missing variables in the processes.

5.8.3 Causes of Sea Level Change

The major cause being proposed at the moment is that of rise in atmospheric temperature, which ultimately heats up the surface of the sea, covering 70% of the globe. This can only influence the surface, with this heat budget being transmitted downward by various physical mechanisms, but mainly by turbulence during wave action. The water-particle motions down the water column cease at about half the deepwater wave length, which for the major waves in a fully fledged storm is around 150 m (Silvester 1974). This, it is contended, is the limit of warming for the oceans since below this level the colder water will remain virtually static due to stratification. Even though tidal oscillations may cause motion below this depth, the orbital motion is only horizontal due to the tides being extremely shallow water waves. The vertical component of the tide in deep water would be no greater than 0.5 m and the water above and below 130 m depth would be acting in concert. Therefore, there would be very little shear stress with no vortex motions resulting therefrom. Thus, only the 130 m thickness of sea should be used in any calculations of volume change due to increased temperature.

As noted already, most tide gages are associated with harbors and are therefore located at the mouths of estuaries, large rivers, or deep embayments, where natural shelter is available. Most large ports of the world are in such locations. However, these sites can suffer greatly from siltation, not necessarily in a uniform manner, but by fluctuations as infrequent river floods transport large volumes of material out to sea.

Thus, any tide gage fixed in some depth of water may have that depth, over a wide area, changed by some reasonable percent. This, in turn, affects the characteristics of the tidal wave propagating through the area by having its crest or high tide level enhanced and its trough or low level decreased. This results in a higher mean level. This hypothesis is based on the theory for shallow water waves, but no quantitative figures can be provided at this time of the likely rise for a given change in widespread depth. This is a worthwhile topic for an applied mathematician versed in long-period wave theory.

As noted in Chapter 3, humps of sand can traverse the coast at time intervals of some 15 years, which could affect the tidal oscillations and indicate a rise or even a fall in sea level over such periods. Besides such fluctuations, as noted in Section 3.8, estuaries are noteworthy for silting up as sediment has no where else to go due to inward transport by swell or storm waves. These are locations of some of the major tide recording instruments so that the possible changes in tidal oscillations need urgent attention, as this may explain some of the anomalies discussed above.

Taking a more macroscopic look at siltation, worldwide in fact, it has been suggested in Section 4.3 that the continents have grown in area by accreting sediment horizontally, resulting from the spreading of material from rivers along the coast to some trap. It was noted that 75% of the land masses consist of sediment of which 75% was deposited under marine conditions. Thus, the shape of the oceanic basins, as well as the volume in which its water is contained, have varied over geologic time. It is this deposition of sand as accreting landmass, plus the building out of continental shelves, that could have partly caused the rise of sea level and still is, even though more slowly than previously.

Looking at the areas of basic rock structure in many continents, which was the limit of their landmasses at that time, and the vast areas now, it can be surmised that the tidal oscillations in the oceans would have changed. Whether this has had any effect on the MSL over time is difficult to say, but the reduced volume to contain the sea water has decreased and so sea level increased. This topic needs to be proved or disproved by adequate data or discussion, as it might explain some of the discrepancies in data already collected.

5.8.4 Engineering Impacts

Bruun (1990) foreshadowed that: "A normal beach recession will be about 0.3 m/yr. If sea level rises 1.5 m by the year 2100, shoreline recession will be about 200 meters." There are many implications for coastal engineers and managers to consider from sea level changes, as suggested by Gordon (1988). These include (1) impact on marine structures and set-back distances, (2) coastal recession both on the open coast and in estuaries, (3) flooding of river and deltaic plains, and (4) changes in the salinity regime of estuaries and water tables.

With respect to sandy shores, Bruun (1962) proposed a model to explain the erosion he foresaw due to sea level rise. This two-dimensional, close-system approach considered that, due to the increased depth, material would be spread out in the offshore region at the expense of that in the nearshore, and causing the berm to retreat, where the outer limit of deposition was not specified too definitely. He assumed that the parabolic shore profile would remain the same due to a similar wave climate at any site. Schwartz (1967) termed this model the *Bruun Rule.*

Beach erosion has almost become a norm and is very widespread worldwide. Indeed, Bruun's proposal (1962) may be useful, but as Bird (1988) explains: "On the other hand, beach erosion is not necessarily an indication that sea level rise is in progress. The modern prevalence of beach erosion is due to several factors that operate in varying associations around the world's coastline: a relative sea-level rise (i.e., coastal submergence) is only one of these. Beach erosion has been noted even on parts of the Scandinavian coast where the land has been rising."

Bruun (1990) has observed that: "The *Rule* has sometimes been used rather indiscriminately without realizing its limitations. One should always remember that it is basically two dimensional, but it is almost applied three-dimensionally." This is one aspect that issue can be taken with this proposition, because longshore transport of material will continue as long as material is available to be shifted by oblique waves. Sediment will be brought into an area and will be retained on the coast by the persistent swell. Infrequently it will be spread offshore in the form of a bar, but will be quickly replaced in the berm. If the water level is higher, the resulting berm will also be higher because of the extra run-up or reach of the waves when depositing this refurbished material. The only extra material demanded will be in the extra thickness of berm, which changes in width greatly even today with a constant sea level.

Gordon (1988) has disagreed with Bruun (1962) by stating: "Everts (1985) recognized the essentially two-dimensional nature of the Bruun approach. He argued that to analyze site-specific response to sea-level rise, it is necessary to consider both onshore/offshore and longshore mechanisms utilizing a sediment budget approach." More thought is required on this question which cannot be proven by model studies.

Muddy coasts will produce different problems from sandy coasts because of their inability to form berms and dunes above MSL. A rise in sea level will cause higher levels in the swampy zones associated with such shorelines. It is these areas, and river plains of even sandy shorelines, that will suffer the most drastic effects of sea level rise. Mounds constructed around deltaic islands to protect farmers working on them will have to be raised. The present ones appear insufficient as they are designed on past storm surges which are often exceeded.

5.8.5 What to Do—If Anything

One suggestion has been made by Bruun (1990): "Theoretical and practical results favor placement of material, not on the beach, but along the entire cross section, all at one time, with grains of various sizes placed *exactly* where they belong in the profile. It is unquestionable that this causes higher stability and less material loss alongshore as well as offshore." He cites a case on the Danish coast where this was carried out, but the long-term effects have not been reported. The expense of this proposal would be prodigious for the tens of thousands of sandy shorelines to be considered the world over.

The alternative is to do nothing, especially with the speculative nature of the problem being faced. Cocks et al. (1988) have questioned: "Can formal decision analysis be applied to situations of such high uncertainty, i.e., to contingencies which do not have an associated probability distribution? ... It may in fact be both a natural and a rational reaction to a highly uncertain and distant threat to do nothing."

Where structures may be involved their life span needs to be taken into consideration, as noted by Whalin and Houston (1987): "Coastal engineers do not currently need to consider accelerated sea-level rise for most projects. Even if a major acceleration starts around 2100, it will be at least 50 years before sea-level has risen one-half meter. The extra cost of designing most structures to accommodate accelerated sea-level rise, they suggested, will generally be greater if one assumes that the structure would have to be rebuilt in 50 years anyway." Stark (1988) supports this view: "Engineers and the whole community are reliant on accurate data and scenarios. Over-reaction could seriously damage national economies and should be avoided."

There is certainly a need to collect tidal records from a network of gages around the world in order to observe similarities or otherwise. Any differences in trend should be examined in the light of perhaps some of the variables outlined in this section and others already accepted as valid. In decades to come, an authoritative trend may emerge from which plans can be made for coping with an acceptable rise in sea level.

REFERENCES

Ahrens, J. P. 1990. Dynamic revetments. *Proc. 22nd Inter. Conf. Coastal Eng., ASCE 2*: 1837–50.

Anderson, G. L., Hardway, C. S., and Gunn, J. R. 1983. Beach response to spurs and groins. *Proc. Coastal Structures '83, ASCE* 727–39.

Artificial Beach Renourishment 1991. (Eds. J. Van der Graaff, H. D. Niemeyer, and J. Van Overeen) *Coastal Eng., 16*: 1-163.

Ashley, G. M., Halsey, S. D., and Farrell, S. C. 1987. A study of beachfill longevity: Long Beach Island, N. J. *Proc. Coastal Sediments '87, ASCE 2*: 1188–1202.

Baba, M. and Thomas, K. V. 1987. Performance of a seawall with a frontal beach. *Proc. Coastal Sediments '87, ASCE 1*: 1051–61.

Bakker, W. T. 1968. The dynamics of a coast with a groyne system. *Proc. 11th Inter. Conf. Coastal Eng., ASCE 1*: 492–517.

Bakker, W. T., Klein Breteler, E. H. J., and Roos, A. 1970. The dynamics of a coast with a groyne system. *Proc. 12th Inter. Conf. Coastal Eng., ASCE 2*: 1001–20.

Balsillie, J. H., and Berg. D. W. 1972. State of groin design and effectiveness. *Proc. 13th Inter. Conf. Coastal Eng., ASCE 2*: 1367–83.

Balsillie, J. H., and Bruno, R. O. 1972. Groins: an annotated bibliography. U.S. Army Corps of Engrs., *Coastal Eng. Res. Center,* Misc. Paper No.1–72.

Barcelo, J. P. 1970. Experimental study of the hydraulic behaviour of inclined groyne systems. *Proc. 12th Inter. Conf. Coastal Eng., ASCE 2*: 1021–40.

Bascom, W. N. 1954. The control of stream outlets by wave refraction. *J. Geology 62*: 600–10.

Battjes, J. A. 1974. Surf Similarity. *Proc. 14th Inter. Conf. Conf. Coastal Eng., ASCE 2*: 466–80.

BERG, D. W. and WATTS, G. M. 1965. Variations in groin design. *Proc. Santa Barbara Specialty Conf. on Coastal Eng., ASCE,* 763–97.

BIJKER, E. W., and VAN DE GRAFF, J. 1983. Littoral drift in relation to shoreline protection. In *Shoreline Protection,* Proc. Instn. Civil Engrs., Thomas Telford London, 81–86.

BIRD, E. C. F. 1984. *Coasts: An Introduction to Coastal Geomorphology,* 3rd ed. Canberra: Australian National Univ. Press.

———. 1988. Physiographic indications of a sea-level rise. In *Greenhouse: Planning for Climate Change,* Proc. Greenhouse 87 Conf., ed. G. I. Pearman. Divn. of Atmospheric Research, CSIRO Australia, Canberra, 60–73.

BISHOP, C. T. 1983. A shoreline protection alternative: artificial headlands. *Proc. Canadian Coastal Conf.,* 305–19.

BRUUN, P. 1962. Sea-level rise as a cause of shore erosion. *J. Waterways and Harbors Div., ASCE* 88: 117–30.

———. 1972. The history and philosophy of coastal protection. *Proc. 13th Inter. Conf. Coastal Eng., ASCE 1:* 33–74.

———. 1990. Worldwide impact of sea-level rise on structures. In *Handbook of Coastal and Ocean Engineering, Volume 1, Wave Phenomena and Coastal Structures,* ed. J. B. Herbich. Houston, Texas: Gulf Publishing Co. 647–72.

CAREY, A. E. 1907. The protection of seashores from erosion. *J. Soc. of Arts,* 650–63.

CARTER, R. W. G. 1988. *Coastal Environments.* London: Academic Press, Harcourt Brace Jovanovich.

CHAPMAN, D. M. 1980. Beach nourishment as a management technique. *Proc. 17th Inter. Conf. Coastal Eng., ASCE 2:* 1636–48.

CHATHAM, C. E. 1972. Movable-bed model studies of perched beach concept. *Proc. 13th Inter. Conf. Coastal Eng., ASCE 1:* 1197–1216.

CHEW, S. Y., WONG, P. P., and CHIN, K. K. 1974. Beach development between headland breakwaters. *Proc. 14th Inter. Conf. Coastal Eng., ASCE 2:* 1399–1418.

CHOU, I. B., POWELL, G. M., and WINTON, T. C. 1983. Assessment of beach fill performance by excursion analysis. *Proc. Coastal Zone '83, ASCE 3:* 2361–77.

COCKS, K. D., GILMOUR, A. J., and WOOD, N. H. 1988. Regional impacts of rising sea levels in coastal Australia. In *Greenhouse: Planning for Climate Change,* Proc. Greenhouse 87 Conf., ed. G. I. Pearman, Divn. of Atmospheric Research, CSIRO Australia, Canberra, 105–20.

DALLY, W. R., and POPE, J. 1986. Detached breakwaters for shore protection. U.S. Army Corps of Engrs., *Coastal Eng. Res. Center,* Waterways Expt. Station, Vicksburg, Miss., Tech. Rep. CERC-86-1.

DEGUCHI, I., and SAWARAGI, T. 1986. Beach fill at two coasts of different configuration. *Proc. 20th Inter. Conf. Coastal Eng., ASCE 2:* 1032–46.

DETTE, H. H., and GÄRTNER, J. 1987. Time history of a seawall on the island of Sylt. *Proc. Coastal Sediments '87, ASCE 1:* 1006–22.

DOUGLASS, S. L., and WEGGEL, J. R. 1986. Performance of a perched beach—Slaughter Beach, Delaware. *Proc. Coastal Sediments '87, ASCE 2:* 1385–98.

Effective Use of the Sea. 1966. Panel on Oceanography of the President's Science Advisory Committee, Washington, D.C.

EMBLETON, C. 1982. Mean sea level. In *The Encyclopedia of Beaches and Coastal Environments*, ed. M. L. Schwartz. Strousburg: Hutchison Ross Publ. Co., p. 541.

EVERTS, C. H. 1979. Beach behaviour in vicinity of groins—two New Jersey field examples. *Proc. Coastal Structures '79, ASCE 2*: 853–57.

———. 1985. Sea level rise effects on shoreline position. *J. Waterway, Port, Coastal and Ocean Eng., ASCE 3*: 985–99.

EVERTS, C. H., DE WALL, A. E., and CZERNIAK, M. T. 1974. Behaviour of beach fill at Atlantic City, New Jersey. *Proc. 14th Inter. Conf. Coastal Eng., ASCE 2*: 1370–88.

FISHER, J. S., and FELDER, W. N. 1976. Cape Hatteras beach renourishment. *Proc. 15th Inter. Conf. Coastal Eng., ASCE 2*: 1512–31.

FOHRBÖTER, A. 1974. A refraction groyne built by sand. *Proc. 14th Inter. Conf. Coastal Eng., ASCE 2*: 1451–69.

GIESE, G. S., and AUBREY, D. G. 1987. Bluff erosion on outer Cape Cod. *Proc. Coastal Sediments '87, ASCE 2*: 1871–76.

GORDON, A. D. 1988. A tentative but tantalizing link between sea-level rise and coastal recession in New South Wales, Australia. In *Greenhouse: Planning for Climate Change*, Proc. Greenhouse 87 Conf., ed. G. I. Pearman. Divn. of Atmospheric Research, CSIRO Australia, Canberra, 121–34.

GROVE, R. S., SONU, C. J., and DYKSTRA, D. H. 1987. Fate of massive sediment injection on a smooth shoreline at San Onofore, California. *Proc. Coastal Sediments '87, ASCE 1*: 531–38.

HICKS, S. D. 1973. Trends and variability of yearly mean sea level, 1893–1971.

NOAA 1973. U.S. Department of Commerce, Washington, D.C., Tech. Mem. No.12.

HOBSON, R. D. 1977. Review of design elements for beach-fill evaluation. U.S. Army Corps of Engrs., *Coastal Eng. Res. Center*, Waterways Expt. Station, Vicksburg, Miss., Tech. Rep. No. 77-6.

HORIKAWA, K., and SONU, C. J. 1958. An experimental study on the effect of coastal groins. *Coastal Eng. in Japan 1*: 59–74.

HSU, J. R. C. 1979. Short-crested water waves. *Ph.D. thesis*, Dept. of Civil Eng., The University of Western Australia.

HSU, J. R. C. and SILVESTER, R. 1989. Comparison of various defense measures. *Proc. 9th Austral. Conf. Coastal and Ocean Eng.*, 143–48.

HSU, J. R. C., SILVESTER, R., and XIA, Y. M. 1989. Applications of headland control. *J. Waterway, Port, Coastal & Ocean Eng.*, ASCE *115*(3): 299–310.

HUGHS, E. P. C. 1957. The investigation and design for Portland Harbour, Victoria. *J. Instn. Engrs. Aust. 29*: 55–68.

IBE, A. C., AWOSIKA, L. F., IHENYEN, A. E., and IBE, C. E. 1989. Erosion management strategies for the Mahin mud beaches, Ondo State, Nigeria. *Proc. Coastal Zone '89, ASCE 1*: 821–35.

INMAN, D. L., and FRAUTSCHY, J. D. 1965. Littoral processes and the development of shorelines. *Proc. Santa Barbara Specialty Conf. on Coastal Eng., ASCE*: 511–36.

Institution of Civil Engineers. 1985. *Coastal Engineering Research*. Maritime Eng. Group. London: Thomas Telford Ltd.

IRRIBARREN, R. 1938. Una formula para el calcula de los diques de escollera. *Rivista de Obras Publcas*, Madrid.

JAMES, W. R. 1974 . Borrow material texture and beach fill stability. *Proc. 14th Inter. Conf. Coastal Eng., ASCE 2*: 1334–44.

————. 1975. Techniques in evaluating suitability of borrow material for beach nourishment. U. S. Army Corps of Engrs., *Coastal Eng. Res. Center*, Waterways Expt. Station, Vicksburg, Miss., Tech. Mem. No. 60.

KRAUS, N. C. 1987. The effects of seawalls on the beach: a literature review. *Proc. Coastal Sediments '87, ASCE 2*: 945–60.

KUHN, G. G., and SHEPARD, F. P. 1983. Beach processes and sea cliff erosion in San Diego County, California. In *Handbook of Coastal Processes*, ed. P. D. Komar. Florida: CRC Press, 267–84.

LEHNFELT, A., and SVENDSEN, S. V. 1958. Thyboroen channel—difficult coast protection problem in Denmark. *Ingenioeren 2*: 66–74.

LILLEVANG, O. J. 1965. Groins and effects—minimizing liabilities. *Proc. Santa Barbara Specialty Conf. on Coastal Eng., ASCE*: 749–54.

LISITZIN, E. 1974. *Sea-Level Changes*. Amsterdam: Elsevier, Elsevier Oceanography Series 8.

LOVELESS, J. H. 1986. Offshore breakwaters: some new design considerations. *Instn. Water Congress and Scientists 40*: 511–22.

MACDOUGAL, W. G., STURTEVANT, M. A., and KOMAR, P. D. 1987. Laboratory and field investigations of the impact of shoreline stabilization structures on adjacent properties. *Proc. Coastal Sediments '87, ASCE*: 961–73.

MAGOON, O. T., and EDGE B. L. 1978. Stabilization of shorelines by use of artificial headlands and enclosed beaches. *Proc. Coastal Zone '78, ASCE 2*: 1367–70.

Manual on Artificial Beach Nourishment. 1986. Rijkswaterstaat Center for Civil Eng. Res., Codes and Specifications, Delft Hyd. Lab., The Netherlands.

MIDUN, Z. B., and LEE, S. C. 1989. Mud coast protection—the Malaysian experience. *Proc. Coastal Zone '89, ASCE 1*: 806–20.

MONI, N. S. 1970. Study of mud banks along the southwest coast of India. *Proc. 12th Inter. Conf. Coastal Eng., ASCE 2*: 739–50.

MOSKOWITZ, L., PIERSON, W. J., and MEHR, E. 1962. Wave spectra estimated from wave records obtained by O. W. S. Weather Explorer and O. W. S. Weather Reporter, 1, 2 N. Y. Univ. Tech. Rep. 3.

MUIRWOOD, A. M. 1970. Characteristics of shingle beaches: the solution to some practical problems. *Proc. 12th Inter. Conf. Coastal Eng., ASCE 2*: 1059–75.

NAGAI, S. 1956. Arrangements of groins on a sandy beach. *J. Waterways and Harbors Div., ASCE 82*(WW2), Paper 1063.

Our Nation and the Sea. 1968. Commission on Marine Science, Engineering and Resources (plus 3 volumes of panel reports), Washington, D.C., U.S. Govt. Printing Office.

PELNARD-CONSIDÈRE, R. 1954. Essai de théorie de l'évolution des formes de rivages en plages de sable et de galets. *Quatième Journées de L'Hydraulique,* Paris, 13–15 Juin. (In French)

PILARCZYK, K. W., VAN OVEREEM, J., and BAKKER, W. T. 1986. Design of beach renourishment scheme. *Proc. 20th Inter. Conf. Coastal Eng., ASCE 2*: 1456–70.

PILKEY, O. H. SR., PILKEY, W. D., PILKEY, O. H., JR., and NEAL, W. J. 1983. *Coastal Design: A Guide to Builders, Planners, and Home Owners*. New York: Van Nostrand Co.

PINCHIN, B. M., and NAIRN, R. B. 1986. The use of numerical models for the design of artificial beaches to protect cohesive shores. *Proc. Symp. on Cohesive Shores,* Nat. Res. Council Canada, 196–209.

PIRAZZOLI, P. A. 1986. Secular trends of relative sea level (RSL) changes indicated by tide-gauge records. *J. Coastal Res. 1*: 1–26.

RIEDEL, H. P., and FIDGE, B. L. 1977. Beach changes at Swan Island, Victoria, 1886–1976. *Proc. 3rd Aust. Conf. Coastal and Ocean Eng.*, 163–69.

ROSSITER, J. R. 1967. An analysis of annual sea level variations in European waters. *J. Royal Astron. Soc. Geophys 84*: 259–99.

SAYRE, W. O. 1987. Coastal erosion on the barrier-islands of Pinellas County, West-central Florida. *Proc. Coastal Sediments '87, ASCE 1*: 1037–50.

SCHWARTZ, M. L. 1967. The Bruun theory on sea level rise as a cause of shore erosion. *J. Geol. 75*: 79–92.

SCOTT, J. R. 1968. Some average sea spectra. *Q. Trans. Roy. Inst. Nav. Archit. 110*: 233–39.

SEIJI, M., UDA, T., and TANAKA, S. 1987. Statistical study on the effect and stability of detached breakwaters. *Coastal Eng. in Japan 30*(1): 131–41.

SILVESTER, R. 1965. Coral reefs, atolls and guyots. *Nature 207*: 681–88.

———. 1974. *Coastal Engineering, 1*. Amsterdam: Elsevier.

———. 1976. Headland defense of coasts. *Proc. 15th Inter. Conf. Coastal Eng.*, ASCE 2: 1394–1406.

———. 1984. Littoral drift caused by storms. In *Seabed Mechanics*, Proc. Symp. IUTAM, ed. B. Denness. London: Graham & Trotman Ltd., 217–22.

SILVESTER, R., and HO, S. K. 1972. Use of crenulate shaped bays to stabilize coasts. *Proc. 13th Inter. Conf. Coastal Eng.*, ASCE 2: 1347–65.

———. 1974. New approach to coastal defense. *Civil Eng., ASCE 44* (Sept): 66–69.

SILVESTER, R., and SEARLE, M. 1981. Headland control to prevent cooling water sand incursion. *Proc. 5th Aust. Conf. Coastal and Ocean Eng.*, 135–38.

SILVESTER, R., and VONGVISESSOMJAI, S. 1971. Characteristics of the jet-pump with liquids of different density. *Proc. 3rd World Dredging Conf.*, 293–315.

SORENSEN, R. M., and BEIL, N. J. 1988. Perched beach profile response to wave action. *Proc. 21st Inter. Conf. Coastal Eng.*, ASCE 2: 1482–92.

STARK, K. P. 1988. Designing for coastal structures in a greenhouse age. In *Greenhouse: Planning for Climate Change*, Proc. Greenhouse 87 Conf., ed. G. I. Pearman, Div. of Atmospheric Research, CSIRO Australia, Canberra, 161–76.

SUNAMURA, T. 1973. Coastal Cliff erosion due to waves—field investigations and laboratory experiments. *J. Faculty Eng., Univ. Tokyo*, Vol. 32.

SUNAMURA, T., and HORIKAWA, K. 1972. A study using aerial photographs of the effect of protective structures on coastal cliff erosion. *Coastal Eng. in Japan 15*: 105–11.

TAIT, J. P., and GRIGGS, G. B. 1990. Beach response to the presence of a seawall. *Shore and Beach 58*(2): 11–28.

THIELER, E. R., BUSH, D. M., and PILKEY, O. H. 1989. Shoreline response to hurricane Gilbert: lessons for coastal management. *Proc. Coastal Zone '89, ASCE 1*: 765–75.

THOM, R. B., and ROBERTS, A. G. 1981. *Sea Defence and Coast Protection Works*. London: Thomas Telford Ltd.

TITUS, J. G. 1990. Greenhouse effect and sea-level rise. In *Handbook of Coastal and Ocean Engineering, Volume 1: Wave Phenomena and Coastal Structures*, ed. J. B. Herbich. Houston, Texas: Gulf Publishing Co., 673–702.

TOYOSHIMA, O. 1974. Design of detached breakwater system. *Proc. 14th Inter. Conf. Coastal Eng., ASCE 2*: 1419–31.

———. 1976. Changes of sea bed due to detached breakwaters. *Proc. 15th Inter. Conf. Coastal Eng., ASCE 2*: 1572–89.

———. 1982. Variation of foreshore due to detached breakwaters. *Proc. 18th Inter. Conf. Coastal Eng., ASCE 3*: 1873–92.

VALLIANOS, L. 1970. Recent history of erosion at Carolina Beach, N.C. *Proc. 12th Int. Conf. Coastal Eng., ASCE 2*: 1223–42.

VAN DE GRAAFF, J., and BIJKER, E. W. 1988. Seawalls and shoreline protection. *Proc. 21st Inter. Conf. Coastal Eng., ASCE 3*: 2090–2101.

VAN DE PLASSCHE, O., ed. 1986. *Sea-Level Research: A Manual for the Collection and Evaluation of Data.* A contribution to Projects 61 and 200, Geo Books, Norwich.

VAN DER MEER, J. W. 1988. *Rock Slopes and Gravel Beaches under Wave Attack*, Delft Hyd. Communication, No. 396.

VAN DER MEER, J. W., and PILARCZYK, K. W. 1986. Dynamic stability of rock slopes and gravel beaches. *Proc. 20th Inter. Conf. Coastal Eng., ASCE 2*: 1713–26.

———. 1988. Large verification tests on rock slope stability. *Proc. 21st Inter. Conf. Coastal Eng., ASCE 3*: 2116–28.

VAN HIJUM, E. 1974. Equilibrium profiles of coarse material under wave attack. *Proc. 4th Inter. Conf. Coastal Eng., ASCE 2*: 939–57.

WALDEN, H. 1963. Comparison of one-dimensional wave spectra recorded in the German Bight with various theoretical spectra. *Proc. Conf. Ocean Wave Spectra*: 67–81.

WALKER, J. R., CLARK, D., and POPE, J. 1980. A detached breakwater system for beach protection. *Proc. 17th Inter. Conf. Coastal Eng., ASCE 2*: 1968–87.

WARD, D. L., and AHRENS, J. P. 1991. Laboratory study of a dynamic berm revetment. U. S. Army Corps Engrs., *Coastal Eng. Res. Center*, Waterways Expt. Stn.

WENZELL, D. 1978. Morphologic effects of Westerland beach renourishment 1972. *Proc. 16th Inter. Conf. Coastal Eng., ASCE 2*: 1859–72.

WHALIN, R. W., and HOUSTON, J. R. 1987. Implications of sea-level rise to coastal structure design. In *Preparing for Climate Change,* Washington, D.C.: Climate Institute.

WIEGEL, R. L. 1964. *Oceanographical Engineering*, Prentice Hall. New Jersey: Englewood Cliffs.

WINTON, T. C. 1983. Prediction of post-fill beach response. *Proc. 6th Aust. Conf. Coastal and Ocean Eng.*, 246–50.

WINTON, T. C., CHOU, I. B., POWELL, G. M., and CRANE, J. D. 1981. Analysis of coastal sediment transport processes fom Wrightsville Beach to Fort Fisher, North Carolina. U. S. Army Corps of Engrs., *Coastal Eng. Res. Center,* Misc. Pap. 81–6.

WONG, P. P. 1981. Beach evolution between headland breakwaters. *Shore and Beaches*, July, 3–12.

YU, G., and BAO, S. 1987. Erosion and erosion control of a silt-muddy beach. *Proc. Coastal Zone '87, ASCE 3*: 2453–60.

Applications of Headland Control

It was probably in the eighteenth century that man thought about shelter for his boats and started in earnest to occupy places on the coast. This would have started within river mouths and deep indentations, but later would have extended to more exposed sites. As he commenced building dwellings and roads close to the shoreline, he demanded that Nature cease shifting it. But it was not till the industrial revolution in the nineteenth century that he had the capacity to construct large rubble-mound structures to achieve his goal, or large dams to conserve drinking water. His vessels became larger which then required deeper channels to his ports. All these influences began to be felt on the coast where the constant transmission of sediment alongshore was intercepted and erosion occurred.

In the mid-nineteenth century, engineering was recognized as a profession and various institutes were formed, which then published descriptions of achievements in various fields. These concentrated on land-based activities for almost a century before coastal works found their way into the literature. This was aided by the needs of nations during wars, it not being until World War II that coastal engineering became acknowledged as a recognized facet of civil engineering. Since that time it has flourished, with the multiplication of technical journals dealing solely with this topic.

Any operation connected with the ocean or its margins is exceedingly costly in research, design, construction, and maintenance. Hence, savings had to be made in this chain of events and unfortunately these were effected in the areas of research and design. Ad hoc problems were tackled without due knowledge of the phenomenon involved, or the consequences of the structures installed on the coast. This aspect has been partly overcome by many large scale research institutions being established. But still many governments and private consultants are not keeping up with the literature, where new thoughts are being expressed. There has been a tendency to copy previous solutions without knowing of their long-term effects.

Many coastal problems revolve around sedimentation in general or erosion and siltation in particular. The long-term reduction in material available for the waves arriving obliquely to the shoreline is the basic problem to be solved. Structures must be devised to retain sand, shingle, or silt, where it is needed, without depleting the supply elsewhere. Many of the attempts to do this in the past have failed and, as a consequence, have multiplied the problems. Even the continued addition of sediment to the coast has not succeeded economically to abate this erosive trend.

However, by observing how Nature has reacted to longshore movement in the presence of headlands or fixed points on the coast, a glimpse may be obtained of a means to retain sand in locations, where it is wanted. The bays that form from wave action over centuries have a shape that is repetitive the world over. These minimize the movement alongshore of the mobile small grained elements that make up the beach berms that separate sea from land. The crenulate shape that they adopt can now be predicted, if only empirically, which involves the most persistent waves in the area, that of the swell. This arrives normal or near normal to the beach around the whole bay periphery, implying that the longshore component of breaking-wave energy is tending to zero and that littoral drift becomes negligible. Even when storm waves usurp the berm and place its contents offshore, the normal accreting swell subsequently returns it directly back onshore. The result is an unshifting shoreline.

The equation employed by early workers to describe the crenulate shape, namely the logarithmic spiral, was difficult to apply and has since been found erroneous for the downcoast or straight segment of this feature. The parabolic (or more precisely quadratic) function, only recently derived, and described fully in this tome, should permit a wider acceptance of this stabilization concept. It should certainly be included in any cost estimates of alternatives for erosion-abatement contracts. There may be secondary problems that are perceived or feared, but these authors would be willing to assist any authorities that forecast difficulties in implementing the procedures suggested.

An ever present concern is with the zone downcoast should littoral drift be retained along some segment of shoreline. This concern is to be admired, for it has not always been the case. It will depend on how the beaches upcoast of each headland inserted are to be accreted. If it is to be achieved by natural means, as with a groin system, then erosion is bound to occur beyond the field of headlands. As will be noted in this chapter, this is not necessarily deleterious, since some sections of coast can readily be lost to another farther downcoast, or silting at river or harbor mouths can be prevented. However, a status quo can be maintained by simultaneous reclamation and headland control, after which normal littoral drift will traverse the field and supply material, to retain the waterline in the downcoast region. The resulting bays will not be in static equilibrium as so designed, but all major infrastructure should be built shoreward of the storm beach profile of this stable shape, because drift fluctuations can cause these limits to be reached.

6.1 DESIGN OF HEADLANDS

For convenience of discussion headlands have previously been depicted as parallel to
the original shoreline, as in Fig. 6.1. These would receive persistent swell obliquely and
hence suffer scouring, as noted in Section 5.3.1 and in more detail in Chapter 7. An
alternative to this is to orient the headland parallel to the expected shoreline, as shown
in Fig. 6.1a. This would receive the full impact of storm waves and swell which would
generate either partial or full standing waves. Vortices will also occur at the downcoast
extremity which, as will be discussed in Chapter 7, can cause subsidence. The third
possibility, shown in Fig. 6.1c, is that of angling the headland into the newly formed
beach. However, during storms, the material fronting the structure will be removed with
insufficient volume to construct a full scale protective bar. High waves can thus impact
the structure over a reasonable length.

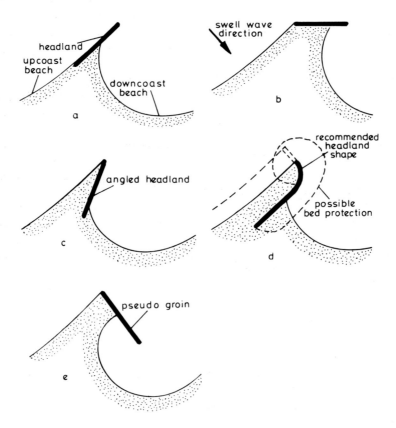

Figure 6.1   Various orientations and shapes of headlands

The best solution, and that strongly recommended by the authors, is shown in
Fig. 6.1d, where the downcoast end of the headland is curved seaward. This will accrete

a beach along the bulk of the straight section of the structure. If its berm is sufficiently wide, the worst storm waves may not remove it completely (see Section 3.4.3 for the estimation of beach and berm recession). But even if they do they must traverse the offshore bar and hence be greatly attenuated before impacting the structure. Thus, the straight segment need be of only modest height and cross section and its armor units can be smaller. It is suggested that its soffit be at the height of the beach berm or approximately 1.5 m above HWM, thus reducing the visual impact on the coast, a necessary condition at pleasure resorts. As soon as the milder or swell waves arrive, this beach is reformed in a matter of days ready for the next onslaught.

Another advantage of this version (Fig. 6.1d) is the slower diffraction of waves around the curve face which minimizes vortex formation and concomitant scouring. The extremity of the curved tip will have to be higher and armored by larger units as it can receive the full impact of storm waves. Also, should storm or intermittent swell waves arrive obliquely to this curved face then reflection from it could scour the bed, necessitating some protective measure, as indicated in the figure. A wave climate derived for the area will determine this need, or it could be left until after monitoring the offshore depths.

The alternative depicted in Fig. 6.1e has not been proven by model study, although it has been observed in some prototype situations. It could be installed with concurrent reclamation in order that the beach is formed at its seaward tip. This is somewhat similar to a groin offset from an original shoreline but oriented parallel to the orthogonal of the incoming swell in that depth of water. The downcoast bay curves to the same side as the upcoast beach, requiring a larger than normal berm which cannot be completely removed during a storm. If it disappears the resulting scour may cause subsidence, but only experience would tell the extent of this. The fronting beach would quickly reform once the swell waves arrived.

A new type of construction for headlands will be discussed in Chapter 8 involving the use of flexible membranes filled with a slurry mortar cast in situ. These could reduce substantially the cost of such structures, particularly if they are to be installed offshore prior to a tombolo forming and the beach accreting to its tip. They have been widely applied in Central and South America (Porraz et al. 1977, Porraz and Medina 1977) where the simple method of construction can employ much unskilled labor.

Overall, the technical considerations in relation to the design of headlands may be summarized as follows:

1. Properties of headlands
   - type, length, shape, distance offshore, spacing, orientation to shore
2. Construction procedure
   - construction sequence and protection
3. Dimensions of bay
   - geometry: size required, indentation, and degree of erosion to the existing shoreline
   - method of formation: naturally or aided by reclamation
4. Aesthetic aspects

Pilot studies of headland control could be carried out on sites where erosion is severe to prove to relevant authorities the validity of the approach. Experience could thus be gained to embark upon a major stabilization program which could save millions of dollars in beach facilities and dredging of channels suffering from siltation. The land gained from the sea could pay for the remedial measures.

6.2 SALIENT BEHIND SINGLE OFFSHORE BREAKWATER

Sand accumulation occurs in the lee of a breakwater placed offshore due to wave diffraction and refraction, aided by nearshore current circulation during its formation. The salient or tombolo produced depends mainly on the length of breakwater and its distance from the original shoreline. Hsu and Silvester (1990) have demonstrated that the distance of the salient apex from the breakwater is preferred to that from the original shoreline, since it relates more to the wave energy input. Previous results with the latter have exhibited unacceptable scatter, whereas the former provides a single curve with practically no deviation, despite the mixed use of model and prototype data. It has also been illustrated that the curved beaches so formed can be related to the static equilibrium shape of bays forming between headlands. The final waterline for some distance either side of the breakwater is more landward than the original, implying beach erosion in this vicinity.

In the case of a salient formed behind a multiple offshore breakwater system, there is no accurate prediction method available at the present stage, despite the same dimensionless ratios, used in the case of single offshore breakwater, having proven useful (Hsu and Silvester 1989a).

6.2.1 Salient Growth

Consider the case when waves diffract into the shadow zone of a sufficiently long breakwater offshore, or a breakwater tip of a harbor, they break at an angle to the original shoreline and hence transport sand toward the centerline of the structure, aided by nearshore current circulations. Initially double salients may form, especially behind a long breakwater, but with adequate wave duration and sediment supply, these will coalesce into a single apex.

This action is displayed in Fig. 6.2, where the initial littoral currents move sediment to form two protrusions with these currents sweeping into the shadow zone from either side. This centerwards drift is offset slightly by waves coming from the opposite extremity and generating a counter current. However, the short-crested system tends to move material into the bay formed between the developing salients, so that ultimately a single apex forms, as illustrated in Fig. 6.2b. There is then a balance between currents along the beaches either side of it. It is possible to enclose large bodies of water between these approaching protrusions as seen in Fig. 6.3 of the tombolo formed leeward of Oga Hanto (peninsula) on the Sea of Japan coast with the large lake of Hachirogata enclosed by barrier beaches on the south and the north.

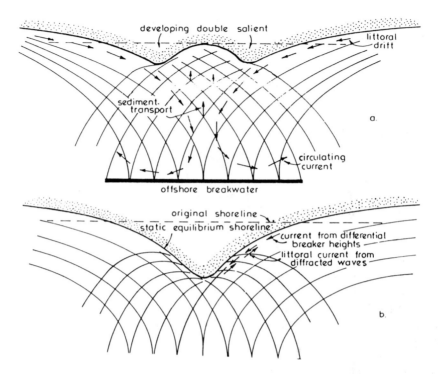

Figure 6.2 Stages of salient formation behind single offshore breakwater, showing: (a) two initial protrusions by littoral currents, and (b) a final single apex

Many workers have studied this phenomenon, assuming waves approaching normally to the structure, using models (Shinohara and Tsubaki 1966, Horikawa and Koizumi 1974, Rosen and Vajda 1982, Mimura et al. 1983, and Uda et al. 1988) or field data (Noble 1978, Dally and Pope 1986), even with numerical experiment (Perlin 1979). For studying the optimum form of sediment deposition in models, the distances to such double apexes should not be used in relationships because insufficient wave duration has been provided for one single apex to form.

A systematic investigation using model studies on this topiç has been reported by Shinohara and Tsubaki (1966), who have presented data of planform with measurement scales normal and parallel to the breakwater distorted, as do many other workers. One such plan has been redrawn to natural scale in Fig. 6.4. It is seen that double apexes existed up to 3 hours duration, but by 5 hours the shape was blunt ended, which was also the case for 6 hours when the test ended. The centerline protrusions from the original shoreline were plotted by the present authors against time. It was seen that at the end of the test the curve was still rising steeply. It was extrapolated to become asymptotic to a line parallel to the time axis, which indicated that for complete equilibrium the apex would have reached the breakwater, so forming a tombolo.

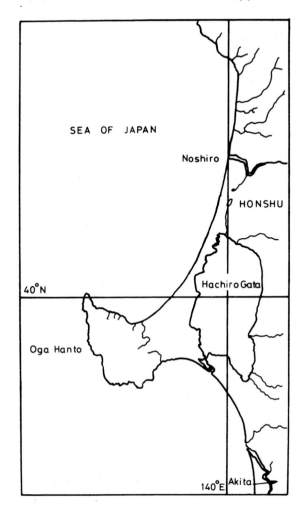

Figure 6.3 Tombolo formed leeward of Oga Hanto (Peninsula) on the Sea of Japan (see Fig. 7.31 for location)

6.2.2 Dimensionless Ratios

A single breakwater parallel to the coast with normally incident waves, as shown in Fig. 6.5, has an optimum apex distance from the original shoreline (Y), and from the breakwater (X). The radius from the diffraction point (i.e., the tip of the breakwater) to the apex at the breakwater centerline is denoted as R_1, at angle α to the breakwater. The length of the breakwater is B and its distance from the original shoreline S. Some workers have monitored the sand volume of the salient, but not the beach degradation either side, even if it was exhibited in the model studies.

  These variables were generally listed and grouped into dimensionless parameters for the breakwater length (B), distance to the original shoreline (S), salient protrusion from the shoreline (Y), water depth (h_B) at the breakwater, the volume of sand accreted, position of the breaker line (X_b) from the original shoreline, and deepwater wave steepness (H_o/L_o).

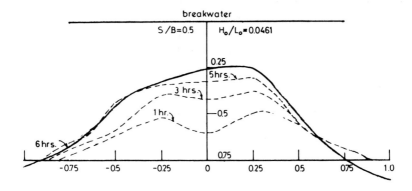

Figure 6.4 Progressive salients as measured over time by Shinohara and Tsubaki (1966)

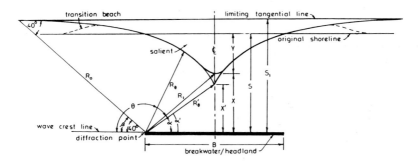

Figure 6.5 Definition sketch of salient formation in lee of offshore breakwater, with normal wave approach

These have resulted in scattering, so the relationships attempted did not inspire confidence in their application. It has been shown by Hsu and Silvester (1990) that all but the first two have little relevance to the problem. The variables used in this reanalysis are listed in Table 6.1 from various papers, together with the equivalent distance ($X = S - Y$) from the salient apex to the breakwater. It is also noted in Table 6.1 whether the data apply to model (M) or prototype (P) situations or numerical simulations (N). All dimensions are in meters. The experimental data provided in Table 6.1 have been taken from Rosen and Vajda (1982), Suh and Dalrymple (1987), and Uda et al. (1988). A total of 46 results, of which 35 are from model tests, 4 from prototype situations, and 7 from numerical simulations (Perlin 1979), have been analyzed in an attempt to establish useful dimensional parameters for engineering purposes. All results analyzed pertain to salients that have not reached the breakwater and hence not formed a tombolo, but curves can be extended to give breakwater offsets that define this condition. Some workers term all salients as tombolos, but this should be reserved for shorelines reaching the structure. Cases for the width of the tombolo at the structure and effect of oblique waves have not so far been treated.

TABLE 6.1   DATA OF ACCRETION BEHIND A SINGLE OFFSHORE BREAKWATER

No.	B	S	Y	X	h_B	X_b	H_o/L_o	Authors	$M/P/N$
1	180	370	170	200	1.8			Dally & Pope (1986)	P
2	610	610	240	370	8.4				
3	4.0	2.0	1.07	0.93				Horikawa & Koizumi (1974)	M
4	1.5	1.8	0.65	1.15	0.09	1.8		Mimura et al. (1983)	M
5	540	555	250	305	7.2			Noble (1978)	P
6	325	1666	10	1656	9.0				
7	100	50	18	32			0.0086	Perlin (1979)	N
8	200	100	26	74			0.0086		
9	300	100	48	52			0.0086		
10	800	400	50	350			0.0086		
11	400	200	96	104			0.0086		
12	200	100	50	50			0.0217		
13	200	100	11	89			0.0300		
14	0.5	3.0	0.16	2.84	0.08	1.36	0.015	Rosen & Vajda (1982)	M
15	1.0	3.0	0.26	2.74	0.08	1.36	0.015		
16	2.0	3.0	0.38	2.62	0.08	1.36	0.015		
17	0.5	2.0	0.21	1.79	0.05	1.36	0.015		
18	0.5	1.0	0.24	0.76	0.03	1.36	0.015		
19	1.0	1.0	0.56	0.44	0.03	1.36	0.015		
20*	1.0	2.5	0.35	2.15	0.08	1.70	0.025		
21*	0.5	2.0	0.23	1.77	0.07	1.70	0.025		
22*	1.0	2.0	0.42	1.58	0.07	1.70	0.025		
23*	2.0	2.0	0.66	1.34	0.07	1.70	0.025		
24	0.5	1.0	0.26	0.74	0.03	1.70	0.025		
25*	1.0	1.0	0.55	0.45	0.03	1.70	0.025		
26	0.5	2.0	0.05	1.95	0.08	1.00	0.039		
27*	1.0	2.0	0.24	1.76	0.08	1.00	0.039		
28	2.0	2.0	0.23	1.77	0.08	1.00	0.039		
29	1.5	0.75	0.27	0.48	0.05	1.60	0.0192	Shinohara & Tsubaki (1966)	M
30*	1.5	1.50	0.31	1.19	0.10	1.60	0.0192		
31	1.5	2.625	0.35	2.28	0.18	1.60	0.0192		
32	1.5	3.75	0.33	3.42	0.25	1.60	0.0192		
33	1.5	0.75	0.50	0.25	0.05	1.60	0.0461		
34*	1.5	1.50	0.58	0.92	0.10	1.60	0.0461		
35*	1.5	2.625	0.41	2.22	0.18	1.60	0.0461		
36	1.5	3.75	0.20	3.55	0.25	1.60	0.0461		
37*	1.0	1.35	0.32	1.03			0.018	Uda et al. (1988)	M
38*	1.4	1.35	0.33	1.02			0.018		
39	2.0	1.35	0.48	0.87			0.018		
40	2.8	1.35	0.54	0.81			0.018		
41	3.6	1.35	0.58	0.77			0.018		
42	2.6	3.75	0.90	2.85			0.018		
43*	3.2	3.75	1.25	2.50			0.018		
44	4.0	3.75	1.15	2.60			0.018		
45	4.6	3.75	1.10	2.65			0.018		
46	5.2	3.75	1.16	2.59			0.018		

* Salients redrawn to natural scale for comparisons of R_o (Section 6.2.3).

The ratio Y/S has generally been plotted linearly against B/S (Rosen and Vajda 1982, Suh and Dalrymple 1987), but as seen in Fig. 6.6 such a presentation suffers from severe scatter. Even when Y/B is used, results indicate that this Y measurement is not applicable. The alternative distance $X (= S - Y)$ used as X/S also served little purpose. This indicates that apex positions should be related to distances from the structure (X) rather than from the original shoreline (Y), because the beach lines of salients are dictated by the diffraction and refraction of waves leeward of the breakwater. Such salient distances (X) divided by breakwater length (B) plotted against B divided by original shoreline distance (S) results in a single curve with very little scatter (as seen Fig. 6.7). This occurs in spite of data being used from models with varying wave and sediment characteristics and also prototype situations.

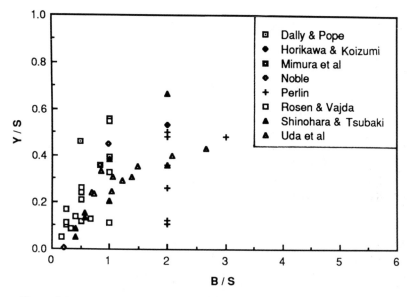

Figure 6.6 Relationship between Y/S versus B/S from 46 data of model studies and prototype conditions for salient behind single offshore breakwater, as given in Table 6.1

Other parameters that have been assumed of importance by previous workers have also been proven to be quite scattered. These include the depth of water (h_B) at the breakwater expressed as h_B/B versus Y/S or X/B, the value of sand accumulation divided by the volume of water leeward of the breakwater, the wave breaking distance X_b from the original shoreline as the ratio of X_b/S against Y/S and X/B, and using the deepwater wave steepness (H_o/L_o) plotted with X/B or B/S.

Not mentioning the difficulty of assessing wave steepness in prototype conditions, the fact that wave steepness appears irrelevant is because at static equilibrium the diffraction, with almost negligible refraction, will not be affected by it. This steepness (H_o/L_o) or wave energy certainly has an effect on the speed at which stability is reached, but at the final stages it is of no consequence on salient size. The same principle applies to the nature of the sediment used in models. As long as the waves in all areas of the model

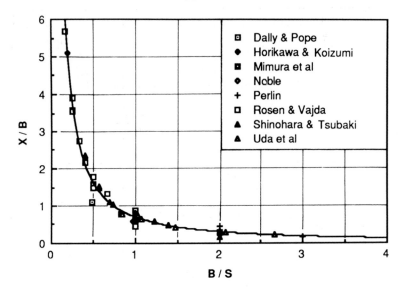

Figure 6.7   Relationship between X/B versus B/S from data in Table 6.1

are sufficient to move the grains, they will be shaped into a stable beach. Lighter or finer particles will be moved more swiftly, but at static equilibrium all kinds of material, if mobile in the most diffracted waves, will give the same planform. Rosen and Vajda (1982) employed coarse bakelite, whereas all other workers used sand. Despite this variation, plus prototype cases, all the results will be shown to fall on the curves presented with very little scatter.

With only three geometric variables involved in the apex position, namely B, S, and X to be nondimensionalized, one variable will be common, for which B is chosen because it is definitive and the most important. The repetition of a variable in two parameters has little relevance when the relationship is an exponential or a polynomial expression. As shown in Fig. 6.7, a single curve results when the ratio X/B is plotted against B/S, giving an equation:

$$X/B = 0.6784 \ (B/S)^{-1.2148} \qquad (\text{C.R.} = 0.98) \tag{6.1}$$

where C.R. denotes the correlation coefficient of curve fitting. Whether salient variables are plotted against B/S or S/B, similar relationships should result, although a slight discrepancy could occur due to different equations being fitted to the data. Hence, as seen in Fig. 6.8, X/B against S/B results in an equation:

$$X/B = 0.6784 \ (S/B)^{1.2148} \qquad (\text{C.R.} = 0.98) \tag{6.2}$$

or

$$X/B = -0.1626 + 0.8439(S/B) + 0.0274 \ (S/B)^2 \qquad (\text{C.R.} = 0.99) \tag{6.3}$$

For a value of $X/B = 0$, or tombolo formation, $S/B = 0.192$ or $B/S = 5.208$.

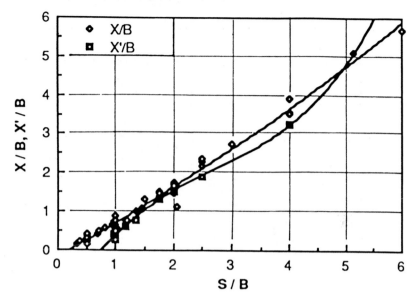

Figure 6.8 X/B and X'/B versus S/B from data in Table 6.1

As seen in Fig. 6.5, the radius R_1 is proportional to X and B by:

$$X/(B/2) = \tan \alpha \qquad (6.4)$$

By plotting $\tan \alpha$ versus S/B, as in Fig. 6.9, it is seen that $\tan \alpha = 0$ when $S/B = 0.19$ or a tombolo forms. Also no salient is formed when $\tan \alpha = 2S/B$ from the geometry. From Fig. 6.5 it is seen that this occurs when $S/B \approx 5$.

6.2.3 Salient Planform

No previous workers have given scientific attention to the shapes of the salient beaches and the recession of the original shoreline beyond the breakwater, where most of the sand accumulated behind the structure came from. This is so despite the apex positions and the accreted sand volumes having been measured. For the case of a single offshore breakwater, it can be shown that the curved waterlines of the saliants are part of a static equilibrium bay whose form can be defined by wave obliquity to a control line joining the tip of the breakwater and the point at the extremity of the bay so formed, as discussed in Section 4.2.3.

In order to prove this relationship, it is necessary to mention the additional variables given in Fig. 6.5, which were not alluded to previously. These are:

1. The radius R_θ from the tip of the breakwater to any point on the beach at angle θ from the wave crest line (through the same diffraction point).
2. The radius R_o (or control line) to the accepted extremity of the bay at angle β to the wave-crest line. This is the same angle as between the control line and what

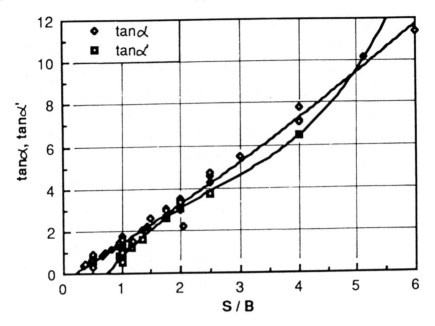

Figure 6.9   Values of $\tan \alpha$ and $\tan \alpha'$ versus S/B from data in Table 6.1

may be termed the downcoast tangent to the bay, which is parallel to the original shoreline and hence also the breakwater. Its distance from the breakwater is S_1.

3. The radius R_θ' is measured to the intersection of the diffracting waves on the breakwater centerline, which is a hypothetical extremity of the beach so formed. This gives a new salient distance from the breakwater of X', which is a starting point for the bays on either side.

Employing the newly developed parabolic form for a bay in static equilibrium (Hsu et al. 1987, 1989a, 1989b; Hsu and Evans 1989), which were presented in Section 4.2.3 and other relevant figures in Chapter 4, the radii (R_θ) are drawn from the tip of the breakwater at angle θ to the wave crest alignment.

For the ideal case of an infinitely long sandy beach, for which no downcoast limit is definable, some assumption with respect to β is required in order to determine the value of R_o. For this purpose a value of $40°$ has been selected. It was found that the variation of R_θ/R_o values, if β was accepted as $30°$, would differ by only 2% when accepting this value of $40°$. To assist in the determination of R_θ, a curve of R_θ/R_o is presented in Fig. 6.10 for $\beta = 40°$ over the range of $\theta = 40°$ to $180°$.

In order to derive R_o the plan forms of salients from the data included in Table 6.1 and Fig. 6.6 were analyzed by redrawing to natural scales (these are noted by asterisks in Table 6.1), in spite of their possible lack of static equilibrium. Even though the apexes may not have arrived at their final positions, the shorelines away from this limit were seen to be in equilibrium by the R_o values for each specific case being very similar, of

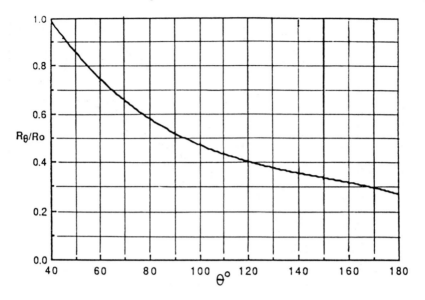

Figure 6.10   Dimensionless ratio of radii R_θ/R_o versus θ for wave obliquity $\beta = 40°$

which an average was taken. A set of 12 data consisting of the averaged distances near R_{90} (normal to the breakwater extremities) were selected on these curved shorelines, for relating R_{90}/R_o by Fig. 6.10 and then graphed as R_o/B versus S/B in Fig. 6.11. The equation of this line is:

$$R_o/B = 0.1737 + 1.683 \ (S/B) \qquad (\text{C.R.} = 1.00) \qquad (6.5)$$

Once R_o is derived, the value of R_θ can be found from R_θ/R_o at any other θ from Fig. 6.10.

The distance S_1 to the tangent of the downcoast limit of the bay is given by:

$$S_1 = R_o \sin 40° \qquad (6.6)$$

and hence S_1/B versus S/B is another line at a slightly different slope as seen also in Fig. 6.11, whose equation is:

$$S_1/B = 0.1112 + 1.082 \ (S/B) \qquad (\text{C.R.} = 1.00) \qquad (6.7)$$

Another point that could be determined on the beach planform is its intersection with the centerline of the breakwater, at distance X' from it, with radius R_θ' from the tip of the breakwater at angle α'. The ratio of X'/B versus S/B has been drawn in Fig. 6.8, together with X/B as in Eq. (6.3). The former is for the 12 intersections of the projected beachlines, while the latter is for 46 data points. The X'/B curve has equation:

$$X'/B = -1.593 + 2.583 \ (S/B) - 0.675 \ (S/B)^2 + 0.083 \ (S/B)^3 \quad (\text{C.R.} = 1.0) \quad (6.8)$$

Equation (6.8) shows that for $X'/B = 0$, a tombolo just forms or the projection of salients from either side intersect the breakwater with $S/B = 0.75$ or $B/S = 1.33$. This should

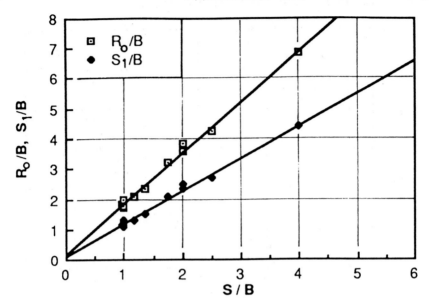

Figure 6.11 R_o/B and S_1/B versus S/B from data in Table 6.1

be compared to $S/B = 0.19$ for the salient apex reaching the breakwater with a finite width of tombolo. It is also seen in Fig. 6.8 that the X'/B and X/B curves approach each other near $S/B = 2$ and then diverge before crossing at $S/B = 5$, or $B/S = 0.20$. This is the value when $X' = X$ or no salient is formed, similar to the case derived above when $\tan \alpha = 2S/B$.

As seen in Fig. 6.5, the radii to the X' and X locations are R_θ' and R_1 respectively. These have been drawn as R_θ'/B and R_1/B in Fig. 6.12. The equation of the former is:

$$R_\theta'/B = 0.6638 \ (S/B)^{1.2115} \qquad \text{(C.R.} = 0.9) \qquad (6.9)$$

An alternative criterion for a tombolo just forming is when $R_\theta'/B = 0.5$. From Eq. (6.9) it gives $S/B = 0.79$ or $B/S = 1.27$, which differs only slightly from $S/B = 0.75$ from the $X'/B = 0$ test.

Also as seen in Fig. 6.5, the angles of these respective radii from the breakwater are α' and α. These can be related to S/B by plotting against the tangent as in Fig. 6.9 above, the equation of which is:

$$\tan \alpha' = -3.1856 + 5.1655 \ (S/B) - 1.3502 \ (S/B)^2 + 0.1656 \ (S/B)^3 \quad \text{(C.R.} = 1.0)$$
$$(6.10)$$

For $\tan \alpha' = 0.0$ this provides the ratio $S/B = 0.75$ or $B/S = 1.33$, when the salient projection coincides with the breakwater.

If the value of $S/B = 0.75$ is to be accepted when the hypothetical beach lines intersect at the breakwater, the value of X/B from Fig. 6.8 or Eq. (6.3) is 0.486 or

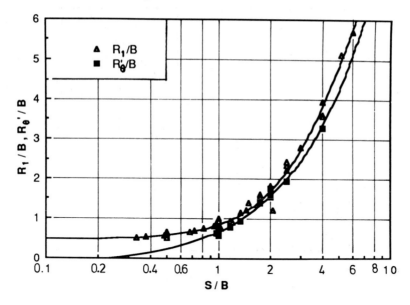

Figure 6.12 R'_0/B and R_1/B versus S/B from data in Table 6.1

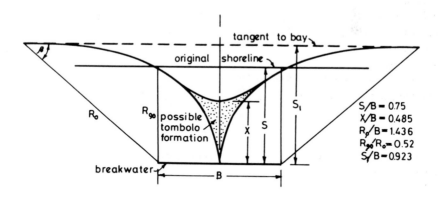

Figure 6.13   Conditions for incipient formation of a tombolo after long duration

$X/S = 0.648$. Any further extension of the salient must take place within the narrow margin between these extrapolated shorelines, as shown in Fig. 6.13, where the stippled area will be very shallow if not completely filled. Even though this involves very small volumes there is practically no littoral drift occurring and hence accretion to smaller X will take excessive duration of wave action. Hence, for $S/B \leq 0.75$ or $B/S \geq 1.33$, the value of X obtained will vary greatly with wave duration. In Fig. 6.7 there is large scatter at $S/B = 1$, due to many tests being conducted at this value, but here $X/B = 0.71$ or

$X/S = 0.71$, which approximates the value derived above. Thus, a critical condition is being experienced which can cause scatter due to insufficient wave duration.

All the results in Figs. 6.7 to 6.12 are presented in Table 6.2, from equations to the curves so derived. This should help in the design of a single offshore breakwater with normally incident waves. Values can be interpolated linearly, between the values so listed. The limiting conditions of $S/B = 0.2$ and 0.75 have been underlined. Between these natural salients may occur, or a tombolo formed after long wave duration, with a finite width at the breakwater. However, this latter situation has not been examined by Hsu and Silvester (1990).

TABLE 6.2   DIMENSIONLESS RATIOS FOR DESIGNING A SINGLE OFFSHORE BREAKWATER

S/B	B/S	X/B	R_1/B	$\tan\alpha$	R_o/B	S_1/B	X'/B	R_θ'/B	$\tan\alpha'$
0.192	5.208	0.000	0.496	0.001	0.497	0.319			
0.20	5.000	0.007	0.498	0.014	0.510	0.328			
0.21	4.762	0.016	0.500	0.032	0.527	0.339			
0.30	3.333	0.093	0.520	0.186	0.679	0.436			
0.40	2.500	0.179	0.549	0.359	0.847	0.544			
0.50	2.000	0.266	0.585	0.532	1.015	0.653			
0.60	1.667	0.354	0.627	0.707	1.184	0.761			
0.70	1.429	0.442	0.674	0.883	1.352	0.869			
0.75	1.333	0.486	0.700	0.971	1.436	0.923	−0.001	0.468	−0.001
0.79	1.266	0.521	0.721	1.042	1.503	0.966	0.067	0.499	0.134
0.80	1.250	0.530	0.726	1.060	1.520	0.977	0.084	0.507	0.167
0.90	1.111	0.619	0.784	1.238	1.689	1.085	0.245	0.584	0.490
1.00	1.000	0.709	0.847	1.417	1.857	1.194	0.398	0.664	0.795
1.20	0.833	0.890	0.985	1.779	2.194	1.410	0.677	0.828	1.355
1.40	0.714	1.073	1.140	2.145	2.530	1.626	0.927	0.998	1.854
1.60	0.625	1.258	1.308	2.515	2.867	1.843	1.150	1.173	2.301
1.80	0.556	1.445	1.488	2.890	3.204	2.059	1.352	1.353	2.703
2.00	0.500	1.635	1.678	3.269	3.540	2.276	1.535	1.537	3.069
2.50	0.400	2.118	2.184	4.237	4.382	2.817	1.938	2.014	3.877
3.00	0.333	2.616	2.718	5.231	5.224	3.357	2.315	2.512	4.630
3.50	0.286	3.127	3.258	6.253	6.065	3.899	2.727	3.028	5.454
4.00	0.250	3.651	3.790	7.302	6.907	4.440	3.236	3.560	6.472
4.50	0.222	4.190	4.303	8.379	7.749	4.981	3.904	4.106	7.808
4.94	0.202	4.675	4.733	9.349	8.489	5.457	4.673	4.597	9.346
5.00	0.200	4.742	4.790	9.483	8.590	5.522	4.793	4.665	9.587
5.18	0.193	4.944	4.959	9.887	8.893	5.716	5.180	4.869	10.360
5.50	0.182	5.308	5.250	10.615	9.432	6.063	5.966	5.236	11.933
5.55	0.180	5.365	5.295	10.729	9.516	6.117	6.101	5.294	12.203
6.00	0.167	5.887	5.685	11.774	10.274	6.604	7.485	5.818	14.970

For assessing the limiting salient conditions or for designing a proposed offshore breakwater, the information in Table 6.2 should be useful. Upon examining this table, the following points can be noted:

1. At $S/B = 0.192$ or $B/S = 5.208$ a tombolo is seen to form since $X/B = 0$, $R_1/B = 0.5$ and $\tan\alpha = 0$. This will eventually develop a substantial length along the breakwater.

2. At $S/B = 0.75$ or $B/S = 1.33$ the extrapolated beach lines meet at the breakwater centerline where $X'/B = 0$. Thus, after long wave duration, a tombolo could form. At least the depths in the vicinity of the breakwater will be very small.

3. At $S/B = 4.94$, $X/B = X'/B = 4.67$, and $\tan\alpha = \tan\alpha' = 9.35$. Also, at $S/B = 5.18$ the value of $\tan\alpha = 2S/B$ which confirms that no salient is formed. Also at this latter stage $X/B = 4.94$ and $X'/B = 5.18$ which are essentially the same. Not until $S/B = 5.55$ does $R_1/B = R_\theta'/B = 5.294$, at which range α' varies from $84.0°$ at $S/B = 4.94$ to $85.3°$ at $S/B = 5.55$.

With respect to (1) and (2), Suh (1985) has stated that B/S "should be greater than 1.5 for a continuously growing tombolo." As inferred in Table 6.2, tombolos may form between $B/S = 1.33$ to 5.21. Since the value suggested by Suh (1985) is close to the smaller value, it is the intersection of the beaches either side on the centerline that is more applicable. Also, Dally and Pope (1986) have concluded regarding tombolo formation that: "The breakwater should be placed offshore a distance between two-thirds and one-half times its length." Thus, $S/B = 0.5$ to 0.67 as compared to $S/B = 0.192$ to 0.75 in Table 6.2. Again the values suggested are closer to the extrapolated beach lines than for the rounded apex proximity.

With reference to (3) above, Inman and Frautschy (1965) have observed: "Experience along the southern California coast indicates that pronounced accretion does not occur if the structure is situated offshore a distance equal to or greater than 3 to 6 times the length of the detached breakwater." The S/B value as given in Table 6.2 range from 4.94 to 5.55 for zero accretion.

6.2.4 Bay Shape Application

Following the relationships of salient planform as presented above, it is now desired to verify the assumption that the salient beach behind a single offshore breakwater is part of a crenulate shaped bay. Employing another planform similar to that given in Fig. 6.4, which was one of the test results obtained by Shinohara and Tsubaki (1966), under the condition of $S = B = 1.5$ m with $H_o/L_o = 0.0461$ and wave duration of 5 hours. The similarity is seen of these salient beaches either side of the apex to the parabolic shape as given by Fig. 6.14, or more generally using the relationships given in Section 4.2.3.

As seen in Fig. 6.14, radii R_θ are drawn at angle θ to the wave-crest alignment at the tip of the breakwater. The value of R_{90} is chosen where this normal to the breakwater tip intersects the beach, which for the RHS is 1.25 m and the LHS is 1.42 m. From Fig. 6.10, $R_{90}/R_o = 0.523$ so that $R_o = 2.39$ or 2.72 m respectively. Other values of R_θ are obtained from Fig. 6.10 for specific angles θ. These have been drawn,

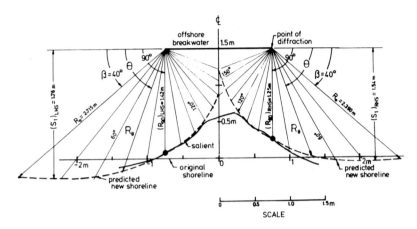

Figure 6.14 Salient planform as provided by Shinohara and Tsubaki (1966) with predicted shoreline shown using specific R_{90} values

giving the static equilibrium shape of the forming bay, which matches the model beaches exceedingly well. There is a degree of asymmetry about this salient from the model test. This was probably due to differing wave heights either side of the breakwater. Although the authors measured waves in both regions, they did not comment on any variations. A larger wave height would expedite accretion in this transient state of salient development. At the final equilibrium stage it would make no difference.

Because of its importance in field applications, coastal engineers should have the ability to estimate the salient planform and beach erosion either side of a breakwater. To predict this situation without recourse to a model, as illustrated in Fig. 6.15, the following relationships are employed from Table 6.2 or respective figures:

$$S/B = B/S = 1.00, X'/B = 0.398 \text{ (Fig. 6.8)}, R_\theta'/B = 0.664 \text{ (Fig. 6.12)},$$

$$\tan\alpha' = 0.795 \text{ (Fig. 6.9)}, R_o/B = 1.857 \text{ (Fig. 6.11)}, S_1/B = 1.194 \text{ (Fig. 6.11)},$$

$$X/B = 0.709 \text{ (Fig. 6.7 or 6.8)}, R_1/B = 0.847 \text{ (Fig. 6.12)}, \tan\alpha = 1.417 \text{ (Fig. 6.9)}$$

From $X'/B = 0.398$, $X' = 0.6$ m; or $R_\theta'/B = 0.664$, $R_\theta' = 1.00$ m; and $\tan\alpha' = 0.795$, $\alpha' = 38.5°$; the starting point for the beach lines either side of the breakwater centerline is obtained. From $R_o/B = 1.857$, $R_o = 2.79$ m from which R_θ/R_o is obtained (from Fig. 6.10) for various angles θ measured from the wave-crest line through the tip of the breakwater. The resulting beach should be asymptotic to a line parallel to the breakwater displaced S_1 from it which should match $S_1/B = 1.194$ or $S_1 = 1.79$ m, implying beach recession of 0.25 m at the bay end of the salient in the model.

To obtain the actual rounded apex distance from the breakwater $X/B = 0.709$, or $X = 1.06$ m, the waterline is drawn asymptotic to the beaches and this normal to the centerline at 1.06 m from the breakwater. This point should be R_1 from the diffraction

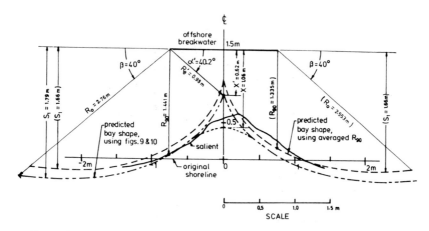

Figure 6.15 Salient planform as provided by Shinohara and Tsubaki (1966) with predicted shoreline shown, using dimensionless ratios as in Table 6.2 or relevant figures in Section 6.2

points as also given by $R_1/B = 0.847$ or $R_1 = 1.27$ m, at angle $\alpha = 54.8°$ or as given by $\tan \alpha = 1.417$.

The resulting shorelines, symmetrical either side of the centerline are shown in Fig. 6.15, but do not match completely with those obtained by Shinohara and Tsubaki (1966) due to their asymmetry. However, they predict the salient very closely and might be used in preference to a small scale model, with limited wave duration. One advantage of this approach is that the beach recession either side of the breakwater is also determined. This would indicate whether it intercepts a seawall that may exist between the original shoreline (S) and this predicted limit (S_1). If large storm waves are not to impact the seawall, the offshore breakwater must be placed so as to provide a full berm width at this recessed location, or even a little more. This width could be gauged from normal beaches in the locality that has suffered these extreme waves.

6.3 STRAIGHT SHORELINE

Consider first the case of a straight shoreline, as in Fig. 6.16, where headlands are so spaced offshore to have a bay penetrate the original shoreline. No net loss of land need be experienced if that gained in the vicinity of each headland balances that removed at the greatest indentation. Beaches are only transposed from one orientation to another, in the knowledge that the new one is more stable. Hence, more use can be made of the bayed beach, to an adequate distance landward of the static equilibrium shape, as distinct from a monstrous set-back line to allow for some unknown future erosion.

Should no loss of the original beach be stipulated, the headlands can be located further offshore or closer together, in order that the greatest indented waterline is asymptotic to this original waterline or even seaward of it, as seen in Fig. 6.17. If this additional land is to be provided from the existing littoral drift, it is obvious that the series of headlands

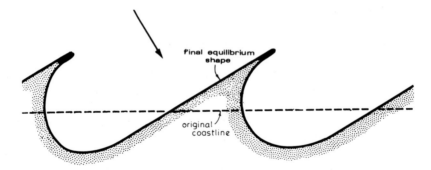

Figure 6.16 Headland control on straight beach for balance of erosion and accretion

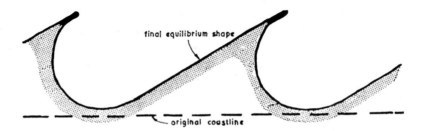

Figure 6.17 Headland control on straight beach for no loss of the original shoreline

must be commenced at the downcoast end. In utilizing drift in this manner, supply of sand downcoast of the last structure could be impeded for many months or even years. This could result in severe erosion between it and the next fixed point farther along the coast. Whereas this may be deleterious in many cases, it can serve a useful purpose where littoral drift needs to be reduced or prevented altogether. Such occasions are the siltation of river or harbor mouths, or water intakes of industrial complexes, which will be discussed in Section 9.9.

With the promotion of beach renourishment as a coastal defense measure, an admirable opportunity is provided for retaining this material in place by inserting headlands a short distance offshore from the new beachline, to place the fill between compartments, as will be discussed in Section 6.8. These massive volumes of sand will then be transposed to form a scalloped shoreline while being retained in place rather than expedited downcoast as noted in Section 5.3.5. Any littoral drift still arriving will be transmitted through the bay system at the original rate. This is unlike a groin field where sand is removed from each cell during a storm and has subsequently to be refilled. All material is retained within each bay segment except for that available from upcoast. In computing volumes required for the initial formation of bays, either in static or dynamic equilibrium, cognizance must be taken of the milder offshore slopes in the downcoast region of the bay due to the more normal approach of the milder or swell-type waves.

6.4 ERODING EMBAYMENT

The cases discussed in Section 6.3 refer to straight lengths of coast, which could be considered applicable if short lengths of shoreline are to be stabilized. There are other cases where natural bays are unstable and receding back to static equilibrium due to reductions in sediment supply. Where the predicted static equilibrium shape is landward of an existing shoreline, the bay can be considered as unstable, inferring that with any decrease in sand supply the bay could erode back to static equilibrium limit but not farther. If this loss of land is unacceptable, headlands could be installed to break the initial bay into compartments, whose stable shapes are predictable. Since the original bay was receiving sand, due to its unstable condition, these intermediate bays will have milder indentations, to start with at least.

Such a situation is depicted in Fig. 6.18, where two headlands (A and D) encompass a large unstable bay. This could erode back to the shape shown dotted, predicted from the value of β at point D. The loss of land so indicated may be unacceptable due to commercial or recreational facilities already erected on the shoreline. Groins, seawalls, and offshore breakwaters not only are costly but also would exacerbate the problem and create more erosion downcoast from them. Beach renourishment may provide protection for 3 years at the most, after which it would have to be repeated.

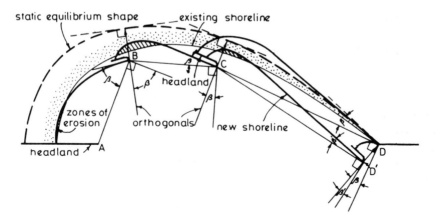

Figure 6.18 Stabilization of an existing bay (full line) shown to be unstable by static equilibrium shape (dotted) with headland control to prevent erosion

A headland control solution is indicated in Fig. 6.18 with a headland installed at B, thus giving a new control line AB. The persistent waves approaching the coast in this region will have a nearshore orthogonal normal to the dotted static equilibrium shoreline. Its angle with the normal to the control line equals β for the new bay. A quick evaluation of its plan outline can be obtained from the indentation ratio a/R_o, from Fig. 4.33 or Eq. (4.5). With $R_o = AB$ the indentation (a) can be drawn as a line parallel to the control line. The static equilibrium shape can then be sketched, starting at B parallel to the tangent of the static equilibrium curve for the major bay and becoming asymptotic to the indentation tangent. The waterline can be continued into the lee of headland A by a log-spiral type curve. Of course, the exact shape could be drawn by the use of Fig. 4.24 or Eq. (4.4) by drawing radii (R) at angles (θ) to the wave-crest line through B, which is parallel to the tangent of the dotted static equilibrium shape of the major bay at the wave orthogonal point. However, this more time-consuming procedure could be delayed until the location of headland B is more firmly fixed.

If the resulting waterline intersects the existing shoreline, it implies that some erosion will take place. In this context even continued supply of littoral drift from upcoast of headland A will not influence bay AB, since it will be transmitted across bay AD to points between B and D, as seen in Figs. 4.49 and 4.50. However, the cost of this small amount of erosion must be compared to the cost of moving B seaward in overcoming it. It is obvious that the presence of B has drastically reduced the loss of land that would have occurred had it not been installed.

Similar principles are applied to the construction of a headland at C and a new control line BC with much reduced β. The static equilibrium shape so derived may not be applicable immediately since littoral drift is continuing, since bay AD was shown to be unstable or seaward of the dotted curve showing stability. The further around bay AD, the more likely the new bays will be less indented by the passage of littoral drift. But it would be wise to keep major facilities behind the static equilibrium line, which is the limit of erosion should upcoast supply be cut off. If, as before, the predicted shoreline intersects that existing, some erosion will be experienced. If this can not be countenanced, either or both B and C can be moved seaward or closer together.

Between C and D substantial erosion could ensue, even more than the final static equilibrium shape for bay AD. This is where a groin-type structure at D extending to D' could be advantageous. Should bays AB, BC, and CD be accreted naturally, by existing littoral drift, the dearth of sediment beyond D or D' may have adverse consequences. However, if these were to be reclaimed to near their dynamic equilibrium shape, the supply beyond D could be as before with no deleterious effects.

6.5 BARRIER ISLAND AND SPIT

A convex coast such as a spit or barrier beach, as discussed in Section 4.4, varies vastly in width and hence narrow sections can be overwashed by storm waves accompanied by storm surge. Supply of sediment to them has drastically reduced in past decades, and therefore, their defense is paramount. As seen in Fig. 6.19, the deepwater wave orthogonals, coming from a persistent oblique direction, are refracted by different amounts before arriving at various points along the curved coast. This results in changing values of β at successive bays formed between headlands, being greater the farther downcoast. Varying indentations must be taken into account when alongshore spacing and distance offshore is determined for each headland. This implies probably closer spacing near the extremity of the spit, but much wider than the normal gap between offshore breakwaters should they be constructed at this location.

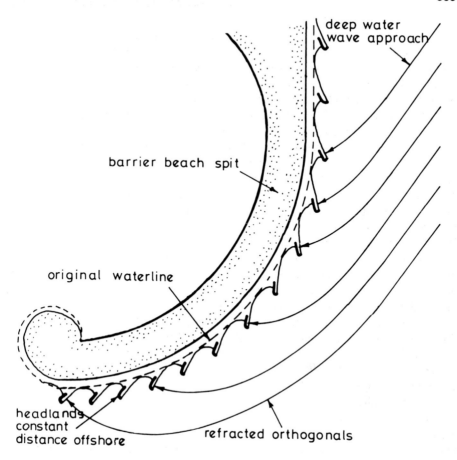

deep water
wave approach

barrier beach spit

original waterline

headlands
constant
distance offshore

refracted orthogonals

Figure 6.19 Stabilization of convex barrier beach or spit by headland control

A suggested system for Spurn Head in the United Kingdom is depicted in Fig. 6.20, where headlands are located at LWM. The two at the southern end are much more seaward than those upcoast. If such a scheme were adopted, construction should commence at the southern end and proceed northward. However, if urgent protection was required at an intermediate point, headlands could be installed with concurrent reclamation. Drift could then pass this location to feed the bays being formed farther downcoast. Similar type man-made barrier beaches have been stabilized by anchor points (Denney and Fricbergs 1979), with an example from Toronto, Canada shown in Fig. 6.21.

6.6 ALTERNATIVE TO GROIN FIELD

The primary variables for groin design are the length and spacing in order to maximize the sand retained. The major difference between groins and headlands is in their orientation,

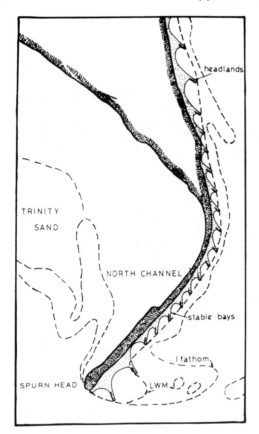

TRINITY
SAND

NORTH CHANNEL

SPURN HEAD

headlands

stable bays

1 fathom

LWM

Figure 6.20 Suggested headland control for Spurn Head, UK

as illustrated in Fig. 6.22, where the line joining the seaward extremities becomes the control line (R_o). For orientation A the full bay is formed for the wave orthogonal shown, but for B, C, and D more curved sections of it will not exist. On the downcoast side, orientations E, F, G make no difference to the shoreline so formed.

The difference between groins and headlands is apparent if orientations C and G are considered as the former. In this case only the tangential portion and part of the curved segment is available as beach. As noted already, storm waves will induce a rip current adjacent to C and dispose of material out to sea. However, if headland A, or even B, were employed no such rip would be formed. Even if it were, sediment would be retained within the cell and be returned to shore by subsequent swell. The transport of sand along the tangential beach during a storm to the most indented section of the bay quickly creates a large offshore bar which is added protection for the zone closest to an original shoreline where development has probably taken place.

The graphs and tables contained in Chapter 4 for determining shapes of static equilibrium bays can be used to predict the shoreline between groins. For this purpose the deepwater direction of persistent swell should be known and this orthogonal refracted to depths between the groin tips. This decides the obliquity β, which is also the angle

Figure 6.21   A typical application of barrier beach for port sheltering (photo permission of the Toronto Harbour Commission)

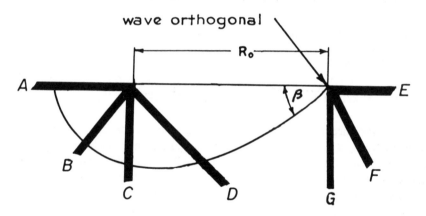

Figure 6.22   Beach orientation produced by groins and headlands

between the tangential downcoast extremity of the partial bay so sculptured and the control line. Since the indentation ratio (a/R_o) is of more relevance, the length of groin from the tip to the waterline of the original shoreline becomes (a), assuming that no existing beach is to be lost. Once β is known and the spacing of the groins (R_o) decided, their lengths (a) can be found from Fig. 5.5, in which groin length is shown as B and spacing as D. For various values of β the ratio of D/B can be ascertained from the figure.

It is seen from Fig. 5.5 that D/B for $\beta = 30°$ to $60°$ ranges over 2 to 3, but for $\beta = 10°$, it may be 7. Such spacing has been examined by Nagai (1956) and Horikawa and Sonu (1965), who recommend 2 to 3 groin lengths. Dunham (1965) has suggested the use of long groins in order to achieve a segment of a bay in static equilibrium. "For this reason, they have sometimes been labelled 'barrier groins,' but regardless of designation, they are intended to serve as artificial headlands."

The headland control principle can be applied to an existing groin field, as illustrated in Fig. 6.23, where an extension parallel to the known shoreline is made at the groin tip. A curved tip at its extremity will accrete a beach in front of the extended arm, while the new beach line will accrete along the heel section of the groin. At its greatest indentation the waterline will be seaward of the original shoreline by the amount of the seaward deviation. These structures could comprise new cheap construction of flexible membrane units to be outlined in Chapter 8.

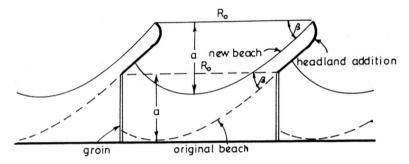

Figure 6.23 ⋯Converting groin into headland to promote the formation of stable bay shape

6.7 MODIFICATION TO OFFSHORE BREAKWATERS

Because of the uneconomic manner of retaining sand on beaches by offshore breakwaters, as outlined in Section 5.3.3, the only suggestion that can be made is to open up the bays so formed to larger proportions. This will obviate the formation of rip currents during storms, which disperses much of the collected material back out to sea for further transmission along the coast by the short-crested wave system existing beyond the breakwaters (see Fig. 5.8).

As seen in Fig. 6.24, as an example, two offshore breakwaters could be removed in order that a crenulate shaped bay forms between those left on either side. Its indentation may be such that the waterline is in close proximity to an existing seawall, without due allowance for the berm to be removed during storms. In this case the alteration is shown of reorienting the downcoast headland, to provide the secondary control line, from which a new bay shape results. This reorientation would also apply to what would be termed the upcoast headland, but is not shown in the figure. The curved shorelines would then intercept the new headland on the same side as the downcoast beach of the bay, as was presented in Fig. 6.1e. There could be much surplus rock or prefabricated concrete blocks for use in other coastal defense structures. For a bay to work in the manner as outlined in this tome, it should be some hundreds of meters in width with an indentation

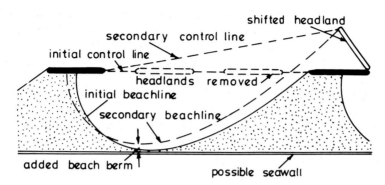

Figure 6.24   Suggested modification of offshore breakwaters to form crenulate bays on a coast

of 200 to 300 meters, in order that any offshore bar formed during a storm is still within the cell. Distances of these bars from the beach were discussed in Section 3.4.3.

6.8 BEACH RENOURISHMENT

Beach renourishment has been accepted as a soft option for coast stabilization schemes, as noted in Section 5.3.5, implying that the beach can defend itself quite readily. Its main disability, besides being expensive, is that fill material is swiftly removed downcoast, so refilling is required at intervals of about 3 years. This not only fails to solve the problem but also creates others by silting harbors and channels downcoast.

There are many variables involved in the erosion process associated with a renourishment project that make it difficult to ascertain which are predominant. These include replenishment length, width of fill or density, sand grain size, method of placement, shoreface slope, inlet proximity, season of pumping, and storm frequency (Pilkey and Clayton 1987, Leonard 1988). All these interact, making comparisons of projects well nigh impossible. Even on the same beach successive renourishment schemes will have a number of variables operating. However, it may be concluded that the ability of beach recovery after storms and the wave obliquity of the persistent swell in relation to the beach are the most important factors which govern the durability of a beach renourishment project. Terminating groins, presumably at the downcoast of a renourished beach, have been adopted to reduce sand loss alongshore. On this account, the influence of storm sequences should not be forgotten, since sediment retained between groins will be deflected seaward by the generation of strong rips. This will result in it being carried much farther downcoast before being welded to the shoreline.

Of all the variables examined on renourishment, storm frequency was found to be the most relevant. Pilkey and Clayton (1987) have questioned: "Why the poor predictability of artificial beaches? By far the most commonly given reason for beach failure is storms." The reason for the imprecise prediction is given by Leonard (1988): "The acquisition of storm data is difficult, however, since storm activity is not consistently recorded from area to area." Of all the reasons given 47% were due to storms. In her comprehensive study of the U.S. East Coast barrier beaches she linked durability with time intervals between project completion and the first storm event.

Leonard (1988) also provided a table of 19 projects with number of storms in the first year and a figure with beach lifetime in months of the same projects. The latter were scaled from this figure in order to produce Fig. 6.25, in which zero storm conditions were omitted. The curve resulted in:

$$t_b = 23.56/N^{0.92} \qquad (\text{C.R.} = 0.78) \qquad (6.11)$$

where t_b is durability in months, N the number of storms in the first year. It could be expected that subsequent years would have a similar wave climate, so that the number of storms on average might be repeated. Since storm activity is so important in the pulsative movement of drift, it is a variable that should be maintained continuously if useful data are to be obtained on replenished beach longevity.

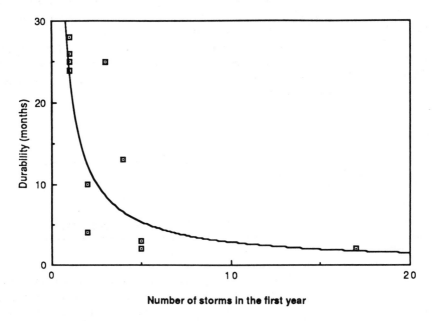

Figure 6.25   Beach fill durability resulting from number of storms in the first year

Substituting in Eq. (6.11) for N, durabilities as in Table 6.3 are obtained, from which it is seen that a storm each second year results in about 4-year life, and one each year in 2 years, and 2 storms per year in 1 year life expectancy. As the number of storms increases beyond 3 per year, as seen in the curve in Fig. 6.25, durability decreases more slowly. However, the difference between 1 or 2 storms has a significant effect on durability, in fact halving it.

Durability values have been provided by Leonard (1988) for 42 out of 280 replenished operations, which she categorized into periods over which 50% of the material was

TABLE 6.3 DURABILITY FROM
NUMBER OF STORMS ANNUALLY

No. of Storms	Durability (months)
0.5	44.6
1	23.5
2	12.5
3	8.6
4	6.6
5	5.4
6	4.5
7	3.9

lost, commenting: "This is a conservative measure since, in most cases, the beach loss of artificial beaches in each of the beach lifetime categories was in considerable excess of fifty percent." The percentages for the Atlantic, Pacific, and Gulf coasts are listed in Table 6.4, which have been graphed in Fig. 6.26 by taking the mean values of durability in years. The similarity of these coasts with widely different wave climates is startling. They indicate that around 60% of the projects had a life expectancy of 3 years. Other pertinent values are that 20% were less than 1 year and 35% were less than 2 years, or more than half were under 2 years duration.

TABLE 6.4 OVERALL PERCENTAGE OF LIFETIME CATEGORIES ON VARIOUS COASTS

Durability (years)	< 1	1–5	> 5	< 2	2–5	Reference
Atlantic	26	62	12	40	48	Leonard (1988)
Pacific	18	55	27	—	—	Leonard et al. (1989)
Gulf	15	75	15	35	55	Dixon & Pilkey (1989)

Because of economical constraints, renourishment is generally carried out in single batch operations, with large volumes deposited very swiftly. As noted above, this results in ready dispersal either side of the fill but mainly downcoast. Since the natural beach erodes more slowly than the protruding filled zone, it would appear preferable to add a little and often in order to preserve natural conditions. This has been recommended by Pilkey and Clayton (1987) and Sorensen et al. (1988).

Upon reviewing the results of four large-volume beach renourishment projects in the United States, Engense and Sonu (1987) have concluded: "As a general guide line within the limited scope of this study, it can be stated that compartmentalization of a project beach has the single most important effect on retention of the placed fill." Terminal structures in the form of groins or jetties have been employed in the past. In a fully compartmentalized beach, longshore sediment transport of the fill can be reduced, but not offshore losses during storms.

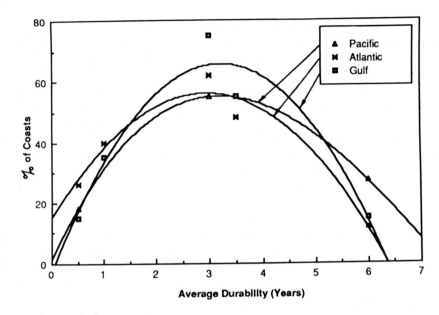

Figure 6.26 Durability expectancies for three coasts of United States

Since the offshore loss of fill material with a field of groins is promoted by rip currents, as described previously, it is beneficial to consider alternative compartmentalization other than using groins, jetties, and offshore breakwaters. Hsu and Silvester (1989b) have discussed various coastal defense solutions, from which a comparison can be made between groins and headlands in keeping sand in position. For beach renourishment the action of headlands in forming stable bays between them deserves more attention.

This approach to coast stabilization has been derived from observing how Nature maintains stable shorelines between naturally occurring headlands. These have remained in position for thousands of years, either in dynamic or static equilibrium. Therefore, the longevity of the beach fill can be ensured and the money spent on any nourishment project will not go down the drain. The ability of the renourished beaches to defend themselves can also be met by the formation of the offshore bar and its subsequent return of sand. The loss of the beach berm, each year or more often, cannot be overcome but the curved waterline so designed can be maintained with bay formation.

6.9 BEACH DOWNCOAST OF HARBOR CONSTRUCTION

Many new harbors are having to be built on coasts where no natural shelter is available such as a coastal indentation, offshore islands, capes, barrier spits, or reefs. This causes a need for long breakwaters which are extended over time as further massive structures are angled to the shoreline, implying that swell waves will be angled to them. Even storm waves can be oblique from many arrival directions. But due to their size they interrupt the littoral drift and so cause erosion of beaches downcoast. This is often met by seawalls and/or groins to safeguard facilities associated with the port. Examples could be cited from around the world.

The method preferred by these authors is to reorient the waterlines parallel to the refracting waves by means of headlands which will form either dynamic or static equilibrium bays. These structures should be constructed prior to the breakwater being installed, since erosion is very swift. In close proximity to a protruding arm accretion may occur, but beyond the swell orthogonal which suffers no diffraction the oblique waves will remove material that is not replaced from upcoast. This is the area requiring stabilization by headland control.

A harbor may be positioned in the lee of a natural headland, because it provides a reasonable source of shelter. However, it might start out as a modest port feature with plenty of littoral drift bypassing it, but this feature may be extended seaward to provide a greater area of calm water or better navigation. These excessive breakwaters interrupt the flow of sediment and so provide new conditions for the downcoast bay, to which it will adapt. It is to be shown how a bay in dynamic equilibrium can be stabilized but the reverse is not possible.

As observed by Komar (1983): "Many occurrences of destructive coastal erosion have resulted directly from the construction of jetties, breakwaters, and other engineering structures." Moutzouris (1990) also comments that: "Coastal erosion and/or accretion are sometimes a result of the construction of harbour or coastal works in an area. Most of the time these side-effects are undesired and engineers have to cope with them. Many cases can be met along the Greek coastline." It will now be shown that similar sized breakwaters placed in slightly different locations can have a beneficial, or at least not a deleterious, effect on a downcoast beach area. This is due to the new diffraction and refraction pattern causing waves to arrive near normal to the existing shoreline of a bay in dynamic equilibrium, which could erode if sediment supply to it is terminated by natural processes or man's intervention. Actual examples of port extensions in Japan will be utilized to explain the procedure.

6.9.1 Straight Shoreline

It stands to reason that a breakwater running out to sea will act very much like a headland leeward of which the most persistent waves, the swell, will diffract and refract to the downcoast beach. This must result in the formation of a bay with concomitant erosion of the beach some distance from the heel of the structure. This deleterious denudation is overcome at great cost by the construction of a seawall in the most affected area, followed possibly by groins and even offshore breakwaters.

Harbors may be sited on long straight sandy coasts, or even in large bays in dynamic equilibrium, where the shoreline for a reasonable distance is linear. A typical case is depicted in Fig. 6.27, where the main breakwater is angled to the shore and intercepts a large volume of drift. The secondary breakwater at A provides shelter from the predominant waves diffracting around the tip of the first and also storm waves that

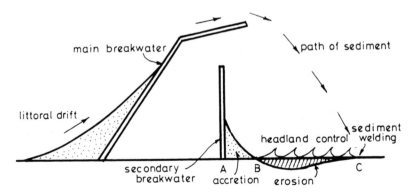

Figure 6.27   Harbor construction on a straight sandy coast, showing potential erosion downcoast and possible headland control

can arrive from a wide fan of directions. The path of sediment bypassing the main breakwater is shown where it welds to the coast at C some distance downcoast from A, where in fact some accretion can be expected as a bay commences to form. It is in this region that headland control is recommended in order to maintain the shoreline in its original position. This is preferred to groins, seawalls, or offshore breakwaters because all these have deficiencies in dealing with lack of sediment supply, as outlined in Chapter 5. Headlands should be installed prior to the construction of the main breakwater since erosion is very swift. Otherwise, some concurrent reclamation may be required, which is expensive.

As noted in Fig. 6.27, the shoreline between points B and C will attain a permanent set-back due to the curved nature of the crests arriving in that region. This is the start of a crenulate shaped bay that could extend to the next headland downcoast, or to the straight beach that originally existed. As seen in this figure, the predicted eroded section BC could have headlands installed to form bays, whose greatest indentation matches the original shoreline. They will approach the static equilibrium shape, for which the equations presented in Section 4.2.3 apply, because no further drift will take place there. These headlands should be inserted prior to harbor construction and reclamation carried out to the limit of the stable shape of the bays. Initially they will not be so indented because the littoral drift will still exist, but will later assume static equilibrium as sediment supply is cut off by the harbor complex.

6.9.2 Bay in Dynamic Equilibrium

A modest port complex, in the form of breakwaters AB and CD in Fig. 6.28, is built at a natural headland upcoast of bay AE which is in dynamic equilibrium. By transferring the orthogonal normal to the downcoast beach tangent at E across to A, the wave-crest line is angled to the control line AE at 32° as shown. The stable bay shape for no port structures can thus be drawn as "static bay for AE." This displays the erosion that could

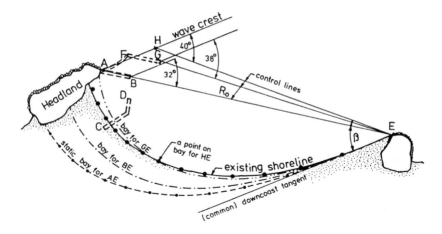

Figure 6.28   Effects of positioning breakwater tip in a dynamically stable bay on beach erosion downcoast of a harbor

take place should littoral drift from upcoast of the natural headland at A cease altogether. However, the existing shoreline, as shown, indicates that sufficient sediment is bypassing point A to maintain it.

When breakwater AB is constructed, the new diffraction point (B) causes a stable shape further seaward, marked as "bay for BE." This again is for a situation of no littoral drift. An alternative of breakwater FG provides a new value of $\beta = 38°$ and R_o from which the static bay, denoted as "bay for GE," forms. This, it is seen, matches the existing shoreline fairly well, with slight accretion to the shoreline adjacent to CD and slight erosion downcoast of it. It should be remembered that in time sediment would be bypassing point G so that the bay would again be in dynamic equilibrium, resulting in a shoreline seaward of this static equilibrium outline of GE. However, it can be seen that the initial construction of FG instead of AB has almost solved any erosion problems of the bayed coast. These two structures are of similar length and except for differing depths should cost much the same. Its obliquity to the wave crests will aid transmission of sediment beyond G without forming shoals, as will be discussed in Section 9.8. The curved structure between A and F could serve to retain sand and build a straight beach between A and F, oriented parallel to the wave-crest line. This structure will not be attacked by waves so need rise only to MSL, and its units could be modest in size just to retain the sand. In fact, sand could just accrete in this zone and provide natural protection.

Should it be desired that the coast from A to E be in static equilibrium, because of imminent interception of littoral drift upcoast from headland A, due possibly to an access channel being dredged for a port in that region, then point H could be chosen as the diffraction point. As seen, $\beta = 40°$ and with the new R_o of HE, the static bay as shown by larger dots results. This matches the existing shoreline exactly, inferring that erosion would not be experienced at any point around its periphery even with no sediment input.

An actual example of such shoreline evolution is given for Oarai, Japan, located as in Fig. 7.31, which faces the Pacific Ocean. As seen in Fig. 6.29, two angled breakwaters formed a harbor in 1916 which were quickly silted up. The stable shoreline south of the site, for no further input of sediment, is shown for $\beta = 37°$ and $R_o = 3.0$ km. However, because of littoral drift the shoreline was seaward of this, implying the existing beach was in dynamic equilibrium. An extension to the harbor, as in Fig. 6.30, established a new diffraction point at A, with $\beta = 45°$ and $R_o = 1.92$ km, resulting in the predicted shoreline which matches the existing one very well.

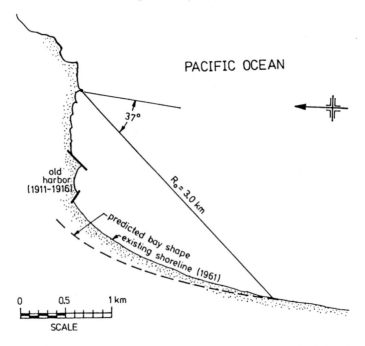

Figure 6.29   Shoreline downcoast of Oarai harbor, Japan (see Fig. 7.31 for location), showing dynamic bay shape in 1961

A further extension of this breakwater from A to B, completed in 1981, and seen in Fig. 6.31, gives $\beta = 35°$ and $R_o = 3.48$ km. The new static equilibrium shape is shown dotted together with the 1976 shoreline. During 1983 to 1985 a detached breakwater CD was constructed for which a new stable bay shape for CE is depicted plus the existing shoreline of 1988 (Mizumura 1982, Kraus et al. 1984). It is seen that the shoreline in 1988 even with two long groins inserted had not accreted out to this new limit for zero littoral drift. Material collected as a salient in the lee of breakwater CD has meant that sand for this accretion has had to come from downcoast, resulting in erosion from 1 to 3 km south of the harbor. This has initiated the installation of seawalls, groins, and concrete block revetments. If and when normal drift conditions are established in the future the bay shape for CE could eventuate, with even further accretion to a dynamic equilibrium shape.

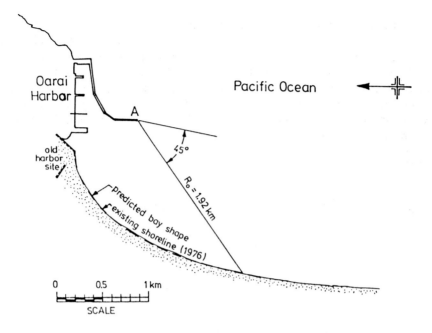

Figure 6.30 Shoreline downcoast of Oarai harbor, Japan, showing a static stable bay shape in 1976

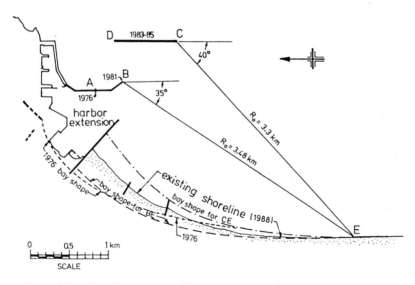

Figure 6.31 Shoreline downcoast of Oarai harbor, Japan, in 1988 showing the salient formation behind the breakwater extention in 1985

6.9.3 Bay in Static Equilibrium

Although, as shown previously, a bay in dynamic equilibrium can be made stable by placing breakwaters at the upcoast headland in a suitable position, this is not necessarily the case for a bay already in static equilibrium. The shoreline in the vicinity of Iwafune harbor, Japan, located as in Fig. 7.31, was originally a small river mouth delta, with a 50 m wide mouth, receiving sediment from it and also a river 5 km to the north. As seen in Fig. 6.32, an upcoast groin and detached breakwater were completed in 1929 to prevent siltation. These were connected in 1934, with an inner revetment and wharf added in 1957 (Haruta 1961), which became the outer breakwater for this river mouth harbor.

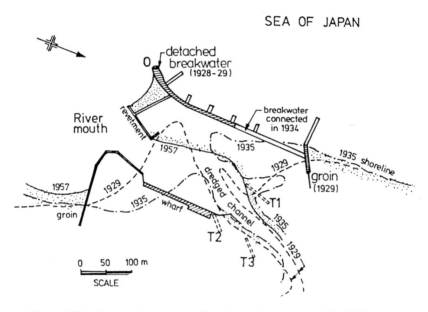

Figure 6.32   First stage development of Iwafune harbor, Japan (see Fig. 7.31 for location), from 1928 to 1957

Aerial photographs of 1965 show reasonably wide beaches upcoast of the main breakwater, since most sediment comes from the north. As seen in Fig. 6.33, some downcoast position had to be located as the limit of the bay formed south of the harbor at point B. The tip of the breakwater (A) and wave-crest line, parallel to the tangent at B, provides β at $30°$ and $R_o = 1.35$ km. The static equilibrium shape fits the actual shoreline as at 1965 and hence no further erosion could be expected.

Mammoth breakwater extensions were carried out, as in Fig. 6.34, resulting in new stable beach shapes, which caused shoaling of the channel from the port, with massive dredging required. In this case two points D and E along the same sandy beach were selected with similar β values of $40°$ to the two breakwater tips B and C, to obtain the bay shapes. The stable shorelines are seen to be well seaward of the 1988 shoreline with its many groins. Material for this addition must come from downcoast, which instigated the

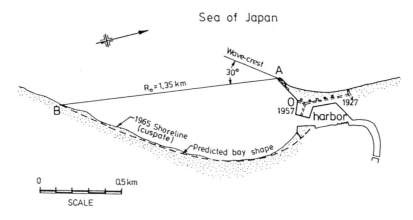

Figure 6.33 Shoreline downcoast of Iwafune harbor in 1965, showing a stable bay shape in static equilibrium

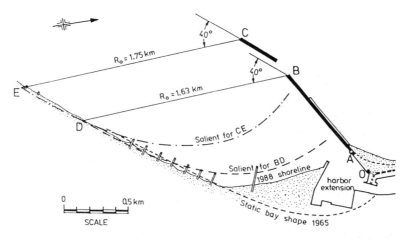

Figure 6.34 Iwafune harbor in 1988, showing salient growth in its lee and beach erosion downcoast due to further breakwater extension

groin construction, and even with offshore breakwaters beyond the downcoast control point (not shown in this figure). These will not prevent the recession necessary for Nature to reshape the beachline to that applicable to the boundary conditions. Where a bay downcoast of a harbor complex is already stable, any modifications to it, by diffraction points being established from breakwater extensions, must involve erosion of one segment to feed the other, in order to produce the salient predicted in Section 6.2.

6.10 ARTIFICIAL AND RECREATIONAL BEACHES

The economic impact of industrial development worldwide has seen many natural beaches being replaced by port facilities and factories. As a result, artificial beaches have been constructed for recreation and tourism, as well as for the restoration of eroding beaches. One basic requirement for this theme is to construct suitable protective structures for preventing sediment loss from the artificially designed beaches, with an attenuated wave environment. Combinations of offshore breakwaters and groins have been employed in all the cases so far reported in the literature (Sato and Tanaka 1980, Berenguer and Tamago 1987, Berenguer and Enríquez 1988, MacIntosh and Anglin 1988, Kuriyama et al. 1988, Rouch and Belleslla 1990, and Spataru 1990). The beaches so designed, either fully or partially enclosed cells, are subjected to mainly diffracted waves and with minimum wave refraction behind the offshore structures, hence the term *pocket beaches* by Berenguer and Enríquez (1988). They have observed that: "In the recent years, the increase of pocket beaches has been remarkable. In Mediterranean countries where tidal effects are insignificant, especially in Spain, Italy, France and Israel, this system of beach regeneration has been widely used because of its rapid results and its economical nature. Japan also has a lot of experience in this kind of work. In Spain, there are over 40 beaches of this kind. They are usually situated on stretches of coastline where there is a lot of tourism and in which there is also a fairly rapid erosional process ... "

The main considerations for an artificial beach are the design of protective structures, layouts of the structures and beach planforms, beach profiles, and wave conditions. Sediment properties and retention are also important. Although the structural dimensions in relation to the geometry of sediment accumulation (length, distance from the original beach, gap between structures, and size of salient) have been suggested by many researchers (Berenguer and Enríquez 1988, Kuriyama et al. 1988, and Spatara 1990), only physical models and empirical relationships are available to assist the design. However, it is possible to numerically calculate the wave conditions in the vicinity of this structure-beach cell using a wave transformation scheme, once a layout is set.

The conventional approaches, in which offshore breakwaters and groins are used, have produced enclosed and partially enclosed beach systems, in which natural wave action is restricted. Generally speaking, two types of beach cells are common. First, the groin is placed on the centerline of the gap between two offshore breakwaters, thus forming a convex seaward salient behind them, see Fig. 6.35A, and second, the groin is located behind an offshore breakwater at each end of the cell, resulting in a concave landward beach between two groins, as shown in Fig. 6.35B. Sato and Tanaka (1980) have compared design criteria of these two types of arrangements. In practical applications, many different shapes of groin have been used, such as a straight trunk with or without an additional attachment at its seaward tip (i.e., T-shaped, inverse L-shaped, Y-shaped, or ring), or curved trunk. In orientation, most groins are perpendicular to the original beach, with some inclined to the latter. Terminal groins are installed at the two extreme boundaries, so forming a closed beach system overall. Offshore breakwaters and groins are often in the form of a permeable rubble-mound.

Artificial beaches have received popular patronage from various beach-goers in summer. However, because they are essentially a closed system, Sato and Tanaka (1980) have reported that: "pollutants produced by sea-bathing people or discharged from the land is likely to stagnate in the vicinity of the shoreline on account of such structures." Besides this environmental drawback, the potential scour seaward of the offshore struc-

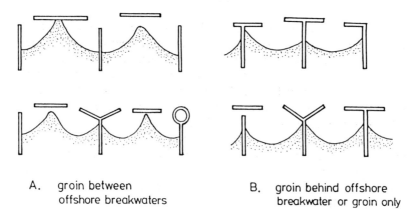

A. groin between   B. groin behind offshore
    offshore breakwaters   breakwater or groin only

Figure 6.35 Two general layouts of offshore breakwaters and groins used in creating artificial beaches

tures themselves needs to be investigated, as noted in Chapter 7. It is for these reasons that an open beach system, which can be produced by headland control, is recommended by the authors. The open system has the benefits of producing aesthetic waterlines of the beach in harmony with its surroundings, plus a stable shoreline after storm attack and subsequent swell action, so preventing the accumulation of pollutants. Also the headland structures, if designed as in Fig. 6.1d, need not rise above the beach berm level.

The concept of headland control was first applied in Singapore (Silvester and Ho 1972), where reclaimed land has been kept in position with added benefit of beautiful beaches for recreation. So far some 1,500 hectares of valuable new land on the southeastern coastal margin has been reclaimed since early 1970 (Wong 1981; Chew et al. 1974, 1986). The stable bay shapes considered in the previous sections in this chapter should be referred to for this purpose, that is, for straight shoreline, see Section 6.3, eroding embayment, Section 6.4, and barrier beach and spit in Section 6.5.

One of the major benefits of producing a bayed beach is to provide variable wave conditions and beach slopes along its complete periphery, from the curved portion upcoast to the straight section downcoast, to suit the physical ability of various age groups. Waves arriving at the coast of a bay after diffraction and refraction produce different wave heights around its whole perimeter. The curved shadow zone behind the upcoast headland has relatively calm water, which is suitable for the beginners and aged, while the straight tangential section of the bay will provide higher waves for good swimmers. In an idealized coastal embayment, either in static or dynamic equilibrium, various beach stages can be observed around the bay (Short 1978), with different breaking conditions. A straight reach of swimming beach can not provide this kind of variety, nor a closed beach system in the conventional artificial beach.

Another benefit of a bayed open beach is its recreational surfing possibilities near the headland. As reported by Walker et al. (1972), conditions for recreational surfing require a certain combination of bottom bathymetry which promotes peeling breakers to sustain lengthy rides. Silvester (1975) has analyzed factors influencing a good surfing site, from the angle of economic and engineering aspects, concluding that "stabilization and recreation might be served by the same headland approach."

The curved bottom contours at such a site are indicated in Fig. 6.36, showing swell approaching them at a slight angle. At the appropriate depth they start to break, with this motion continuing along the crest in a peeling action, which is most desired by a surfboard rider. This continues until the breaking contour becomes parallel to the wave orthogonal or normal to the wave crest. This is the end of the ride, but the surfer can return seaward outside the breaker region to the starting position once more. Natural headland locations are preferred sites for surf riding and hence more can be created while at the same time stabilizing the coast.

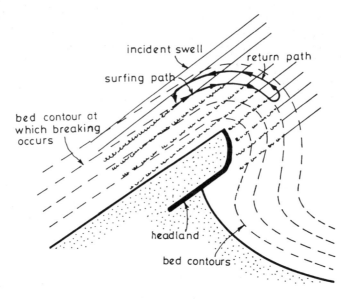

Figure 6.36 Surfing conditions around a headland of a crenulate shaped bay (Silvester 1975)

6.11 MAN-MADE ISLANDS

Nature's method of providing beaches at points around islands is by means of crenulate shaped bays, which are stable because there is no littoral drift due to no river systems supplying sediment to the coast. Such a picture is given in Fig. 6.37, where the particular shapes of these beaches match the crests of swell waves persistently arriving. Although these margins would have suffered storm attack through the ages they are still in place, because the offshore bars have been returned to shore with no longshore loss.

This certainly should provide a clue as to what should be done when artificial islands are being conceived. The use of headlands, and the predictable bay shapes between them, permit the full use of the coastal margin that is maintaining the inner land mass. This is unlike the approaches so far, which are virtually seawalls inviting Nature

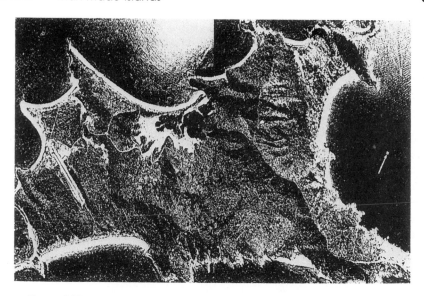

Figure 6.37 Keppel Island on the Queensland coast, Australia, showing stable bay shapes (Reader's Digest 1983)

to do her utmost to dismantle man's puny attempts to hold her at bay. It is a fact of life that she will always have the upper hand. Silvester (1982) has addressed this matter.

A typical approach is displayed in Fig. 6.38, where a semi-circular stone revetment faces the incoming waves, with a port complex on its leeside. It is seen from the

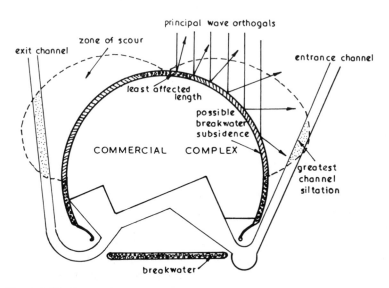

Figure 6.38 Normal circular seawall protection for a man-made island showing zones of bed scour and channel siltation

orthogonals drawn that waves are reflected at various angles, which will scour the seabed to the limits shown dotted. This material will be moved parallel to the curved wall, as will be discussed in detail in Chapter 7. It could well find its way into the access channels to and from the harbor, so creating two tasks, that of revetment wall maintenance and that of continual dredging for navigation purposes.

To obtain maximum area for minimum protection the circular shape is preferred. Even here the use of headlands and bays can be employed to maintain the desired boundary. As seen in Fig. 6.39, the incoming swell waves are refracted differentially across the circular bed contours, with those arriving closest to the leeward side suffering the greatest change in direction. This implies that their obliquity (β) to the bay in that location is larger than at the seaward face of the island. However, the changing shapes of the bays around the island periphery can be predicted once the refraction pattern can be established. Each will be in static equilibrium once the beach and offshore profiles have reached stability, which may take a year or two of storm wave action.

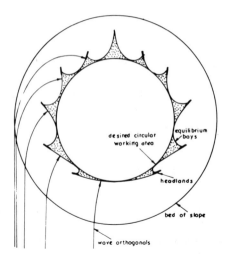

Figure 6.39 Headland stabilization of circular island showing wave orthogonals

Some islands are being constructed as airports, with a suggested layout shown in Fig. 6.40, where the preferred "walking stick" shape of the headland is included. The bays sculptured between these are of vastly different size, but each serves its purpose in providing the working areas required for runways, air terminal, cargo handling, and workshops. The bays on the leeward side are well protected and can serve as refuges for ferries. Entrances to tunnels connected to the mainland will suffer no scouring as no littoral drift will be taking place. Storm waves will not have ready access to these leeside bays.

Herbich and Haney (1982) consider that: "Artificial islands appear to hold great potential for the future. More research is required in all phases of planning and construction ... " One of the best known studies on artificial islands is contained in the *Bos Kalis Westminster* report of the North Sea Island Study Group, published in 1975 (Van der Burgt 1976), which represented the Dutch Government in the conceptual design of

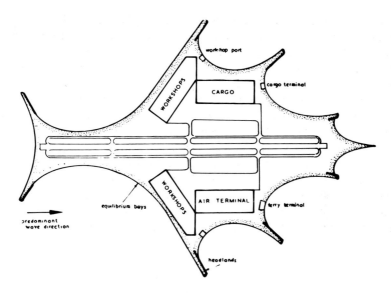

Figure 6.40 Suggested stable shape for an island airport

a 3,300 hectare island. This multi-purpose industrial complex could eventually generate 2% of the total Dutch national income. It is hoped that the headland concept presented above will be given fullest consideration in any deliberations.

REFERENCES

BERENGUER, J. M., and ENRÍQUEZ, J. 1988. Design of pocket beaches: the Spanish cases. *Proc. 21st Inter. Conf. Coastal Eng., ASCE 2*: 1411–25.

BERENGUER, J. M., and TAMAYO, C. 1987. Beach restoration at Malaga urban area: Pedregalejo case study. *Proc. Coastal Zone '87, ASCE 4*: 4667–81.

CHEW, S. Y., WONG, P. P., and CHIN, K. K. 1974. Beach development between headland breakwaters. *Proc. 14th Inter. Conf. Coastal Eng., ASCE 2*: 1399–1418.

CHEW, S. Y., HO, S. K., WONG, P. P., and LEONG, Y. Y. 1986. Beach development between headland breakwaters in a low wave energy environment, Pasir Ris, Singapore. *Proc. 20th Inter. Conf. Coastal Eng., ASCE 2*: 1016–31.

DALLY, W. R., and POPE, J. 1986. Detached breakwaters for shore protection. U.S. Army Corps Engrs., *Coastal Eng. Res. Center*, Waterways Expt. Station, Vicksburg, Miss., Tech. Rept. CERC-86-1.

DENNEY, B. E., and FRICBERGS, K. 1979. The use of anchored beach systems in metro Toronto. *Proc. Coastal Structures '79, ASCE 2*: 237–57.

DIXON, K. L., and PILKEY, O. H., 1989. Beach replenishment along the U.S. coast of the Gulf of Mexico. *Proc. Coastal Zone '89, ASCE 3*: 2007–20.

DUNHAM, J. W. 1965. Use of long groins as artificial headlands. *Proc. Santa Barbara Specialty Conf. on Coastal Eng., ASCE*, 755–62.

Engense, A. K., and Sonu, C. J. 1987. Assessment of beach nourishment methodologies. *Proc. Coastal Zone '87, ASCE 4*: 4421–33.

Haruta, T. 1961. Recent coastal processes in Niigata Prefecture. *Coastal Eng. in Japan 4*: 73–83.

Herbich, J. B., and Haney, J. P. 1982. Artificial islands. In *The Encyclopedia of Beaches and Coastal Environments*, ed. M. L. Schwartz. Stousburg: Hutchison Ross Publ., 64–66.

Horikawa, K., and Sonu, C. 1965. An experimental study on the effect of coastal groins. *Coastal Eng. in Japan 1*: 59–74.

Horikawa, K., and Koizumi, C. 1974. An experimental study on the function of an offshore breakwater. *Proc. 29th Annual Conf.*, Japanese Soc. Civil Engrs., 85–87.

Hsu, J. R. C., and Evans, C. 1989. Parabolic bay shapes and applications. *Proc. Instn. Civil Engrs.*, Pt. 2, 87, 557–70.

Hsu, J. R. C., and Silvester, R. 1989a. Salients leeward of multiple offshore structures. *Proc. 9th Austral. Conf. Coastal & Ocean Eng.*, 347–51.

———. 1989b. Comparison of various defense measures. *Proc. 9th Austral. Conf. Coastal & Ocean Eng.*, 143–48.

———. 1990. Accretion behind single offshore breakwater. *J. Waterway, Port, Coastal & Ocean Eng., ASCE 116*(3): 362–80.

Hsu, J. R. C., Silvester, R., and Xia, Y. M. 1987. New characteristics of equilibrium shaped bays. *Proc. 8th Austral. Conf. Coastal & Ocean Eng.*, 140–44.

Inman, D. L., and Frautschy, J. D. 1965. Littoral processes and the development of shorelines. *Proc. Santa Barbara Specialty Conf. on Coastal Eng., ASCE*, 511–36.

Komar, P. D. 1983. Coastal erosion in response to the construction of jetties and breakwaters. Chapter 9 in *CRC Handbook of Coastal Processes and Erosion*, ed. P. D. Komar. Florida: CRC Press, 191–204.

Kraus, N. C., Hanson, H., and Harikai, S. 1984. Shoreline change at Oarai beach: past, present and future. *Proc. 19th Inter. Conf. Coastal Eng., ASCE 2*: 2107–24.

Kuriyama, Y., Irie, I., and Katoh, K. 1988. Follow-up of artificially nourished beaches. *Coastal Eng. in Japan 31*(1): 105–20.

Leonard, L. A. 1988. An analysis of replenished beach design on the U.S. East Coast. *M.S. thesis* (Geology), Duke University, Durham, N.C.

Leonard, L. A., Clayton, T. D., Dixon, K. L., and Pilkey, O. H. 1988. U.S. beach replenishment experiences: a comparison of the Atlantic, Pacific, and Gulf coasts. *Proc. Coastal Zone '89, ASCE 2*: 1994–2006.

MacIntosh, K. J., and Anglin, C. D. 1988. Artificial beach units on Lake Michigan. *Proc. 21st Inter. Conf. Coastal Eng., ASCE 2*: 2840–54.

Mimura, H., Shimizu, T., and Horikawa, K. 1983. Laboratory study of the influence of detached breakwaters on coastal changes. *Proc. Coastal Structures '83, ASCE 2*:740–52.

Mizumura, K. 1982. Shoreline change estimates near Oarai, Japan. *J. Waterway, Port, Coastal and Ocean Div., ASCE 108*(WW1): 65–80.

Moutzouris, C. I. 1990. Effect of harbour works on the morphology of three Greek coasts. *Proc. 27th Inter. Nav. Congress*, 17–21.

Nagai, S. 1956. Arrangements of groins on a sandy beach. *J. Waterways and Harbors Div., ASCE 82*(WW2), Paper 1063.

NOBLE, R. M. 1978. Coastal structures' effects on shorelines. *Proc. 16th Inter. Conf. Coastal Eng.*, *ASCE 3*: 2069–85.

Our Nation and the Sea. 1968. Commission on Marine Science, Engineering and Resources (plus 3 volumes of panel reports), U.S. Govt. Printing Office, Washington D.C.

PERLIN, M. 1979. Predicting beach planform in the lee of breakwater. *Proc. Coastal Seds. '77, ASCE 2*: 792–808.

PILKEY, O. H., and CLAYTON, T. D. 1987. Beach replenishment: the natural solution. *Proc. Coastal Zone '87, ASCE 2*: 1408–19.

PORRAZ, M., MAZA, J. A., and MUNOZ, M. L. 1977. Low cost structures using operational design systems. *Proc. Coastal Seds. '77, ASCE*, 672–85.

PORRAZ, M., and MEDINA, R. 1977. Low cost, labour intensive coastal development appropriate technology. *Sea Technology*, Aug. 19–24.

Reader's Digest. 1983. *Guide to the Australian Coast,* ed. R. Pullan. Sydney, Australia, Reader's Digest Publ. Co.

ROSEN, D. S., and VAJDA, M. 1982. Sedimentological influences of detached breakwaters. *Proc. 18th Inter. Conf. Coastal Eng., ASCE 3*: 1930–49.

ROUCH, F., and BELLESLLA, B. 1990. Man-made beaches more than 20 years on. *Proc. 22nd Inter. Conf. Coastal Eng., ASCE 2*: 2394–2401.

SATO, S., and TANAKA, N. 1980. Artificial resort beach protected by offshore breakwaters and groins. *Proc. 17th Inter. Conf. Coastal Eng., ASCE 2*: 2003–22.

SHINOHARA, K., and TSUBAKI, T. 1966. Model study on the change of shoreline of sandy beach by the offshore breakwater. *Proc. 10th Inter. Conf. Coastal Eng., ASCE 1*: 550–63.

SHORT, A. D. 1978. Wave power and beach-stages: a global model. *Proc. 16th Inter. Conf. Coastal Eng., ASCE 2*: 1145–61.

SILVESTER, R. 1975. What makes a good surfing beach. *Proc. 2nd Aust. Conf. Coastal & Ocean Eng.*, 30–37.

———. 1982. Planning for artificial islands. *Symp. Eng. & Marine Environ.*, Brugge, 2, 115–22.

SILVESTER, R., and HO, S. K. 1972. Use of crenulate shaped bays to stabilize coasts. *Proc. 13th Inter. Conf. Coastal Eng., ASCE 2*: 1347–65.

SORENSEN, R. M., DOUGLAS, S. L., and WEGGEL, J. R. 1988. Results from the Atlantic City, N. J. beach renourishment monitoring program. *Proc. 21st Inter. Conf. Coastal Eng., ASCE 3*: 2806–17.

SPATARU, A. N. 1990. Breakwaters for the protection of Romanian beaches. *Coastal Eng. 14*: 129–46.

SUH, K. 1985. Modelling of beach erosion control measures in a spiral wave basin. *M. Eng. thesis,* University of Delaware.

SUH, K., and DALRYMPLE, R. A. 1987. Offshore breakwaters in laboratory and field. *J. Waterway, Port, Coastal and Ocean Eng., ASCE 113*(2): 105–21.

UDA, T., OMATA, A., and YOKOYAMA, Y. 1988. On the formation of artificial headland beach using detached breakwater. *Proc. 35th Japanese Conf. Coastal Eng.*, 26–30. (In Japanese)

VAN DER BURGT, C. 1976. The significance of artificial islands. *Proc. Symp. on The Present-Day Challenge of the Sea*, Netherlands Ship Model Basin and Delft Hyd. Lab., NSMB Pub. No. 515, v.1–v.20.

WALKER, J. R., PALMER, R. Q., and KUKEA, J. K. 1972. Recreational surfing on Hawaiian beaches. *Proc. 13th Inter. Conf. Coastal Eng., ASCE 3*: 2609–28.

WONG, P. P. 1981. Beach evolution between headland breakwaters. *Shore and Beach 49*(3): 3–12.

Effects of Maritime Structures

Structures built in the sea are extremely expensive by any standard of civil engineering since they must withstand tremendous forces which are only now becoming predictable with any degree of certainty. While dissipating waves, they also reflect them which causes almost a double application of energy to the seabed, which can result in scouring. This can cause subsidence of the structure due to liquefaction of the sedimentary bed adjacent to them. Means need to be devised to prevent such erosion, particularly in zones where obliquely reflected waves propagate. Similar problems are found with massive structural units below the water surface since orbital water motions close to the bed equate to those where reflecting faces protrude above it.

First, the reflection of waves from normal rubble-mound structures, with or without precast concrete armor units needs to be discussed. The scour due to partially or completely reflected waves arriving normal to a breakwater is then outlined. The kinematics and mass transport due to short-crested waves from oblique reflection has been presented in Chapter 2, but their influence on a sedimentary bed will be discussed in this chapter. Model studies verifying this phenomenon will then be discussed, followed by observations of scour in the field. This erosive situation has only recently become of concern to the coastal engineering fraternity, even though its possibility was published more than 20 years ago.

Rubble-mound structures, especially those faced with large precast concrete monoliths to just below the trough level of storm waves, appear to dissipate waves very effectively, as observed by the turbulence within voids of these units. However, these larger than normal blocks are not taken down the full depth of the face for economic and construction reasons. For a reasonable proportion of the submerged face, smaller stone is placed where water particle orbits are almost horizontal. Since their voids are already filled with water, there will be little exchange as crests and troughs arrive so that the face reacts almost like a glass finish. Thus, while the upper fraction of the wave (admittedly the most active) disposes of its energy, the lower part could be almost completely reflected. This applies to normal and oblique incidence alike, although most model tests are conducted in flumes implying the former approach.

Tests have been carried out on breakwaters using different materials (Gunback 1976, Sollitt and Cross 1972, Anonymous 1970) and summarized by Losada and Giménez-Curto (1981). The reflection coefficient was connected with the Irribarren number $I_r = \tan\alpha/(H/L_o)^{0.5}$, where α is the face slope, H is the local wave height, and L_o is the deepwater wave length. The resulting curves, taking the maxima of the data points rather than the mean, are shown in Fig. 7.1. These data do not include the size of the units involved, which could make a difference to the degree of reflection. Substituting some reasonable values of $\tan\alpha = 0.5$, $H = 1$ m (swell), $T = 12$ sec gives $I_r = 7.5$. It is seen that at this value in the figure $C_r \approx 0.8$.

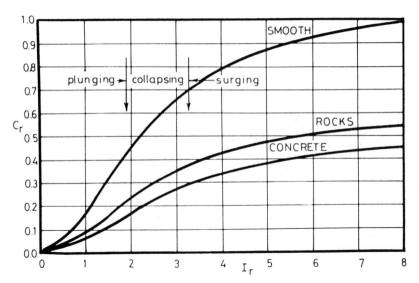

Figure 7.1   Reflection coefficient (C_r) versus Irribarren Number (I_r) showing wave type (Losada and Giménez-Curto 1981)

It should be noted that Gunback (1976) has armor thickness 0.18 of the water depth, whereas Sollitt and Cross (1972) had this ratio at 0.36. Also, the former author had 0.82 of the depth taken up with these units, whereas the latter used 0.57. As noted above, prototype structures can have significantly varying proportions of the face covered by armor units or smaller stone and hence reflection coefficients could differ greatly. There is need for more tests to account for realistic breakwater cross sections, where a large proportion of the submerged face comprises smaller stones forming a mound on which the larger armor units rest at their base. The size of these units needs to be properly proportioned to give correct reflection coefficients in modeling.

7.1 SCOUR DUE TO NORMAL WAVES

Scour from waves approaching normally has been studied by Herbich et al. (1965), Xie (1981), and Irie et al. (1986). Xie used fine (0.10 mm) and coarse sands (0.15–0.78 mm) in his tests. As seen in Fig. 7.2, these were slightly spaced apart, but not sufficiently to differentiate on this basis. A curve of scour (S/H) has been drawn through 0.10 mm points as prototype conditions would be reproduced better by this fine sand. Since the larger waves will effect the greatest scour, up to incipient breaking, a value of H_{max}/d for any given d/L can be derived to obtain the ratio S/d for storm conditions (Goda 1970). This curve is also indicated in Fig. 7.2, which obviates the need to hindcast wave height (H) in close proximity to the wall. These waves are reflected from the structure, for which Herbich et al. (1965) found scouring capacity the same for wall slopes of 45° and 90°.

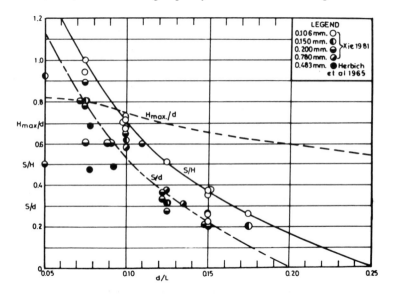

Figure 7.2   Scour due to normal waves reflected from a structure

Assuming a water depth adjacent to the wall is 3 m and the incident wave period 10 s, then $d/L = 0.056$ and $S/d = 0.8$. Hence, a trench will be scoured 2.4 m deep at 39 m from the face. The centers of trenches will occur at odd multiples of quarter wave lengths from the reflecting wall. Xie (1981) recorded the profiles of these troughs which were similar to sinusoids. These have their steepest slope about halfway between the face and the greatest depression. This increased slope in the vicinity of the wall will permit greater pore pressure buildup than for a flat bed and can therefore contribute to subsidence of a rubble-mound supporting heavy loads of similar material or caisson-type structures.

It has been found by Irie et al. (1986) that if rubble berms extend to about a quarter wave length the toe readily subsides into the trough forming on the bed. Their test rig is exhibited in Fig. 7.3, together with profiles along three lines normal to the vertical face.

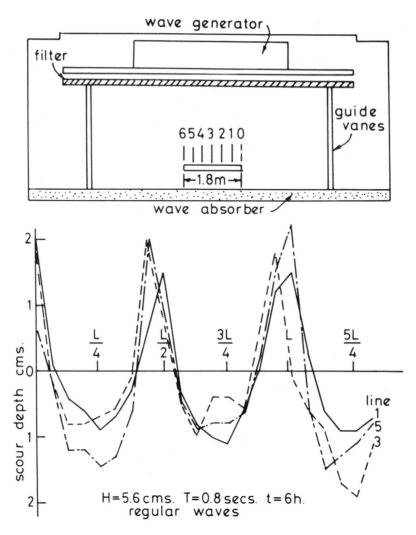

Figure 7.3   Test rig and results of Irie et al. (1986) for $L = 78.1$ cm and $d = 22$ cm

It is seen that scour occurs at the nodal zones of $L/4$, $3L/4$, and $5L/4$, where accretion occurs at the antinodes of 0, $L/2$, and L. Since the wave length in this case is 78.1 cms, the slopes of these troughs near the apexes approximate 22°. These make a strong propensity for pore pressure to cause a collapse seaward.

7.2 SCOUR DUE TO SHORT-CRESTED WAVES

When waves arrive obliquely to structures, they generate short-crested wave systems, as noted in Section 5.3.1. The kinematics of such systems have been presented in Chapter 2, which may be summarized as follows:

1. The directions and magnitudes of water-particle orbits differ across planes normal to the face but are similar along alignments parallel to it. These alignments are defined by fractions of the *crest length* (Z/L'). When this is 0, 1/2, 1, ... orbits are parallel to the face and at $Z/L' = 1/4, 3/4, 5/4, ...$ they are normal to it. Halfway between them $(Z/L' = 1/8, 3/8, 5/8, ...)$ the orbits are elliptical and angled to the vertical within the water column, but in a near horizontal plane near the bed.

2. These orbital motions have increased velocities and accelerations in horizontal and vertical directions due to the doubling of the wave height and increase in wave length. Their attenuation with depth differs from the progressive incident waves.

3. The vortex motion produced at $Z/L' = 1/8, 3/8, 5/8, ...$ alignments produces increased macroturbulence which keeps sediment particles suspended for longer than the case for incident waves.

4. The mass transport in short-crested wave systems is both normal to and parallel to wave propagation, being maximum at alignments $Z/L' = 0, 1/2, 1, ...$ The tendency is to sweep material toward these alignments. The increase in suspension and this mass transport cause ready scouring of the bed.

5. Model studies of water-particle motions have verified the theory derived for such kinematics both in terms of orbital motions and mass transport.

6. Reflected waves suffer diffraction as soon as they leave the reflecting wall. This reduces their height along the ribbon of waves contained within the orthogonals through the wall extremities. The narrower this band, the quicker this reduction in height.

The point raised in (6) above could well be expanded upon, as it points out that waves are not uniform along the reflecting wall. As seen in Fig. 7.4, the narrow band of reflected waves of width B will have a diffraction $K = 0.5$ along its boundaries (Silvester and Lim 1968). The energy of the waves between them will be reduced as

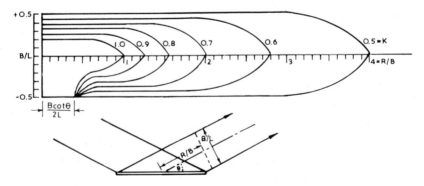

Figure 7.4 Diffraction coefficients of waves reflected from a wall for various ribbon widths (B/L) and distances (R/B) from the center of the wall

waves propagate from the wall, so that on the centerline at $R/B = 4$, the height is reduced by half. Isopleths of other coefficients are shown which indicate that reflected height equal to incident only occurs out to $R/B = 1.0$ and for only part of the ribbon width B. From the wall center to its downcoast tip the full reflected wave height exists, which in terms of wave propagation along the centerline gives $R/B = B \cot \theta / 2L$, where θ is the orthogonal obliquity to the wall. In spite of these differentials in wave height, the orbital motions are still very complex and energetic and can put sediment into suspension, even beyond the $K = 0.5$ limit shown.

7.2.1 Model Evidence of Scour

The oscillation and subsequent removal of sediment from the floor takes place in three stages. First, the shear of water particles causes dunes to form. Second, the sweeping effect of mass transport produces mounds to form parallel to the reflecting wall for the case of 100% reflection. For a lesser reflection coefficient these may be slightly angled to the structure. As Lin et al. (1986) have concluded: "The characteristics of water particle motions under the short-crested wave system enforced the sediment transport and form into special bed forms such as troughs, holes, triangular bars and longitudinal bars." Their tests were conducted with a caisson-type breakwater and a rubble-mound support which complicated the reflection pattern. Third, the whole bed is eroded within the area affected by reflected waves, even outside the limits of complete reflection.

When rectilinear oscillations of water particles take place at the bed, the dune crests are normal to them. At $Z/L' = 0$, 1/2, 1, ... these are at right angles to the face, whereas at $Z/L' = 1/4$, 3/4, 5/4, ... they are parallel to it. Where vortices are formed (at $Z/L' = 1/8$, 3/8, 5/8, ...) sediment particles are ejected radially to form almost circular mounds, which gives complex bed formations. All these forms were exhibited in Fig. 5.2, for waves angled 45° to a wall.

Depending on the alignment that has the greatest orbital velocities at the bed, which in turn is dictated by the angle of incidence, troughs are formed along the reflecting wall. For $\theta > 45°$ they occur at $Z/L' = 0$, 1/2, 1, ... , but with $\theta < 30°$ mounds form in this location. In Fig. 7.5, initial erosion is displayed at nodal alignments ($Z/L' = 1/4$, 3/4, 5/4, ...) where dunes were parallel to the wall for this case of $\theta = 20°$. This topic will also be discussed in Section 9.8. Monochromatic waves produced this result, whereas in nature swell waves would be changing period with time, so producing general denudation of the floor.

Tests carried out for 60 hours, for waves angled 30° to a wall in 25 cm depth with wave period 1.0 sec, produced a result as in Fig. 7.6 on a sand bed 5 cm thick. It is seen that for this limited duration material was in transit from the RHS, so causing accretion to the left. Should this experiment have had longer duration this shoal would have disappeared, leaving the whole area within the ribbon of reflected waves, and even beyond it, devoid of material. At the present time the limiting depth of scour within a specific short-crested wave system cannot be predicted.

Tanaka et al. (1972) conducted a similar test for $\theta = 30°$, in 5.5 cm depth with 0.52 sec waves and 2.8 cm height. Progressive erosion and siltation is shown in Fig. 7.7,

Figure 7.5   Differential erosion occurring at quarter crest lengths from the wall

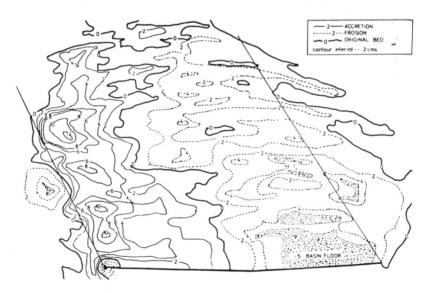

Figure 7.6   Scour of bed to the concrete floor after 60 hours duration

where again material was removed outside the limiting orthogonal through the upcoast tip
of the breakwater. The variations in bed level in this figure are based on the assumption
that a profile recorder was 25 cm from the original bed in the data provided. Irie and

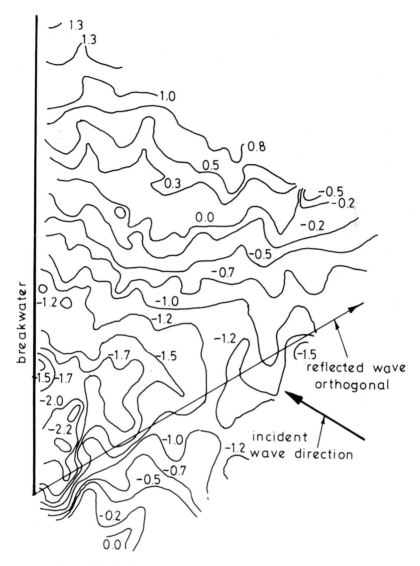

Figure 7.7 Bed contours after 10 hours duration (Tanaka et al. 1972)

Nadaoka (1984) have conducted model tests on a breakwater where the bed was sloping upward toward the downcoast end, as seen in Fig. 7.8. It is seen that zones of erosion and accretion vary across the normal to the structure. This was similar to a prototype structure with waves at similar incident angle.

Later, Irie et al. (1985) conducted tests with angled waves of both regular and irregular form. These have been reported more fully in Irie et al. (1986). Their model basin is depicted in Fig. 7.9, where it is seen that waves approached from either 30° or

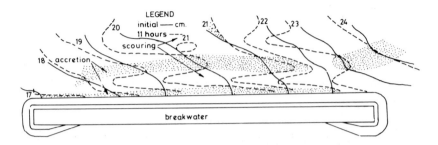

Figure 7.8 Tests showing differential erosion and accretion for a duration of 11 hours with irregular waves (Irie and Nadaoka 1984)

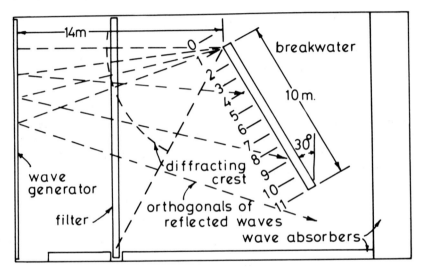

Figure 7.9 Model basin used by Irie et al. (1986) for scour tests by obliquely reflected waves

45°. The sandy bed sloped 1:100 lengthwise along the basin from the wave generator. This implies a slope of 1:115 normal to the breakwater. Thus, the water depth adjacent to the breakwater varied along it, being 6.7 cm deeper at alignment 1 than at alignment 11 for the 30° obliquity. The breakwater was of caisson-type with a trapezoidal rubble berm adjacent to it, as illustrated in Fig. 7.10. Its top surface was maintained constant along its length, causing a reduced width from alignment 1 to 11 of Fig. 7.9, and a rise in sand level from the bed slope previously alluded to.

The incident and limiting reflected orthogonal at 30° are shown with a single diffracting crest in Fig. 7.9. Other orthogonals within this zone of diffraction are also shown which are reflected from the wave generator, the resulting reflected orthogonals are also included. All these wave rays have not been refracted across the sloping bed for ease of presentation. The height of these doubly reflected waves will vary along the breakwater and hence contribute differing amounts of energy along its face.

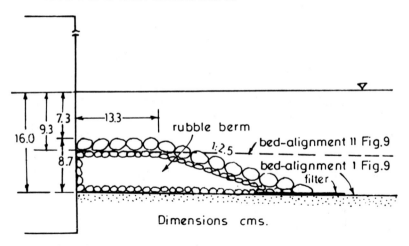

Figure 7.10   Cross section of breakwater as used by Irie et al. (1986) in Fig. 7.9. The actual bed slope normal to the structure was 1:115

The report by Irie et al. (1986) contains many bed profiles, both normal to and parallel to the breakwater, measured over time intervals. These have been simplified and further diagrams drawn by the authors (Hsu and Silvester 1989a), in order to gain a different perspective on this phenomenon of scour due to short-crested waves. The variables employed are listed in Table 7.1.

TABLE 7.1   VARIABLES USED IN TESTS BY IRIE ET AL. (1986)

Test	Scale	Model wave height (cm)	Prototype wave H (m)	Model period(s)	Prototype periods	Mean sed. dia (mm)	Obliquity $\theta°$	Type of waves
1	1/75	8.0	6.0	1.15	10.0	0.14	30	regular
2	1/75	8.0	6.0	1.39	12.0	0.14	30	regular
3	1/75	8.0	6.0	1.30	11.3	0.14	30	irregular
4	1/75	8.0	6.0	1.30	11.3	0.14	45	irregular

Results for run No. 1 (Table 7.1) with 11 hours duration are presented in Fig. 7.11. Distances normal to the reflecting face are shown as fractions of (L'), the distance at which island crests propagate parallel to it, these are at $Z/L' = 0$, 1/2, 1, ..., which can be equated to antinodes of standing waves for normal approach. It is seen that greater scour occurred at $Z/L' = 1/4$ and 3/4 or the nodal zones. At the antinodes horizontal orbital velocities were minimal, so resulting in smaller scour during this limited duration. The numbers on the curves represent profiles as noted in Fig. 7.9. The greatest scour occurred at lines 3 and 4 and the smallest at 7. During this transition period material removed from upcoast (3 and 4) is still traversing the area farther along the wall (7), so making for smaller depths.

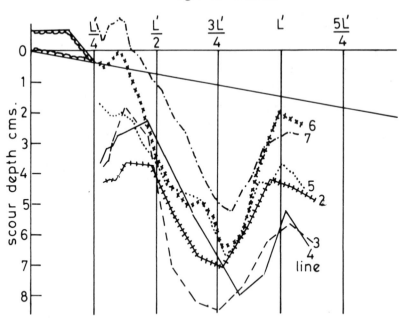

Figure 7.11   Bed profiles normal to the breakwater for lines shown in Fig. 7.9 along its length for regular waves. Lengths L' represents 169 cm. The rubble berm is not to scale (Irie et al. 1986)

Since these quarter crest lengths ($L'/4$) are 43 cm, the mean slope of the trough centered on $3L'/4$ can be derived. At most points along the breakwater these are similar, at about 1:10 or 5.7°. The steepest slope is on line 3 at 1:3.8 or 14.8°. The mean diameter of sand was 0.14 mm, which is similar to prototype conditions, but the waves were reduced to 1/75 of normal size. It is possible that under prototype conditions pore pressure could be built up sufficiently to cause slumping of this trough. This could undermine the rock berm, especially as it is taken out to $L'/4$ from the face of the main structure.

The profiles obtained for run 3 with irregular waves of almost the same period are more uniform in character as seen in Fig. 7.12. This is because each wave component has its own particular nodal and antinodal distances from the wall. The L' alignments have been defined for the peak wave in the spectrum, with greatest scour occurring close to the $3L'/4$ line, similar to that for regular waves in Fig. 7.11. Also, maximum scour occurs at lines 3 and 4 and minima at lines 5 and 6. Accretion appears to have taken place on $L'/4$ at this duration of 11 hours. There is little doubt that, with continuation of waves, scour depths would have become more uniform and greater. Even at this limited duration the bed slope for lines 2 and 3 is 1:3 or 18°, which is steeper than in Fig. 7.11. In practice this could have slumped, particularly during a storm period.

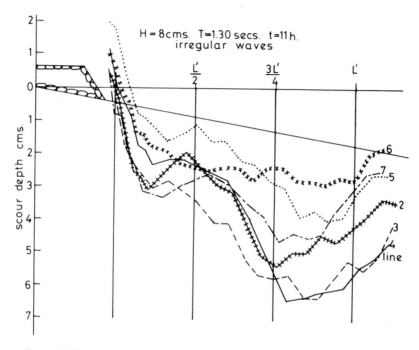

Figure 7.12   Bed profiles normal to the breakwater for irregular waves. Lengths L' represents 212 cm. The rubble berm is not to scale (Irie et al. 1986)

The ratio of short-crested wave height to that for incident waves in the Irie et al. (1986) model is shown along the length of the breakwater in Fig. 7.13, for alignments $z/L' = 0$, 1/2, and 1. It is seen that adjacent to the wall ($Z/L' = 0$) this is above unity at line 1, indicating the presence of reflected waves from the wave generator. By line 2 the incident and reflected waves from the wall have built to over 2, showing again the addition of energy from wave generator reflection. It then fluctuates below and above 2 along the wall but becomes steady at 2 from line 8 to line 10. At alignments $Z/L' = 1/2$ and 1 the waves do not reach their optimum of 2 until line 4 because of diffraction of the reflected wave. They then decrease from lines 6 to 8, but either remain constant after that, or for $Z/L' = 1$ exceed 2 before approaching 2 at line 10. As seen in Fig. 7.9, the reflected energy from the wave generator is minimal at the downcoast end of the breakwater, in the vicinity of lines 8 to 11 values of 2 are achieved. Also included in Fig. 7.13 are scour depths along alignments $Z/L' = 1/4$ and 3/4. These have maxima at lines 1 and 2 respectively, decreasing along the length of the wall at this wave duration of 11 hours. In fact, for $Z/L' = 1/4$ accretion above SWL is exhibited, indicative of material still being transitted.

Results from runs 1 and 2 (Table 7.1), as depicted in Fig. 7.14, show the influence of wave period with regular waves. It is seen that the longer period effects greater transport, removing more material from lines 1 to 4 and adding more from 8 to 11. The scour for alignment $Z/L' = 3/4$ is greatest at line 8 for the 1.39 sec wave, whereas it

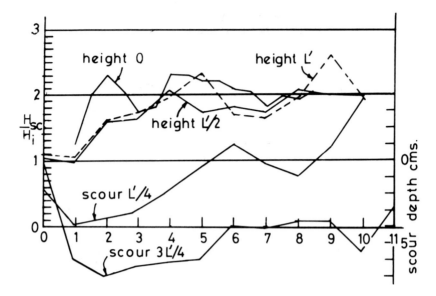

Figure 7.13 Wave height variation and scour along the breakwater for regular waves for laboratory rig shown in Fig. 7.9. Line numbers refer to that figure (Irie et al. 1986)

occurred at line 3 for the 1.15 sec wave. The reasons for the peaks midway along the breakwater are difficult to explain, since more uniform scouring lengthwise would be expected. It may be due to wave variation as exhibited in Fig. 7.13, or refraction over temporary offshore shoals.

Run number 3 for irregular waves gave lengthwise scour as seen in Fig. 7.15, with the same 11-hour duration. Here again more scour was displayed for $Z/L' = 3/4$ than for 1/4 with shoals appearing near the center of the breakwater. Overall erosion was less than for regular waves as presented in Fig. 7.14. Irie et al. (1986) also measured scour over varying durations both normal to and along the breakwater, the latter being for $Z/L' = 1/4$ and 3/4. From these data a graph has been prepared, as in Fig. 7.16, of scour versus duration. As for most sediment scouring phenomena erosion is seen to be swift initially but decreases in rate with time. By the slopes of the lines at duration 11 hours it is observed that the erosion process was still continuing. Only for line 8 at $Z/L' = 3/4$ for the 1.39 sec wave is equilibrium seen to be approached.

Tests conducted by the same authors for approach angles of 30° and 45° (runs 3 and 4 in Table 7.1) are recorded in Fig. 7.17. The scour for 45° appears to be less than for 30° over this 11 hours duration. The deepest scour for line 4 has shifted from the nodal alignment $Z/L' = 3/4$ for 30° toward the antinodal of $Z/L' = 1.0$ for 45°. It could be surmised from Fig. 7.9 that waves reflected from the breakwater (at 45° to the incident waves) would have a limiting reflection orthogonal through line 1 at the end of the structure which is parallel to the wave generator. The diffraction of reflected waves would thus result in lower heights for those waves being reflected by the generator blade

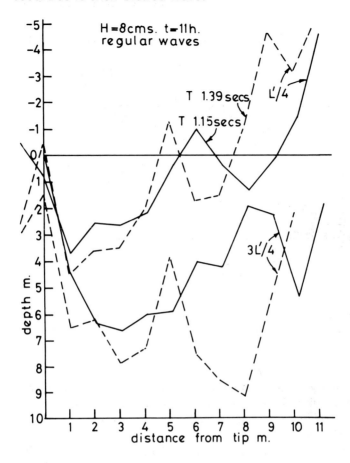

Figure 7.14 Comparison of scour along the breakwater after 11 hours duration for waves of 1.15 and 1.39 s period (Irie et al. 1986)

back to the breakwater. The wave energy near the structure is thus reduced and may have resulted in the more modest erosion.

The scour of irregular waves over time is illustrated in Fig. 7.18, which is similar to that for regular waves in Fig. 7.16. However, the rate of scouring is less, it being about 0.5 cm after 0.75 hours for irregular waves of 1.30 sec period, but 1 cm for regular waves 1.15 sec and 2 cm for 1.39 sec waves, all for the $Z/L' = 1/4$ alignment. The line locations have been chosen for maximum scour and therefore differ in the two figures. The scour for the $Z/L' = 3/4$ alignment is more scattered between the two figures, but the trend is the same. After 11 hours the scour for this alignment with 30° approach is around 4 cm for irregular waves (Fig. 7.18), while it ranges from 6 to 9 cm for regular waves (Fig. 7.16). Although all these have the same significant wave height, there will be more time in the spectral waves when heights less than this occur. With regular waves all heights are the same, arriving at the frequency specified by the wave period.

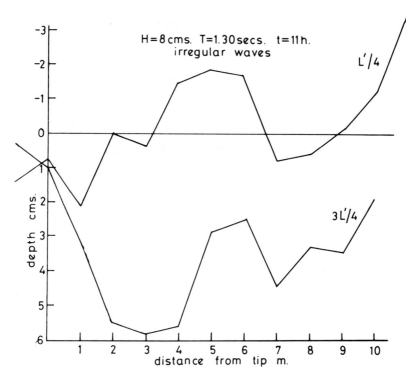

Figure 7.15 Lengthwise scour for irregular waves after 11 hours duration (Irie et al. 1986)

Although a scale of 1:75 is indicated in Table 7.1 for tests by Irie et al. (1986), caution should be exercised in extrapolating scour dimensions to prototype situations. The waves are substantially reduced in size, hence the macroturbulence is disproportionately less, while the sand size is similar. In these tests the duration of 11 hours is seen to be insufficient for final profiles to have been reached. Time scales with actual breakwaters are lacking because most surveys available have been carried out just after construction was completed, thus not allowing for equilibrium conditions to have been reached. There is a need for sequential surveys, plus model studies of varying scales.

The comprehensive series of tests by Irie et al. (1986) are worthy of this reanalysis because they bring out many facets of such model studies, the main one being the need for long enough duration for equilibrium to be observed. The great concern being experienced by Japanese engineers for the consequences of such scour should point the way for coastal engineers throughout the world. Many breakwaters in Japan, angled to the coast at ports, are now being developed along sandy shorelines, most normal embayments having already been used for this purpose. Field measurements of scour will be presented in the next section. No structures should be built which are likely to be angled to persistent waves without the construction of a three-dimensional model to predict the consequences. If scour is indicated, then some protection should be afforded the bed

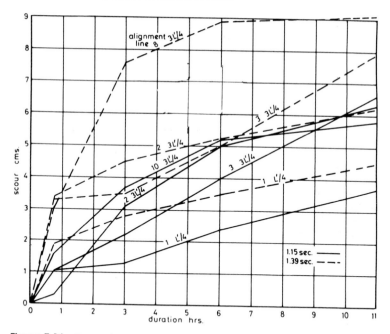

Figure 7.16 Progressive scour at certain lines along the breakwater for regular waves on alignments of $L'/4$ and $3L'/4$ (Irie et al. 1986)

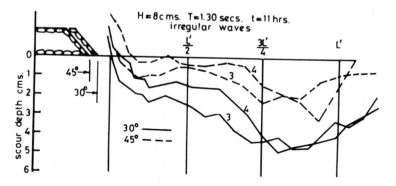

Figure 7.17 Profiles normal to the breakwater at lines 3 and 4 for irregular waves with initial approach angles of 30° and 45°. Rubble berm is not to scale (Irie et al. 1986)

whenever reflected waves are present. This means more than the provision of berm toes extending only a matter of meters from the breakwater face (Hsu and Silvester 1989b).

7.2.2 Field Evidence of Scour

It is contended that the main energy source for bed scouring is obliquely reflecting waves, predominantly swell that is persistent in its duration and direction. As noted already,

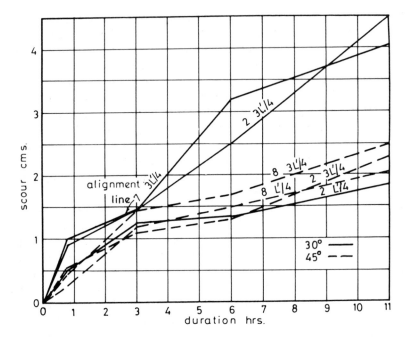

Figure 7.18 Progressive scour at certain lines along the breakwater for irregular waves on alignments of $L'/4$ and $3L'/4$ with initial approach angles of $30°$ and $45°$ (Irie et al. 1986)

these low height long period waves are reflected very effectively from rubble-mound structures. In so doing they generate short-crested systems whose orbits include vortices that expand to the bed even in relatively deep water. The result is the construction of troughs and mounds parallel to the face which are large in magnitude while material is transported in the direction of the island crests so formed.

Examples of bed removal will be cited from around the world, which phenomenon has only been recognized fully in the last two decades. As noted in Section 7.2.1, Japanese engineers have made a concerted attack on this problem, due to the many breakwaters angled to the coast in their country. Much more research is called for in three-dimensional models with mobile beds to verify the theory on short-crested wave systems, with adequate duration for the complete effects to be examined. Prototype cases will be presented for Europe, Japan, the United States, and Africa.

Europe. A number of port structures in Europe have suffered erosion adjacent to them, which has been dealt with in a traditional manner, of covering the affected area by matting or stones. In other cases the failure of the breakwater was explained by instability of the armor units rather than denudation of the bed supporting the structure. The ports to be discussed are Sines in Portugal in great detail, Hirtshalls in Denmark, Rotterdam in the Netherlands, Zeebrugge in Belgium, Thorshavn in the Danish Foeroes, and Ashdod in Israel.

Sines.   This is a prime example of subsidence of the breakwater face which has been widely publicized because of its catastrophic nature. Many international agencies have attempted to provide an answer as to the cause, including a major report by ASCE (Edge et al. 1982). As accepted in the report: "other failure scenarios can be proposed from the available data"; even though the opinion was expressed that "the causes of failure and the sequence of events leading to the failed layer during the storm of February 26, 1978 will never be completely defined." As noted by the references already listed for this section, both for model studies and field measurements, the bed erosion could well be due to waves arriving obliquely to structures and being reflected from them. Silvester (1986), Silvester and Hsu (1987), Hsu and Silvester (1989a), Silvester and Hsu (1989) have summarized some of these prototype occurrences and reanalyzed results from model studies.

Sines breakwater is oriented N–S at latitude 38° N and hence receives the incessant swell from storms continually passing across the Atlantic Ocean from west to east between latitudes 40° to 60°, with winds circulating in an anti-clockwise direction. The storm waves generated by these winds are optimal from the W–NW directions as wind fields follow waves already produced. These disperse as swell to arrive in the deep water off the Portuguese coast from a NW direction. This is confirmed by percentages of time for swell in the east Atlantic Ocean as given in Table 7.2 (HMSO 1959). The most appropriate latitude to accept is 30°–40°, where the duration from NW is over four times that from the SW. Even if these values are averaged with those in the 40° to 50° latitudes, over twice the swell waves arrive from the NW quadrant.

TABLE 7.2 PERCENTAGE TIME FOR
SWELL IN EASTERN ATLANTIC

Latitude	NW	SW
30°–40°	49.2	11.25
40°–50°	37.6	25.5
30°–50°	43.4	18.4

Note: These do not add up to 100%
due to percentages from NE and SE
directions being omitted.

Refraction across the continental shelf may cause the approach angle to the Sines breakwater to be about 30°, as depicted in Fig. 7.19. This is ideal for eroding the bed in the form of trenches parallel to the straight segment of the breakwater at spacings of 1/4, 3/4, and 5/4 of the crest length, which is dictated by the wave period and depth. Hence, the centerlines of the bed depressions change with the period or wave length of the swell, which vary continually. The sand placed in suspension by the vigorous water-particle motions of the short-crested system, including strong vortices directly on the bed, are transported southward by the mass transport of the waves. This implies that the zone near the head may have still been receiving sand on the date of the collapse, with the bed possibly higher than the original. The region to the north would have been scoured

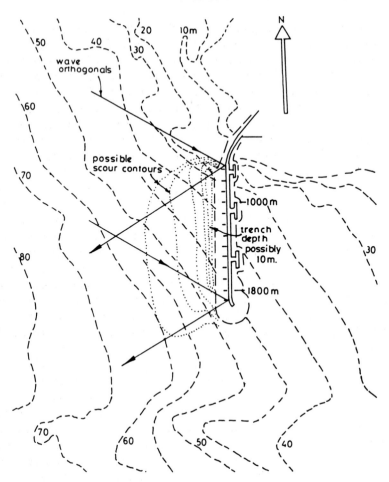

Figure 7.19 Plan of Sines showing wave orthogonals and hypothesized contours of scour

first and become deeper than farther toward the south. As seen in Fig. 7.19, erosion could take place many crest lengths from the breakwater, encompassing a triangular area between it and the reflection orthogonal through its upcoast limit. It becomes obvious that a purported protective toe of a few meters width will do little to prevent such scouring.

When the seabed is deepened adjacent to a breakwater, the floor slope up to the structure is increased and can almost reach its angle of repose, as seen in the model studies of Figs. 7.11 and 7.12. The consequences when storm waves arrive include a buildup of pore pressure in the soil with possible collapse of its face, or certain reduction of its ability to withstand the load imposed on it. As noted by Markle (1986): "It is generally thought that toe scour is the significant problem after major storms. Bedding layers slough off into the scour holes, and this damage migrates back to the toe of the primary armor. The resulting instability of the armor stone toe leads to downslope migration of the onslope armor and eventual deterioration of the structures."

It is thus believed by these authors that the Sines collapse was due to bed failure, which caused smaller stones at the foot to subside and with it the layers of armor units being supported by this mound of rock. Cross sections of the Sines breakwater in Edge et al. (1982) show it resting on base rock, but it was resting on 12 m of sand: "It was concluded that it would not be necessary to remove any bed material before placing the rubble mound." Although many laboratory tests have been conducted on most aspects of this collapse, they have tended to concentrate on the size of dolosse units used and their placement or misplacement on the face. Neither the consultants in the first place nor researchers since have, to the authors' knowledge, conducted tests with a mobile bed using obliquely arriving waves. The authors were told by one consultant that a three-dimensional model test was not carried out because it was too expensive.

Wave conditions at Sines reported in Edge et al. (1982) included a table showing percentage of time that swell from N and NW exceeded 3.7 m for each quarter of the year. These were 15, 8, 2, and 10, respectively. These percentages would be exceeded greatly by lower wave heights, which could be just as effective in scouring the bed due to their high reflectivity. These authors referred to the frequent "Azores High" pressure system which causes cyclones to pass across the Atlantic from west to east north of the coast of Portugal. They noted that: "For certain wave conditions, concentration of waves occurring from W to NW directions may occur in the vicinity of the breakwater." With respect to storm conditions at Sines, Edge et al. (1982) stated: "When this high pressure region is displaced, low pressure disturbances may move directly toward the coast of Portugal." But with the anti-clockwise circulation of the winds in these centers the maximum waves still arrive from the WNW (Mettam 1976). For the actual damaging storm in February 1978 it was reported: "At noon on the 24th, the second center was about 48° N 37° W." With Sines at 38° N this implies that the waves were arriving from the NW quadrant. Thus, the majority of the storm waves and swell are oblique to the N–S oriented structure 1.2 km in length and can thus induce severe scour.

During the storm of February 26 the waves were not to design height, but prior to this date waves of 9 m and greater were experienced as seen in Table 7.3, which lists the number of weeks during which waves exceeded 6, 7, 8.5, and 9 m during the previous 3 years. This has relevance to the accepted mode of failure, of dolosse being picked up by the uprush and falling on others during the downrush and hence causing breakage and final collapse. From the statistics in Table 7.3 this action should have occurred in 1975 and 1976 when large waves were more prevalent.

Also listed in Table 7.3 are the number of weeks during which swell waves exceeding 2 m were experienced. It is seen that these ranged from 43 to 46 when only 48 were considered for the year, 4 in each month. This duration is almost continuous, with these waves reflecting almost fully from an oblique angle. It is this persistence that could be the culprit in the erosion of the bed, with the storm incidence and concomitant liquefaction acting as a symptom of the real problem or "mal de mer."

As noted in Chapters 2 and 5, a wall receiving oblique waves will have the reflected waves moving out within a ribbon bounded by the orthogonals through its limits, as in

TABLE 7.3 RECORD OF DOLOSSE PLACEMENT AND NUMBER OF WEEKS WITH CERTAIN WAVES BEING EXCEEDED

Year		1974 Oct.–Dec.	1975	1976	1977	1978 Jan.–Feb. 26
No. Dolosse placed		1,489	5,563	6,786	5,908	56
Wave Heights (m)	9	—	3	3	1	—
	8.5	1	4	4	1	—
	7	2	7	6	3	1
	6	2	8	10	8	2
	2	12	46[1]	44[2]	43	7

[1] Four weeks were listed for each month, giving 48 for the year

[2] No record available for October

Fig. 7.4. At the ribbon edge the wave height is half that of the incident wave (for complete reflection) due to diffraction taking place, which absorbs energy from within the band. However, for the central half of the ribbon and for about one width distance, measured along it from the breakwater center, the resulting combined height is double that of the incident wave. This also applies from the center to the downcoast tip of the reflecting wall, or the southern head in the case of Sines. Hence, waves in this region are the largest that can be produced in this short-crested system.

The material removed from the upcoast limit of reflection, or the northern section of Sines breakwater, has to move along the structure. The trench so dug is deepest first in this region and progressively deepens along the reflecting face. It is contended that at the downcoast or southern limit accretion occurred over a temporary period, but given sufficient duration the complete length would have been eroded. As noted in the previous section, trenches are scoured at either $Z/L' = 1/4$, 3/4, 5/4, or $Z/L' = 0$, 1/2, 1, with the greatest depths at 1/4, 1/2, or 3/4 crest length from the face. However, scour can occur for a number of crest lengths out to sea, reducing due to the loss of energy in the reflecting wave from diffraction. With the change in wave length, as swell periods alter over time, the erosion takes place over the whole bed. It is difficult, due to scaling problems, to predict rates of scour from model studies and none appear to have been conducted for Sines anyway. The hypothesized scour distribution at the time of failure is depicted in Fig. 7.19, with the possibility of 10 m depth along the centerline of the closest trench.

Although it is stated in Edge et al. (1982) that: "Subsequent to the failure of the breakwater, extensive and detailed bathymetry surveys were conducted by the contractor," these have not been generally available. As the wave climate in 1977 was more moderate, with little storm action (see Table 7.3), the littoral drift would have been minimized, which could have set the scene for severe scour in front of Sines. A survey available for 1982 showed accretion seaward of the breakwater indicating an abundance of sand available. Between 1978 and 1982 there were many storms which enhanced the rate of longshore drift, as explained in Chapter 3. It is contended that erosion adjacent to the breakwater would have commenced as soon as part of the N–S section of the structure had been completed. It expanded as construction proceeded, with material being transported southward. By the time of the storm on February 26, 1978, a large proportion of the N–S arm at the southern end would have been accreted, or possibly at its original level (see Figs. 7.13, 7.14, and 7.15).

It is pertinent to note that the northern extremity of the breakwater curves eastward toward the mainland (see Fig. 7.19). This provided an indentation which would have retained more sediment than farther southward because waves were more normal to the structure. This section of bed, therefore, could not have suffered the same degree of scour as along the N–S segment, or even may have been unaffected by any sediment transport. This section was not damaged in spite of the dolosse armor units being smaller.

It is submitted that, due to slumping of the bed adjacent to the breakwater, or buildup of pore pressure that reduced its shear resistance, the base mound supporting the dolosse units plus the toe subsided, as depicted in Fig. 7.20. A similar concept has been presented by Smith and Gordon (1983), except that they implied collapse of the small sized rubble at the floor of the slope due to pore pressure within its almost unstable face. "Whilst the Dolosse Slope is 1.5 to 1 for what are completely interlocking units both toe zones are proportioned at a steeper 1.33 to 1 slope. Since the normal angle of repose of a mixed boulder population is usually only 1.25 to 1, these boulders are almost stressed to their limit merely holding themselves in place." This combination of bed and toe failure has been stressed previously, as discussed by Markle (1986).

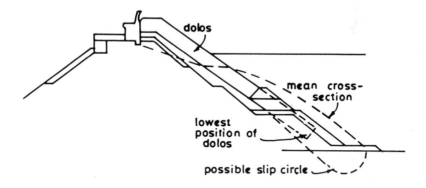

Figure 7.20   Suggested subsidence mechanism to fit observations subsequent to failure

This sudden collapse of the toe would have caused the armor units to slide over the underlying 3–6 T stone down their 1.5 to 1 slope. This would have resulted in breakage of dolosse at the base and within the mass of these units, but not necessarily those at the top, because they had no concentrated point loads applied to them. This penetration of rubble, and even perhaps some dolosse, into the bed means that the cross sectional area lost from the top would not match the added cross section of the mass at the base, as measured from the mean bed level.

It is significant that the collapse occurred over the bulk of the N–S segment of the breakwater, indicating that similar unstable conditions existed for this 1.2 km length of structure. Only at the southern head, and in the northern region near where the breakwater curved eastward, did subsidence not take place during this event. Discussion will now be presented on the following topics:

(a) Subsidence prior to February 26.

(b) Breakage of dolosse at depth.

(c) Resulting cross-sections.

(d) Damage at the breakwater head.

(a) Subsidence prior to February 26.   In Appendix B of Edge et al. (1982), the following information (partially quoted) was from the Portuguese Investigation Committee to the Societa Italiana Condotte d'Acque: "In January 1978, after the storms occurred during the winter months of 1977, an inspection along the west breakwater took place, by boat, of the condition of the dolosse armor units. In the zone between the progressive 828 and the progressive 870 there were also some broken or displaced dolosse at the water level, in a certain place of that zone, rocks of the enrockment underlying the dolosse armor unit could be seen, an indication that, in that place, dolosse were broken or displaced in a large quantity. ... The Inspectorate recommended to Condotte, orally, to carry out the reloading of this zone between the progressives referred to above, which Condotte did—the remaining 20 dolosse calculated by the Inspector as necessary in order to make a complete reloading, could not be placed before the storm of February 26, 1978, because of sea conditions."

In a further letter it was stated: "Information from another origin disclosed that this was a well localized spot—susceptible of creating alarming suspicions to an attentive and experienced observer. ... Was Condotte conscious of that or did it underestimate the occurrence, as it seems can be inferred from the transcript above? ... An atmosphere of apprehension for what had happened did not evolve—an atmosphere foretelling other similar phenomenons, the traces of which began appearing along the outside slope in zones where, as a consequence of the storms at the end of February, greater destruction of the dolosse armour unit took place. ... Was there not felt a need for a minute observation of all the outside slope of the breakwater, both in the emerged and the underwater parts?"

In reply Condotte stated: "The 'alarming suspicions' were groundless insofar as the same zone was concerned is demonstrated by the fact that between progressives 828 and 870 the effects of the February storms were not felt ... the consideration that the trunk at issue had been built in August 1975 and, therefore, if there was any kind of defect it could have manifested itself long ago, all contributed to the thought that the phenomenon was not to become generalized but rather to be locally remedial."

This section of breakwater, at the northern extremity of the straight section, would have suffered bed erosion first and hence suffered subsidence even before the general collapse. As noted by the above quotation, the underlying rock face was exposed, presumably above the waterline, which caused such sliding motion of the mass of armor units. Once this collapse had occurred the buried stone with new material above it was able to withstand the onslaught of the February 26 storm. Unless observers were aware of this bed supporting deficiency they could not suspect its generalized applicability. As it turns out, there was insufficient time for anything worthwhile to be done along the full length of the breakwater.

(b) **Breakage of dolosse at depth.**   The inspection by divers at a point about one third of the straight length of breakwater from the head was reported by Edge et al. (1982) as follows: "Essentially the inspection by the panel members corroborated the data reports ... by the Navy divers that virtually all the underwater units were broken, irrespective of the state of the dolos above the water level." This implies that many of the armor units remaining above or at the water level were undamaged, despite the later observation: "In the early stages of the storm, when the significant wave height reached 6 m, some dolos units began to move in the vicinity of the mean water level. ... As the wave height and period increased, the movement became more severe and the unit itself accelerated to the velocity of the uprushing waves."

It was also reported: "It must again be emphasized that although the armor units were intact above the water level in some locations, the armor units were by-and-large damaged below the water level." This was the case even though, "in many sections of the breakwater the dolos armour above the water level was completely removed, leaving the primary filter layer, and in some cases the core material, exposed." Further support for this differential damage is supplied in Appendix E of Edge et al. (1982), where the consultants provide a "Rationale for unreinforced dolos on Sines." They state: "During site visits (before the storm of 26 February) we have seen very few broken dolos above water level. We have also noted a number of dolos which have moved a considerable distance during the storm from their initial position without breaking. We therefore consider that there is evidence that the dolos had sufficient strength to resist normal movements in service. ... We have no information regarding the number of breakages recorded by underwater inspection of the dolos which have been moved considerable distances during the collapse which occurred on 26th February. We would, however, expect such breakages as a consequence of the major displacements which have occurred."

All the above quotations confirm to the reader, it is hoped, that the layers of dolosse slumped down the supporting filter layer. This action broke the bulk of units at the base of this mass, which is readily accepted when Edge et al. (1982) observed: "There is also a point where a unit located on the bottom layer of a breakwater armour section cannot support its own weight plus those units resting above." It would not take much of a subsidence to overstress dolosse in either compression or shear and so cause breakage.

(c) **Resulting cross sections.**   Also in Appendix B of Edge et al. (1982) a letter to Condotte d'Acque raised the following query: "In the cross-sections surveyed ... between September/October 1975 and January 1976, substantial differences are detected between the project cross sections and the executed cross sections insofar as the slope of the outside breakwater is concerned. These differences are observed in the cross sections corresponding to the following distances to the origin (m): 1069.09; 925.09; 949.09; 969.09; 989.09; 1009.09; and 1029.09." As seen in Fig. 7.19 these refer to the

northern regions of the breakwater. This appears to be further evidence of early scour
and subsidence with concomitant bulging of the profile at depth as rock and armor units
were received from above.

Cross sections of the final profiles were provided by Edge et al. (1982), many of
which were incomplete, in that the upper envelope did not clearly define the top surface
of the dolosse or the filter material beneath, and the lower section of the envelope was
not always carried down to the bed level, as seen in Fig. 7.21. The similarity of these
profiles from 729 to 1705 m (see Fig. 7.19) indicates the uniform nature of the subsidence
that occurred. However, a rough estimate was made of the area in the upper segment
to that of rock and armor units gained near the bed. The percentage of lower addition
to upper subtraction is listed in Table 7.4, which indicates that some of the volumes of
displaced material were buried in the trench, or fell into the liquefied bed as created by
the excess pore pressure. Only at Station 930 was the accumulated base material greater
than that lost at the top. This was within the area added to the previously collapsed
section prior to the storm of February 26. By the percentages shown in Table 7.4 it is
seen that part of the volume removed from the upper sections had been absorbed in the
bed, indicating a massive burial of rock and dolosse. Any buildup of the bed material
would have been immediately suspended and transported southward.

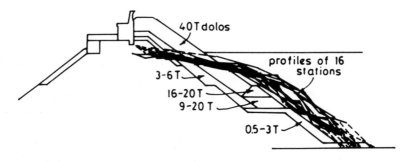

Figure 7.21   Cross sections after collapse as recorded by Edge et al. (1982). Extrapolated surfaces are shown dotted

It is significant that in the main cross section shown by Edge et al. (1982) the bed
is shown as rock, although in the text it is noted: "The deposits consist of sand and
gravel in thickness of up to 12 m." Committee members dived to inspect the bottom:
"The divers made a descent along the face of the breakwater near caisson number 81
... to a depth of −30 m where the bed rock is at a depth of −45 m." This reference
to bed rock rather than sand again appears to cloud the issue of bed erosion. It is
hard to perceive that the bed could be inspected, especially as much sand would be in
suspension due to the constant swell action. They would not be able to observe the
trench suggested, particularly as this would spread for some tens of meters from the
face. Without the hydrographic data supposedly obtained immediately after the collapse,
no proof is available of the erosion as suggested above.

TABLE 7.4 PERCENTAGE OF AREA GAINED IN
CROSS SECTION NEAR THE BED TO THAT LOST
ABOVE ZERO HORIZONTAL MOVEMENT

Station	Lower addition/upper loss (%)
720	74
750	39
780	56
905	58
930	130
955	56
1085	15
1115	40
1145	59
1175	87
1205	76
1235	74
1250	46
1605	98
1645	70
1705	47
Avg. Omitting 930	60

(d) Damage at the breakwater head.   In Edge et al. (1982) it was observed: "At the time of the February 26, 1978 storm some construction still remained to be done near the head and at berths 1 and 2. The superstructure was incomplete at the head (the wave wall had not yet been poured) and some 3–6 ton stone and dolos were missing ... Several dolos around the head were marked before the storm to measure settlement of the structure.... After the storm, the units were located horizontally and vertically once again on 10th March 1978. However only units $d1$ to $d7$ could be found." These are noted in Fig. 7.22, where it is seen that only $d6$ and $d7$ were on the seaward face of the structure and hence subject to the full impact of the waves in some 47 m of water. The actual movements of these two marked units is listed in Table 7.5.

TABLE 7.5 MOVEMENTS OF MARKED DOLOSSE NEAR
THE HEAD (SEE FIG. 7.22)

Dolosse No.	Seaward (m)	South (m)	Vertically (m)
$d6$	0.21	0.05	0.16
$d7$	0.01	0.11	0.40

The minor amount of movement, as indicated in the table, produced the following statements: "Near the head, as indicated previously, some settlement and a few broken

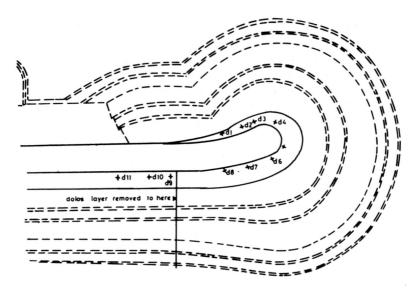

Figure 7.22 Positions of marked Dolos prior to the storm and subsidence zone marked

units were observed," and "The head which was placed at 1:2 slope was much less damaged during the storm than the trunk." In fact, it was only the 1/2–6 T stone below the supporting berm for the dolosse that was decreased in slope on the seaward side of the breakwater, as indicated by cross sections. The dolosse right at the tip were at a smaller angle. Thus, the conditions at $d7$ at least were quite similar to dolosse near the superstructure elsewhere along the trunk. Even so, the movements can be considered negligible, despite the depth being the greatest and the short-crested waves being the largest, as noted earlier. This indicates that the bed in this region was as original, if not actually accreted, during this storm. This implies that if the bed had remained static along the straight length of the breakwater, the collapse may not have occurred.

Three sets of model tests were conducted in an attempt to reproduce failure of the dolosse face. The first series (Morais 1974) by the Laboratorio Nacional de Engenharia Civil of Portugal "was based on the hypothesis that broken dolosse in the armour were responsible for the failure" (Edge et al. 1982). Various percentages in horizontal bands were broken by hand to test their effect on stability, but "only when 50% of the armour units were already broken did the breakwater fail under the assumed storm conditions of February 26."

The National Research Council of Canada, in concert with the Public Works Department, carried out tests with the structure as constructed (Mansard and Ploeg 1978). It is stated: "With 8.5–9.5 m waves (estimated peak of the storm) most units below the water level moved substantially. The units immediately below the mean water level were moved to another location further down the slope." If the inference is that these movements would break the dolosse and thus cause the final failure, the question must be raised as to why this had not happened much earlier. From Table 7.3 it is seen such storm conditions existed on 17 occasions prior to the storm of February 26, 1978. However, this had not caused the sudden type of failure experienced by that event except at the northern end, where removal could be associated with bed subsidence as discussed previously. Thus, it would appear that should prototype dolosse be lifted temporarily by larger waves it was insufficient to displace or break them in large enough numbers to create a general subsidence. When the face was undermined at the bed the general vertical movement could have broken the bulk of them, particularly those at the bottom and mid-depths of the layers, as observed.

The third series of tests was conducted by the Laboratoire Central d'Hydraulique de France (Langlais and Orgeron 1978). "When broken dolos were placed in the armour layer no significant increase in the failure process was observed" (Edge et al. (1982). With regard to movement of dolosse it was stated: "A result of the tests showed that for the conditions of the 26 February, 1978 storm the wave uprush could have caused vertical forces sufficient to lift a dolos from the armour layer." If this meant only those at the top of the incline, since dolosse are supposed to be interlocking, why were many of these units intact after the collapse?

All the model studies were conducted in flumes, which assumed that the storm waves arrived normal to the face, which was not the case, since they arrived from at least WNW if not more northerly than this. The island or intersecting crests were thus traversing the face southward and would have orbital velocities parallel to it besides normal to it. Even so, units were not sufficiently disturbed to create a general collapse during the 3 years of high waves prior to February 1978 (see Table 7.3).

It is a pity, in hindsight, that no three-dimensional models were built with movable beds to study scouring possibilities of oblique waves to this structure. It must be recognized that this type of occurrence at the time was not well known by the profession, despite the matter being publicized previously (Silvester 1972, Silvester 1974). This oversight might be blamed on the concentration on seawalls where reflection can be accepted as full, plus the belief that storm waves are well dissipated on rubble-mound slopes with precast concrete armor units. The role of swell waves, with modest heights but with large periods, has been discounted in the past, even though their durations are almost continuous.

Of wave heights experienced at Sines (Edge et al. 1982), all weeks of the 3 years showed some wave height, those with 2 m or greater being listed in Table 7.3, to indicate their persistence. Swell of 1 m was recorded for 48 weeks in 1975–76, which equates to continuous. When reflected obliquely these are able to generate the vortex action and complex orbital water motions similar to those experienced in storm waves, but reduced somewhat in intensity. It would seem appropriate, or even essential, in all future developments with breakwaters angled to swell or storm waves, that such three-dimensional studies be made with sedimentary beds. Even though scale factors would preclude an assessment of rates of bed erosion, the zones of initial and greatest scour could be determined, which should be complemented by frequent hydrographic surveys of the prototype (De Rouck and De Meyer 1987).

It appears, to these authors at least, that the reasons derived for the failure of the Sines breakwater cannot explain its suddenness or completeness. "The damage consisted of the complete loss of some two thirds of the armour layer of 42 ton dolos units" (Edge et al. 1982). The explanation by Smith and Gordon (1983) gets closest to the truth: "Any wave less than the design wave that could initiate any toe fluidizing action, thus, could trigger an irreversible progressive sliding failure in any breakwater, and the added effect of only a small handful of $H_{\max}$ waves could be catastrophic." The above discussion has taken the liquefaction argument down to the sandy bed itself, which causes a subsidence of the breakwater toe with similar consequences. As questioned by the same authors: "How otherwise, is it possible to explain the extreme abrasion evident in the Dolos and Dolosse pieces salvaged from the damaged sections of Sines?"

The evidence of scour being brought to light by Japanese workers, referenced previously, both from model studies and prototype measurements, to be considered below, is providing a due warning of consequences from oblique wave reflection. The paper by De Rouck and De Meyer (1987) outlines the erosion measured and predicted for breakwaters of equal magnitude to Sines. Even though these authors tend to blame strong tidal currents for this scouring, they are unlikely to exert great influence without the stirring action provided by wave action, particularly that from short-crested systems.

Wearne (1985) has stated: "The important general conclusion from the reports of important failure or lesser damage is that none were caused by hitherto unknown physical phenomena that acted without prior warning. All were caused by not knowing or using properly existing information. Therefore they were due to problems of perception or communication." This certainly is the case with respect to Sines, since knowledge of scouring propensity has been around for some decades, but perhaps not emphasized sufficiently in quarters where designs were being carried out. It is hoped that this will not continue into the future.

Markle (1986) recognized the need to design toe buttressing and toe berm material to withstand the movement from wave action and acknowledged that these "are susceptible to damage and failure when placed on an erodible bottom material" which "can result in the undermining and displacement of stones that were otherwise able to withstand the wave and flow environment but failed because of undermining induced displacement."

As with structures on land, there is more to be learned from failures than from structures that have withstood all environmental conditions, as they may have been overdesigned. There is much to be gained from views expressed on every aspect of catastrophies, whether they are novel or not. It is a pity if these must be restricted due to sub-judice provisions in the case of imminent litigation. In the case of Sines no firm conclusions have been reached, so there is a need to pursue the matter further, until confidence can be placed in any one or two explanations. Only then can future collapses be avoided.

Hirtshalls.   This port, at the northern tip of Denmark, has breakwaters as seen in Fig. 7.23. A new Eastern breakwater was under construction in 1973 when it was damaged. The blame for this was placed on the extension of the W breakwater from 1972 to 1974 (De Rouck and De Mayer 1987). A section normal to the E breakwater showed a 13 m scour for some 250 m offshore. Data points closer than 25 m were not

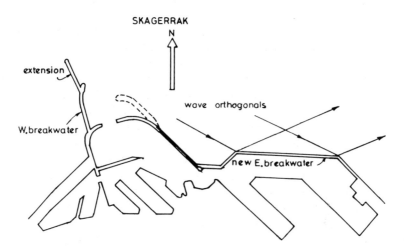

Figure 7.23   Hirtshalls harbor, showing E breakwater which collapsed during construction

available and hence the proximity of erosion to the structure could not be ascertained. This removal took place over 2 years whilst the structure was still being built, which highlights the swiftness of this action. As seen in the figure, the predominant WNW waves derive from the passage of cyclones passing eastward north of Denmark. They are quite oblique to the E breakwater and hence could generate intense short-crested systems along its length. The extension of W breakwater would have had little influence on this scouring mechanism since the supply of sediment had already been interrupted by the existing structure and the dredged channel to the port.

Rotterdam.   Euro-Poort was expanded at Rotterdam by the construction of a curved breakwater as depicted in Fig. 7.24. Erosion has occurred over a 2.5 km width in this relatively shallow water. De Rouck and De Mayer (1987) explained this as follows: "After completion of the new entrance, the depth in front of the bed protection can be expected to increase due to the concentration of the tidal currents and large scale dredging." It is difficult to comprehend how the tidal currents were increased more than by the N breakwater which already existed.

The dredging of the access channel, which must have taken place long before the S breakwater construction, would have interrupted bed transport by the predominant NNW waves. No mention is made of the possible wave reflection from the curved face, which as seen in Fig. 7.24 is at varying angles around the periphery. The short-crested systems established by this action, it is contended, could be the major cause of the said erosion. De Rouck and De Mayer (1987), in fact, did allude to breaking of progressive waves in the scouring mechanism. A blanket covering the bed, not specified, was subsequently eroded at its edge. This material would have been placed in suspension and carried southward by the interaction of reflected and incident waves, aided by tidal currents and mass transport due to the wave system.

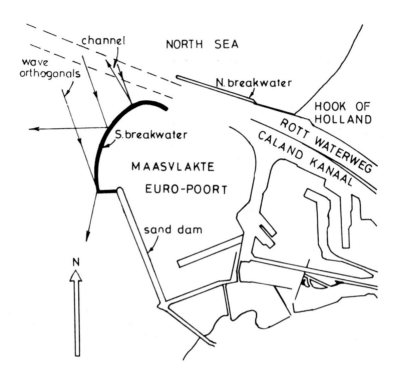

Figure 7.24 The plan of Euro-Poort, Rotterdam, where scour occurred seaward of the S breakwater

Zeebrugge.   The ultimate development of this harbor is depicted in Fig. 7.25, where it is seen two major breakwaters (NW and NE) enclose a large basin. The original version was a curved mole constructed between 1901 and 1910. This suffered erosion on its seaward face (Migniot et al. 1983) which will be discussed. Prior to the NW and NE breakwaters being constructed the LNG terminal was installed in 1981, which comprised an almost rectangular reclamation with a seawall surround. Erosion took place along the northern face soon after it was commenced and continued along the protective breakwater to its west. This initial erosion will be outlined. Finally, the results of model tests on the scour adjacent to the NW breakwater will be presented.

As seen in Fig. 7.26, the erosion at the curved mole over 9 years produced scour holes greater than 7 m deep along the main body and east of the tip of this structure. As noted by Migniot et al. (1983): "Materials were periodically dumped in the trench and protected the structure toe against the erosion which moved offshore along as the breakwater construction progressed. The maximum erosion depths were 8–9 m." Besides this alignment a second trench was oriented east–west. "These two types of erosion respectively correspond to flood and ebb current actions." There is no doubt this was aided by the suspension action of the short-crested wave action due to the oblique reflection of the incident waves as indicated in Fig. 7.26.

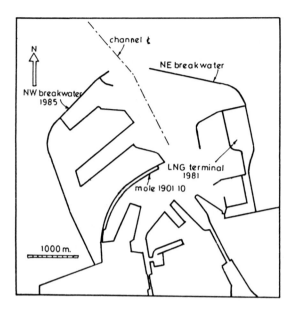

Figure 7.25 The ultimate outline of Zeebrugge harbor, showing the NW and NE main breakwaters

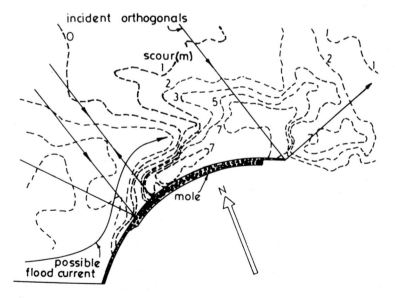

Figure 7.26 Scour along the curved mole of Zeebrugge harbor, which received oblique incident waves

The LNG construction suffered the full impact of swell and storm waves, but it was the northern arm that received these waves obliquely, as depicted in Fig. 7.27. Referring to this erosion trench, Migniot et al. (1983) stated: "It suddenly appeared after the February 1981 spring tides, created by strong flood currents whose velocity reached about 3 m/s. Then it progressively extended until July 1981 when the deepest point reached −20 m." It should be realized that at this time of the year wave action would also have been excessive. As the breakwater (as seen in the figure) was extended, so the trench also expanded to the west, with material so removed transitting the former depression and so reducing its depth temporally. This structure is now in calmer water, as seen in Fig. 7.25, and hence is not in jeopardy.

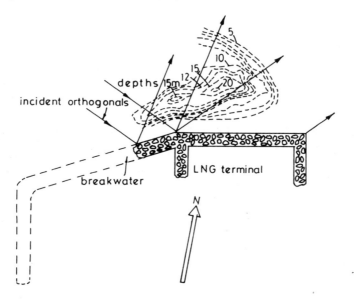

Figure 7.27 Trench erosion outside the breakwater protecting the LNG terminal at Zeebrugge harbor

Migniot et al. (1983) report that several models were constructed to predict erosion along the NW breakwater. "Natural sandy sediments were represented by material of appropriate density ($\rho = 1.22$) giving a correct similitude in the reproduction of sedimentary movements." The final stages of erosion are indicated in Fig. 7.28, about which the authors concluded from the tests: "In particular, they forecast the bed evolution around the north-western breakwater bend where a 700 m long, 250 m wide, at depth −17 m trench will appear during the structure construction. Then this trench will spread and reach 10 years after completion of the north-western breakwater 2000 m in length, 350 to 400 m in width with a maximum depth below −20 m." Whether this has occurred in the prototype is not known to these authors.

Thorshavn.   The Danish Faeroes could be considered as a port of Europe, therefore the port of Thorshavn will be described here. Sorensen (1985) has written on

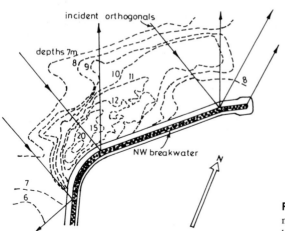

incident orthogonals

depths 7m

8 9

10 11

8

12

15

20

NW breakwater

N

7

6

Figure 7.28 Erosion indicated by a model along NW breakwater of Zeebrugge harbor, showing trench erosion

breakwater failures and has cited ten modes of damage of which four relate to undermining of the seabed with concomitant collapse of the toe and berm supporting larger armor units. Of this harbor he comments: "This example further documents how the damage to a structure is an integrated problem. Once the support of the armour layer fails, the armour layer slides down resulting in more severe wave impacts on the crown wall and erosion under the superstructure. The result was severe damage to the wall that would probably otherwise have been limited if the 'chain reaction' had not occurred."

The plan of the harbor and its site location are shown in Fig. 7.29, where it is seen that all major waves must approach the breakwater from the east, which is oblique to it. This aspect is not discussed by Sorensen (1985). He gave the reason for the failure of the main armor layer as the lack of a suitable berm below level −5.0 m. Model tests had shown that the 5.0 t weight of this berm stone was adequate for the design wave height of 4.0 m, when in fact collapse occurred at 3.7 m height. This indicates to the authors that bed erosion must have taken place due to the persistent swell waves prior to the storm, much like the Sines breakwater previously alluded to.

Ashdod.   This port in Israel will also be considered as part of the European community. Migniot et al. (1983) report serious erosion of a trench adjacent to the Ashdod breakwater during its construction in 1962–63. The depth changes between 1957 and 1963 are shown in Fig. 7.30, where it is seen that along the central portion of this 600 m long structure erosion reached about 4 m, "this value being eventually greater during storms to reach 6 to 7 m." Migniot et al. (1983) continue this comment as follows: "The sudden and local scouring of bottoms led to partial damaging of the structure south slope, the toe having sled in the trench and the then non supported tetrapod armour having subsided several metres, jeopardizing the breakwater overall stability." They suggest damage could have been less if a 10 m wide toe had been placed adjacent to the structure, but this would be disputed by these authors.

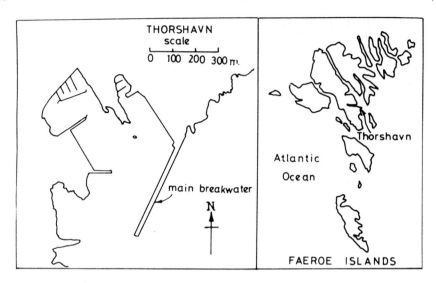

Figure 7.29   Site plan of Thorshavn harbor, showing breakwater orientation

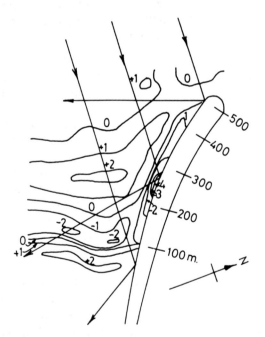

Figure 7.30   Trench scour which
occurred at Ashdod harbor, Israel

Japan.   Takeyama and Nakayama (1975) reported on breakwater failures in Japan, of which 900 had occurred from 1965 to 1972, as noted in Table 7.6. Of these 440 were due to subsidence and 259 were explained by bed scouring, or about 30%. Many of these had new armor units added at great expense.

TABLE 7.6 NUMBER OF BREAKWATER FAILURES RECORDED IN JAPAN (TAKEYAMA AND NAKAYAMA 1975)

Year	No. collapsed	No. due to subsidence	No. due to scouring	Scouring/collapsed (%)
1965	259	140	61	23
1966	71	27	23	32
1967	56	12	23	41
1968	94	53	36	38
1969	39	5	14	36
1970	212	141	59	28
1971	91	51	26	28
1972	78	11	17	22
Total:	900	440	259	Avg: 31

Several ports will now be discussed, all of which are oblique to the coast and hence to the predominant waves arriving. Some are on oceanic margins while others bound the Sea of Japan where persistent swell does not exist. In this latter case, the number of storms in any year is an important variable. The ports to be discussed are Naoetsu, Fukui, Niigata, Tagonoura, and Kashima. These are noted in Fig. 7.31.

Naoetsu harbor.   Long-term shoreline changes in the vicinity of Naoetsu harbor, on the coast of Sea of Japan, have been discussed by Tsuchiya et al. (1976). The harbor has been formed mainly by a breakwater running almost 45° to the coast which faces north. Only this structure is included in Fig. 7.32, where it is seen that it has progressed over some 15 years and has since been elongated, actually up to 1985. Prior to 1960 the Seki River discharged on the eastern side of the developing breakwater but was then diverted to the western side.

The bed contours of 1961 in the figure protrude seaward in two places, probably due to the influence of these two river exits. The interaction of the incident waves from NW to NNW with their reflective components has straightened out these contours, resulting in both accretion and erosion by 1975. Scouring is most pronounced in proximity to the breakwater, where a maximum of 1.5 m has occurred at the 10 and 11 m contours of 1975. Contours in 1961 only extended to 14 m and hence erosion up to the tip of the 1975 structure could not be determined.

The breakwater was realigned in 1978 and completed in 1985, as indicated in Fig. 7.33, when another survey was made. The contours for 1975 and 1985 are shown in this figure, from which zones of scour can be observed. The limit of erosion is shown which is near the bend in the breakwater. The most severe removal was 1.5 m where the 16 m contour of 1985 meets the structure. Remedial measures have already been undertaken by adding 32 tonne blocks seaward of the original caissons, making up a volume equal to that of the caissons. With the continued arrival of waves from the NW and NNW further erosion is possible, for which some bed protection by mortar filled mattresses may be called for.

Figure 7.31 Map of Japan showing harbors and sites discussed in the text

Fukui port. Fukui is also located on the Sea of Japan coast south of Naoetsu harbor, beyond the Noto peninsular, as seen in Fig. 7.31. It similarly consists of a straight breakwater at 45° to the coast, oblique to the predominant storm waves which are generated by cyclones traveling eastward across the Sea of Japan. Swell is of short duration. As indicated in Fig. 7.34, the wave crests are approximately 30° to the breakwater.

Scouring has been observed by Irie and Nadaoka (1984) in prototype situations. The breakwater was completed in 1978 when a hydrographic survey was conducted, which can be compared to contours obtained in 1976, as in Fig. 7.34. Some 3 m has been removed over half the length of this breakwater, which took place in a matter of months (Silvester 1986). The limit of erosion (contour 0 m) extends for all but 200 m of its length. It ran from the tip to 500 m from the face, the zone enlarging toward the toe where reflected waves propagate.

Silvester and Hsu (1987) have further reexamined this site, with a comparison of 1978 to 1982 contours, as seen in Fig. 7.35. Here a second limiting scour line for the latter year is shown. It is seen that the erosion has widened from the tip for half the

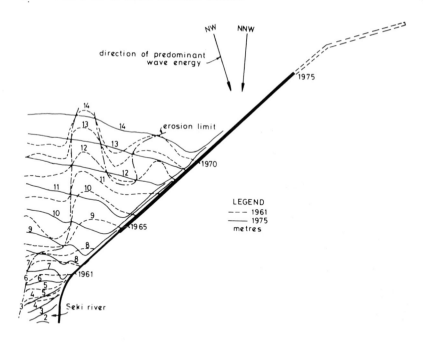

Figure 7.32   Main breakwater of Naoetsu harbor with bed contours for 1961 and 1975

breakwater length. The material so removed has caused temporary accretion toward the toe, but the area of erosion beyond this has increased.

The centerline of maximum scour is shown, which is 35 m from the breakwater face. The depth of this trough is shown in the side elevation where it is apparent that scour has increased to 4 m for one third of the length. For another third the scour approximates 3 m, which indicates a progressive deepening in these 4 years. Material will still be transported toward the heel, so that scour depths are less there at present. The rate of removal in this situation depends on the number of storms experienced during that period, which may have been more or less than normal.

Although the contours available did not allow for accurate horizontal measurements, their distances from the breakwater face were measured to give an approximate half trough profile as in Fig. 7.35 at the 1,300 m point. It is seen that the average slope is 1:5, but that adjacent to the breakwater could be in the order of 1:2.3, which approximates the angle of repose for the sediment. Irie et al. (1985) experienced similar steep slopes to the trough in model studies of only 11 hours duration, both for monochromatic and irregular waves. In the model, slopes were 1:3.4 where waves were small with respect to the sediment size. The propensity for pore pressure buildup could result in tilting of the caisson-type units, or burial of rubble of armor material.

Niigata harbor.   This harbor is also situated on the Sea of Japan, north of Naoetsu (see Fig. 7.31). Sato et al. (1969) gave contours in 1966, presumably soon after comple-

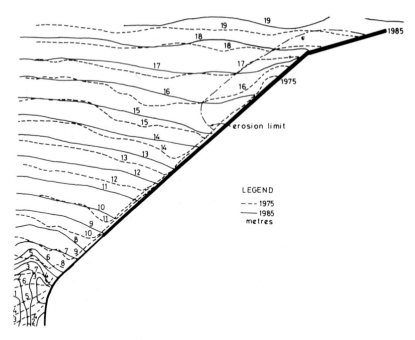

Figure 7.33   Naoetsu harbor with bed contours for 1975 and 1985

tion of the breakwater. These can be compared to those measured in 1963 and provided by Sato and Irie (1970). The two sets are included in Fig. 7.36, with accretion zones shown stippled. It is seen that near the corner a depression of 11.4 m has been scoured in an original depth of 6.5 m. Sato and Irie (1970) showed contours for a slight extension of the breakwater in 1967 and the installation of a rubble-mound base for caisson units to the limit shown dotted in the figure. There was some accretion in the region previously scoured, probably due to the transmission of sediment along the submerged mound during this short time period. It would be instructive to see hydrographic data some years later.

Tagonoura port.   This port is situated in Suruga Bay on the south coast of Honshu facing the Pacific Ocean (see Figs. 7.31 and 7.37a). As seen in Fig. 7.37b, the harbor is formed by a breakwater at about 45° to the shoreline, to which waves from typhoons, arriving generally from the SSW, can arrive through a fan of directions at 33° to 77° to the breakwater. Ichikawa (1967) has described the scouring and resultant damage to the caisson units detailed in Fig. 7.37c. As seen in the figure, these comprise both rectangular and circular units, some offset and others angled to the remainder. The joints between most of these massive units were filled with concrete, which became displaced when the caissons tipped seaward after the bed eroded. The openings so formed enhanced the sediment removal by swashing. The bed profile along the breakwater for the years 1962 to 1965 are exhibited in Fig. 7.37d, where it is seen that depth increases in the order

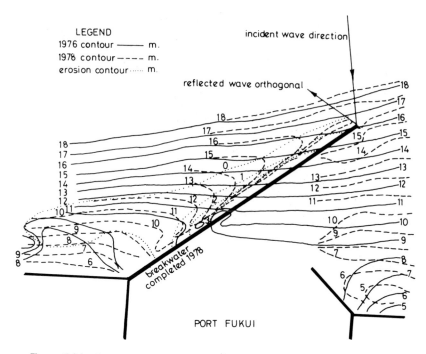

Figure 7.34 Contours at the port of Fukui, Japan in 1976 and 1978, showing zones of scour (Irie and Nadaoka 1984)

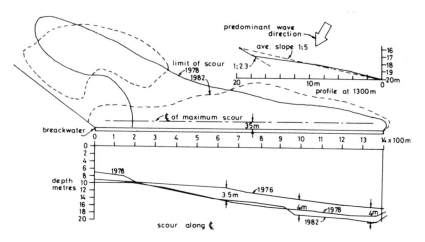

Figure 7.35 Details of scour in the vicinity of the port of Fukui, Japan, in 1978 and 1982

of 5 to 6 m occurred. In some cases this was to the toe level of the caissons, which varied as indicated in Fig. 7.37c, causing subsidence, probably aided by pore pressure enhancement.

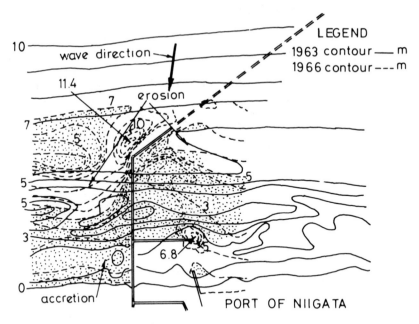

Figure 7.36 Erosion and accretion at Niigata between 1963 and 1966 (Sato and Irie 1970, and Sato et al. 1969)

Ichikawa (1967) conducted tests in a flume (implying normal approach for all waves) for two cases of bed slope 1:5 and 1:10. Scour was measured after only 40 mins. duration, with results as plotted in Fig. 7.38, using parameters HL/dd_w and S/d_w, where H is the wave height in 0.9 m depth d, L is the wave length in the same depth, S is the scour, and d_w is the depth adjacent to the wall. Some scatter is apparent, more so for slope 1:10 than for 1:5 for which a relationship:

$$S/d_w = 0.033 \ (HL/dd_w)^2 \tag{7.1}$$

applies. A line parallel to this for maxima in the envelope of the 1:10 slope data gives the constant in Eq. (7.1) as 0.011, implying that the scour ratio tripled. In the prototype the wave obliquity would permit greater scour adjacent to the structure and thus enhance removal of sediment. No comparison could be made between model and prototype because wave characteristics could not be derived for a similar approach depth at the floor adjacent to the units.

Kashima port. A large complex has been constructed on a relatively straight sandy beach on the Pacific coast east of Tokyo (see Fig. 7.31). Again this consists of a large straight breakwater comprising caissons at about 45° to the coast. Waves arrive from both the NE and SE quadrants, the latter from typhoons which travel northward during the summer. The orientation of the breakwater was selected to protect the harbor from

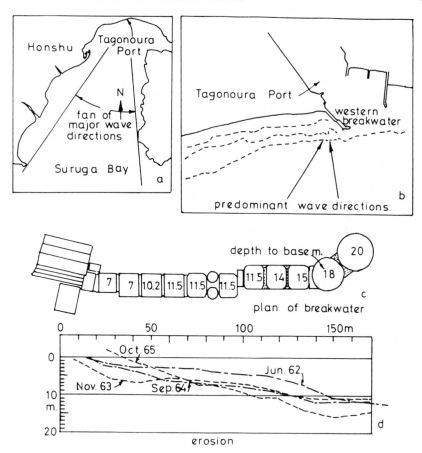

Figure 7.37 Tagonoura port showing: (a) location plan, (b) site plan, (c) breakwater details, and (d) bed profiles

these severe wave conditions. However, a large proportion of the wave energy emanates from extra-tropical cyclones passing eastward across the northern Pacific Ocean.

Swell data available (Monthly Meteorological Chart 1956) is listed in Table 7.7, where it is seen that the monthly mean from the NE is 23.1% and from SE is 16%. Thus, approximately 50% more swell arrives from the NE than from the SE and these predominate over 9 months. The typhoon season from June to August contains mainly SE swell, although the percentage of heavy swell is minimal during these months. The data in Table 7.7 distinguish only quadrants of direction, but further information (Kawakami 1976) was obtained by the Ministry of Transport, Japan, for 1971–74. Directions are listed in Table 7.8, for all waves and those greater than 2 m. It is seen that those in the NE quadrant are predominant which, as seen in Fig. 7.39, the NE and ENE waves are quite oblique to the breakwater, whilst those from the E will create standing waves. Besides the swell, waves can be generated locally by winds in the vicinity of the port.

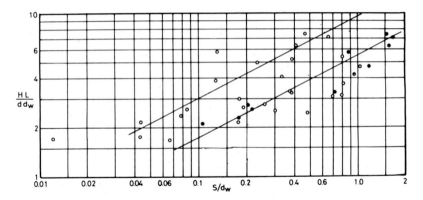

Figure 7.38 Scour adjacent to vertical wall with normal wave approach (Ichikawa 1969). Slope 1:5 closed dots, slope 1:10 open dots

TABLE 7.7 PERCENTAGES OF SWELL FOR 5° SQUARE OF LATITUDE-LONGITUDE ADJACENT TO KASHIMA COAST

Month	J	F	M	A	M	J	J	A	S	O	N	D	Avg.
NE	15	28	26	21	33	25	9	17	31	25	28	19	23.1
SE	6	8	8	17	18	30	30	22	22	15	12	4	16
Heavy Swell all directions	14	23	14	8	5	3	2	4	8	4	18	22	—

TABLE 7.8 PERCENTAGES OF WAVE DIRECTIONS IN VICINITY OF KASHIMA PORT

Dir.	All	Height > 2.0 m
NE	21	7
ENE	38	8.5
E	37	4.5
ESE	3.5	—

These will already be short-crested traveling at various angles to each other, but will create more complex systems on reflection. The percentage of winds (Kawakami 1976) is listed in Table 7.9, from which the significance of the NNE direction is apparent.

The angled segment of the breakwater in Fig. 7.39 was commenced in 1966 and was 88% of its final length in 1970, it being completed in 1975, when the latest hydrographic data became available. Contours of 1970 and 1975 are included in the figure, showing erosion east of the breakwater and accretion to the west of the dredged channel. The line of limiting erosion expands from the tip toward the toe where the shoreline has receded. A maximum of 2 m adjacent to the breakwater has scoured near the 14 m contour of

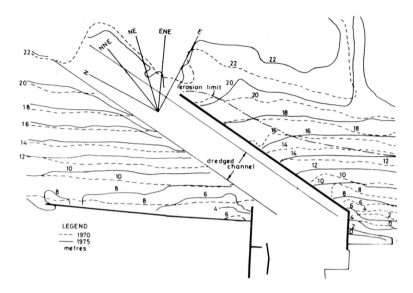

Figure 7.39   Kashima port with contours for 1970 and 1975

TABLE 7.9   PERCENTAGE OF WINDS IN
VICINITY OF KASHIMA PORT

Wind	Avg.	> 5 m/s	> 10 m/s
N	10	4	1
NNE	14	9	3
NE	11	7	2
ENE	8	3	1
E	5	2	0.5

1975, but it must be remembered that sediment would have been still traversing this region, while being removed from the tip. More current data might indicate the dangers due to continual wave reflection.

United States.   Concern is also being felt in the United States for the stability of rubble-mound structures as exemplified by Markle (1986), in which questionnaires of 21 Corps of Engineers districts were analyzed, with 12 reporting problems at 42 sites. These related to undermining of toes, berm and toe displacement, and scour of bed material. Solutions discussed included addition of precast blocks, excavation of bed material to rock bottom, filling of scour holes with core material, enlargement of berms, and placement of asphaltic mats (which proved unsuccessful). These band-aid remedial measures are not likely to provide a permanent solution, as admitted by the author: "It is concluded that design guidance is seriously needed on the proper sizing and placement configurations needed to provide adequate buttressing stone and toe berms for rubble mound coastal breakwaters and jetties." No mention is made of the real culprit of bed erosion and means of overcoming this.

In the discussion Markle (1986) states: "A scouring bottom is a problem in itself. No matter how well a toe is designed, if the local bottom materials (sands, silts, clays etc.) are exposed to sufficient energy levels for scour to occur, the toe structure is doomed to failure unless the toe berm is extended out to a point where the energy levels are below those which initiate scour. In most cases this is not practical or feasible." As noted on many occasions throughout this book, the short-crested systems, resulting from oblique reflection, can enhance scour capability tremendously. Slumping of additional berm material into the trough along the face adds to the reflecting surface and exacerbates the problem. As also noted by the author: "Some repairs have been successful thus far, while other areas require frequent repair work." Just two cases will be cited here, those of Grays Harbor and Newburyport.

Grays Harbor.   A plan of Grays Harbor, Washington is presented in Fig. 7.40, showing the south jetty which was scoured on the northern or channel side. The outer 1,700 m of this structure is now below mean low water, with no repairs being contemplated. The most persistent waves at this site would be from the NW, for which orthogonals are shown in the figure. These impinge at 45° to this side of the structure. The short-crested waves would move material eastward, aided by the flood flow through

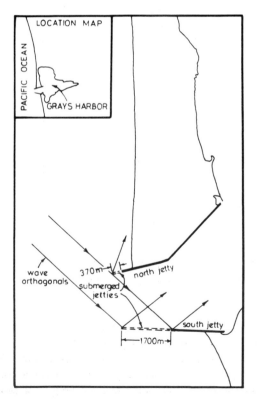

Figure 7.40   Grays Harbor, Washington, U.S.A., showing wave reflection

the entrance. Ebb flow might transport partially suspended sand out to sea, but is not the sole cause of erosion.

Newburyport.   This port, located in Massachusetts, is depicted in Fig. 7.41, where both the north and south jetties have been damaged by subsidence. Although the south structure could be undermined by channel flow, this action would be aided greatly by the oblique waves. Since both jetties have collapsed, blame cannot be placed on outflow at the entrance alone. Markle (1986) has remarked: "In most instances, instability (failure) of the structure's toe does not become evident until it has resulted in damage to the primary armor which has progressed up to or above the still water level (SWL)."

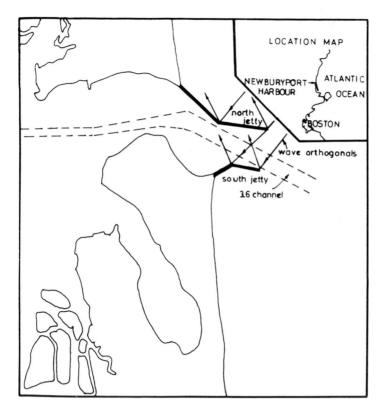

Figure 7.41   Newburyport harbor, Massachusetts, U.S.A., showing oblique wave orthogonals

Such subsidence of large dolosse armor units due to collapse of the supporting mound and toe, consisting of smaller rock, has been highlighted in the preceding discussion of the Sines breakwater. This erosion scenario is supported by Markle (1986): "It is generally thought that toe scour is the significant problem after major storms. Bedding layers slough off into the scour holes, and this damage migrates back to the toe of the primary armor. The resulting instability of the armor stone leads to downslope migration of the onslope armor and eventual deterioration of the structures."

Africa.   Only two ports will be considered here, both on the Mediterranean Sea, namely Tripoli in Libya and Arzewel Djedid in Algeria.

Tripoli.   This harbor complex has been referred to by Barony et al. (1983), Lindo and Stive (1985), and Sorensen (1985). Its plan is shown in Fig. 7.42, where it is seen that the 2.2 km SW section was constructed between 1973 and 1977 and the NE segment between 1976 and 1980. Two storms in 1981 caused the following damage to Stage I (Lindo and Stive 1985): "Over hundreds of meters the parapet wall collapsed and many tetrapods were fractured, while toe erosion was observed along several places. Further more tetrapods had been broken along the total of the breakwater—in general above MSL more than below." Whether breakages at depth were adequately observed is difficult to assess. The mode of collapse can be ascertained from a remark by Barony et al. (1983): "Movement of tetrapods left traces on the seaside face of the parapet wall." The fact that this wall was left clear of tetrapods indicates to the authors that subsidence had occurred. It seemed to have received the full impact of the waves since whole sections had "blown away."

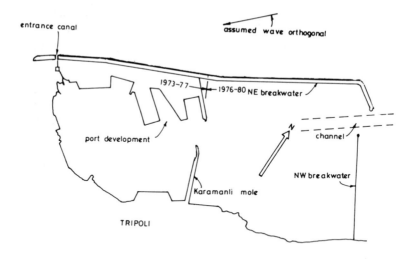

Figure 7.42   Site plan of Tripoli harbor, Libya

Sorensen (1985) discussed this breakwater: "Serious overtopping occurred after the collapse when venting and suck holes appeared in the reclamation of the working area, causing settlement of the backfill material adjacent to the breakwater. Fine reclamation material should be separated from the underlayer and armour units by suitable screens."

Arzewel Djedid.   This port has been discussed by Sorensen (1985), where the main breakwater suffered severe damage in 1980. The original and final profiles are given in Fig. 7.43, where it is seen that subsidence must have taken place. Sorensen (1985) blames the collapse on "insufficient structural strength of the large tetrapods on the very steep slope of 1:1.33." He also suggests that compaction of the armor units occurred for which model structures were conducted. On this aspect he states: "It is very important to notice that most often the compaction occurred without any individual unit being displaced from the armour layer. The model did not simulate the fragility of the units and consequently the model did not reproduce the severe damage observed in the prototype." This model would most likely have been conducted in a flume with a solid bed so the wave bed interaction of scour and pore pressure buildup could not be checked.

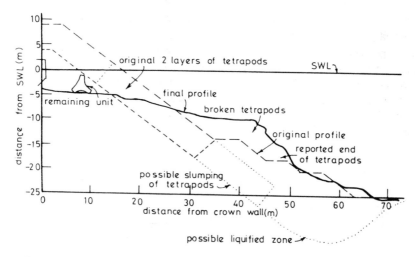

Figure 7.43   Arzewel Djedid, Algeria, showing the original and final profiles of main breakwaters

As seen in Fig. 7.43, the area of collapsed debris is smaller than that removed from the segment to 28 m from the centerline. This percentage is similar to those given in Table 7.4 for the Sines breakwater. It appears therefore to the authors that pore pressure enhancement due to scouring caused slumping of the sedimentary bed and ultimate subsidence of the face.

7.2.3 General Comments

One disconcerting aspect of the scouring reported for prototype conditions is its swiftness. Before breakwaters are complete the obliquely reflected waves, either of storm wave or swell character, have created an unbalanced situation with respect to energy application to the floor. Large volumes of material are thrown into suspension and carried horizontally by the mass transport of the short-crested wave system parallel to the breakwater face. This pseudo-current is aided and abetted by tidal currents in the region where influence diminishes as the bed is deepened.

The trench so dug along the breakwater length produces steep floor slopes adjacent to it, which during a storm build up high pore pressure. These can result in collapse of the steep slopes or liquefaction, which reduces the support capacity of the bed and hence results in subsidence of the rubble components into this liquid mass. The armor units supported by the berm at the base of the breakwater face thus slide swiftly down and in so doing are broken by the earthquake type impact. Although all units are subject to breaking from this action, those at the bottom of the pile are more likely to be fractured, while also being buried. Inspection of these components after the event is very difficult.

Before any major design is carried out for a port involving breakwaters, model studies to various scales should be implemented with respect to their stability. Not only does this involve the structure itself, but also its contact with the bed or its foundation. This could be the weakest link in the chain of reactions due to wave action. As concluded by Sorensen (1985): "The cost of hydraulic model tests is very small compared to the cost of construction, design and supervision etc. and is not the item to economize on in a breakwater project." Emphasis has been placed previously on the role of movable bed models in a three-dimensional situation where either persistent swell waves or those of storm character arrive obliquely to the structure.

Another aspect of these events, where the environmental factors have had influences not properly predicted and hence resulted in failure, is that of their publicity. Engineers learn probably more from collapsed structures than they do from standing edifices that may be overdesigned. As stated by Sorensen (1985): "In this respect the failures, although unacceptable and costly for the owners, are of such great importance for the engineering community that it is moral obligation to publish such events internationally."

7.3 SCOUR AT TIP OF BREAKWATER

When waves are reflected obliquely from a breakwater, the short-crested system so established causes island crests to propagate parallel to the face at distances $Z/L' = 0$, $1/2, 1 \ldots$ from it. That at $Z/L' = 0$ causes waves, with heights about double the incident, to move along the structure and arrive at its downcoast tip. When a crest reaches this point, the seaward water level is above the almost MWL on the leeward side, which causes water to follow around the breakwater tip in the form of a vortex. Similarly, when the trough arrives waterlevel on the seaward side is less than in the leeward region and hence a second vortex of opposite rotation is formed. This action is enhanced when the breakwater is narrow, as for example, with caisson-type structures.

These two vortices occur at frequencies of one wave cycle and expand down to a fixed boundary, either to the supporting rubble mound or the seabed. In so elongating their rotational velocity is increased, which magnifies their sucking capacity on any sedimentary floor (Migniot et al. 1983). They thus have a great capacity for removing material at the toe of the structure which can result in its collapse. This action has been investigated by Chang (1975, 1977) by means of a three-dimensional model, as illustrated in Fig. 7.44. It is seen that monochromatic waves are reflected at 45° by a

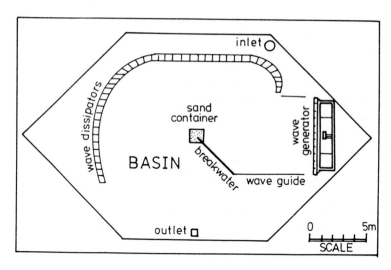

Figure 7.44   Model basin of Chang (1975, 1977) for testing scour at breakwater tip

thin breakwater, with the island crests proceeding to its tip, where a sand chamber is located beneath the basin floor.

Wave dissipators were located around the boundary of the wave basin to minimize the influence of secondary reflections in the test area. After screeding the sand in the chamber to floor level, waves commenced. At this initial stage large clouds of sediment were seen to rise as each vortex applied its energy to the bed. A hole soon scoured which was accompanied by a circular mound whose height and width varied around its periphery. Its dimensions were monitored over time by means of a pointer gauge as seen in Fig. 7.45. Tests with various wave heights and periods resulted in scour versus time relationships, as exhibited in Fig. 7.46, which shows swift erosion at first, with it becoming asymptotic to some maximum scour value. At the final stage the vortices drew very little material into suspension and any so removed was thrown onto the sides of the conical depression and hence recirculated.

In Fig. 7.46 the curves A, B, C, and D refer to various wave characteristics. The smallest scour for D had an incident wave height of 7.93 cm and period of 1.25 s, whereas the largest scour for A is for the smaller height of 3.51 cms but the largest periods 2.03 s. It is thus seen that the vortex strength, or its scouring propensity, is more affected by wave period than height. This highlights the influence of swell which is of this nature, besides being more persistent than storm type waves.

The scour can be nondimensionalized by dividing by incident wave height (S/H_i), graphed against $d/(gT^2)$ in Fig. 7.47, the latter being equivalent to d/L_o. As depth ratio decreases, so scour increases to an apparent limit of $S/H_i = 7.0$. Where $(d/gT^2)10^3 = 16$ [or $d/L_o = 0.1$] the scour approaches zero, which for 10 s waves implies a depth of 15.6 m. The conical hole dimensions can be nondimensionalized by dividing depth and horizontal distance by the depth at the center, as illustrated in Fig. 7.48 for varying

Figure 7.45　Pointer gauge for measuring scour hole depth at breakwater tip

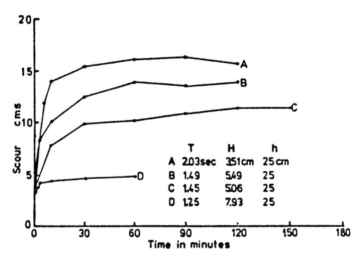

	T	H	h
A	2.03 sec	3.51 cm	25 cm
B	1.49	5.49	25
C	1.45	5.06	25
D	1.25	7.93	25

Figure 7.46　Measured scour depth versus time on linear scale

wave characteristics and duration. This geometrical similarity is independent of wave characteristics and duration. The bed profile is closely associated with the angle of repose for the sediment, which for this case was 30° for this 0.3 mm sand.

　　Markle (1986), from his observation of structures in the Portland district of the Pacific coast of the United States, was led to comment: "In some instances, scour at jetty heads has been so severe that it was not economically feasible to try to fill and stabilize the scour holes. The best approach in these cases was to abandon the outer 200 to 300 ft of the jetty heads and rehabilitate the remainder of the structure." It would appear to these authors that prior protection by means of mattress type containers filled with slurry mortar in situ for a reasonable radius around the tip would have provided a solution. This procedure is discussed in more detail in Section 8.3.

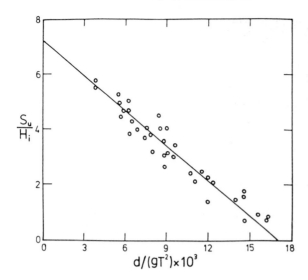

Figure 7.47 Dimensionless scour depth S/H_i versus equivalent water depth d/gT^2

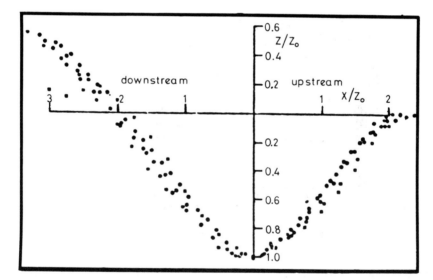

Figure 7.48 Dimensionless scour depth Z/Z_o with width X/Z_o for all wave conditions and durations

Although narrow breakwater widths have been utilized in the tests discussed above, the prototype structures observed by Markle (1986) comprised heads of conical rubble mound material armored by large precast concrete units. It would appear that the strength of these vortices was such that they could penetrate large thicknesses of such material and withdraw sand from beneath them. This again calls for three-dimensional models in which the complete structure is involved, with movable beds and oblique persistent waves of swell proportions.

REFERENCES

ANONYMOUS. 1970. High island water scheme, Hong Kong. *Hyd. Res. Stn., Wallingford*, UK.

BARONY, S. Y., MALLICK, D. Y., GUSBI, M., and SEHERY, F. 1983. Tripoli harbour NW breakwater and its problems. *Proc. Inter. Conf. Coastal and Port Eng. in Developing Countries*, 808–19.

CHANG, H. T. 1975. Scour at end of breakwater with oblique incident waves. *Proc. 2nd Aust. Conf. Coastal and Ocean Eng.*, 180–84.

CHANG, H. T. 1977. Vortex scour due to wave action at a breakwater tip. *Proc. 6th Aust. Hydrau. and Fluid Mech. Conf.*, 348–51.

DE ROUCK, J., and DE MEYER, C. 1987. Port structures their care and maintenance. *PIANC Bull.*, No. 58, 133–68.

EDGE, B., BAIRD, W. F., CALDWELL, J. M., FAIRWEATHER, V., MAGOON, O. T., and TREADWELL, D. O. 1982. *Failure of the breakwater at port Sines, Portugal*. Coastal Eng. Res. Council, ASCE.

GODA, Y. 1970. A synthesis of breaker indices. *Trans. Japanese Soc. Civil Engrs.* 2(2): 227–30.

GUNBACK, A. R. 1976. The stability of rubble-mound breakwaters in relation to wave breaking and run down characteristics and to $\xi = \tan\alpha/(H/L_o)^{0.5}$ number. Div. Port and Ocean Eng., *Norwegian Inst. Tech.*, Report No. 1.

HERBICH, J. B., MURPHY, H. D., and VAN WEELE, B. 1965. Scour of flat sand beaches due to wave action in front of sea walls. *Proc. Santa Barbara Specialty Conf. on Coastal Eng.*, ASCE, 705–26.

HSU, J. R. C., and SILVESTER, R. 1989a. Model test results of scour along breakwaters. *J. Waterway, Port, Coastal and Ocean Eng., ASCE 115*(1): 66–85.

———. 1989b. Berm toe protection of breakwaters. *Proc. 9th Aust. Conf. Coastal and Ocean Eng.*, 96–102.

ICHIKAWA, T. 1967. Scouring damages and vertical wall breakwaters of Tagonoura port. *Coastal Eng. in Japan 10*: 95–101.

IRIE, I., and NADAOKA, K. 1984. Laboratory reproduction of seabed scour in front of breakwaters. *Proc. 19th Inter. Coastal Eng. Conf., ASCE 2*: 1715–31.

IRIE, I., KURIYAMA, Y., and ASAKURA, H. 1985. On the protection methods of scour in front of breakwaters. *Proc. 32nd Japan Conf. Coastal Eng.*, 445–49. (In Japanese)

———. 1986. Study on scour in front of breakwaters by standing waves and protection methods. *Rep., Port and Harbour Res. Inst.*, Japan 25(1): 4–86. (In Japanese)

KAWAKAMI, T. 1976. Deputy Director Kashima Construction Office, Ministry of Transport, Japan (personal communication).

LANGLAIS, C., and ORGERON, C. 1978. Synthese des resultats des essais en canal realises Jusgu'an 10 Aout, 1978. Maisson-Alfort, LCHF Rep. No. 3.

LIN, M. C., WU, C. T., LU, Y. C., and LIANG, N. K. 1986. Effects of short-crested waves on the scouring around the breakwaters. *Proc. 20th Inter. Conf. Coastal Eng., ASCE 3*: 2050–64.

LINDO, M. H., and STIVE, R. J. H. 1985. Reconstruction of main breakwater in Tripoli harbour, Libya. *Dredging and Port Construction*, Feb., 19–27.

LOSADA, M. A., and GIMÉNEZ-CURTO, L. A. 1981. Flow characteristics on rough permeable slopes under wave action. *Coastal Eng. 4*: 187–206.

MANSARD, E. P. D., and PLOEG, J. 1978. Model tests of Sines breakwater. Hyd. Lab., *Nat. Res. Council Canada*, Rep. No. LTR-HY-67.

MARKLE, D. G. 1986. Stability of rubble-mound breakwater and jetty toes; survey of field experience. U.S. Army Corps Engrs., *Coastal Eng. Res. Center,* Waterways Expt. Station, Vicksburg, Miss., Tech. Rep. REMR-CO-1.

METTAM, J. D. 1976. Design of main breakwaters at Sines harbours. *Proc. 15th Inter. Conf. Coastal Eng., ASCE 3:* 2499–2518.

MIGNIOT, C., MANOUJIAN, S., WENS, F., and NEYRINCK, L. 1983. Erosion near harbour structures under wave and current action, how to foresee and remedy. *Proc. Inter. Harbour Congr.,* pp. 1.55–1.67.

Monthly Meteorological Chart. 1956. Western Pacific, Mo. 484, HMSO London.

MORAIS, C. C. 1974. Irregular wave attack on a dolos breakwater. *Proc. 14th Inter. Conf. Coastal Eng., ASCE 3:* 1677–90.

SATO, S., TANAKA, N., and IRIE, I. 1969. Study on the scouring at the foot of coastal structures. *Coastal Eng. in Japan 12:* 83–98.

SATO, S., and IRIE, I. 1970. Variation in topography of seabed caused by the construction of breakwaters. *Proc. 12th Inter. Conf. Coastal Eng., ASCE 2:* 1301–19.

SILVESTER, R. 1972. Wave reflection at seawalls and breakwaters. *Proc. Instn. Civil Engrs. 5:* 123–31.

———. 1974. *Coastal Engineering, 2.* Amsterdam: Elsevier.

———. 1986. The influence of oblique reflection on breakwaters. *Proc. 20th Inter. Conf. Coastal Eng., ASCE 3:* 2253–67.

SILVESTER, R., and HSU, J. R. C. 1987. Scouring due to reflection of oblique waves on breakwaters. *Proc. 8th Aust. Conf. Coastal and Ocean Eng.,* 145–49.

———. 1989. Sines revisited. *J. Waterway, Port, Coastal and Ocean Eng., ASCE 115(3):* 327–44.

SILVESTER, R., and LIM, T. K. 1968. Application of wave diffraction data. *Proc. 11th Inter. Conf. Coastal Eng., ASCE 1:* 248–72.

SMITH, A. W., and GORDON, A. O. 1983. Large breakwater toe failures. *J. Waterway, Port, Coastal and Ocean Eng., ASCE 109(2):* 253–55.

SOLLITT, C. K., and CROSS, R. H. 1972. Wave transmission through permeable breakwaters. *Proc. 13th Inter. Conf. Coastal Eng., ASCE 3:* 1827–46.

SORENSEN, T. 1985. Experience gained from breakwater failures. *Proc. Conf. on Breakwaters,* 103–18.

TAKEYAMA, H., and NAKAYAMA, T. 1975. Disasters of breakwaters by wave action. *Port and Harbour Res. Inst.,* Japan, Tech. Note No. 200. (In Japanese)

TANAKA, N., IRIE, I., and OZASA, H. 1972. A study on the velocity distribution of mass transport caused by diagonal, partial standing waves. *Rep., Port and Harbour Res. Inst., Japan 11(3):* 112–40. (In Japanese)

TSUCHIYA, Y., SHIBANO, T., and NAKANISHI, T. 1976. Long-term shoreline change of the Naoetsu harbor. *Coastal Eng. in Japan 19:* 109–20.

WEARNE, S. H. 1985. General lessons of recent failures. *Proc. Instn. Civil Engrs. 78(1):* 1451–52.

XIE, S. L. 1981. Scouring patterns in front of vertical breakwaters and their influence on the stability of the foundations of breakwaters. *Delft Univ. of Tech.,* Report of Coastal Eng. Group.

Alternatives to Normal Breakwaters

Normal here connotes the rubble-mound structure covered on its relatively steep face with large armor units consisting of stone or precast concrete blocks of a variety of shapes. Research and development has revolved mainly around the stability of these massive units that sit precariously on the underlying stone of smaller size. The unfortunate failure of some major breakwaters has caused a rethink of their design criteria, but this has concentrated on the stability of the two-layer thickness of these armor units. Little thought has been given to the sedimentary bed supporting the structure, which can also become unstable under the action of waves. Pore pressure buildup can produce liquefaction adjacent to or even under the toe of the steep slope supporting the armor units, so causing it and them to subside, leaving the underlayer, or even core, at the mercy of the storm waves.

This has led to the rejuvenation of the berm breakwater which is comprised solely of quarry-run stone of size between the previous core material and the armor units. It is closer to that of the underlayer of the traditional breakwater. As the title implies, waves will oscillate these elements to form a beach berm, by throwing some above MSL to form a crest, and swashing others to enlarge the berm below water. The stones so accreted on the seaward face will rest at the angle of repose which is approximately 30°. This was the original method of constructing breakwaters because equipment was not then available to handle rock of large dimension or weight. One or two of these structures are still standing after many decades of severe storms, for which their designers deserve credit.

The berm-like profile formed near SWL serves to break the waves, much as occurs on a normal beach. Thus energy is dissipated, except for that which is reflected seaward, mainly at the base. Since these smallish stones are mobile, to some degree, any waves that arrive obliquely to the structure can effect their transport along it. This needs attention both along the trunk and at the heads of berm breakwaters.

It is not much stretch of the imagination to extend this principle to a breakwater consisting solely of sand. As already noted, the best defense of land against the sea is the beach itself, which forms an offshore bar during storms to limit the erosion. Thus, a barrier beach structure to serve as a breakwater must make allowance for this transient loss of berm during storm sequences. This type of action is not present in berm breakwaters, but design procedures can be developed to cope with it. Also the sand is more mobile along the beach so formed, so this littoral drift must also be prevented, or reduced to a minimum if upcoast supply is still available. It will be shown how these requirements can be met.

Another recent development that could reduce the cost of breakwater construction is geotextile containers filled with grout. These are not unlike the precast concrete units developed to provide the weight and shape to achieve stability on steep slopes. However, they differ in two major aspects: first, they are cast in situ and so do not need to be handled in placement, and second, they have intimate contact with their neighbors so do not rock or suffer dynamic forces under wave action. These both lead to a reduction in the physical strength requirement of these units, which are sausage-like in character, but more oblate spheroidal in cross section. The sand/cement mortar mix can have admixtures which reduce the cost substantially.

The final alternative structure to be presented is mobile and is akin to the floating breakwater that has received much attention in recent years. However, it is recommended that it be submerged a small distance to get it out of the large wave forces that are predominant at the sea surface. By so doing the waves transmitted leeward of the breakwater are reduced over a greater range of the storm wave spectrum. Such structures can be used to attenuate waves reaching dredgers operating in the open sea, or to serve as temporary headlands whilst bays are reclaimed to their desired shape, prior to fixed structures being installed. They can also serve as marina protection, for which only the floating breakwater has been used to date.

8.1 BERM BREAKWATERS

These have received the attention of experts in the field (*Berm Breakwaters: Unconventional Rubble-Mound Breakwaters* 1988) after some failures had occurred due to misunderstandings in design. Prior to this the new design was considered a breakthrough (Baird and Hall 1987, Fairweather 1987) even though the concept had been used in the last century. These were to replace the traditional breakwater which has generally two layers of armor stone, one or more filter layers of smaller rock, and then a core of quarry-run material.

The armor unit size in the traditional structure is a function of slope and wave conditions. Should these not be available from the quarry, large precast concrete blocks of varying shape have been devised to provide stability. Some of these need rigorous placement in order for them to interact and so restrict lifting. This requires strict supervision under conditions of poor visibility, deep water, and stormy weather. The uprush and downwash occurring around these units produces large velocities and forces which are demanding ever increasing weights.

Baird and Hall (1987) have suggested the use of quarry-run rock of smaller dimensions which reduces unit cost and simplifies construction. They state: "This alternative approach is not really 'new.' Many historical breakwaters were built using much smaller stones." They cite the case of Plymouth breakwater, built in 1855, which is still standing the rigors of Atlantic storms: "They will not fail in a catastrophic manner because the permeable berm of armour stones consolidates into a nested surface.... The berm profile is quite regular through the water line, typically in the order of 1:5." Hall and Wilkinson (1987) confirm this action: "The structures built in the early 1800s were constructed by tipping the entire yield of a quarry into the sea and letting the action of the waves redistribute and sort the stones into a more stable structure.... By allowing the breakwater to rearrange itself into a more effective shape to dissipate wave energy, it is possible to use much smaller sized material. The same processes occur constantly on our beaches."

Van der Meer (1988a), in the introduction to his treatise on rock slopes and gravel beaches, states: "A more economical solution can be a structure with smaller elements, where profile development is being allowed in order to reach a stable profile." He admits: "The alternative structures with high economical potential, such as S-shaped and berm breakwaters, required new design techniques." He added that such material can stand up to storm wave attack, but omitted discussion on longshore transport potential. Thus, with the failure of some traditional breakwaters, new designs are now under consideration, but as Fairweather (1987) has stated: "Breakwaters are notably fragile structures. So the decision to try a new design is always a difficult one. Nonetheless, several berm breakwaters have been built since 1983, and these are mostly success stories."

8.1.1 Berm Breakwater Profiles

Many tests have been conducted (*Berm Breakwaters: Unconventional Rubble-Mound Breakwaters* 1988) in models to determine the stable shape of breakwaters comprising small stone. Their equilibrium profiles may be considered as in dynamic or stable state, depending on the degree of motion of the surface material. In the tests, an initial smooth slope was adopted which was allowed to erode to various profiles, dictated by the wave energy in each monochromatic wave (proportional to H^2T^2) as seen in Fig. 8.1 (Burcharth and Frigaard 1988).

Seaward of the milder *berm slope* the face steepens, as noted by Van der Meer (1988b): "The lower slope is often steep and close to the natural angle of repose of the armour." Hall (1988) includes a table for 8 prototype structures with loose rubble and slopes of 1:1 to 1:2 (vertical: horizontal), averaging 1:1.6. These are of the same order as normal rubble breakwaters for the smaller stones supporting, as a mound, the larger armor units further up the face.

A typical cross section of stable shapes is given in the inset of Fig. 8.2, where d_c/H for steep waves approximates 0.8 (Hall 1988), in depth d at the toe of the breakwater. The ratio of $(H/d)_{max}$ can be derived from Goda (1974), from which the curve is drawn

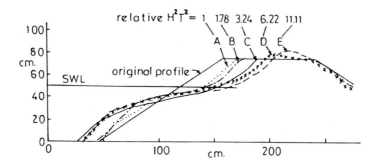

Figure 8.1 Profiles obtained after 3,000 waves with values of significant wave height and spectral peak wave period used to compute relative wave energy per wave, H^2T^2 (from Burcharth and Frigaard 1988)

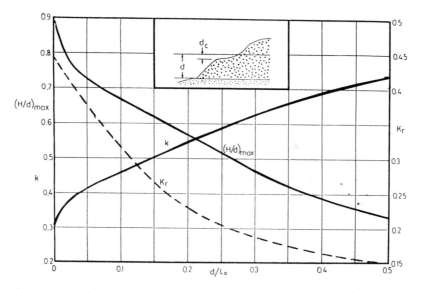

Figure 8.2 Typical berm breakwater shape with graphs of maximum wave height, fraction of reflection surface, and reflection coefficient

in the figure. The fraction (k) of the steep breakwater slope which could provide full reflection in this part of the water column is thus:

$$k = (d - 0.8H)/d = 1 - 0.8(H/d)_{max} \tag{8.1}$$

which curve is also plotted. It is seen that for $d/L_o = 0.15$ this fraction is 0.5 and enlarges as the toe depth increases. Thus, as these breakwaters are taken into greater depths, so a larger percentage of its face is providing almost 100% reflection. In this context the water-particle motions in this part of the water column are essentially horizontal. Water already fills the spaces between rocks and hence there is little exchange due to these horizontal oscillations. Because of this, the face acts as almost a glass finish and hence the efficient reflection. More research is required into reflection coefficients from berm breakwaters in varying depths of water.

Accepting that the situation can be equated to a vertical wall rising to depth d_c from the MSL, the tests of Dick (1968) and theory of Dean (1945) and Ogilvie (1960) are applicable, to obtain a reflection K_r as presented by Silvester (1974). The resulting curve is included in Fig. 8.2, where it is seen to rise from 0.15 at $d/L_o = 0.5$ to approximately 0.45 for very shallow conditions, when all orbital motions are virtually horizontal. For toe depths of less than 20 m and wave period of 12 seconds, $d/L_o < 0.1$ which gives $K_r > 0.3$. This will generate a partial standing wave or its equivalent in a short-crested system for oblique approach, so that vortices and macroturbulence will be available to scour bed material.

The profile of the berm section will be similar to those obtained from tests on shingle beaches as given in Section 5.5, with Figs. 5.30 and 5.31 being particularly pertinent. From these cross sections the volume of material per unit length of structure required to form a safe berm can be determined by equating eroded and accreted areas by templates in Fig. 5.29. With more research the relationships derived to date can be refined and modified into useful form.

8.1.2 Other Considerations

Mansard (1988) discusses the waves to be used in models of berm breakwaters and concludes that more research is required in wave transformation during shoaling and that prototype waves require measurement before they can be used in models. Ahrens (1988) comments: "Because of their high porosity, reef breakwaters are surprisingly stable to wave attack and, at the same time, can dissipate wave energy effectively." He introduced a *spectral stability number* that included a term H^2L^2 in it by which the required volume could be obtained to give a desired crest height of the stable profile. Jensen and Sorensen (1988) addressed the topic of stone mobility: "More research is required to quantify this aspect and to define acceptable limits and parameters for description thereof." In describing a berm breakwater constructed in Iceland, Baird and Woodrow (1988) observed: "The study demonstrated that by varying the geometry of a breakwater cross section a design can be prepared that makes full use of the yields of a quarry and may use armor stones weighing five times less than the stones required for a conventional design. At this location this design approach achieved cost saving, compared to that for a conventional design, in the order of 40 per cent." Anglin et al. (1988) expressed concern about rock movement along the structure by oblique waves, but this was not included in the recommended research program. Neither was the possible scouring of the sedimentary bed at the toe of the steep outer face.

8.2 BARRIER BEACH BREAKWATERS

The progression from conventional to berm breakwaters appears natural, it at least employs Nature in the construction process. As noted above, Hall and Wilkinson (1987) have observed that the beach profiling is the same as occurs on beaches. Extension of

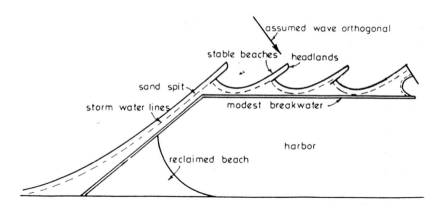

Figure 8.3   Typical breakwater layout showing application of stable bay beaches instead of a normal structure

the concept to using sand as the bulk material for such structures would appear logical, since beaches can defend themselves so readily. However, the excessive mobility of this material in the presence of oblique waves demands some provision to prevent removal. From what has been presented throughout this book, the obvious approach is headland control or fixed points along the breakwater length between which stable bays can form as suggested in Fig. 8.3. The cost of these structures plus reclamation must be weighed against the advantages that sand *barrier beach breakwaters* provide, four of which will now be discussed:

1. Waves are completely dissipated, either on the swell-built profile or the offshore bar during storms. This obviates overtopping and the creation of spray, even for the largest waves. Sufficient beach width should be provided for the temporary loss of the berm, plus a little more for safety, during the severest storms. The direction and duration of waves from these low pressure centers must be assessed, because differential scour can occur around the periphery of the bays, as discussed in Section 4.2.8.

2. Areas are provided around the seaward boundary of a port for commercial, industrial, or other purposes. Land is at a premium in the vicinity of harbors, where most activities are carried out on the landward perimeter. By providing zones around the whole periphery, designs of these mammoth facilities become more flexible. Many port complexes usurp shorelines previously used by the public for leisure or small-craft launching, hence the additional beaches could be a bonus.

3. The material for construction, namely the sand, is readily available on the site, either from adjacent land areas or as spoil from dredging the harbor basins or access channels to them. The costs of quarrying stone, hauling it to the coast, and placing it in position are all eliminated, except for the headland construction which is small in volume compared to the alternative rubble-mound structure. Most port contracts include dredging items, for which removal seaward is normally a necessary and costly component. Using the spoil to construct a breakwater nearby

can achieve two goals simultaneously. Any shortfall in sand supply could be made up from an adjacent offshore zone, so long as the normal bed profile is maintained to support the beaches being formed. It could be taken from a zone which will be filled quickly by littoral drift accumulating against the new structure.

4. Barrier beach breakwaters obviate reflection from the structure which, as discussed in Chapter 7, causes severe scouring of the adjacent sedimentary bed, sometimes resulting in structural subsidence. The berm breakwater, as discussed in Section 8.1, has a steep underwater face that can reflect a portion of the incident wave, so even these structures can suffer from such scour. The beach version, on the other hand, transposes automatically to a wave breaking profile as soon as a storm arrives.

8.2.1 Prevention of Sand Removal

There is little to elaborate on the predictable shape of the bays which will form between the headlands spaced along the line of the proposed breakwater. They should be designed for static equilibrium, since sand supply from upcoast can become minimal at times. It is preferable to retain all beach material in each bayed compartment, even when it is removed offshore during a storm. Although the subsequent swell waves should return this bar material directly back onto the beach, it is possible for those arriving just after a storm to be slightly angled downcoast. This could cause a loss of the downcoast tip of this bar, which may need to be prevented. This can be accomplished by extending the curved tip of the headland seaward in the form of a submerged groin, as seen in Fig. 8.4.

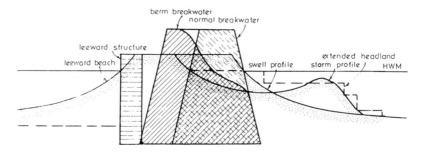

Figure 8.4 Schematical cross section of breakwater alternatives showing submerged groin extension at tip to prevent loss of sand from offshore bar

This figure also shows the relative cross sections of a plain sand feature and that of a leeward wall to provide larger depths adjacent to it, for berthing ships perhaps. The storm beach profile (as calculated from Section 3.4.3) dictates the heights or submergence of the required groin extension to prevent sand loss into the next bay or past the final headland. The preferred construction for these protruding structures is with geotextile units, to be presented in the next section, since the dumping of stones on line in deep water is extremely expensive. If littoral drift is still arriving from upcoast, the resulting bays will not be indented to the design values and so some sand transmission past the submerged groins will take place.

8.2.2 Existing Cases

There are two locations where anchored beaches have been employed to provide calm water for ports. These are at Saldanha Bay in South Africa, completed in 1976, and Toronto in Lake Ontario, Canada, commenced in 1982, although smaller versions were commenced in 1970. At Saldanha material was dredged from the bay, while in Toronto earth fill and construction debris were utilized.

Saldanha Bay. This port is some 120 km NW of Cape Town on the southern Atlantic coast, where continuous swell is experienced from SSW (Zwemmer and Van't Hoff 1982). The site plan in Fig. 8.5 displays a breakwater running from Hoedjies Point to Marcus Island, some 2 km in length. The startling fact is that this structure consists of sand in the form of a crenulate shaped bay. To derive this stable shoreline the authors conducted model tests to refract waves with periods from 10 to 18 s, recording their crest patterns in the correct location. They then took an average of these in the knowledge that: "The horizontal alignment of a beach in general, must be designed in such a way that the predominant wave direction will always impinge perpendicular to the beach in order to prevent lateral transport of material.... As other kidney-shaped beaches exist in the area, it was quite a challenge to try to design an artificial beach which could withstand the coming ocean swell having its origin in the 'roaring forties'." Any applications of this concept will be aided by this treatise and the many references contained in it.

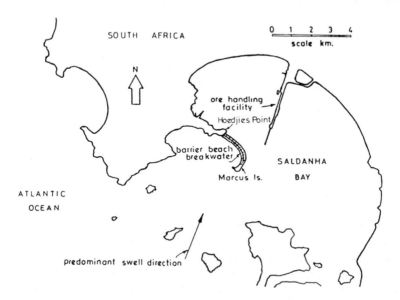

Figure 8.5 A sand breakwater installed at Saldanha Bay, South Africa

The difficulties of placing the base material and subsequent berm and dune reserve in position have been fully described by Zwemmer and Van't Hoff (1982), which is recommended reading. Although the continuous swell waves swashed sand around the tip of the spit as it proceeded from Hoedjies Point, it was readily available for redredging for its extension. The first stage of reclamation consisted of 19 million m^3 up to 6.60 m below M.L.W.S., of which some 1 million were lost outside the profile. The second stage involved the placement of 1.3 million for the beach itself. The closure was effected within 3 months, which is quite an achievement.

It stands to reason that such a man-made bay must adjust to the actual wave climate, which averages things out much better than man's statistical capabilities. This may cause some erosion in one place and perhaps addition in another. Also the assumed offshore profiles may not match the design slopes due to deficiencies in the wave climate data or presumed relationships between sediment movement and orbital motions due to wave action. However, bed contours should become steady within 3 years, so that the final beach shape should be established within this period.

Because of these adjustments, surveys have been conducted (Schoonees et al. 1990) which displayed an average scouring of around 0.5 m/yr from 1978 to 1986 over the whole bayed profile to 20 m depth. Such erosion was not uniform. As noted by the authors: "The beach profiles showed either erosion or remained basically stable. The most severe beach erosion occurred at profiles 2 and 36." As seen in Fig. 8.6, these were at the extremities of the bay. The erosion at profile 2 could be due to the rocky protuberance just west of it which may have caused reflected waves to enter the sediment dynamics. The bay shape predicted by methods in Section 4.2.3. is shown dotted in the figure (using a figure of Schoonees et al. 1990) and is seen to indicate erosion propensity in the region from profile 30 to 38. These need only the addition of sand, if in fact there is not enough already there to prevent overtopping. The designed offshore slope of 1:35 had gone to 1:45, which accounted for most of the erosion. This material modification should not greatly affect the volume of sand above MSL.

The apparent remedial measures could not include sand replenishment because the volumes required were not available. "It was therefore recommended that a suitable rock protection be provided along the spending beach breakwater" (Schoonees et al. 1990). This was to involve geotextiles covered with stone, the volume of which was approximately 200,000 m^3. The cost of this should be compared to 500,000 m^3 of sand calculated to be needed for beach realignment. It appeared strange to these authors that sand would not be available within Saldanha Bay for this modest volume, especially as it could serve the purpose of dredging larger turning basins within the port. This could be deposited on the berm or dune system to allow for the severest storms of the future. Nature could do the spreading offshore in creating the stable profiles.

Toronto.   Along the metropolitan shoreline of Toronto reclamation has been carried out since 1970, to provide recreation areas and sheltered water for marinas (Denney and Fricbergs 1979). Material has been provided from land leveling and excavation of building sites. In fact, contractors had to pay a fee for disposal of debris, which was deposited in lower layers and later covered with sand or cobbles. It is a good illustration of stone-type material serving as a base for a beach. Headlands were constructed to anchor bays formed between them, as seen in Fig. 8.7 as also in Fig. 6.21.

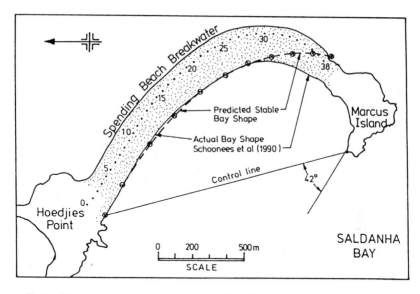

Figure 8.6   Predicted and actual bay shape of sand breakwater at Saldanha Bay shown in Fig. 8.5

Figure 8.7   Anchored beaches for reclaimed land at Bluffers Park system, near Toronto, Canada (photo permission of the Toronto Harbour Commission)

A more comprehensive scheme was embarked upon in 1982 (Fricbergs 1985), in which a 4 km long cobble beach was constructed to shelter a larger water area for a port. The complete complex is illustrated in Fig. 8.8, in which even an airport has been included in this large reclamation project. The major wave input in this elongated lake is from the east, so requiring the main leg to be oriented N–S. The sand and cobbles were sorted by the waves, the former being retained on the beaches and the latter moved offshore. Pinchin and Nairn (1986) have reported on this venture: "A 3.3 km long revetment armoured with quarry stone was considered in 1977 with an estimated cost of

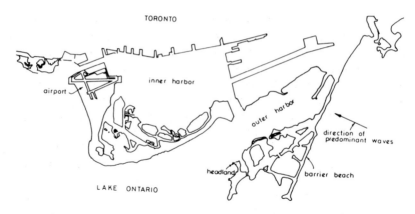

Figure 8.8 Sedimentary beaches with headland control serving as breakwaters for the port of Toronto, Canada

$4.8 million. A later design, however, consisting of pebble, gravel and cobble beaches anchored between 8 armoured hard points and covering more than 4 km in length had an estimated cost of $1.9 million in 1982 dollars (Fricbergs 1985). This new design utilized scrap rubble and demolition material from Toronto." The relative volumes of this subsidized material and normal excavation soils is not known, but the 60% saving speaks for itself.

At the extremities of the long spit changes in shoreline orientation are coped with by headlands between which stable bays formed, similar to those exhibited in Fig. 8.7. This type of beach sheltering can be accomplished on oceanic margins once the wave climate, particularly the direction of swell just after cyclones, is known. The demand of the beach berm width by the largest storms, while forming the protective offshore bar, can be determined by inspecting similarly exposed beaches in the vicinity, and confirmed from relationships given in Section 3.4.3. The curved tips of the headland must be able to withstand the brunt of the storm waves, but their trunks need suffer little attack (see Section 6.1).

8.2.3 Suggested Barrier Beach Breakwaters

Some major ports will now be discussed, all of which have been alluded to by De Rouck and De Meyer (1987) in terms of failure. Four will show possible protection by beaches for existing breakwaters in place, whilst two will indicate headland control, should the original breakwaters not have been constructed. With most damaged complexes new armor units have been added, either retaining the same profile or reducing the slope near the waterline. For example, Fullalove (1983) has stated: "The proposed new profile of Sines is perhaps the most significant indication of a return to more traditional engineering. It bears a striking resemblance to the 150 year old Plymouth breakwater which designer John Rennie based on the shape assumed by unarmored breakwaters after wave action." Even so, there could still be oblique reflection from the steeply sloped submerged section (see Section 8.1) of these S-shaped structures, which is conducive to scouring. The alternative to be discussed, of providing frontage beaches, overcomes this problem.

In each case the direction for the most predominant waves, swell on oceanic margins or locally generated milder waves for others, has been assumed. A more rigorous analysis of the wave climate would be required, with knowledge of cyclone paths, should the concept be applied. In this respect the approach angle of the more moderate waves directly after a storm is required, since these will determine the bay shape more than some annual average, as discussed in Section 3.4.5. All waves need to be refracted to the depths between headland tips, or along the control line. The spacing and distance offshore of headlands is infinite, but those chosen should serve to explain the principle of providing sufficiently wide beaches for the spasmodic demand of the storm waves. The bay shapes presented have been derived roughly by using the indentation ratio of Section 4.2.6, but finally the more accurate method of Section 4.2.3 should be applied.

Hirtshalls.   This port is located on the northern tip of Denmark, and the scour which took place adjacent to the eastern breakwater (see Fig. 8.9) has been outlined in Section 7.2.2. The surmised orthogonal of the swell type waves is shown in the figure, which result from storms passing eastward north of Denmark. As seen, three headlands plus a spur groin are suggested, in order that three deeply indented bays are sculptured between them. They could be positioned closer to the breakwater to reduce the beach width, but leaving enough when the storm waves from the north remove the berms. Due to the direction of these erosive waves more berm material will be demanded adjacent to the headlands, which will determine the extent of encroachment in utilizing this beach area. Also some storm waves could arrive from NE and reflect from the curved and back faces of these headlands, so demanding bed protection, as discussed in Section 6.1.

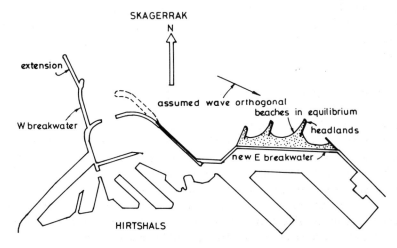

Figure 8.9   Plan of the port of Hirtshals, Denmark, showing the east breakwater damaged during construction

It is assumed that the east breakwater remain as built, although some of its rock material could be utilized in constructing the headlands, as it will now be protected by a sandy shoreline. Besides protecting the remains of this structure, the beaches can provide more working areas for commerce or pleasure, or both. If the bays had been provided initially the eastern breakwater could have been very modest in height, up to working level for industrial purposes since wharves could have been constructed along it, particularly if geotextile units were used.

Tripoli.   The scour problems at this port have been reported by Barony et al. (1983) and Lindo and Stive (1985), as discussed in Section 7.2.2. The alternative to the expensive remedial measures adopted, of using beaches to protect this 4.5 km long structure, is depicted in Fig. 8.10. The assumed swell orthogonal is from the NE which makes it very oblique to the breakwater. Other physiological features on the coast point to a general SW littoral drift and hence the given orientation of the bays. For this concept to be applied, the directions of both storm waves and the more moderate waves after storms would need to be ascertained.

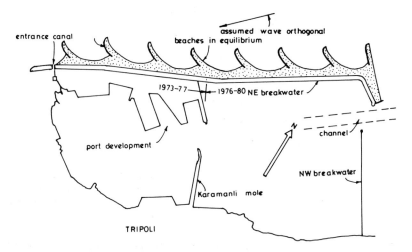

Figure 8.10   Plan of Tripoli harbor, Libya, showing an alternative treatment of the major breakwater

The costs of reclamation and headlands should be compared to the mammoth investment in the berm breakwater proposed, discounting the likely damage to the latter. One great advantage of the beach solution is the extra working area along the NE segment of this structure, plus the additional space adjacent to the already reclaimed wharf area. Overtopping and spray generation could be overcome and the beaches themselves could be used for fishing and bathing. It is not suggested that this alternative be contemplated unless further damage ensues, but it serves to show possible approaches for similar port sites.

Rotterdam.   The erosion that has occurred offshore from the curved south break-water at Euro-Poort, Rotterdam, was discussed in Section 7.2.2. The three headlands suggested in Fig. 8.11 will form deeply indented bays for the wave orthogonal shown. Their placement determines the width of shoreline at the greatest indentation region and hence the area available for commercial and other activities. With the shoals available at this site, much sand is available for reclamation. Therefore, the headlands could be sited further seaward to provide larger areas landward these stable beaches.

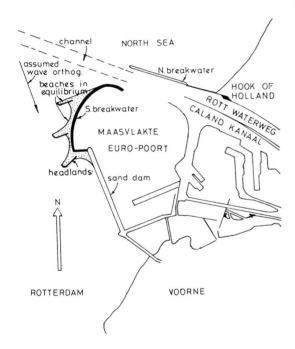

Figure 8.11   Plan of entrance to Euro-Poort, Rotterdam, The Netherlands, exhibiting headland control to reduce wave reflection

Sines.   The failure of this large straight breakwater (Edge et al. 1982) has been discussed in detail in Section 7.2.2. The present interest is centered on an alternative method of protecting the existing structure against further damage. As noted in Fig. 8.12, two headlands and a spur groin are suggested, between which two bays could be formed. More fixed points at closer spacing and less distance seaward could be adopted. Because of the excessive depths the tips of the headlands may need to be extended under water, in the form of a submerged groin, to maintain the offshore swell and storm profiles of the beaches. These must extend down to 30 m in the north to 50 m in the south. Such extensions are shown dotted but are not to scale.

The required reclamation is large by any standards, but some of the material could be gained naturally from littoral drift traveling southward. If the southern groin were installed first a beach would be accreted, whose orientation would indicate the approach

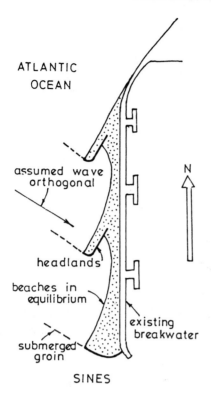

ATLANTIC
OCEAN

assumed wave
orthogonal

N

headlands

beaches in
equilibrium

existing
breakwater

submerged
groin

SINES

Figure 8.12 Plan of the main breakwater at Sines, Portugal, indicating a treatment to protect the modified structure

angle of the persistent swell. Due to the slow accumulation expected at these large depths a normal profile could form to the limit of the offshore bar and a steeper slope (perhaps the angle of repose) beyond it. This is similar to the berm-breakwater concept discussed previously. Once the southern beach had been built up the first headland, near the center of the breakwater, could be constructed, with some sand dumped in front to protect it initially from storm waves. The bulk of the material for the northern beach could be gathered naturally. Subsequent littoral drift would pass through the system, causing bays to be indented less than the designed static equilibrium shape.

Bilboa.   This port is situated on the northern coast of Spain, within the Bay of Biscay. It receives incessant swell from a NW direction, as indicated in Fig. 8.13. The existing breakwater, running in a NE direction, suffered damage in close proximity to its SW end, which prompted the addition of 35 t concrete cubes along its complete length. An alternative to this solution is shown in the form of one headland and a spur groin, with sand fill or natural accretion. Because of the approach angle of the wave orthogonal the bays so formed have a small indentation. This type of protection may not be necessary, but since the original breakwater slope is being maintained with no toe protection further subsidence may be possible.

It is understood that a second breakwater, shown dotted in the figure, is contemplated. If installed it will be much more oblique to the predominant waves than the

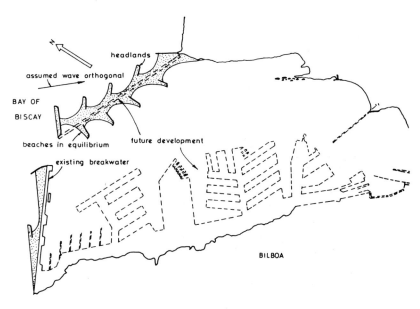

Figure 8.13 Plan of the proposed port at Bilboa, Spain, showing a barrier beach alternative to the planned SE breakwater

existing breakwater. This would result in a short-crested system with severe scour potential. In case the work has not commenced, the alternative shown of several headlands could be considered. These bays would be sculptured by the waves from the curved tip of one headland to the straight segment of the next. Material for this reclamation could be gained within the harbor as dredging will be necessary to provide channels, berths, and turning circles. Greater use could thus be made of this large expanse of calm water. Since bathing beaches are in short supply along this rugged coast, the provision of nine embayments for leisure could be an extra boon to Bilboa.

Zeebrugge.   The difficulties with scouring, both actual and predicted, have been noted in Section 7.2.2. An alternative of using beaches will now be outlined, based on their installation in place of or in addition to the rubble-mound existing complex, as seen in Fig. 7.25 previously. The new concept is depicted in Fig. 8.14, where for the western side a sand spit is formed concurrently with headlands and a groin-type structure at the northernmost tip, near the access channel. This would have provided calm water for the construction of wharf retaining walls which then could have been reclaimed. The submerged groin extensions to the curved tips of the headlands would run in a WNW direction in order to retain sand within each bay after each storm.

On the eastern side of the harbor, headlands and reclamation could have provided a large working area for the LNG terminal once the western margin of the rectangular area plus the SW–NE breakwater had been constructed. This could have consisted of small stone as little or no wave impact would have been exerted on it.

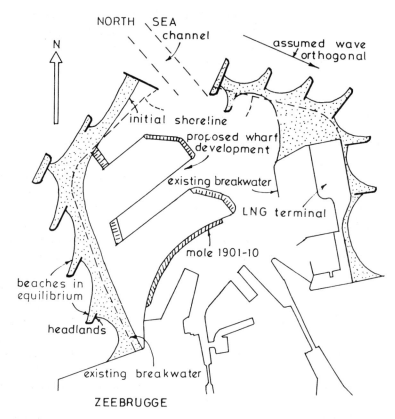

Figure 8.14 Barrier beach version for Zeebrugge harbor, with or without original structures

The assumed wave orthogonal would require more rigorous analysis than provided here, but the principle only is being presented. The source material could come from the access channel and deepening of the harbor proper. Another source could be westward of the main spit, since any depressions there would be quickly saturated by drift from along the coast.

Any cost comparisons with the actual installation should include headlands and dredging. No maintenance costs for the beaches are envisaged, as can be for rubble-mound structures. Since the headlands are protected by beaches, their heights need only be modest, as also the units comprising them. Only at the curved tips would sizable rock be required if geotextile units are not employed. The direction of the fiercest and longest duration storm waves would have to be assessed to determine which sections of the bays will be eroded the most during storm sequences.

The advantages of the scheme presented are similar to those mentioned for previous ports, namely added working areas and reduction of spray, but there are others for this particular case. For example, the access channel will not be silted, as is likely with the waves arriving obliquely to NW–SE arm of the western breakwater (see Fig. 7.28).

Once the beaches are stabilized naturally, which might take up to 3 years for the offshore profiles to reach equilibrium, they will require no maintenance. Due to the shortness of the Belgium coast and its need for protection, pleasure resorts are few and far between, so that the new beaches could be an added tourist attraction.

8.3 GEOTEXTILE CONSTRUCTION

The age-old concept of constructing shoreline structures with rock or precast concrete units, based on weight for any given wave height, has recently come into disrepute (Harlow 1980). The formula of Hudson (1959) has been used widely since its inception, even though it contains only height as the force input. This may have been suitable in shallow water, where celerity of all waves is essentially the same, determined as it is by water depth only. However, as structures have been taken to ever greater depths, the waves strike breakwaters at velocities which are dependent on both wave period and depth. Thus, longer period components of the spectrum approach at greater speed, with resultant higher forces for any given wave height.

There has been a need for larger and larger armor units, which increases the cost prodigiously. If these are just dumped randomly, or are even placed with so-called interlocking action, they still only have point contact with each other. They therefore suffer from rocking motions and can be plucked from their position in the sloping face by high water pressures developed beneath them, or by outward flow when the wave trough arrives. These unreinforced units are not designed to absorb dynamics loads; once cracked and broken their weight is halved, making subsequent removal easier. They can then be thrown around by the waves to break further units.

An aspect of breakwater design which has received little attention is the inner core comprising smaller material. Swashing of water through this rubble-type stone can dislodge it and cause settlement of the armor units resting on it. Smith and Gordon (1983) state: "Unfortunately it seems that so much design attention is normally applied to the behaviour of armour in the actual wave breaking zone that the potential hydraulic forces that must be applied to the submerged armour tend to be largely ignored."

Bruun and Gunback (1978) explained the phenomenon of "resonance" on the slope of rubble-mound material as: "the situation that occurs when run-down is in a low position and collapsing-plunging wave breaking takes place simultaneously and repeatedly at or close to that location." Sawaragi et al. (1982) have shown this to occur when the surf similarity parameter, $\tan\theta/(H/L_o)^{0.5}$, has values between 2 and 3, where $\tan\theta$ is the slope of the face, H is local incident height, and L_o is the deepwater wave length. This is the same as Irribarren's number (Irribarren 1965) which has been related to minimal sized stones for stability (Losada and Giménez-Curto 1979). The importance of this parameter in design has also been shown (Bruun and Johannesson 1976), suggesting an S-shaped profile for breakwaters. This, however, still results in steep underwater slopes which can readily reflect waves.

Reflection of oblique waves from the lower face of breakwaters, even when the upper zone is being dissipated by large armor units, can effect substantial scour of the bed, which results in a trench with side slopes equal to the angle of repose of the sediment. This critical state can not countenance buildup of pore pressure without sudden failure of the face, as near breaking waves approach the structure. The degree of reflection from rubble-mound slopes has been studied (Madsen 1983) which shows its dependence on: (1) porosity, (2) width of material, (3) diameter of stones, (4) water depth, (5) wave period, and (6) incident wave height.

A graph was put in the form of a parameter kW

$$kW = 2\pi (W/d)(d/L_o)/\tanh 2\pi d/L \qquad (8.2)$$

where W is the width of equivalent rubble mass, d the water depth, L the local wave length, and L_o the deepwater wave length, from which the curves of Fig. 8.15 have been derived. As seen in the inset of this figure, an equivalent submerged rectangular volume of stone has been used in the analysis which is believed to be equivalent to the trapezoidal cross section normally existing. The reflection coefficient Cr is based on the wave as a whole, which may involve breaking near SWL. Thus, considering reflection for lower sections of the water column, the reflection component for this would be even greater than indicated in Fig. 8.15. It should be remembered that in many breakwaters the thickness of protective stone layers is not great and many engineers insert fabric just beneath them to reduce scouring of smaller sized material of the core. This concentrates flow of the downwash from the wave which is maximum at the trough. When a steep crest approaches simultaneously, an upward flow in this permeable mass occurs in the same region, which magnifies the outward force on the stones and the possibility of failure.

The recommendations of the various authors quoted can be summarized by the conclusions of Sawaragi et al. (1982), in which they state: "Design formulae in future must take the hydrodynamic forces produced by resonance into consideration as an external disturbing force." All this points to the need for a new approach to the design of breakwaters, which permits large scale voids to be present throughout the whole structure and obviates the use of small sized material to save costs.

As noted already, scouring of the seabed adjacent to breakwaters from waves reflecting obliquely may also be another cause of such slumping of armor units (Silvester 1972, 1977). Waves, particularly the insidious swell, do not necessarily approach normal to a structure, despite the accepted flume tests to determine stability. When they are angled to the face, especially the long-period components of the wave spectrum, short-crested systems are established which are conducive to scouring along the length of the structure. Some engineers believe this cannot be the case in deep water, but the vortices so generated in this system can penetrate to the bed with very little attenuation (Silvester 1972). The macroturbulence associated with this system maintains sediment in suspension which is then carried downcoast by the excessive mass transport in the short-crested system (Hsu et al. 1980).

Thus, if any progress is to be made in the design of coastal structures, the following points should be noted:

1. Armor units should be designed which fit snugly together, giving resistance to movement by friction over a wide surface.

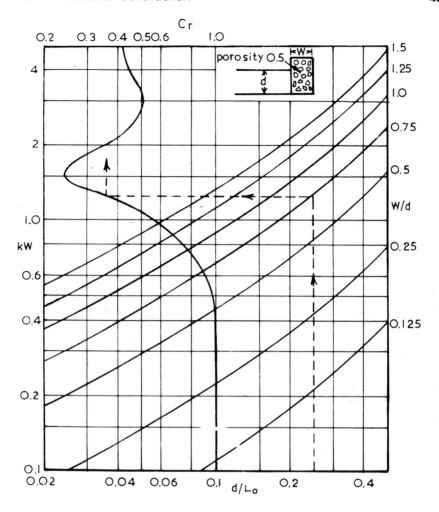

Figure 8.15 Reflection coefficient C_r for given W/d and d/L_o

2. Core material should consist of units equally as large as other armor blocks if they can be manufactured from cheap concrete.

3. Spaces should be provided between elongated units to permit wave absorption, reduce reflection, and attenuate upward pressures.

4. The bed seaward of the breakwater should be protected from scouring by an impermeable mat for either traditional rubble-mound structures or geotextile units.

8.3.1 Characteristics of Geotextile Units

Porraz and Medina (1977, 1978) and Porraz et al. (1977, 1979) have pioneered the development of geotextiles acting as formwork for cement mortar units cast in situ. Their purpose was to utilize local unskilled labor rather than expensive imported technology, especially for developing countries. Although it was proved useful in this situation, this new concept has application in western countries where many marine structures are required at isolated sites, and infrastructure for transport of stone is not always available or must be provided at great cost.

This solution utilizes high strength fabrics which can be sewn to any desired size of armor unit. Since they are not man-handled once fabricated, they need not have the same strength as the precast concrete alternative. Being cast in place, these units will assume the shape of those already cast beneath them. They therefore have a broad contact surface which prevents rocking. Good frictional resistance is provided which can resist the pushing and pulling action of waves. The undulatory nature of the surface contact also provides a saw-tooth resistance to shear forces of one layer against its neighbor beneath and above.

The suggested weak mortar mix need be only of sufficient compressive strength to support the weight above, plus the moment from the side force of the waves. Since the flexible membrane is required to hold the mixture in place until it sets, any subsequent deterioration due to UV rays or other conditions is of little concern. The resulting hard mortar must resist swashing by water, perhaps laden with sediment.

Thus the demands, as set out in items 1, 2, and 3 above, can readily be met. As with any engineering structure, costs of alternatives must be assessed. In such a comparison it is not the unit cost (e.g., cost per m^3) that is important, but the overall expenses of establishment, interruption in construction, maintenance, and relative volumes of material. It will be shown that geotextile units can be cheaper than rubble-mound, with or without precast armor units.

It is believed that longer containers than those used previously have greater advantages. Thus, sausage-type units are preferable to large bags because of their more comprehensive contact with adjacent bodies. Ray (1977) used bags filled with sand but found them very unstable for a number of reasons, even in laboratory conditions. The overall size of each homogeneous mass can, in fact, be made so large that it is inconceivable that any wave could disturb or remove them. The weight of individual sausages can be such, and their frictional resistance with each other so effective, that even the stability coefficient (Hudson 1959) normally used in breakwater design becomes irrelevant. Tests are required to prove this contention but these should not be costly. As for precast concrete armor units, these cast in situ monoliths would not be reinforced; but, as they do not suffer dynamic forces due to rocking, the need for high shear and tensile strength is obviated.

Settlement of the structure, as occurs with randomly placed rubble material which compacts as the load grows, is not likely with mortar filled units since they are added uniformly. Compaction of the sand below the breakwater occurs evenly due to the rectangular nature of the cross section. Should any cracking take place due to such differential settlement, this will be in sausages well down in the structure where wave forces are reduced. Even so, the broken units are still massive compared to the alternative stone components.

8.3.2 Design of Units

Liu (1978) presented a mathematical derivation and carried out an experimental verification of plastic containers filled with water and tested in the atmosphere as seen in
Fig. 8.16. This is equivalent to a mortar mix with $SG = 2$ when submerged in water.
The definition sketch of Fig. 8.17 presents the variables involved: fabric circumference
(S), pressure head (b_1) as measured from the base, height of unit (H), width of unit (L),
contact width at base (L'), height from base to vertical slope of unit (Y), tension in fabric
per unit length of sausage (T), the diameter (D) of the circular area (A) equivalent to that
of the unit $= (S/\pi)$.

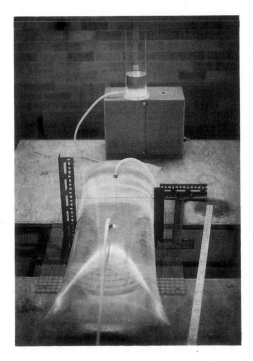

Figure 8.16 Experimental equipment for
testing water filled sausage

The theoretical curves and experimental data for parameters H/D, $2Y/D$, $4A/\pi D^2$,
L/D, L'/S, and $2T/\rho b_1^2$, are presented in Fig. 8.18, where close agreement is observed.
Two conditions are noted of smooth and rough bed beneath the unit, which made little
difference to the verification, except perhaps for L'/S. Two examples of predicted and
measured shape are shown in Figs. 8.19 and 8.20 for differing values of pressure head
(b_1). The dimensionless parameters useful in design are slightly different from those
included in Fig. 8.18, so are presented separately as in Fig. 8.21 (Liu and Silvester
1977). Because many of the useful fabrics are available in 12 ft widths (3.66 m), a list
of variables is included in Table 8.1. Units can be fabricated in multiples of this width
after sewing with longitudinal seams.

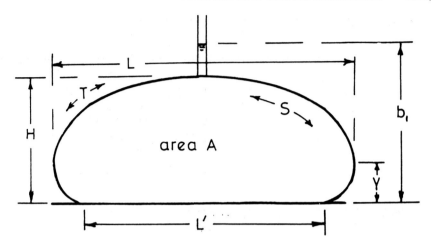

Figure 8.17   Definition sketch of mortar filled sausage

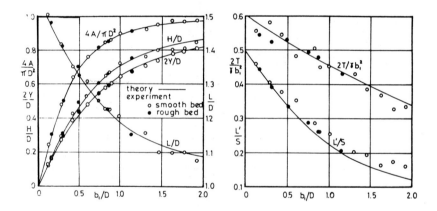

Figure 8.18   Dimensionless parameters from theory and experiment

If the breaking strength of this specific fabric in the warp direction is 1,260 kg/m, the maximum height of sausage possible (cast underwater) is 1.0 m. The increase in cross-sectional area to the equivalent rectangular shape (A/HL) is only 4% from 0.9 to 1.0 m in height and hence this lower value, with much lower tensile stress of 693 kg/m is to be preferred. It should be noted that tensile stress for above-water casting is almost twice that for the submerged state.

Sausages could be laid across the width of the structure with only their ends in contact with the open sea. Except by cracking in tension, these units cannot conceivably be detached from the breakwaters by waves. As each unit should have full support beneath it and full vertical load above it, the resulting face should be almost perpendicular and not less than 4:1. As indicated in Fig. 8.22, the preferable alignment of units is 45° to the axis, in order to make each longer and heavier. The underlying row will provide

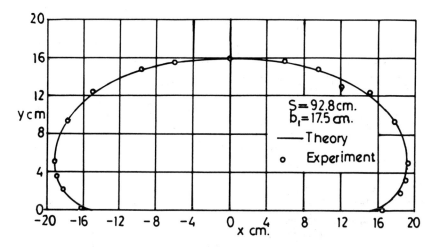

Figure 8.19   Comparison of measured and predicted shape for a pressure head $b_1 =$ 17.5 cm

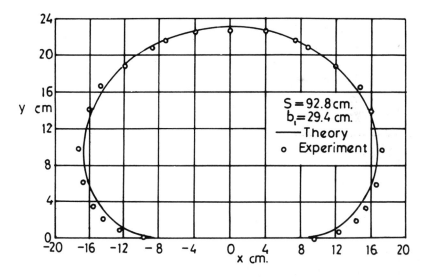

Figure 8.20   Comparison of measured and predicted shape for a pressure head $b_1 =$ 29.4 cm

an undulating surface which the upper unit will follow, so giving a "saw-tooth" contact for good shear strength (see section *BB* of Fig. 8.22).

Positioning of sausages in alternate layers should be carried out accurately to keep them from sitting on the space beneath, which would cause flexure in the units running normally between them. Such positioning can be effected by "nailing" the unfilled membrane to those already in place. At the extremity of the structure a rectangular extension should be provided, as in Fig. 8.23, so sausages of equal length can be positioned at all

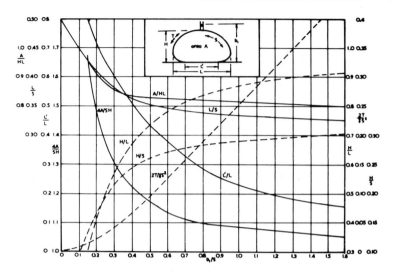

Figure 8.21   Design parameters for mortar $SG = 2$

TABLE 8.1   VARIABLES IN DESIGN OF MORTAR-FILLED SAUSAGES WITH CIRCUMFERENCE OF 12 FT (3.66 m)

H (m)	H/S	b_1/S	b_1 (m)	L/S	L	L'	A/HL	A	$2T/\gamma S$	T (kg/m)	T' (kg/m)
1.0	0.273	0.62	2.27	0.348	1.27	0.46	0.825	1.05	0.130	1,202	2,096
0.9	0.250	0.42	1.54	0.361	1.32	0.62	0.832	0.99	0.075	693	1,209
0.8	0.218	0.31	1.13	0.380	1.39	0.82	0.851	0.95	0.043	397	692
0.7	0.191	0.25	0.92	0.395	1.45	0.94	0.875	0.89	0.031	286	499
0.6	0.164	0.19	0.70	0.410	1.50	1.05	0.905	0.81	0.021	194	338
0.5	0.137	0.16	0.59	0.425	1.55	1.21	0.950	0.74	0.015	139	242

levels. This is a critical region of the structure, where vortices due to wave diffraction are pronounced (Chang 1977), and hence bed protection should be extended around the leeward side.

The suggested spaces would provide channels through which water could oscillate to relieve high pressures on the seaward face. These tubes of corrugated form would have virtually rounded orifices at each end. Jets would discharge on the leeward side of the structure which will be normal to each other at alternate layers. The vortices generated should attenuate each other quickly and reduce wave transmission on the harbor side. The jets' leeward flow would be reduced by the alternative passages where water would proceed to zones of lower pressure. This particularly would be so for waves arriving obliquely to the structure, but storm waves are also multidirectional and so wave forces and pressures will not be synchronous along the whole length. Uplift forces on sausages will be greatly reduced by the provision of such a pressure-relieving mechanism. Tests are required to measure transmitted waves through these novel breakwaters, as well as

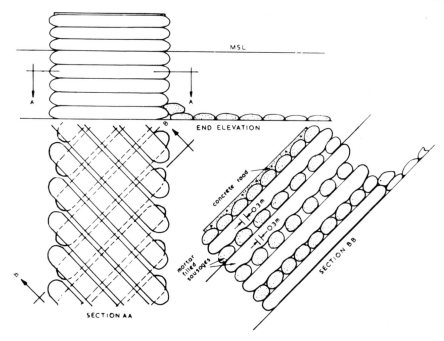

Figure 8.22   Plan and sections of breakwater trunk using sausage-type units

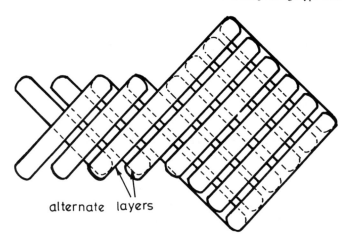

Figure 8.23   Enlargement of breakwater tip to permit units of equal length as in trunk

records of internal pressures. The voids would provide for excellent spawning of small fish and so keep the angler happy.

The cross section of the breakwater will depend on the amount of overtopping that can be permitted during the worst storms. Overturning moments and shear forces must be resisted at all levels, particularly at the bed and the crown. Since the frictional forces to resist horizontal shear are determined by the vertical weight from above, the sausages at the crown must be carried high enough to where forces are sufficiently reduced. This is where the diagonal alignment of units is beneficial since the spatial force of any wave is spread along each sausage progressively and not applied instantaneously over its complete length.

Wave forces are reduced in triangular fashion from the SWL, to the reach of the partial standing wave, as indicated by *AB* in Fig. 8.24. If overtopping to the extent of *BC* is accepted, the horizontal force (as indicated by the hatched area *CD* plus shear from the flow) must be resisted by the top sausage layer through its contact at level *D*. This is provided by both surface friction, through the weight of this layer, plus the undulatory surface with units just below level *D*. Should experience indicate that insufficient allowance has been made, new sausages can readily be cast in any spaces created in the zone *CD* and others added above *C*, where horizontal forces are reducing linearly.

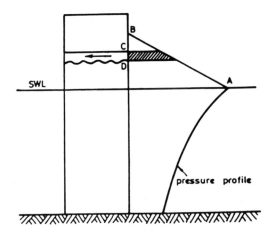

Figure 8.24 Horizontal forces exerted on structure due to partial standing wave

Near the bed internal and leeward jets could promote scouring; therefore, it is suggested that the first three layers from the bed be cast without spaces between. Those at the crown should also be poured adjacent to each other. In this case, alternate units should be cast first so that subsequent sausages clasp them along their sides. This implies that they do not act individually in being lifted by internal pressures, nor sheared separately. A continuous undulatory surface will result at the crown on which a concrete road can be cast for later access (see Fig. 8.22).

As already stated, all breakwaters reflect some energy from incident waves, particularly those of long period. The structures made from geotextiles will be steeper and, in spite of perforations as suggested, may reflect waves readily. This quality could be utilized for bypassing sediment across harbor and river mouths by forming a short-crested system from oblique swell waves, as will be discussed in Section 9.8. In either type of construction, rubble-mound or otherwise, the seabed near these structures should be protected by mortar mattresses.

Sausages or mattresses could be cast over the floor seaward of the breakwater and along the ribbon of reflected waves. The extent of this needs further research and experience in prototype situations, as was discussed in Section 7.2.2. The thickness of these units has not been analyzed to date but may be equal to the depth of the wave trough below SWL for standing waves, which is equivalent to the lift exerted on them. Additions to permit a safety factor should be considered. Alternate sausages laid with spaces between could have later units cast, which would form "tongue and grooved" joints between them. This was illustrated in Fig. 8.22 but is enlarged in Fig. 8.25. By this means the vortices generated within the short-crested wave pattern (Silvester 1972) cannot penetrate to the floor. This is not the case with stone cover (generally adopted) since vortices enter the voids and suck sand from beneath the rubble elements. An alternative is to make these units much wider or more mattress like.

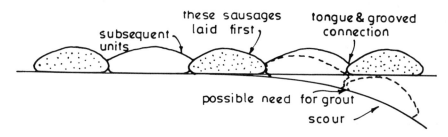

Figure 8.25   Method of casting units to protect the bed

8.3.3 Mortar Mix

Since these monoliths are not required to be handled once they are cast, nor suffer impact forces due to rocking, they can be designed for their major strength requirement, of withstanding compressive loads. This includes the vertical weight of the structure above, plus the compression resulting from bending moments exerted by wave impact. These can be met by mortar strengths much less than those of normal concrete (Goh 1983). An added requirement is resistance to wear by water with suspended sediment. Since limestone meets these demands satisfactorily, a strength similar to this material could be adequate. It should be noted that the mortar, not the geotextile, provides wearability against water flow. The fabric is needed to give support only for the period of mortar setting, after which it can deteriorate from UV rays or tearing incidents of any origin.

Besides these basic strengths, other criteria for the mortar are as follows:

1. It must have 100% slump, or be a slurry, in order to flow to any part of the fabric container without vibration.
2. It must provide a tensile strength equal to that of limestone.
3. It must have a specific gravity equal to or exceeding that of limestone.
4. It must give a water wearing capacity better than that of limestone.
5. It must be as cheap as possible for the cementing material available.

Tests have been carried out by Goh (1983), and Silvester (1983, 1986) which compared various ratios of cement and admixtures to sand, to give compressive and tensile strengths of limestone. Samples of limestone from seven different sites were found to differ greatly in these values so that one, which appeared better by far from the rest, was used for the comparison.

To derive the most economical mix, the additions to Portland cement covered by the tests were: ground granulated blast furnace slag, fly ash, and diatomite. Of these only the first improved the characteristics of the test specimens. Since the bulk of material on any coastal site would be beach sand, this was included, besides sand from terrigenous sources. Also, seawater was used in some tests, instead of fresh, to determine its influence. In order to reduce the water/cement ratio, while maintaining a high slump and easy flow through pumping equipment, various percentages of detergent were added.

Measurements made on all limestone samples were compressive and tensile strengths, specific gravity, and scour by a water jet over a period of 72 hours. The sample showing maximum compressive strength gave 9.2 MPa. The ratio of tensile/compressive strength was found to be a linear relationship with a constant of 0.127, thus giving a tensile strength of 1.17 MPa. The scour as determined by the maximum depth of hole after 72 hours with a constant jet pressure head resulted in the following expression:

$$\text{Compression (MPa)} = 10.4 - 1.91 \text{ scour (mm)} \tag{8.3}$$

implying that scour did not become zero until the compressive strength exceeded 10.4 MPa. For the accepted compression of 9.2 MPa, the scour was thus 0.63 mm. The specific gravity related as follows:

$$SG = 1.64 + (\text{compressive strength}/32 \text{ MPa}) \tag{8.4}$$

which for 9.2 MPa gives $SG = 1.93$.

The traditional approach with mortars is to use a high cement content and low water/cement ratio. This increases the cost and the possibility of cracking, while workability lessens. The object in this case was to have a slurry for placement without vibrators, with a compressive strength and other characteristics similar to the best limestone sample. Admixtures were used to achieve this while at the same time minimizing cracking in what are virtually bulk castings.

Various gradations of sand in mortars have been used (Alvariz et al. 1974, Neville 1973, Yen et al. 1978) results of which are summarized in Fig. 8.26. It is seen that increasing sand/cement ratio decreases 28-day strength, as expected. The fine sand curve (the lowest) indicates that with a ratio of 5:1 a strength of 10 MPa was obtainable, which is slightly greater than for the limestone tested. Since it was anticipated that this would be enhanced by admixtures, this ratio was adopted for all tests. Included in the data of Fig. 8.26 are the water/cement ratios used (w/c), whose increase caused the strength to fall. To obviate this and to obtain a smooth mix for pumping and placing, a detergent known as Dispernex F was added at rates X of 2.5, 5.0, and 6.76 milliliters per kilogram of cementing agent (equivalent to 0.676% by weight approximately).

Granulated ground-blast-furnace slag was the only additive that improved strength. As seen in Fig. 8.27, the optimum 28-day strength came from ratio $s/(s+c) = 0.6$, which

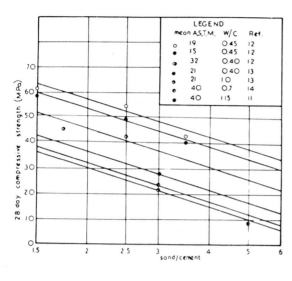

Figure 8.26 Mortar strength versus sand/cement ratio for various sand gradings

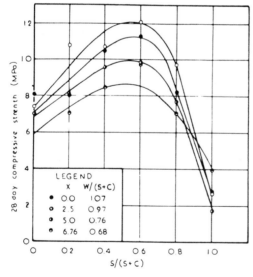

Figure 8.27 28-day strength for ground granulated blast-furnace slag (s), cement (c), laboratory sand and fresh water (W), Dispernex F detergent with X mls/kg of ($s + c$). Ratio sand/($s + c$) = 5.0

is equivalent to slag being 150% by weight of cement. It is also observed that $X = 2.5$ gave better results than zero detergent. Extra specimens were therefore tested for these specific mixes for beach sand and seawater. Compression tests were carried out at 7, 14, and 28 days, as indicated in Fig. 8.28, where it is seen by extrapolation that the 300-day value was 70% greater than for 28 days; this age characteristic has been observed by Bamforth (1980). Also apparent in this figure is the benefit of using beach sand and seawater. Reasons for this could be the increased content of calcium carbonate

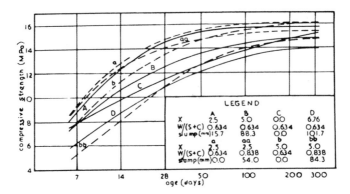

Figure 8.28 Strength versus age, for: (a) Laboratory sand and fresh water (——), (b) beach sand and seawater (- - -). All cases $s/(s+c) = 0.6$, sand$/(s+c) = 5.0$

in beach sand (15.5%) compared to 0.1% in terrigenous material, plus the presence of sodium magnesium compounds in seawater. These have all been found in Portland cement (Montgomery and Dunstan 1981).

Since tests were conducted on specimens with limited slumps it was necessary to extrapolate strengths for complete slump, or 300 mm in this case. In Fig. 8.29 the optimal 300- and 28-day compressive strengths are plotted against slump. It is seen that for a 300 mm slump the highest strength possible is 11.8 MPa, which is higher than the 9.2 MPa obtained from the limestone. The water/cement ratio $[w/(s+c)]$ for this optimum slump is found from Fig. 8.30 to be 1.18 with $X = 2.5$. Further tests are suggested to

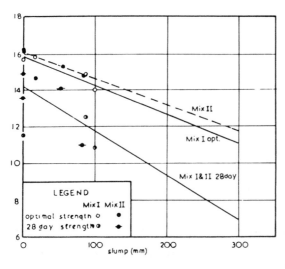

Figure 8.29 Strength versus slump: Mix I with laboratory sand and fresh water, Mix II with beach sand and seawater. $s/(s+c) = 0.6$, sand$/(s+c) = 5.0$

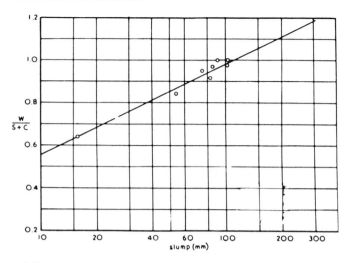

Figure 8.30 Slump versus $W/(s + c)$ for both laboratory and fresh water and beach sand and seawater for $X = 2.5$ mls/kg $(s + c)$

confirm the strengths and slumps as predicted in Figs. 8.29 and 8.30. However, as a first trial on local sands these are worthwhile test values.

Results of the indirect tensile strength (Brazil test) are shown in Fig. 8.31, where again the highest value obtainable was for $s/(s+c) = 0.6$ and $X = 2.5$ ml. This turned

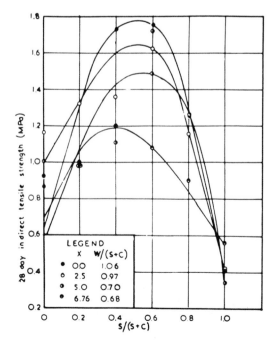

LEGEND

X	$w/(s+c)$
0.0	1.06
2.5	0.97
5.0	0.70
6.76	0.68

Figure 8.31 Indirect tensile strength (by Brazil Test) versus $s/(s + c)$, for $X =$ mls detergent/kg $(s + c)$, and sand/$(s + c) = 5.0$

out to be 1.64 MPa, which is higher than the 1.17 MPa from limestone. The ratio of tensile/compressive strength was 0.129, similar to that for limestone (0.127). Thus, for the optimum compressive strength of 11.8 MPa at 300 days for 300 mm slump a tensile strength of 1.52 MPa can be expected, which still exceeds that of limestone.

Water scour of the cement mix specimens, under the same jet pressure head for the limestone tests, showed that compressive strength in excess of 7 MPa suffered zero scour over 72 hours. This was an improvement on the limestone, which required 10.4 MPa to withstand scour. The specific gravity of the cement mixes correlated with 28-day strength by

$$SG = 1.68 + \text{(compressive strength/30 MPa)} \tag{8.5}$$

which was commensurate with that of limestone, Eq. (8.4). For the accepted strength of 11.8 MPa previously derived, an $SG = 2.07$ was obtainable, which exceeds that for the limestone (1.95).

From these tests the following proportions are suggested for further tests on the desired slump and other characteristics:

Beach sand/(slag + cement) = 5.0

Slag/(slag + cement) = 0.6 by weight

Seawater/(slag + cement) = 1.18 by weight

Detergent = 2.5 mls/kgm of (slag + cement)

The above mix should give a complete slump and provide the following values:

300-day compressive strength = 11.8 MPa

300-day tensile strength = 1.52 MPa

Specific gravity = 2.07

This mortar when set should also provide good resistance against water scour, certainly better than that of limestone, which is widely used in marine structures.

8.3.4 Costs

It is difficult to compare costs of geotextiles units with equivalent rubble-mound construction due to:

1. A specific situation having to be analyzed which may not be appropriate at any other site, where material supply conditions may differ.
2. The methods of costing equipment usage in sausage construction varying greatly.
3. The purchase price of cement and additives (the bulk of the cost) being determined by local suppliers, which can increase swiftly over time.
4. The wave climate of any area being unique.

Even so, it is worth comparing the cost of a known rubble-mound structure with a geotextile alternative for an assumed mode of construction. This has been carried out

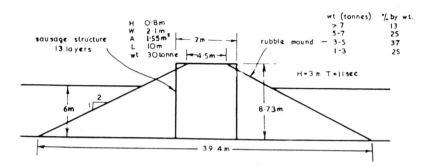

Figure 8.32   Cross sections of breakwaters used in cost comparison

(Goh 1983) for a limestone breakwater in Western Australia for which costs and design waves were known. A structure consisting of mortar sausages was designed and costed. The cross section of the two alternatives are illustrated in Fig. 8.32.

The total length of the breakwater was 1,350 m and contained 300,000 tons of limestone rock, of which the armor units weighed 12 tons. The design wave had height of 3 m in 6 m of water depth and its wave period was 11 sec. The job was completed in 24 weeks and cost $1.125 million in 1978, which excluded the access road. Inflation to 1982, when the comparison was made, was 2.5, giving the comparative cost of $2.8 million.

The design of the sausage structure covered sliding at the base, bearing capacity of the soil, overturning of the structure, and the shear at SWL. A protective mat was to be provided for this vertical faced breakwater, even though bed protection was not provided for the rubble-mound alternative. The sausages comprising this mat were required to be 0.94 m thick in order to resist uplift by standing waves and to be carried out for a quarter wave length of 20 m.

A particular nylon fabric was chosen 5.0 m wide to give a circumference (S) of this dimension. With a filled height (H) of 0.8 m, this gave a cross-sectional area (A) of 1.55 m^2. For a 7 m width of structure, sausages laid 45° to the axis were 9.9 m long, thus giving a total volume for each of 15.35 m^3. These would weigh over 30 tons each. Allowing 0.3 m spacing between units for the central 5.0 m section of the total water column of 8.73 m, the final volume of mortar, including the protection mat, was calculated as 93,000 m^3.

Mortar mixing equipment and accessories for two teams were priced, including standbys in case of breakdown. This total was divided by 5 for 20% depreciation on the job. Fuel and labor costs (at current prices) were added as follows:

Nylon fabric	$157,000
Cement + slag	1,829,000
Detergent	66,000
Machinery + running	104,000
Labor	329,000
Total:	$2,485,000

This was 88% of the limestone version cost. If the protection mat were excluded the total could be $1.79 million or 64% of the rubble-mound alternative.

It is seen that, even with some reservations in the manner of estimating, this novel approach equals or is more likely to be less than that of the traditional method of construction. However, the final cost of any complex should include that for the probable maintenance. In an extreme storm sequence it is normally accepted that rubble-mound structures will require some rebuilding. But with the massive units used in geotextile version, with good surface contact, no rocking or displacement can be envisaged. Even if it does, the addition of extra units will not require as large establishment cost as the rock alternative, involving the use of trucks and cranes. Access to and across the sausage-type structure is much easier than for a breakwater containing large stones or precast armor units. Thus, this new approach should be priced and compared methodically with other designs of breakwaters, groins, or headlands. Once contractors have gained some experience in this type of construction, prices will reduce substantially.

8.3.5 Logistics

One of the great advantages of this constructional concept is the simplicity of the operation. Very little equipment is necessary and this is transportable by trucks of reasonably low capacity. Certainly the cement and additives must be delivered either in bags or in bulk, but it can serve remote sites where there is little infrastructure in terms of roads or water supply. Two concrete mixers, a slurry pump, and a small air compressor for use by divers is all that is necessary on the beach, and perhaps for seawater in mixing and cleaning a pump is required. Sand may need to be shifted a short distance, for which a modest front-end loader should suffice.

Two mixers could be filled by unskilled laborers, using lightweight boxes for measuring volumes. The mixers then feed into the hopper of a slurry pump which delivers through a 75 mm pipe up to 200 m to the point of construction. For longer delivery a second hopper and pump may be necessary. Unskilled workmen can also connect this pipe to inlets of successive sausage units, perhaps without the pump being stopped. Work underwater must be carried out by divers with breathing masks.

The alternative of establishing a ready-mix complex on site will depend on the site conditions and magnitude of the operation. Hoppers for sand and bulk cement or blast furnace slag (or other admixtures) would be required to feed mixer trucks which should pour continually into the pump hopper, which can also be mobile. It is necessary with any method to maintain a continuous flow of slurry since settling in the pipe could make restarting of the pump difficult. From two normal mixers 1 ton of mortar could be poured every 3 min, but longer sausages may require doubling of this rate. Any one unit should be cast within 3 hours to prevent any initial setting prior to complete filling.

There is little doubt, from experience so far gained in the developing countries, that construction by this method is much faster than for the traditional rubble-mound structures. Cessation of operations due to inclement weather, shortage of funds, or labor disputes, does not involve expensive equipment with large overheads. Temporary concrete block anchors with ties may be required to hold geotextiles when the first sausages are poured. Once the base units are in place, membranes for upper layers can be nailed to these. As the container fills, nails pull out, perhaps with slight tears which are of little consequence. As the units become full small tears across the crown permit excess water to drain, leaving solid mortar to the desired thickness of the sausage. There are many other facets of construction that could be alluded to but these can readily be sorted out on the job.

8.3.6 Applications

All types of marine structure, massive or modest in size, can be served by these geotextile units that are cast in situ to any size desired. Even diagonal sausages at the crown cannot be dislodged so that overtopping can be countenanced during severe storm sequences. Koerner and Welsh (1980) have described a number of applications for fabric formwork in marine and fluvial structures. These include support of offshore pipelines, filling of scour holes adjacent to bridge piers, mattress protection of levees against wave action, and causeways for all types of services. They can replace expensive rubble-mound structures in almost all conditions.

Should engineers have greater faith in stone or precast concrete armor units than in the proposed geotextile units being proposed, they may find that mortar filled units of very low strength are cheaper per unit volume than normal quarry run core material. This could arise due to the sand being available on site and, as a pumpable mixture, easier to deliver into place than trucking rock fragments. This alternative structure is depicted in Fig. 8.33A, where the outside protection is afforded by rocks or concrete blocks. This does not preclude the necessity of protecting the bed should swell waves arrive obliquely to the structure.

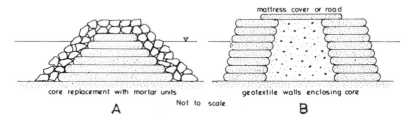

Figure 8.33   Use of geotextiles: (a) to replace core material, and (b) to provide a space for core fill

If the reader accepts the geotextile sausage units as an ideal stable unit he could construct the alternate shape as shown in Fig. 8.33B. This consists of two relatively narrow rectangular structures parallel to each other, between which sand or stones could be dumped. The top would be covered by a relatively thin mattress filled with mortar for ready access, or utilization as a working area, in which case a reinforced roadway would be in order. This may require extra units to prevent overtopping during storms.

The obvious benefit of geotextile units filled with mortar is that the bulk of the material, sand, is available in the vicinity of the site. This overcomes the difficulty of providing large rocks from quarrying, and the expense of crushing and screening small stone in the fabrication of concrete blocks. The necessity for good road networks is overcome plus the cost of fuel in cartage. This may apply more to developing countries, as does the importation of less technology in the service of complex equipment. Rock itself is becoming scarcer or more costly to quarry. In the United Kingdom, for example, it is cheaper to import such material from Scandinavian countries than to quarry it locally. As breakwaters become larger in cross section, as they extend into deeper water, the armor units must grow, so requiring larger cranes and trucks for handling.

The mortars suggested for these geotextile units should consist of lean mixes for which less quality control is demanded. The overall size of the structure, as indicated in Fig. 8.32, to dissipate waves is less than for the rubble-mound alternative, which is dictated more by structural requirements of continuing construction through all weather. Lower overhead and mobilization costs are involved in the case of job interruptions. Semi-skilled laborers are cheaper in all types of economies.

Alvariz et al. (1974), Porraz et al. (1977, 1979), and Porraz and Medina (1977, 1978) report on flexible membrane units used in groins and similar structures. Pilot studies would now be an advantage to see how these structures stand up to normal wave climates. So far only units of bag proportions have been used as distinct from the sausage concept as being proposed. These larger monoliths withstand waves better and can be cast more economically. With the disputes that have arisen over the failure of a major breakwater (Harlow 1980, Edge et al. 1982, Silvester and Hsu 1989) there is a need for a new approach to this construction which entails more than a different shape of precast concrete armor unit.

8.4 SUBMERGED PLATFORM BREAKWATERS

This type of breakwater commenced as one floating on the surface, the first of which was called the "Bombardon" used in the Normandy invasion in World War II (Lochner et al. 1948). This took the form of a Maltese Cross with some arms filled with water to provide mass for long natural periods of oscillation. It damped moderate waves but failed in an unexpected severe storm and was swept away. After the war cheaper alternatives were sought (Bulson 1967, Griffin 1972, Richey and Nece 1974, Hattori 1975, Dattatri et al. 1977), but few have passed the model stage. A surface-plate type was tried as a prototype by Hasler (1974), but to be effective its width had to equal one wave length of the incident wave in depth at the site. Harris and Webber (1968) and Shaw (1973) also tested a plate structure which reduced wave height by 75% when the width was 0.67 the wave length. Others have taken the form of scrap tires (Kowalski 1974), and catamaran pontoons (Miller 1974a, Christensen and Richey 1974, Adee and Martin 1974, Adee et al. 1976).

Submerging the plate a small distance below SWL permits the top portion of the wave to traverse it while the pressure and water-particle motion of the part below the plate is transmitted to its landward side immediately (Patarapanich and Silvester 1984). If the wave above the plate has its trough at this point when the crest reaches seaward end, water flows upward so generating turbulence, which dissipates some wave energy. Besides this, waves are reflected seaward from both edges, and breaking is occasioned by the meeting of crests from the front and rear at some point on top of the plate. These several sources of energy dispersal cause the transmitted wave leeward of the breakwater to be greatly reduced.

The width of the plate (B) must be such as to maximize the energy loss, which from the above description of the action should be about half the length (L') of the wave in shallow conditions above the plate (see definition sketch in Fig. 8.34). What is needed is to attenuate components over a reasonable range of the spectrum of either storm waves or swell. As these propagate over the plate they steepen and due to opposing crests, as mentioned above, will break. It is found that the submerged plate only has to be about one quarter to a third the width of a surface platform to achieve the same result.

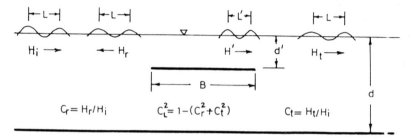

Figure 8.34 Definition sketch of variables involved in the submerged platform breakwater

8.4.1 Theory and Experimental Verification

Patarapanich (1980) has provided a finite element method (FEM) solution to wave action over the submerged platform and has verified it by model tests of his own and of Dick (1968). The reflected wave from the rear end traveling seaward over the plate is again reflected at the front end, but this is not included in the analysis (Patarapanich 1980) since this action is negligible in the overall action.

Figure 8.35 shows the FEM curves and Patarapanich data for various B/L and d/L values showing C_r, C_t, and C_L values plotted against d'/d. It is seen that the experimental curves for C_r at their maxima are below those of the theory (FEM), which is possibly due to the reflected wave from the rear of the plate being partly dissipated in the breaking process with the incident wave propagating over it. The minima of the C_t data are between those of the relevant FEM curves as some of the secondary reflection at the front of the plate is adding to this transmitted energy. The coefficients for energy loss (C_L) have maxima around $d'/d = 0.15$, which is the same for minimal values of C_t. However, this does not match either the theoretical or experimental curves for C_r, indicating that reflection plays a small part in reducing the transmitted wave. The preferred submergence thus appears to be at $d'/d = 0.15$. Patarapanich and Cheong

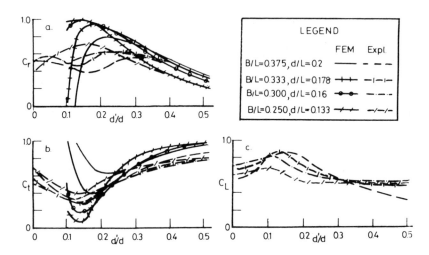

Figure 8.35 Coefficients of reflection (C_r), transmission (C_t), and loss (C_L) for a range of submersion (d'/d) and ratios of width (B/L) and depth at the site (d/L) (Patarapanich 1980)

(1989) conducted further flume tests, in which optimum submergence (d'/d) was found to be 0.05 to 0.15 with regular and random waves. In respect to the latter, they state: "Results of the coefficients of reflection, transmission and energy loss for random waves averaged over the frequency range of the wave spectrum, show similar characteristics to those of regular waves. Optimum submergence is found to be slightly higher at around 0.1–0.2."

Figure 8.36 shows FEM and Patarapanich plus Dick experimental curves for $d'/d = 0.1$ and 0.2 giving C_r, C_t, and C_L against B/L' values. Here again the experimental maxima C_r are less than the theoretical, but the reverse is so for C_t. The maxima for C_L and minima for C_t are for $B/L' = 0.6$ to 0.8, with maxima for C_r at around $B/L' = 0.6$. From this it can be concluded that B/L' should be designed for 0.6 to 0.8 to cope with most wave conditions, and not 0.5 as suggested in the previous discussion. The additional data by Patarapanich and Cheong (1989) showed that the optimum ratio of B/L' was 0.5–0.7 but dependent on the d/L ratio. The FEM solution indicated that wave reflection increases and hence transmission decreases with smaller depth or greater wave length.

The effect of wave steepness was also studied by Patarapanich (1980), who found that incident wave steepness reaches a critical value when it starts to break on the platform. Beyond this steepness the C_L curve exhibits a sudden rise from the higher energy loss caused by turbulence in the wave breaking process. Wave reflection decreases, thus decreasing C_t, which is minimal when C_L is maximum, but not when C_r was maximum. This again indicated that more energy is dissipated by breaking than by reflection seaward. The critical incident wave steepness is a function of d/L, d'/d, and B/L', which occurs when C_L is maximum.

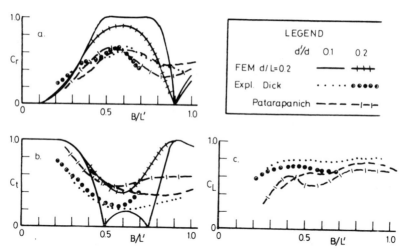

Figure 8.36 Coefficients C_r, C_t, and C_L versus B/L' for FEM and experimental data
(Patarapanich 1980)

8.4.2 Hydrodynamics of Platform

The forces exerted on a submerged platform were derived by Patarapanich (1980, 1984) by the FEM, which compared favorably with the long wave solution (Siew 1976, Siew and Hurley 1977) and the pressure forces under the plate matched those obtained by Ijima et al. (1971). Curves were provided for shallow-water and transitional depths, but only the latter will be presented here. Reasonable depths and wave periods would indicate that d/L values at any breakwater site would approximate 0.1. The vertical force on a submerged platform is positive when upward. The horizontal force is positive in the direction of incident wave propagation. These forces are nondimensionalized by dividing by a hydrostatic pressure such that

$$\overline{F}_y = 2F_y/\omega HB \tag{8.6}$$

where $\overline{F}_y$ is the vertical force per unit length of breakwater

    ω is the unit weight of seawater

    H is the incident wave height

    B is the breakwater width

and

$$\overline{F}_x = 2F_x/\omega HB \tag{8.7}$$

for the horizontal force per unit length of breakwater. Besides the horizontal force due to the orbital motions of the wave acting on the platform, there is an additional component due to thickness of the platform. This was shown to be negligible by Patarapanich (1980). There is also friction drag, which is also inconsequential.

The overturning moment due to the time and spatial variations in the vertical forces is given by

$$\bar{M} = 2M/\omega HB^2 \qquad (8.8)$$

The variations of $\overline{F}_y \overline{F}_x$ and $\overline{M}$ with B/L' are shown in Fig. 8.37 for only two values of d'/d, namely 0.1 and 0.2, as this was the region where optimum wave attenuation occurred. Values for the suggested value of 0.15 could be interpolated from these curves. It can be seen that for the recommended range of $B/L' = 0.6$ to 0.8, maxima values are not reached for any of these dimensionless variables. These curves are presented only for $d/L = 0.1$ which, as already suggested, should apply to most conditions. As seen in Fig. 8.38, the maximum values of $\overline{F}_y$, $\overline{F}_x$, and $\overline{M}$ are shown over a d/L range from 0.01 to 0.2 which should cover most applications of submerged platforms acting as temporary headlands. For $d/L < 0.05$ the forces and moments are essentially constant for a given d'/d value.

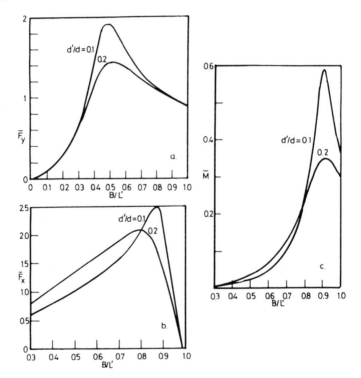

Figure 8.37   FEM curves for $\bar{F}_y$, $\bar{F}_x$, and $\bar{M}$ for $d/L = 0.2$ and $d'/d = 0.1$ and 0.2 (Patarapanich 1980)

8.4.3 Applications

There are various aspects that must be considered when designing a floating or submerged platform breakwater. These include direction of wave approach, irregularity of

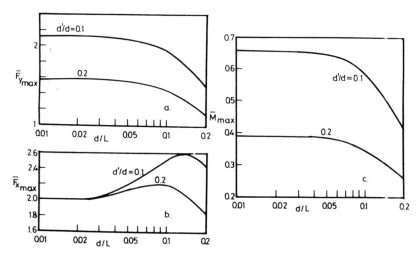

Figure 8.38 Maxima values of $\bar{F}_y$, $\bar{F}_x$, and $\bar{M}$, over a range of d/L for $d'/d = 0.1$ and 0.2 (Patarapanich 1980)

the wave profile, variations in the seabed topography and the tidal range. Besides these environmental variables are those intrinsic to the structure itself, such as method of support either from the bed or the sea surface, method of anchoring, number and size of individual units attached to each other, materials of construction, to name only a few.

Oblique reflection, rather than normal as assumed in all tests, may reduce the reflection capacity of the submerged platform, but this requires further research. All waves in a developing sea will be angled slightly to the mean wind direction so that the crests will be very irregular. This will cause all the types of motions experienced by a ship, but reduced by the degree of submersion. Each component of a wide spectrum of waves will be attenuated according to its wave length on arrival. Transmission levels would need to be expressed in terms of energy or height squared. Since waves will diffract around the ends of a breakwater, the coefficients of diffraction may need to be derived for each period in the spectrum, but at least for the major components. The analysis carried out was for constant depth, so that any undulations of the bed seaward of a breakwater will need to be taken into account. Tide variations will alter d/L and d'/d values, no matter how a platform is supported. The drag due to tidal currents could become greater than the force due to the waves to be attenuated, but possibly not the storm waves to be encountered.

The submerged platform could be buoyant and held in position by cables from anchors or weights on the seabed. It could also be held up by floats at the surface, in which case the whole structure is subject to sway, heave, roll, surge, pitch, and yaw motions, but not to the same extent as a ship on the surface. The forces involved in all these motions will be exerted on lines from either support system. The horizontal forces must be overcome by anchors or large weights tied by long lines to both ends of each breakwater unit. Allowance should be made for some degree of swing of the complete complex due to change in wind direction or tidal flow. The alternative is to anchor from both seaward and landward sides and so retain the breakwater in a fixed location and orientation. The individual units should be made as long as structural design will permit, keeping in mind the large forces in the connecting links between each. There are many new construction materials available as discussed by Miller (1974b).

REFERENCES

ADEE, B. H., and MARTIN, W. 1974. Theoretical analysis of floating breakwater performance. *1974 Floating Breakwater Conf.*, Mar. Tech. Rep. Ser. No. 24, Univ. Rhode Is., 21–39.

ADEE, B. H., RICHEY, E. P., and CHRISTENSEN, D. R. 1976. Floating breakwater field assessment program, Friday Harbor, Washington. U.S. Corps of Engrs., *Coastal Eng. Res. Center,* Tech. Rep. No. 76–17.

AHRENS, J. P. 1988. Reef breakwater response to wave attack. *Berm Breakwaters, ASCE,* 22–40.

ALVARIZ, J. A. M., FARADJI, M., and PORRAZ, M. 1974. Breakwater, rockfill and in situ rocks construction with Bolsacreto system. *Proc. 4th Inter. Ann. Conf. Mats. Tech.,* 1–20.

ANGLIN, C. D., DEAN, K. B., and WILLIS, D. H. 1988. Unconventional rubble-mound breakwaters—concerns. *Berm Breakwaters, ASCE,* 270–78.

BAIRD, W. F., and HALL, K. R. 1987. Breakwater breakthrough. *Civil Eng.,* Jan., 45–47.

BAIRD, W. F., and WOODROW, K. 1988. The development of a design for a breakwater at Keflavik, Iceland. *Berm Breakwaters, ASCE,* 138–46.

BAMFORTH, P. B. 1980. In situ measurements of the effect of particle portland cement replacement using either fly-ash or ground granulated blast-furnace slag on the performance of mass concrete. *Proc. Instn. Civil Engrs.,* Part 2, 777–800.

BARONY, S. Y., MALLICK, D. V., GUSBI, M., and SEHERY, F. 1983. Tripoli harbour NW breakwater and its problems. *Proc. Inter. Conf. Coastal and Port Eng. in Developing Countries,* Columba, 808–19.

Berm Breakwaters: Unconventional Rubble-Mound Breakwaters. 1988. Proc. Workshop, Nat. Res. Council Canada 1987, eds. WILLIS, D. H., BAIRD, W. F., and MAGOON, O. T., American Soc. Civil Engrs.

BRUUN, P., and GUNBACK, A. R. 1978. Stability of sloping structures in relation to $\zeta = \tan \alpha / \sqrt{(H/L_o)}$ and risk criteria in design. *Coastal Eng. 1*: 287–322.

BRUUN, P., and JOHANNESSON, P. 1976. Parameters affecting stability of rubble mounds. *J. Waterways, Harbors and Coastal Eng., ASCE 102*(WW2): 141–64.

BULSON, P. S. 1967. Transportable breakwaters. *Dock and Harbour Auth. 48* (June): 41–46.

BURCHARTH, H. F., and FRIGAARD, P. 1988. On the stability of berm breakwater roundheads and trunk erosion in oblique waves. *Berm Breakwaters, ASCE,* 55–72.

CHANG, H. T. 1977. Vortex scour due to wave action at a breakwater tip. *Proc. 6th Aust. Conf. Hydrau. and Fluid Mech.,* 348–51.

CHRISTENSEN, D. R., and RICHEY, E. P. 1974. Prototype performance characteristics of a floating breakwater. *1974 Floating Breakwater Conf.,* Mar. Tech. Rep. Ser. No. 24, 159–79.

DATTATRI, J., SHANKAR, N. J., and RAMAN, H. 1977. Laboratory investigation of submerged platform breakwater. *Proc. 17th Congress, IAHR 4*: 89–96.

DEAN, W. R. 1945. On the reflection of surface waves by a submerged plane barrier. *Proc. Cambr. Phil. Soc. 41*: 231–36.

DENNEY, B. E., and FRICBERGS, K. 1979. The use of anchored beach systems in metro Toronto. *Proc. Coastal Structures '79, ASCE 2*: 835–52.

DE ROUCK, J., and DE MEYER, C. 1987. Port structures, their care and maintenance. *P.I.A.N.C., Bull.*, No. 58, 133–68.

DICK, T. M. 1968. On solid and permeable submerged breakwaters. *Civil Eng. Res.*, Queen's Univ., Kingston, Canada, Rep. No. 59.

EDGE, B. L., BAIRD, W. F., CALDWELL, J. M., FAIRWEATHER, V., MAGOON, O. T., and TREADWELL, D. D. (1982). *Failure of the Breakwater at Port Sines, Portugal*. Coastal Eng. Res. Council, ASCE.

FAIRWEATHER, V. 1987. Bold new breakwaters. *Civil Eng.*, Jan., 47–48.

FRICBERGS, K. 1985. Shorezone development of rubble mole in Toronto. *Proc. Canadian Coastal Eng. Conf.*, 497–511.

FULLALOVE, S. 1983. Designers becalmed in deep water. *NCE Inter.*, June, 20–21.

GODA, Y. 1974. A synthesis of breaker indices. *Trans. Japanese Soc. Civil Engrs. 2*(2): 227–30.

GOH, P. J. P. 1983. Use of mortar filled containers for marine structures. *M. Eng. Sci. thesis*, University of Western Australia.

GRIFFIN, O. M. 1972. Recent designs for transportable wave barriers and breakwaters. *J. Mar. Tech. Soc. 6*(2): 7–18.

HALL, K. R. 1988. Experimental and historical verification of the performance of naturally armouring breakwaters. *Berm Breakwaters, ASCE*, 104–37.

HALL, K. R., and WILKINSON, D. 1987. Breakwaters the full circle. *Marine Studies Bull.*, Mar. Studies Center, Univ. Sydney 4(6): 21.

HARLOW, E. H. 1980. Large rubble-mound breakwater failures. *J. Waterway, Port, Coastal and Ocean Div., ASCE 106*(WW2): 275–78.

HARRIS, A. J., and WEBBER, N. B. 1968. A floating breakwater. *Proc. 11th Inter. Conf. Coastal Eng., ASCE 2*: 1049–54.

HASLER, H. G. 1974. The 'Seabreaker' floating breakwater. *1974 Floating Breakwater Conf.*, Mar. Tech. Rep. Ser. No. 24, Univ. Rhode Is., 181–91.

HATTORI, M. 1975. Wave transmission from horizontal perforated plates. *Proc. 22nd Japan Conf. Coastal Eng.*, 513–17. (In Japanese)

HSU, J. R. C., SILVESTER, R., and TSUCHIYA, Y. 1980. Boundary-layer velocities and mass transport in short-crested waves. *J. Fluid Mech. 99*(2): 321–42.

HUDSON, R. V. 1959. Laboratory investigation of rubble-mound breakwater. *J. Waterways and Harbors Div., ASCE 85*(WW3): 93–121.

IJIMA, T., OZAKI, S., EGUCHI, Y., and KOBAYASHI, A. 1971. Breakwater and quay wall by horizontal plates and permeable materials. *Coastal Eng. Res. Lab.*, Hyd.-Civil Eng. Dept., Kyushu Univ., Japan, Tech. Rep. No. 2.

IRRIBARREN, R. 1965. Formule pour le calcul des diques en enrochements naturels on elements artificiels. *Proc. 21st Inter. Naval Congress*, Sect. II, Stockholm.

JENSEN, O. J., and SORENSEN, T. 1988. Hydraulic performance of berm breakwaters. *Berm Breakwaters, ASCE*, 73–91.

Koerner, R. M., and Welsh, J. P. 1980. Fabric forms conform to any shape. *Concrete Construction*, 401–05.

Kowalski, T. 1974. Scrap tire floating breakwaters. *1974 Floating Breakwater Conf.,* Mar. Tech. Rep. Ser. No. 24, Univ. Rhode Is., 233–46.

Lindo, M. H., and Stive, R. J. H. 1985. Reconstruction of main breakwater in Tripoli harbour, Libya. *Dredging and Port Construction,* Feb., 19–27.

Liu, G. S. 1978. Mortar sausage units for coastal defense. *M. Eng. Sci. thesis,* University of Western Australia.

Liu, G. S., and Silvester, R. 1977. Sand sausages for beach defense work. *Proc. 6th Aust. Conf. Hydrau. and Fluid Mech.,* 340–43.

Lochner, R., Faber, O., and Penney, W. G. 1948. The 'Bombardon' floating breakwater. *The Civil Eng. in War,* Instn. Civil Engrs. 2: 256–90.

Losada, M. A., and Giménez-Curto, L. A. 1979. The joint effect of the wave height and period on the stability of rubble-mound breakwaters using Iribarren's number. *Coastal Eng. 3*: 77–96.

Madsen, P. A. 1983. Wave reflection from a vertical permeable wave absorber. *Coastal Eng. 7*: 381–96.

Mansard, E. P. D. 1988. Towards a better simulation of sea states for modelling of coastal structures. *Berm Breakwaters, ASCE,* 1–21.

Miller, D. S. 1974a. Floating breakwaters. *Civil Eng., ASCE 44*(3): 77–79.

———. 1974b. Materials and construction techniques for floating breakwaters. *1974 Floating Breakwater Conf.,* Univ. Rhode Is., Mar. Tech. Rep. Ser., No. 24, 247–61.

Montgomery, D. C., and Dunstan, M. R. H. 1981. A particular use of flyash in concrete-rolled concrete dams. *Civil Eng. Trans., Instn. Engrs. Aust.,* Vol. CE *23*(4): 227–33.

Neville, A. M. 1973. *Properties of Concrete.* Bath: Pitman Press.

Ogilvie, T. F. 1960. Propagation of waves over an obstacle in water of finite depth. *Univ. California,* IER Ser. 82, Rep. 14.

Patarapanich, M. 1980. Submerged Platform Breakwater. *Ph.D. thesis,* University of Western Australia.

———. 1984. Forces and moment on a horizontal plate due to wave scattering. *Coastal Eng. 8*(3): 279–301.

Patarapanich, M., and Cheong, H. F. 1989. Reflection and transmission characteristics of regular and random waves from a submerged horizontal plate. *Coastal Eng. 13*: 161–82.

Patarapanich, M., and Silvester, R. 1984. Optimum conditions for a submerged platform breakwater. *Proc. 4th Congress Asian and Pac. Div., IAHR,* Thailand, 539–50.

Pinchin, B. M., and Nairn, R. B. 1986. The use of numerical models for the design of artificial beaches to protect cohesive shores. *Proc. Symp. on Cohesive Shores,* Nat. Res. Council Canada, 196–209.

Porraz, M., Maza, J. A., and Munoz, M. L. 1977. Low cost structures using operational design systems. *Proc. Coastal Sediments '77, ASCE,* 672–85.

Porraz, M., and Medina, R. 1977. Low cost, labour intensive coastal development appropriate technology. *Sea Technology,* Aug., 19–24.

———. 1978. Exchange of low-cost technology between developing countries. *ECOR General Assembly,* Washington, D.C.

PORRAZ, M., MAZA, J. A., and MEDINA, R. 1979. Mortar-filled containers, laboratory and ocean experiences. *Proc. Coastal Structures '79, ASCE*, 270–89.

RAY, R. 1977. A laboratory study of the stability of sand-filled nylon bag breakwater structures. U.S. Army Corps of Engrs, *Coastal Eng. Res. Center*, Misc. Rep. No. 77–4.

RICHEY, E. P., and NECE, R. E. 1974. Floating breakwaters—state of the art. *1974 Floating Breakwater Conf.*, Mar. Tech. Rep. Ser. No. 24, Univ. Rhode Is., 1–19.

SAWARAGI, T., IWATA, K., and KOBAYASHI, M. 1982. Conditions and probability of occurrence of resonance on steep slopes of coastal structures. *Coastal Eng. in Japan 25*: 75–90.

SCHOONEES, J. S., KLUGER, J. W. J., and ZWAMBORN, J. A. 1990. Causes of damage to Saldanha sand breakwater. *Proc. 22nd Inter. Conf. Coastal Eng., ASCE 3*: 2416–29.

SHAW, P. 1973. Test on scale model floating breakwater. *Dock and Harbour Auth. 54*, No. 636, (Oct.): 200–03.

SIEW, P. F. 1976. Some boundary value problems in fluid mechanics. *Ph.D. thesis*, Univ. of Western Australia.

SIEW, P. F., and HURLEY, D. G. 1977. Long surface waves incident on a submerged horizontal plate. *J. Fluid Mech. 83*(1): 141–51.

SILVESTER, R. 1972. Wave reflection at seawalls and breakwaters. *Proc. Instn. Civil Engrs. 51*: 123–31.

———. 1974. *Coastal Engineering, 1*. Amsterdam: Elsevier .

———. 1977. The role of wave reflection in coastal processes. *Proc. Coastal Sediments '77, ASCE*, 639–54.

———. 1983. Design of in situ cast mortar filled armour units of marine structures. *Proc. 6th Aust. Conf. Coastal and Ocean Eng.*, 289–92.

———. 1986. Use of grout-filled sausages in coastal structures. *J. Waterway, Port, Coastal and Ocean Eng., ASCE 112*(1): 95–114.

SILVESTER, R., and HSU, J. R. C. 1989. Sines revisited. *J. Waterway, Port, Coastal and Ocean Eng., ASCE 115*(3): 327–44.

SMITH, A. W. S., and GORDON, A. D. 1983. Large breakwater toe failure. *J. Waterway, Port, Coastal and Ocean Eng., ASCE 109*(2): 253–55.

VAN DER MEER, J. W. 1988a. Rock slopes and gravel beaches under wave attack. *Delft Hyd. Lab.*, Comm. No. 369.

———. 1988b. Application of computational model on berm breakwater design. *Berm Breakwaters, ASCE*, 92–103.

YEN, T., SU, C. F., and CHANG, M. F. 1978. A possibility of increased mortar strength for ferro-cement. *Inter. Conf. on Mats. Construction for Developing Countries*, Bangkok, 665–83.

ZWEMMER, D., and VAN'T HOFF, J. 1982. Spending beach breakwater at Saldanha bay. *Proc. 18th Inter. Conf. Coastal Eng., ASCE 2*: 1248–67.

Bypassing Mechanisms

Although it has been implied in the previous chapters that littoral drift could be prevented by treating the whole coastline with headlands, between which stable bays could be formed, this is not a realistic possibility. There will always be tens, hundreds, or even thousands of kilometers of beach from which material can be moved to any point. Perhaps, at the upcoast end of a physiographic unit, where net movement is indicated, the bays could reach stable equilibrium and littoral drift become zero, but these sections will be limited in extent. So taken over all, there will always be a need to bypass drift across river mouths and any harbors connected to the coast.

The problem of shoals and sand spits at harbor and river mouths is ubiquitous (Nichol et al. 1990). Because of the pulsative nature of littoral drift, as discussed in Section 3.5, no adequate solution has been found. The calculations of annual rates of transport do not take into account the large fluctuations during the winter and summer seasons. The accretion in some calm area in proximity to a protruding breakwater or dredged channel could be an order of magnitude greater over short periods than these longer-term computations. This can result in shoals remaining in place for very long periods before they can be removed by mechanical means. In the meantime they are continually shifting position, making for navigation problems and the possibility of litigation should a vessel strike them.

The requirement of dealing with this continual flow of sediment across channels, natural or man-made, is a multimillion dollar dredging business. It has been sufficient to devise very sophisticated machines to remove vast volumes of material from the bed and to pump it downcoast, or into barges for disposal out to sea. With the former some semblance of continuity of drift is maintained, but with the latter, downcoast beaches will suffer from erosion. An alternative to sea dredging is a land operation where cranes or bucket equipment from a gantry digs holes in the beach berm and near shore which are filled from subsequent littoral drift.

Where long distances are involved in shifting intercepted material to sites along the coast, trucking may be the only resort. This may be adopted in recirculating drift from a downcoast point back to the upcoast limit of the beach to be retained. Many inlets or outlets have jetties constructed on either side to maintain them in position. The size of the opening is controlled to vary the tidal velocities through them, with a view to giving deeper channels through the submerged bar formed at the entrance.

Because of the difficulties in open sea dredging, provision has been made by having sand swashed over sills into calmer water within the inlet, from whence it is pumped elsewhere. These structures are known as *weir jetties*. A novel mode of doing the same thing is the jet pump, which is driven by seawater and sucks sediment, which is then transported in a pipe to where it is required. These form conical holes which are then filled by incoming drift. Again, by fluidizing the bed by jets of water beneath the bed during an ebb tide, sediment is carried out to sea, there to be taken downcoast by oblique swell waves.

Two new methods to be introduced consist of a natural means of utilizing the wave energy at the site and another involving long-term accretion upcoast of the waterway to be treated. Since the moderate waves, particularly for a few days after a storm, bring the bulk of the material to the opening, they and the long-term swell could be used to expedite its transmission downcoast without forming shoals. Retention of sand before it arrives will prevent shoaling at an entrance. It implies that the downcoast region is starved of this drift, so resulting in beach erosion. This can be dealt with by stabilization methods already alluded to, if in fact it is necessary.

The introduction of these two innovative ideas fits the observation of Parker (1979) in his discussion of weir jetties: "Laboratory research is going to help but it alone is unlikely to provide the solution or even a substantial part of it. History and tradition say that most of the input for the ultimate solution to the bypassing problem will come from people who observe the sea a lot, think a lot about what they see, and dream a lot." Research will never replace intuition, it can only confirm or disprove it.

All the above modes of bypassing sediment across mouths of waterways debouching to the sea will now be discussed, some in more detail than others as this book is more concerned with innovative concepts. If sand from such an operation is delivered to a downcoast beach by pipeline or barge, it should be dumped within the region of bar formation during the course of an annual storm. This will permit the accretionary waves, which push bar material back onto the beach, to refurbish the berm with the added stock.

9.1 DREDGING

There is little need to expand on this topic as many books, journals, and articles have been written which deal with this mechanical means of winning sand, silt, or shingle from the sea for many purposes. Huston (1968) has discussed the fundamentals and Silvester (1974) outlines the operation. There are many types of dredge the choice of which depends on the material to be removed, the distance for disposal, the size of the job, and the time required for completion. Ultimately, the availability of units may be the deciding factor. Contracts generally extend over months or years, even shifting a dredge from one site to another may take months, with due allowance for weather. Other factors entering the choice of dredge are proximity to fuel and spare parts, and wave climate (especially if pipelines are involved).

9.2 LAND DREDGING PLANTS

When persistent swell or frequent storms occur in the area, sea dredging on a continuous basis may be impossible. This is where removal by land-based equipment may become economical. It mainly concerns the digging of large depressions in the foreshore upcoast of the entrance, which it is presumed are filled by subsequent drift. As has been discussed in Chapter 3, the bulk of longshore movement takes place just after each storm as the bar is moved shoreward. This implies that much material is moved along offshore, in the region of the bar itself. By the time the bar reaches the berm the drift has reduced drastically, to that rate for the final swell-built profile which is minimal. The dredged depression may thus receive only a small fraction of the total littoral drift arriving at the site.

9.3 TRUCKING

This becomes necessary where the investment in a dredging operation is not warranted, or the distances are so great that pipelines are out of the question. The wave climate or the economics may not permit barge transport. The large volumes involved will require a mammoth fleet of vehicles with road infrastructure to suit. Such a scheme was adopted by Yajima et al. (1983), where as seen in Fig. 9.1 the barrier beach spit at Amanohashidate in Japan received sand from two ports to the north and recycled material from its southern tip. This beach has groins for its full extent,which may or may not expedite the travel of sediment to the southern end, depending on the frequency of storms in the elongated bay. As noted in Section 5.3.2, groins tend to move material farther offshore than a normal beach during such sequences.

Dumping sand by truck in widening a beach means that the beach face will initially be steep, but the waves will soon sort out the various grain sizes and provide an equilibrium swell-built profile. At least this procedure supplies sand more slowly than with a massive dredging operation and therefore it will last longer in the system, as noted in Section 5.3.5. All engineering jobs are judged on the basis of economics and hence a continuous cost of recirculating or bypassing material becomes a financial burden on rate payers, much like the interest paid on borrowed money.

9.4 EFFECT OF JETTIES

Natural sandy inlets will have widths and depths or waterway cross-sectional areas commensurate with the tidal prism involved. The bars formed offshore by the interaction of littoral drift and the ebb flow are similar to those at controlled inlets, except the latter

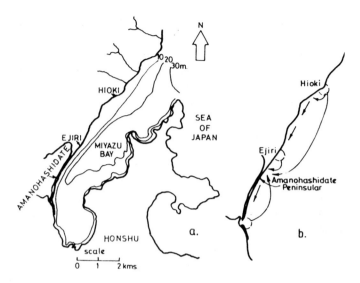

Figure 9.1 Amanohashidate Peninsular, Japan, showing: (a) site plan, and (b) sand transfer system

tend to have higher seaward velocities. The purpose of jetty construction is to move the bar farther seaward in order to maintain a deeper channel for navigation. This results in a larger volume of material being retained that is more difficult for waves to return to the downcoast shoreline, this occurring farther along the coast. The beaches between this welding of the bar and the entrance suffer greater erosion.

Richardson (1991) discussed these two results in management of inlets and states: "An artificial bypassing system will have a higher probability of success if it is designed to address one of these purposes rather than both simultaneously. Although beach erosion and channel shoaling often are related, the optimum approach to dealing with each by artificial bypassing may be quite different. A system that attempts to do both may become a compromise that does neither very well." Nichol et al. (1990) have also commented on this conflict of interest: "Trapped material can result in troublesome channel shoals while the loss of this and other material from the natural littoral system often causes erosion of downdrift beaches." This interruption of sediment transport is also alluded to by Fitzgerald (1982) .

9.4.1 Effect of Littoral Drift and Size of Bar

The bar size and depths over it at any time will be determined by the volume of drift coming from upcoast and the amount of sediment discharge from the river or inlet. When a storm arrives (as discussed in Chapter 3) the littoral drift is optimized, with the majority of it being moved alongshore just landward of the bar returning to the shore. The more severe the storm the further seaward does this silting take place, and the more modest this sequence the closer is the resulting bar to the entrance. Generally rain accompanies the strong winds in the form of a cold front passing across the coast. This produces a flood in the river or inlet system which furnishes extra sand to feed to the bar. The volume proportions from each source are difficult to assess, but the larger bar volumes on these occasions are generally blamed on river flow when it could well be from littoral drift. In either case, it takes 2 or 3 days for the full effect of each to be felt. The combination of large flood discharge and severe storm waves causes the bar to be forced farther from the entrance.

On the dimensions of bars Fitzgerald (1982) has stated: "The morphology of the bar complexes is highly variable with widths ranging from 40–300 m and lengths from 300 to over 1,500 m. Generally, the size of the bar complexes increases as inlet size increases and as the rate of longshore sediment transport increases." These volumes will vary with the distance offshore of the bar crest, when this is greater (due to a strong ebb current) the more it acts like a groin to impede littoral drift. This is why controlled inlets with jetties can cause more downcoast erosion than uncontrolled inlets, because larger volumes are held for longer away from the coast. However, where the tidal flow is small the bar is closer to the entrance with consequences as noted by Bruun (1991): "In practice, it is known that inlets with relatively small flows tend to close up during the most severe storm tide conditions." This should perhaps be amended to read "*after* the most severe ...,*" as the excessive littoral drift will not arrive for a day or two after the sequence.

Richardson (1991) notes the need to ascertain the longshore transport rates: "Artificial sand bypassing systems often are sized to operate on a relatively regular basis; therefore, their proper and economical design depends on knowing not only the maximum rates at which local processes may move sand to them, but also the probable time sequence of different transport rates." It can be taken for granted, from Chapter 3, that these maxima will occur only a few days after any storm. The absolute optimum follows the most severe storm. It is unfortunate that it is during the storm season that the bar requires the most attention in regard to mechanical deepening of channels through them. Nichol et al. (1990) have commented: "Sometimes emergencies would occur after winter storms when the entrance shoal would return shortly after it was dredged." It is not always possible to put off dredging for navigation reasons, but soon after a storm is the most economical.

The volume remaining in a bar for a lengthy period will depend on the wave energy continually arriving to move it shoreward (Walton and Adams 1976). Thus, the full wave climate of storm waves and swell should be assessed when trying to predict the size and motions of shoals at mouths of tidal and fluvial waterways. These fluctuating volumes in bars are supposedly fed by some uniform littoral drift which is calculated over a yearly period. From the discussion above and in Chapter 3, this "river of sand" concept is outdated, but at the moment cannot be replaced with realistic figures for the transient rates during excess periods of supply. It has resulted in large formulae variations on this matter as noted by Nichol et al. (1990): "Surprisingly, even after more than 20 years experience there exist opinion differences of the annual drift quantity at Santa Cruz. One view advocates that the current entrance dredge quantity of 150,000 to 190,000 cubic meters per year reflects the total drift volume while other views suggest that an additional 75,000 to 150,000 cubic meters may be naturally bypassing the entrance each year."

As noted in Section 4.4.3, the presence of an arcuate shaped bar at an entrance can accumulate much material on its upcoast side. As Fitzgerald (1982) has observed: "The dominant direction of longshore sediment causes preferential deposition on the updrift side of the ebb-tidal delta. This results in a downdrift migration of the main ebb channel such that it will tend to parallel the downdrift inlet shoreline (Bruun and Gerritsen 1959)." This sharp curvature in flow erodes the shoreward side of the bar and may breach it. "This process may occur gradually during a six to twelve months period or it may happen catastrophically during a single storm when tidal currents are stronger."

9.4.2 Effect of Tidal Flow

Fitzgerald (1982) quoting Bruun and Gerritsen (1959) states that bypassing can take place by three main methods:

1. through wave action around the periphery of the arcuate shaped bar,
2. through tidal currents being deflected downcoast,
3. through migration of shoals in a downcoast direction.

Bruun (1990) has used a ratio T_p/M_t where T_p is the tidal prism based on a spring tide and M_t is the total amount of material carried to the entrance per annum, both in the same dimensions. Bruun et al. (1978) examined inlets from a number of countries from which Table 9.1 was derived. It is seen that as the tidal prism increases or the littoral drift decreases, so navigation conditions improve. This is one reason why at controlled inlets the exit is narrowed to provide higher ebb flow velocities, which have a similar effect to higher tidal prisms. However, this entrance condition may raise flood levels upstream.

TABLE 9.1 EFFECT OF TIDAL PRISM ON INLET
ENTRANCE CONDITIONS

T_p/M_t	Entrance conditions
> 300	no bar but shoals seaward
150–300	small bar
100–150	low bar, no navigation problem
50–100	wider and higher bar, navigation problem
20–50	wide and shallow bar, navigation difficult
< 20	very shallow bar, navigation very difficult

When a storm abates the following swell waves transport this inlet bar material shoreward, much like the normal bar is refurbished to form a new berm, as discussed in Section 3.4.5. Two major differences occur in this situation, that of wave refraction and that of flood tide assistance to the waves in carrying material to the entrance. Because of the arcuate shape of the bar, waves refract toward the opening causing a protuberance of the shoreline adjacent to it.

The action is concentrated on the downcoast side of the inlet because the bar extends farther in that direction than on the upcoast, where oblique swell will be arriving almost normal to the bar. Fitzgerald and Hayes (1980) have observed: "Wave refraction around the ebb-tidal delta causes a longshore transport reversal along the downdrift barrier." Where a downcoast jetty protrudes from the shore at a controlled inlet, wave diffraction of the oblique swell waves also causes accretion adjacent to it, as observed by Nichol et al. (1990): "The net result often is an accumulation of sand on the downdrift beach immediately adjacent to the entrance followed by an area of erosion farther downdrift."

The second mechanism has been described by Fitzgerald (1982): "The net movement of sand onshore has also been attributed to increased wave suspension during the flood cycle than during the ebb cycle (Oertel 1972, Hubbard et al. 1977, Fitzgerald and Levin 1981)." This process was outlined in Section 3.7, where following currents were shown to lengthen waves while steepening their crests, thus increasing their mass transport at the bed.

The sand on the bar is being continually moved by the oblique swell which is supplying material from upcoast. The waves are steepened over this submerged mound so that transport is magnified. It stands to reason, therefore, that if a channel is dredged through the bar it will readily be filled on its upcoast edge, while its downcoast face will be scoured. The channel thus changes position, making navigation hazardous. This action has been described by Bruun and Gerritsen (1959) as follows: "A submerged bar in front of an inlet or harbour entrance on a littoral drift coast will often function as a 'bridge' upon which sand material is carried across the inlet or entrance. Every channel dredged through the bar will, therefore, be subjected to deposits."

9.4.3 Bar Welding and Beach Recession

It is important to know at what location the bar welds to the downcoast beach because that is where normal supply is again maintained. This will vary from year to year or even from one storm to another, as Fitzgerald (1982) has stated: "The location where the bar complexes move onshore and the length of barrier island shoreline that will be directly affected by the process is controlled by (1) inlet size, (2) wave versus tide dominance, and (3) channel orientation." The inlet size sets the length of the bar, particularly that downcoast of the entrance, so establishing how close to it this material will move shoreward. Where a bar is formed farther offshore, due to a strong ebb tide, the oblique swell waves will move it a greater distance before it reaches the beach. Where the tidal flow is deflected downcoast, the large vortices generated from it and consequent bars will be located more downdrift than if discharge is normal to the coast.

The time taken for these inlet bars to coalesce with the beach will be much longer than for the offshore bar formed during a storm on a normal beach which, as in Chapter 3, was seen to be in the order of days. Fitzgerald (1982) has addressed this question: "The frequency of bar welding events at mixed energy inlets varies from 3–7 years." Not only are the distances longer for the bar as a body to be shifted but the depths are greater, making sediment movement more difficult for the existing waves. However, Fitzgerald and Hayes (1980) also comment: "The effects of these processes can be measured over a period of several years or in as little time as a few hours." This will depend upon the strength and duration of the swell following the formation of the bar.

The process of downcoast accretion, through this natural bypassing mechanism, occurs sporadically due to the ever changing energy input of the swell waves, which as noted in Chapter 3 are changing in wave height and period almost hourly. As observed by Fitzgerald (1982): "This means that along mixed energy (tide dominated) coasts sand is not continuously transferred past inlets but rather it is added to the downdrift shoreline in discrete packets." This supports the thesis, pronounced in Chapter 3, that littoral drift is pulsational, with large amounts added to these bars directly after each storm. The transient storage in these features makes for different frequency of transfer downcoast, but the large fluctuations are recognized. The sudden influx of material as a submarine bar welds to the coast can contain water in the form of a small lagoon as its crest rises above sea level to form a new waterline. As noted by Fitzgerald (1982): "When the bar welds to the inter-tidal beach its cuspate nature usually results in a small ponded water region being formed in front of the bar."

From the zone of bar welding to the entrance the beach suffers a lack of sand, and hence is denuded. This erosion takes place up to that adjacent to the downcoast side of the opening, where accretion occurs due to refraction of waves across the bar, as noted previously. Also, diffraction may take place around the tip of a jetty located at this point. The rates of shoreline loss have been addressed by Walton and Adams (1976) as follows: "In Florida, shoreline recession rates in the near vicinity of inlets are one to two orders of magnitude higher (10–70 feet per year) than average shoreline recession rates away from the influence of inlets (1–3 feet per year) (U.S. Army Corps of Engineers 1971). It is apparent that these inlets act as sand sinks in their capacity to absorb tremendous quantities of sand in both their outer bars and their inner shoal areas."

9.5 WEIR JETTIES

Parker (1979), in the introduction to his discussion of weir jetties, reports of an emperor some 2,200 years ago who was wrongly advised to construct a mole for a harbor which soon silted up. He also recalls that at the 100th anniversary of the ASCE in 1952, coastal engineering was not represented. In 1961 the Chief of Engineers of the Corps of Engineers in an address at Los Angeles listed the major needs to be addressed: "High on his list was an economical means of moving sand on our beaches and across the inlets piercing those beaches." Parker (1979) goes on: "Problems arising from the sea's stubborn insistence on rearranging the world's shoreline have plagued man since the dawn of time. They are still largely unsolved." Much progress has been made in mathematics and theory in the last two decades, but not enough openminded observation of how Nature works. "Harbors shoaled up and had to be abandoned. Fortifications were undermined by the sea and collapsed." It is hoped that this tome provides the overall view needed for engineers to be innovative and work from their intuition rather than from data gathering and analysis. It is a case of harmonizing with nature rather than harnassing her.

In respect to the topic of this section, Parker (1979) states: "Weir jetties are a product of engineering art, not engineering science. Their roots are empirical, not rational." He believes that if properly designed they can effectively bypass sand: "I make this point with full knowledge of the criticism that has been levelled at weir jetties and clear understanding that problems have generally been the order of the day where weir jetties have been constructed." However, he adds: "Weir jetties are not the alpha and omega of sand bypassing. There are other alternatives."

This type of structure evolved in the United States through the Corps of Engineers, which is responsible for maintaining dredged channels. Parkers (1979) reports three circumstances that led to the weir jetty. These were:

1. the ingenuity and perception of Hodges who in 1952 observed the accretion on the leeside of a reef at Hillsboro (Hodges 1955),..
2. the obsolescence of the Corps' hopper dredgers which could not operate effectively in even a mild wave climate,
3. the economies of transposing to pipeline dredges which require calm conditions for good operation.

Parker (1979) admitted that knowledge was limited on this concept. "Perdido, East Pass, Ponce de Leon, and Masonboro can properly be called the first generation." Richardson (1991) has confirmed the experimental nature of these structures: "Experience to date with weir jetties has been mixed, although the original, a naturally occurring rock reef at Hillsboro Inlet, Florida, has functioned well for several decades." It will be noted in Fig. 9.2 that this reef was essentially parallel to the upcoast shoreline, with the seaward edge dictating its alignment. As will be shown, all the subsequent designed weirs were angled to this upcoast beach from 45° to 90°. The action envisaged was that they serve almost as a permeable groin for the weir section close to shore, and impermeable at the seaward end.

9.5.1 Action of Weirs

According to Weggel (1981), the weir should be located in the surf zone of the upcoast shoreline, in order that the sediment placed in suspension is carried over it, to the calm zone created by the seaward extension of the jetty which is of normal proportions. His experiments showed that: "Most of the sand transported over the weir moves across in a relatively narrow region close to where the weir, beach, and waterline intersect." This means that for greater efficiency the waterline at all stages should be restricted to the length of the weir. This implies that this should be known prior to design and construction, which is difficult to predict because the structure acts like a groin, in this case angled to the shoreline.

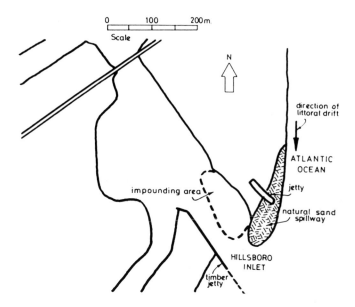

Figure 9.2   Site plan of Hillsboro Inlet, Florida, U.S.A., showing natural reef weir jetty

Murells Inlet, depicted in Fig. 9.3, was controlled in 1980, after which shoreline changes as shown occurred after only 2 years. The new and old ebb tidal shoals show their respective positions and aerial extent, which complements the discussion in Section 9.4.2. With respect to this installation Parker (1979) has stated: "At Murells Inlet, the shoreward

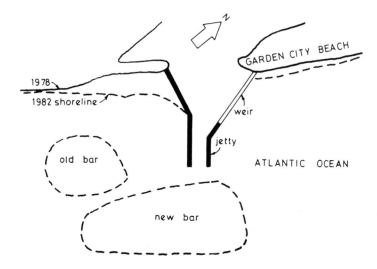

Figure 9.3   Weir jetty at Murells Inlet, Florida, U.S.A., showing shoreline and bar changes after 2 years

end of the weir was fixed at a point selected to define the desired post construction permanent shoreline." This is difficult to do, as noted by Douglass (1987): "The growth of the Garden City spit from 1957 to 1963 correlates with an anomalous period of strong southerly sand transport. Since the area may experience a balanced long-term net transport rate, the use of the weir in the north jetty could be reconsidered."

Weggel (1981), in his suggestion that weirs extend to the breaker line, notes that this changes during storm sequences with a bar forming offshore. In this event he recommends that the weir extend to this new edge of the surf zone. As noted in Chapter 3, the location of the bar is dictated by the strength of the storm, of which some annual mean value serves little purpose. Although he states that seasonal changes should be taken into account, this is not a criterion that can specifically be used in design. Even for the swell-built profile, which exists for a longer period of time than the storm version, the breaker line shifts with the height of the incident wave. As noted in Chapter 3, swell heights and related periods will vary continually while various segments of the original storm spectrum are arriving at the coast. The location of breaking is also affected by the tidal stage.

Where sediment is being moved alongshore it must be accompanied by a littoral current which, when it is intercepted by a structure protruding from the coast, will be deflected seaward. As noted earlier, all the weir jetties constructed so far are angled downcoast, making this deflection easier. Thus, any material in suspension close to the weir will be moved seaward as readily as it is attracted by overtopping water across this lip, set just above MLW level. Seabergh (1983), in his model tests noted this occurrence: "The offshore movement discussed here is related to the presence of a jetty, and the offshore movement occurs due to an oceanward deflection of the wave-generated longshore current by the jetty system even with the presence of a weir (which might be conceptually expected to entirely capture this current)."

It has already been noted that the natural reef weir at Hillsboro Inlet (see Fig. 9.2) was parallel to the upcoast shoreline and hence no deflection seaward of the littoral current could take place. The sand arriving at the reflective seaward face of this reef would cause a short-crested wave system as the incident waves were slightly oblique to it. The suspension capacity of these waves are such that the overtopping waves on this relatively wide weir were fed with an abundance of sediment, which is readily transmitted across it. Parker (1979) lists four actual weir structures with the comment: "These systems have not generally enjoyed the same measure of success as at Hillsboro." *The story of weir jetties may have been vastly different if this orientation and width had been adopted.*

Also, because of the obtuse angle of these weir jetties to the upcoast shoreline, the persistent swell waves will arrive obliquely to them, so generating short-crested waves adjacent to them. These aid the suspension of material at the bed which the deflected littoral current carries seaward. This will cause deepening of the bed in the vicinity of the structure and perhaps its subsidence during a storm, especially at the outer end beyond the weir. Seabergh (1983) noted this in his model studies: "Some sediment moved offshore due to two mechanisms. First, longshore currents for the larger wave conditions were deflected offshore along the ocean side of the jetty, moving sediments with them.

Second, reflected waves from the upcoast jetty interacted with incident waves forming a short-crested wave field upcoast of the weir jetty." This current action will continue past the tip of the structure so causing scour in that region. Snetzer (1969) has observed this: "... At East Pass, some erosion has occurred for several hundred feet seaward from the ends of the jetties and accretion beyond that point, indicating a seaward movement of the marine bar." This bar must refer to the ebb-tide shoaling close to the entrance.

9.5.2 Modeling

Like any modeling involved with sediment movement, that of weir jetties will suffer similar scale effects. While bed load is transmitted over the weir, since it is concentrated at the junction of the waterline and lip, there should also be a proportion of suspended material that overflows seaward of this point. This latter fraction is unlikely to be present in the model due to waves being disproportionately small in respect to the sediment size. Weggel (1981) comments: "There is also some evidence that farther from shore in deeper water near the breaker region, sand moves across the weir in suspension. In the laboratory the amount in suspension appears relatively small in comparison with the amount carried over as bed load near shore."

Another major difference between model and prototype is the use of monochromatic waves in the former and the existence of spectra in the latter. For swell conditions this is of little import, but the inability to produce realistic storm sequences in models, with offshore bar formation, inhibits the production of true fluctuations in supply conditions. As noted in Chapter 3, littoral drift is pulsative with vast volumes arriving at the jetty in a matter of hours. This could smother the weir section, resulting in the waterline forming seaward of it. Seabergh has stated: "The model provides a relatively uniform environment of monochromatic waves and smooth slopes in contrast to the broader spectrum of waves and varying bathymetry of the prototype.... Also, offshore bars would form in nature which would protect the beach somewhat from further erosion, but these bars do not always form on the underlying concrete model bed." He observed that a short-crested wave system was observed from an aerial photograph of Masonboro Inlet weir jetty as in his model studies of similarly oriented jetties.

9.5.3 Prototypes

Walton et al. (1985), in their measurements of sand accumulation over a timber weir at Rudee Inlet, have concluded that: "It was found that the predominant portion of the longshore transport was confined to the upper beach profile, as in earlier laboratory tests, but that the transport over the weir did not necessarily correspond or correlate well with the calculated longshore transport." This would have applied to the swell-built profile when, as noted in Chapter 3, the drift would have been minimal and not to the expectations of a computation based on accretion over a year, which would have included pulses after each storm. The major deposition at the jetty after such events would be close to the bar moving shoreward, which would be swept out to sea by the deflected littoral current. This led to the observation that: "These values suggest the prototype weir was significantly less effective in trapping sand in its deposition basin than the laboratory weir."

The same authors found that distributions of proportionate sediment discharge over the weirs of both model and prototype were similar, but efficiency in terms of the fraction of longshore transport delivered to the basin differed greatly in the two cases. "Satisfactory quantification of efficiency factor was not found. Qualitative reasons have been given for such differences, but remain unproved."

9.6 JET PUMPING

The water jet pump is a device, as illustrated in Fig. 9.4, whereby a high velocity jet entrains water in a mixing tube through a suction nozzle and delivers it through a diffuser to a pipe downstream at a pressure head between that of driving and suction lines (Silvester 1960, 1961; Silvester and Mueller 1968; Silvester and Vongvisessomjai 1971; McNair 1976; Richardson and McNair 1981). The hydraulic characteristics for driving and driven discharges can be obtained from these references. Those for water alone will differ from that for suction fluid containing sediment, which makes for a difference in density. These pumps are very adaptable as they have no moving parts and can operate in any submerged condition.

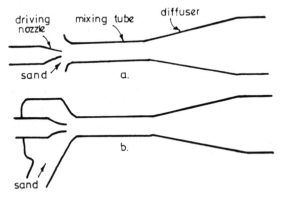

Figure 9.4 Details of the water jet pump, for: (a) open sand entry, and (b) a *Y* suction entry

When used for pumping sand it is advisable to have some water enter the suction nozzle with it, in order not to clog the mixing tube and pipeline. The relative diameters of driving nozzle and mixing tube determine the suction delivery for a given discharge. These pumps can even commence and continue operation when completely buried in sand. According to Pekor and Griffin (1978), they can deliver a sand slurry for 120 m before a booster pump is required. The sand can be drawn in an open nozzle or be supplied via a *Y* connection upstream of it (see Fig. 9.4).

Because of the relatively small clearances between the driving nozzle and suction flare and within the mixing tube, any large sized debris will prevent pumping. This could consist of large shells, timber pieces or bark, drink cans, rags or vegetation, to name a few. Pekor and Griffin (1978) have addressed this problem: "Various proposed solutions have included a recoverable basket strainer, crushers, and wipers attached to the suction. Skin divers have been used." Coughlan and Robinson (1990) have fabricated "a large trial jet pump which will be used as a 'cleaning' pump for removal of debris from the sand trap." This refers to a fixed plant where ten pumps are in line normal to the beach spaced some 30 m apart. Over its 3 years of operation pumping capacity has declined by over 50%. "The reduction in pumping rates is due to operational problems caused by debris being transported into the sand trap by large seas."

Due to the high velocities in the mixing tube and diffuser, the water laden with sand is extremely abrasive, so that hard rubber lining has been suggested. These segments of the pump should be designed for ready replacement. Rambo and Clausner (1989) have stated: "Depending on such factors as eductor wear, amount and type of debris accumulation, and sand availability, deployment and retrieval may be required as often as daily or weekly. An effective method is needed to accomplish this task." These costs should be recognized when planning such a bypassing operation.

As can be gathered from the above discussion, jet pumps can be mobile or fixed in position. A mobile version has been outlined by U.S. Army Corps of Engineers (1984) for Oceanside City, California, whereby some 11 eductors would eventually be installed across the entrance and one upcoast of the main breakwater. Sand is pumped via a fixed booster station to downcoast beaches. Pipes with hole outlets would aid fluidization in the pumping operation. The conical craters so formed would have slopes equal to the angle of repose of around 30°, unless wave action decreased this. The pumps for supplying the driving jet are mounted on floating platforms. It was anticipated that some sediment could bypass the trap during storm events. The initial cost was not to exceed US$5 million.

The fixed system, noted previously, has been described by Coughlan and Robinson (1990), where a single line of jet pumps runs 300 m from the original shoreline. Whether this is sufficient to intercept drift arriving just after a storm of large magnitude, or even of normal intensity, is questionable. Much of the drift moves alongshore just inside the bar which is proceeding shoreward. By the time it arrives at the newly formed berm the rate would have decreased to practically zero. They comment: "Aerial photography indicates that sand can still leak past the seaward end of the bypass trestle despite the reduction in accumulated sand volume, and further excess pumping will be required to achieve additional shoreline recession."

9.7 SAND FLUIDIZATION

The concept of placing a pipe with exit holes beneath a planned channel zone was first conceived by Haygard et al. (1969). Water flow through the bed creates liquefaction of the sand which is then readily removed by an overflowing current. So far it has been envisaged only for entrances to small harbors such as marinas, where relatively small channel depths are involved. Further research was carried out at Scripps Institution of Oceanography, California (Inman and Harris 1970, Bailard and Inman 1975) and at Lehigh University, Pennsylvania (Kelly 1977, Murray and Collins 1978, Weisman and Collins 1979). Subsequent field tests were carried out at Corsons Inlet, New Jersey (Weisman et al. 1980, 1982; Parks et al. 1983). A full scale installation was then carried out at Anna Maria, Florida (Collins and Parks 1987).

The discharge required will depend on the grain size of the sand and its packing, plus the lateral extent of fluidization required. When the system is turned on during an ebb tide, material will be carried out the entrance to feed downcoast beaches. This is likely to be only a fraction of that required for beach stability, the rest coming from bar formation in the natural bypassing action discussed in Section 9.4. However, the goal here is to maintain a minimal depth channel within the entrance. It could be more economical than dredging such small volumes, as a large part of the cost is mobilization at sporadic intervals.

9.8 WAVE REFLECTION

With the knowledge of kinematics in short-crested wave systems in Chapter 2, and their consequences on a sedimentary floor in Chapter 7, the choice becomes apparent of using the persistent swell, bringing drift to an inlet entrance, to help transmit it downcoast. This natural bypassing mechanism, by the double application of energy to the floor, could reduce the dredging bill around the world by many millions of dollars per annum.

The pulsative nature of littoral drift, as discussed in Chapter 3, means that large volumes of material arrive in the offshore bar across an entrance very quickly, but will take some time, even months, to disperse downcoast. This massive input is accompanied by the stronger swell just after a storm, which can be used in reflection to move much downcoast immediately. Silvester (1975, 1977, 1985) has addressed this matter in some detail, but it is worth some reiteration here. The spatial distribution of the short-crested waves will be discussed first, followed by model tests for bypassing sediment. The scouring potential of these waves will then be addressed, leading to structural characteristics in prototype situations.

9.8.1 Wave Phenomena

Consider waves reflecting obliquely from a limited length of straight wall as in Fig. 9.5, where the orthogonals through the tip are presented for clarity. As seen in the figure, the reflected waves are similar to those entering a breakwater gap, which immediately start to diffract at these bounding orthogonals. The wave crest shapes are shown dotted. The energy so spreading into the curved segments of the crests is at the expense of that contained in the ribbon between the limiting orthogonals shown in the figure. It has been shown by Silvester (1978) that along these alignments the wave height is constant at half

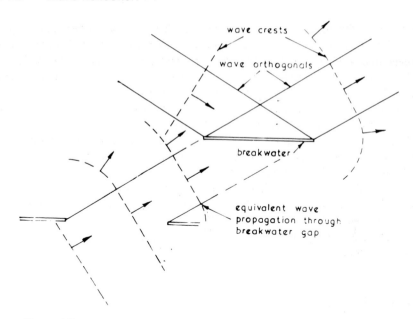

Figure 9.5 Crests and orthogonals of obliquely reflected waves compared to those passing through a breakwater gap

that of the incident wave at the gap itself (see Fig. 2.72). In the same way heights of reflected waves at both edges of this ribbon are half those of the incident waves before reflection, as was seen in Fig. 7.4.

   Thus, it is only over a fraction of the ribbon of reflected waves that the motions as depicted in Fig. 2.47 will occur. However, in other zones, where the two angled wave components are of unequal heights, orbital motions will still be complex and contain strong vortex motions and hence macroturbulence, which is so conducive to scour. The net motion near the bed in the said $K = 1$ zone will be parallel to the wall, but outside it will be deflected to the orthogonal alignment of the incident wave because of its greater energy.

   The suspension capabilities of the short-crested system existing in the ribbon of reflected waves is illustrated well in Fig. 9.6. This is an aerial photograph of waves reflecting obliquely from a limestone reef with a rough vertical edge. This reef is covered in sand for most of the year but after winter storms it can reflect incoming swell for a short period before it rebuilds the berm. It is seen by the light coloring of the sea that sand is suspended to the limit of the orthogonals through the reef extremities (not shown). The apparent concentration at one edge could be due to a surface current from shear of wind coming from the left of the figure. This bed disturbance extends many wave lengths from shore. The original colored print showed large concentrations of sand in a boiling action at the face of the reef which contained many significant potholes. This suspension process would be similar to that at Hillsboro Inlet, where efficient transmission of sand over the reef took place, as discussed in Section 9.5.

Figure 9.6 Sediment suspension along ribbon of reflected swell produced by a natural limestone reef

9.8.2 Model Tests

The dune formation on a model bed under short-crested waves was shown in Fig. 5.2. Another is displayed in Fig. 9.7, where reflected waves could travel much farther from the wall. The various dune alignments parallel to it occur over several crest lengths. These crests of incident and reflected waves are seen to match the dune orientations. For example, at their intersections the water orbits are parallel to wall and the dune crests normal to it, and at the quarter crest lengths the motions are normal to the wall resulting in dunes parallel to it.

The angle of incidence to a structure, for waves arriving from deep water, will depend on their refraction across the continental shelf to the depth adjacent to it. Assuming that the nearshore bed contours are parallel to the beach and the reflecting wall, the angle of obliquity can be obtained, from which Table 9.2 has been prepared. It is seen that even for the large oblique approach angle $\alpha_o = 45°$ in deepwater, a wave of 8 s period (which suffers the least refraction of those included) has an incident angle to the wall of only 29°. As noted in Section 2.2.3, the most predominant swell waves in oceanic conditions are around 12 s, so that with the same deepwater $\alpha_o = 45°$, the crests would be angled at 20° in 30 ft (9.15 m) of water. Structures closer to shore would have smaller obliquities to walls, either natural or man-made, so that these types of angles should be used in model tests.

There are certain experimental difficulties in studying long duration of waves in models with mobile beds with such small incident angles. As seen in Fig. 9.8, reflections from basin boundaries may occur and hence many secondary oscillations could build up after a reasonably short time. The layout in the figure (Silvester 1977, 1985) optimizes the bed area under the influence of the short-crested pattern, with minimal reflection

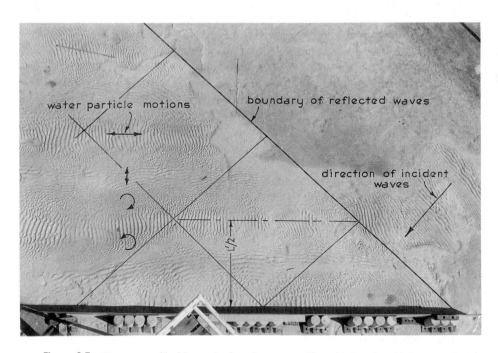

Figure 9.7 Wave crests of incident and reflected waves matching with dune formation on a sandy
bed

TABLE 9.2 WAVE ANGLES AT
STRUCTURE (α) AFTER REFRACTING
FROM DEEP WATER

d (m)	T (s)	α_o (deg.)	α (deg.)
10	14	45	17
10	12	45	20
10	10	45	24
10	8	45	29

problems. However, as noted by the wave-crest diffraction pattern, some waves will be
reflected normally from the wave generator blades, to the right of the figure. These might
set up standing waves which could influence sediment movements in the test area. By
sending waves in batches and stopping the wave machines periodically, this secondary
wave interference can be overcome.

Silvester (1985) conducted tests with walls at angles of 20° and 30° to the wave
crests, some of which were either side of an entrance, which it was desired to maintain
free of shoals. The basin as used is as depicted in Fig. 9.8, where 0.3 mm mean
diameter sand was spread to a uniform thickness of 15 mm. A viewing tower 20 m

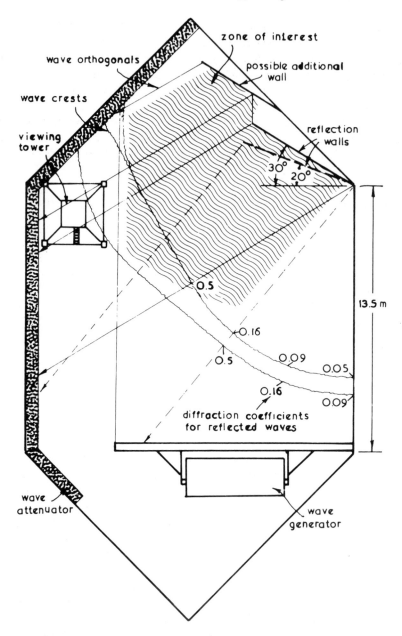

Figure 9.8 Basin layout used to study bed scour from waves reflecting at 20° and 30° to a wall

high was available within the basin, so permitting photographs of the dune formation and zones completely scoured of the sand. The operable water depth was 24 cm and wave period 1.2 s, giving $d/L_o = 0.147$ and a wave length $L = 1.63$ m. The 15 mm sand thickness reduced this length to 1.6 m and if shoaling of an equal amount occurred during transmission of sediment from one area to another, the depth ratio d/L could change to 0.135. This was considered negligible for refraction to occur of either incident or reflected waves. No sand was added during the experiment so that the bed eroded to the limit of the concrete floor. Details of the tests conducted are listed in Table 9.3 and presented in Fig. 9.9. The durations of the waves for each test, as included in the table, excluded the periods of intermission used to overcome wave reflection from the generator blade. After each test the basin was drained and the sand bed photographed from the tower. The concrete floor was painted a color to distinguish it from the white sand, so that areas of erosion were readily identified.

TABLE 9.3   LENGTHS OF WALL AND ANGLES OF WAVE APPROACH WITH VARIOUS ENTRANCE OPENINGS AS IN FIG. 9.9 (SILVESTER 1985)

Test	Upcoast length (m)	α (deg.)	Downcoast lengths (m)	α (deg.)	Opening (m)	Duration (hrs)
a	1.83	20	—	—	—	94
b	3.66	20	—	—	—	138
c	5.50	20	—	—	—	466
d	1.83	20	3.66	—	set back 0.82	50
e	1.83	20	3.66	—	in line 0.82	50
f	1.83	20	3.66	35	set back 0.82	50
g	1.83	30	3.66	20	set back 0.82	26
h	1.83	50	3.66	20	set back 0.82	50
i	1.83	65	3.66	20	set back 0.82	50
j	1.83	30	3.66	30	set back 0.82	26

Figure 9.10 shows the conditions after 94 hours for wave angle as shown in Fig. 9.9a, showing the zones suffering the swiftest scour. These were where the bed dunes are parallel to the wall, or crest distances $Z/L' = 1/4$ and 3/4 from the wall (see Fig. 5.2). Here the water-particle orbits are rectilinear oscillations parallel to the bed and are stronger than in other alignments parallel to the wall, because the wave is approaching that of a standing wave due to the small $20°$ approach angle. For this limited duration the sand has remained in zones where dunes are normal to the wall, or equivalent to the antinodes of these standing waves. In a natural setting, swell wave periods will change with time so that scour over the whole region will occur.

This scouring extended over five batches of dunes parallel to the wall which is equivalent to 2.5 crest lengths. This crest length is $\lambda \cos 20°$ where λ is the length of the incident wave of 1.6 m. Thus scour has taken place for a distance of about 4 m from the wall or about twice the length of the wall. In Fig. 7.4 it is seen that at this $R/B = 2$ the centerline $K = 0.7$, while at its edges it is 0.5. Even so, scour has occurred in Fig. 9.10 to the edge of the ribbon of the reflected waves.

In Fig. 9.11 the wall length has been doubled to 3.66 m (see Fig. 9.9b) and after 138 hours the scour is pronounced. The extent of bed scour has occurred much farther

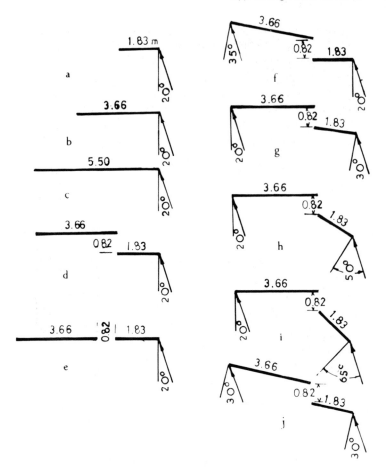

Figure 9.9　The combinations of reflecting walls used to study sediment removal due to oblique wave reflection (Silvester 1985)

out. The conditions for Fig. 9.9c were exhibited in Fig. 7.5, where the wall is 5.5 m long and the duration 466 hours. Sand again remained in alignments $Z/L' = 0$, 1/2, 1 ..., or along island crest propagation. This indicates very little water-particle motion and minimal mass transport. In a complete standing wave, to which this 20° approach angle approaches, there would be zero water movement. This is of little consequence when changing wave periods arrive, as occurs naturally.

For case (d) in Table 9.3 and Fig. 9.9, where two walls are offset to form an opening, the bed conditions are displayed in Fig. 9.12 after 50 hours duration. The obliquity is 20° for both reflecting structures which has occasioned sand to remain at the entrance even though it has been removed for many crest lengths from the walls. Once a channel had been dredged through this modest shoal, it should not be silted as any bars formed from the continued littoral drift would form downcoast of the mouth. Many

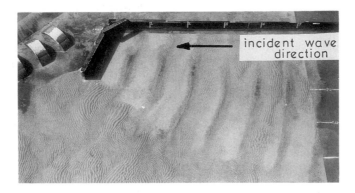

Figure 9.10   Bed scour from 1.83 m wall with 20° wave obliquity after 94 hr duration

Figure 9.11   Bed scour from 3.66 m wall with 20° wave obliquity after 138 hr duration

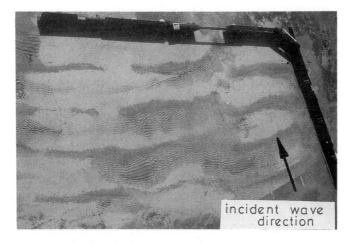

Figure 9.12   Bed scour from walls either side of an opening to represent inlet mouth

factors would not to be taken into account in any prototype situation, but the principle seems to be proven of the effectiveness of using the reflecting bypass system.

The reader is referred to Silvester (1985) for details of other tests as outlined in Table 9.3 and Fig. 9.9, including those with inlet openings between the walls, with other obliquities. This points to the need for conducting three-dimensional models for any breakwaters or other structures that are likely to receive swell waves obliquely. These must have sedimentary beds that can be moved by the modest waves over the whole area. In natural conditions the water orbital motions will be able to disturb the bed in all regions, so this must be the case in any model.

9.8.3 Applications

As indicated in Chapter 7, approach angles in the region of 30° to 60° are most effective in reflecting waves for the purpose of scouring the bed within or just outside the ribbon of reflected waves. The greater the obliquity for a given breakwater length the narrower this band of active removal. Thus, in creating a zone of short-crested waves the structure or structures must be so oriented as to concentrate their attention on shoals predicted to occur seaward of an entrance. This would necessarily be in line with any proposed access channel through the bar.

One suggestion is illustrated in Fig. 9.13, where the main breakwater acts as a headland, which controls the position and orientation of the upcoast beach. The new structure may in fact realign the beach, as indicated in the figure. The littoral drift from this source then comes under the influence of the short-crested waves within the reflection ribbon, with this material traveling parallel to the face of the breakwater. Beyond the tip of the structure a shoal will occur due to the sediment arriving in a zone of only incident wave energy. This cannot handle the sedimentary load unless depths decrease to give waves a better grip on it. Should such a mound be undesirable, particularly if in close proximity to the navigation channel, a second breakwater is also depicted in the figure.

There is an infinite variety of lengths and orientations that could be adopted, some of which should be tested in a three-dimensional mobile bed model. At present the maximum scour depth for a given wave input and water depth cannot be predicted for zero input of sand. Such a theoretical or even empirical solution would serve little purpose as there will always be fluctuating injections of sand, since the problem is one of bypassing material across the mouth. But, there must be some trust in the mechanism that this double application of wave energy to the floor must improve conditions. There is a great need for prototype studies of this mechanism, perhaps by pilot studies just for research.

Another perhaps more realistic case is illustrated in Fig. 9.14, where an inlet is already controlled by jetties either side of it. Depending on the strength of the issuing velocities, from tidal or fluvial action, so the arcuate shaped bar will form close to or away from the opening. The most important location of the bar will be that when it is maximized in size, which will be just after the more severe storms. These may be accompanied by rain which results in flooding and hence pushing of the bar farther offshore. This may not be the case, so the timing of precipitation and high winds should

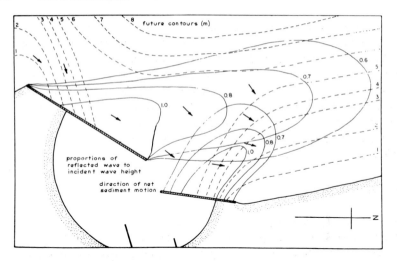

Figure 9.13   Suggested reflecting structures to transmit sediment downcoast without forming shoals in a navigable channel

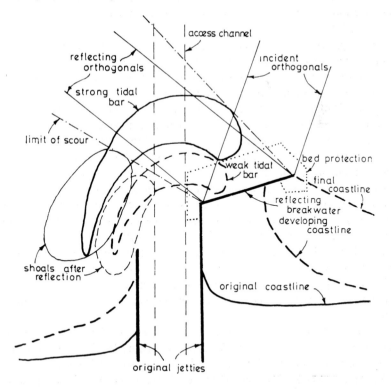

Figure 9.14   Reflecting wall angled to upcoast jetty to concentrate wave energy on bars at location of access channel

be determined in detail. But, do not rely on a uniform input of sediment to the bar, because long periods of moderate waves acting on a swell-built profile will provide minimal littoral drift, as discussed in Chapter 3.

The length and orientation of the reflecting walls in Fig. 9.14 should be decided on the arrival direction of the persistent swell, since these waves can act for long periods in getting rid of material from the desired zones. Whether the strong swell waves just after a storm arrive from this same direction is of little consequence, as their duration will be short. They will be supplying the pulse of material that enlarges the bar swiftly and will be effective in transmitting some segment of the shoal downcoast, but the long-term swell achieves the major goal in respect to the channel area. Also as seen in the figure, a vast accretion will result upcoast from this reflecting wall which will improve navigation temporally, but also erode downcoast beaches. Stabilization by the headland control method is recommended *prior* to construction of the remedial structure.

For these reflecting structures to be most effective they should contain near vertical faces, even though rubble-mound breakwaters can reflect long period swell waves quite effectively, as shown in Chapter 7. The geotextile version, discussed in Section 8.3, is particularly applicable to the wave reflection task discussed above. They need only extend to a height slightly above HWM since these structures, if properly designed, can withstand massive overtopping without units being dislodged. They must be of a width to withstand the impact of incipient breaking waves, in the depths to which they extend.

It is obvious that the bed within the ribbon of reflected waves, particularly that adjacent to the structure, must be protected from scouring, in order to prevent its collapse. For this purpose the geotextile approach is again strongly recommended, as these mattress-like units prevent the orbital vortices from penetrating to the bed and so suspending sand grains for lateral removal. The required thickness of these units for given storm wave characteristics and depth of water has not been determined as yet. The limit would be the depth of the wave trough below SWL, but it is believed it can be smaller than this. As soon as a large flat slab tends to move upward from the floor a suction is created because water cannot readily flow in the minimal space between it and the bed. Mattress units cast in situ follow the bed undulations perfectly and hence restrain water intake very effectively. This phenomenon needs some theoretical and experimental attention.

The suggested area of floor protection is indicated in Fig. 9.14, where it is seen to cease in proximity to the proposed access channel and the existing bar. Of particular importance is the downcoast tip of the reflecting structure, since as noted in Section 7.3, large scale vortices are generated, as crests and troughs of the island waves reach it. The upcoast tip should also be treated because storm waves can arrive from a wide fan of directions, so that similar deep scour could occur at that end. This protective mattress should be installed before the breakwater since scour is very swift. The outer edges of the mattress may suffer some undermining since some scouring is expected beyond its limits. Seams could be sewn lengthwise along these edges, which would permit the subsidence of the outer segment into the trough, but maintaining continual flexible contact with the remainder of the mattress (see Fig. 8.25). The sloping portion then protects the edge against water infiltrating beneath it as the wave troughs tend to lift it from the floor. These are but minor problems that can readily be overcome by a little thought and research.

9.8.4 Effect on Navigation

One of the immediate questions raised when wave reflection is suggested for bypassing material across a river or harbor mouth is that of its effect on navigation of small craft. The short-crested system so generated has island crests double the height of the incident wave. The purported choppy sea is considered to be a hazard to such craft. The island crests should be identified as they propagate parallel to the reflecting wall. A boat handler should therefore be able to avoid these peaks and to traverse the body of water between them. They occur on all alignments at set distances from the wall whilst between these alignments the surface is relatively calm or oscillating vertically very little.

As seen in Fig. 9.15, the craft may commence moving normal to the reflecting wall at the nodal point, or one quarter of the wave length behind crest A. It proceeds across the $L'/2$ distance, which is relatively calm, to arrive at the next island crest orthogonal (EBD) one quarter wave length behind crest C, which was in position B when the craft was at its initial position. Depending on available speed, some other nodal point may

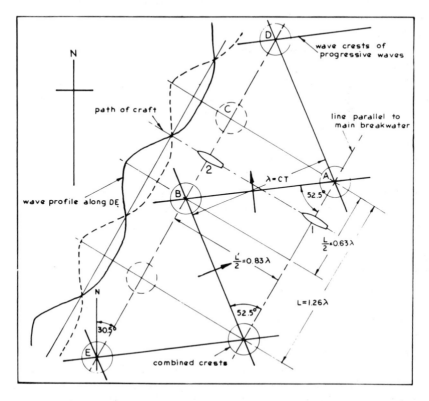

Figure 9.15 Passage of small craft across a short-crested wave system

be chosen. For swell waves these high water mounds should be clearly visible and will be moving parallel to the wall. As the craft proceeds seaward, the crest heights will be reducing and normal long-crested swell will be encountered.

A more realistic picture is presented in Fig. 9.16, where surface contours are presented (Hsu 1979) for two waves angled at 120° to each other, similar to Fig. 2.48. The boat path is shown for stationary waves, but is near normal to the wave orthogonals in the case of wave propagation. It can be seen that the craft crosses troughs to $0.2H_{sc}$ below SWL and heights of $0.2H_{sc}$ above it. It therefore experiences a total vertical motion of $0.3H_{sc}$ or 0.6 of the incident wave height. Even these are spread over a longer time than in riding incoming swell waves. Thus, it is contended, a smoother passage is obtained through such a system than by having to encompass succeeding crests of progressive waves.

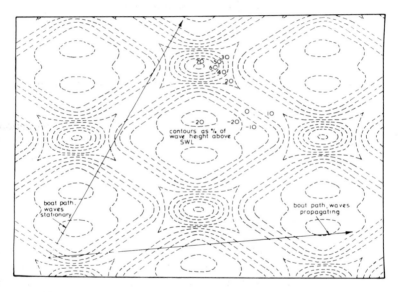

Figure 9.16   Water surface contours in waves angles 120° to each other, showing position of craft when moving normal to the island crest orthogonal

An alternative method of proceeding to the open sea where a breakwater is reflecting swell waves obliquely is illustrated in Fig. 9.17. This is to turn sharply parallel to the structure and travel along the alignment three quarters of a crest length from the face. This will be a zone of minimal vertical movement but will suffer horizontal oscillations, with the passage of the island crests either side. This will cause the craft to traverse a zig-zag course of reasonably small amplitude. This could be unnerving at first but a little education as to what to expect would promote safety. At the upcoast end of the structure, where reflected waves cease, the craft is turned seaward to encounter the incident swell waves, which it must at some stage.

Thus, navigation through these angled wave trains should not be any more difficult than a single train of incident waves. It should not be equated to storm conditions where

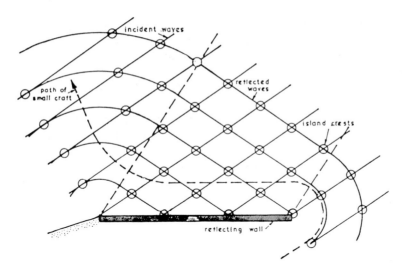

Figure 9.17 Alternative passage for small craft proceeding seaward through a short-crested wave system

waves are also traveling in various directions. In this situation waves of all periods are traveling at all angles so that prediction of peaks in the next instant of time is almost impossible, besides strong winds. With short-crested waves from swell, the island crests are consistent and travel in a predictable path parallel to the seawall.

9.9 UPCOAST ACCRETION

One way to prevent shoaling at an inlet, river, or harbor entrance is to stop the arrival of sediment from upcoast. This can be achieved by means of headlands as detailed throughout this book, but particularly in Chapters 4 and 5. This necessarily means that regions downcoast of such entrances will be devoid of drift and hence will erode. This again may require the treatment of headland control unless they are rocky in nature, can be allowed to recede because of nondevelopment, or have silting problems that can be improved by less input of sand. The complete length of shoreline should be considered, not just the shoaling problem at the entrance.

As seen in Fig. 9.18, a series of headlands can be installed consecutively upcoast of an entrance to form bays in static equilibrium, where shapes are predictable as in Chapter 4. The one nearest to the entrance is constructed first, updrift of which a near triangular shape of beach will be accreted. The next upcoast headland is then inserted, forming a tombolo, which then accumulates another near triangular zone, with a curved beach in its lee. These additions can stop siltation at the entrance for many years, to the limit of coast under the jurisdiction of the agency with the problem. Some port authorities have control over only a limited section of shoreline, while local councils manage small fractions of the sea margin that should be considered in its entirety.

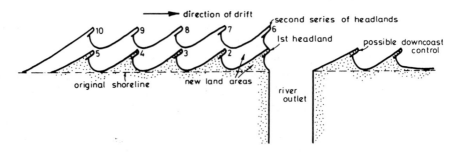

Figure 9.18   Consecutive construction of headlands to accrete littoral drift upcoast from a navigable inlet

Once these geographic limits have been reached a second row of headlands can be inserted seaward of the first, to continue the process for more decades. The costs incurred in their construction are thus spread over many years and hence can be construed as an annual charge. However, this cost must be compared to that for other structures or mechanical means of bypassing, and the fluctuating load of material creating the problem. There is a credit side to this account because valuable land is being gained from the sea, in a location where it can be used. The large areas so accreted can be sold or leased for industrial, residential, or leisure purposes, in the knowledge that the beaches are stable. Even if a small fraction of the previous littoral drift still traverses the headland field the static equilibrium shapes of the bays are known, behind which major infrastructure is kept. In this regard the berm demand during the fiercest storm possible should be kept in mind.

The shoaling entrance may be the mouth of a river which supplies a significant sediment supplement to the downcoast area. It may therefore suffer little denudation from the long-term accretion suggested above. In this regard any possible future reduction in river flow by damming or flood control measures should be known. This, of course, would create its own coastal problems separate from the intervention proposed here.

One example of a possible application of this method of reducing siltation of a port entrance is that of Kaohsiung harbor in Taiwan, as seen in Fig. 9.19. The main port

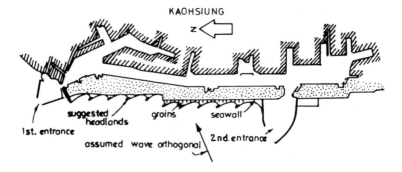

Figure 9.19   Plan of Kaohsiung harbor, Taiwan, showing suggested headland control to stabilize the coast and prevent siltation of the port entrance

area is formed by a barrier beach which has been purportedly protected by a seawall and later groins. However, deepening of the offshore area has occurred due to reflection from the wall. By inserting headlands the resulting stable bays could add land area, which is in such great demand for commercial purposes. If these are allowed to form by progressive construction from the port mouth with natural filling, the silting problem could be overcome for many years. When the field of headlands is complete, to the other port entrance, a second line of the series could be inserted. There are plans to reclaim large areas from the sea to add working area to the port, so the proposed use of headlands with concomitant bays could serve this purpose extremely well.

REFERENCES

BAILARD, J. A., and INMAN, D. L. 1975. Analytical model of duct-flow fluidization. *Proc. Symp. Modelling Tech., ASCE*, 1402–21.

BRUUN, P. 1990. Coastal inlets. *Handbook of Coastal and Ocean Engineering*, Vol. 1, ed. J. B. Herbich. Houston, Texas: Gulf Publ. Co., 829–66.

BRUUN, P., and GERRITSEN, F. 1959. Natural by passing of sand at coastal inlets. *J. Waterways and Harbors Div., ASCE 85*: 75–107.

BRUUN, P., MEHTA, A. J., and JONSSON, I. G. 1978. *Stability of Tidal Inlets: Theory and Engineering.* Amsterdam: Elsevier.

COLLINS, A. G., and PARKS, J. M. 1987. Anna Maria, Florida: case study of sand fluidization for channel maintenance. *Shore and Beach 55* (Apr.): 43–48.

COUGHLAN, P. M., and ROBINSON, D. A. 1990. The Gold Coast seaway, Queensland, Australia. *Shore and Beach 58* (Jan.): 9–16.

DOUGLASS, S. L. 1987. Coastal response to jetties at Murrells Inlet, South Carolina. *Shore and Beach 55*: 21–32.

FITZGERALD, D. M. 1982. Sediment bypassing and mixed energy tidal inlets. *Proc. 18th Inter. Conf. Coastal Eng., ASCE 2*: 1094–1118.

FITZGERALD, D. M., and HAYES, M. O. 1980. Tidal inlet effects on barrier island management. *Proc. Coastal Zone '80, ASCE 3*: 2355–80.

FITZGERALD, D. M., and LEVIN, D. 1981. Hydraulic morphology and sediment transport patterns at Pamet River Inlet, Truro, Mass. *Northeast Geol. 3*: 216–24.

HAYGARD, T., GILMOUR, I. A., and MOTTRAM, W. D. 1969. A proposal to remove sand bars by fluidization. *New Zealand J. Sci. 12*: 851–64.

HODGES, T. K. 1955. Sand bypassing at Hillsboro Inlet, Florida. *Bull. Beach Erosion Board*, U.S. Army Corps of Engrs., Vol. 9 (2).

HSU, J. R. C. 1979. Short-crested water waves. *Ph.D. thesis*, The University of Western Australia.

HUBBARD, D. K., BARWIS, J. H., and NUMMEDAL, D. 1977. Sediment transport in four South Carolina inlets. *Proc. Coastal Sediments '77, ASCE*, 582–601.

HUSTON, J. 1968. Dredging fundamentals. *J. Waterways and Harbors Div., ASCE 93*(WW3): 45–69.

INMAN, D. L., and HARRIS, R. W. 1970. Crater sink sand transfer system. *Proc. 12th Inter. Conf. Coastal Eng., ASCE 1*: 919–33.

KELLY, J. J. 1977. Fluidization applied to sediment transport. I. *Lehigh Univ., Bethlehem, Penn.,* Fritz Eng. Lab. Rep. 710.1.

MCNAIR, E. C. 1976. A sand by-passing system using a jet pump. *Proc. 15th Inter. Conf. Coastal Eng., ASCE 2*: 1342–69.

MURRAY, W. A., and COLLINS, A. G. 1978. Fluidization applied to sediment transport. II. *Lehigh Univ., Bethlehem, Penn.,* Fritz Eng. Lab. Rep. 710.2.

NICHOL, J. M., RICHARDSON, W., and FIRE, S. 1990. Sand bypassing at harbor entrances. *Proc. 27th Inter. Nav. Congress 2*: 145–54.

OERTEL, G. F. 1972. Sediment transport of estuary entrance shoals and the formation of swash platforms. *J. Sed. Petrology 42*: 857–63.

PARKER, N. E. 1979. Weir jetties—their continuing evolution. *Shore and Beach 47* (Oct.): 15–19.

PARKS, J. M., WEISMAN, R. N., and COLLINS, A. G. 1983. Fluidization applied to sediment transport (FAST) as an alternative to maintenance dredging of navigation channels in tidal inlets. *Wastes in the Ocean*, Vol. II, eds. D. R. Kester, B. H. Ketchum, I. W. Duedall, and P. K. Park. New York: John Wiley & Sons.

PEKOR, C. B., and GRIFFIN, E. D. 1978. Jet pump bypass system. *Proc. Coastal Zone '78, ASCE 3*: 2761–71.

RAMBO, G., and CLAUSNER, J. E. 1989. Jet pump sand bypassing, India River Inlet, Delaware. *Dredging Research*, U.S. Army Corps of Engrs., Inform. Exch., Bull., Vol. DRP-89-2.

RICHARDSON, T. W. 1991. Sand-passing. *Handbook of Coastal and Ocean Engineering*, Vol. 1, ed. J. B. Herbich, Gulf Publ. Co., Houston, Texas, 809–28.

RICHARDSON, T. W., and MCNAIR, E. C. 1981. A guide to the planning and hydraulic design of jet pump remedial sand bypassing systems. U.S. Army Corps of Engrs., *Coastal Eng. Res. Center*, Waterways Exp. Stn., *Rep.* HL-81-1.

SEABERGH, W. C. 1983. Weir jetty performance: hydraulic and sedimentary considerations. U.S. Army Corps of Engrs., *Coastal Eng. Res. Center*, Waterways Exp. Stn., Hyd. Lab. Tech. Rep. HL-83-5.

SILVESTER, R. 1960. The water-jet pump—its uses in hydro-electric schemes. *Water Power (London) 12*: 176–80.

———. 1961. Characteristics and applications of the water-jet pump. *La Houille Blanche 16*: 451–60.

———. 1974. *Coastal Engineering, Vols 1 & 2*. Amsterdam: Elsevier.

———. 1975. Sediment transmission across entrances by natural means. *Proc. 16th Congress IAHR 1*: 145–56.

———. 1977. The role of wave reflection in coastal processes. *Proc. Coastal Sediments '77, ASCE*, 639–54.

———. 1978. Diffraction through a breakwater gap. *Proc. 4th Aust. Conf. Coastal Eng.*, 128–31.

———. 1985. Sediment bypassing across coastal inlets by natural means. *Coastal Eng., 9*: 327–45.

SILVESTER, R., and MUELLER, M. H. G. 1968. Design data for the liquid-liquid jet pump. *J. Hyd. Res., IAHR 6*: 129–62.

SILVESTER, R., and VONGVISESSOMJAI, S. 1971. Characteristics of the jet-pump with liquids of different density. *Proc. 3rd World Dredging Conf.*, 293–315.

SNETZER, R. E. 1969. Jetty-weir systems at inlets in the Mobile Engineer District. *Shore and Beach 37* (Apr.): 28–32.

U.S. Army Corps of Engineers. 1971. *Natural Shoreline Study*. Regional Inventory Report, South Atlantic-Gulf Region, Jacksonville District, Jacksonville, Florida.

————. 1984. *Experimental sand-bypass system at Oceanside Harbor, Oceanside, San Diego, California*. Phase 3 Final Rep. prepared by Moffat and Nichol, Engineers, contract DACW09-82-C-0038.

WALTON, T. L., THOMAS, J. L., and DICKEY, M. D. 1985. Cross-shore distribution of sedimental transport at a weir jetty. *Proc. 7th Aust. Conf. Coastal and Ocean Eng. 1*: 525–36.

WALTON, T. L., and ADAMS, W. D. 1976. Capacity of inlet outer bars to store sand. *Proc. 15th Inter. Conf. Coastal Eng., ASCE 2*: 1919–36.

WEGGEL, J. R. 1981. Weir sand-passing systems. U.S. Army Corps of Engineers., *Coastal Eng. Res. Center*, Waterways Exp. Stn., Special Rep. 8.

WEISMAN, R. M., and COLLINS, A. G. 1979. Stabilization of tidal inlet channels - design recommendations. *Lehigh Univ., Bethlehem, Penn.*, Fritz Eng. Lab. Rep. 710.3.

WEISMAN, R. M., and COLLINS, A. G., and PARKS, J. M. 1980. Stabilization of tidal inlet channels—Corson Inlet field test. *Lehigh Univ., Bethlehem, Penn.*, Fritz Eng. Lab. Rep. 710.4.

————. 1982. Maintaining tidal inlets by fluidization. *J. Waterway, Port, Coastal and Ocean Div., ASCE 108*(WW4): 526–38.

YAJIMA, M., UEZONO, A., YAUCHI, T., and YAMADA, F. 1983. Application of sand bypassing to Amanohashidate beach. *Coastal Eng. in Japan 26*: 151–62.

What Direction
Coastal Engineering

From the indictment of scientific experts made in 1966, quoted in Chapter 5 (*Effective Use of the Sea* 1966), the present-day progress must be examined to see where research effort should be placed in the future. There are certain proven principles regarding beach processes and offshore structures that can guide and indicate new measures to be taken. The need is to keep open minds and not throw out any novel suggestion until all aspects of feasibility and economics have been examined. There are data collection programs and modeling procedures that should be questioned for their applicability to the problem being solved. This chapter discusses some issues with a view to making clients and consultants alike more conscious of the dollars spent in research and development in this fast developing field.

10.1 INTRODUCTION

There is a continual tendency for neophytes in any professional field to copy their superiors in thinking and in applying knowledge to practical problems, even when they are not being forced so to do. Introducing new concepts takes a great deal of stamina and self-conviction. It is easier to argue a standard procedure than to establish a novel one. But in following the easier path ingenuity, embodied in the title engineer, may be abdicated. In coastal engineering today there is a dire need for novelty and experimentation.

There are some basic principles which emerge from research and theory, and even from common observation of natural features, that may be accepted under many meteorological, geological, and hydrographic conditions. Such axioms should aid further investigations. These might be stated as follows:

1. Along all shorelines there is a direction (generally oblique) from which most wave energy arrives.

2. The longshore component of the energy, as in 1 above, can transport a specific load of sediment both inside and beyond the surf zone.

3. Littoral drift and offshore transport is not a continuous process but fluctuates seasonally and over longer terms.

4. Storm wave transfer of sand from the beach berm to an offshore bar, and its return by swell, creates a continual change in beach profiles which varies significantly between oceanic and enclosed sea margins.

5. The infrequent erosion of the beach berm, as in 4 above, is an event that cannot be prevented by man, but is only transient, with the berm being present for the majority of the year, on oceanic coasts at least.

6. The formation of the offshore bar, as in 4 above, serves as a protection in limiting the erosion that takes place during a storm sequence.

7. Short-crested waves created by oblique reflection, and even in the generation process, expedite the transport of sediment through magnified orbital water particle velocities and macroturbulence.

8. The more normal the approach of waves to a beach, the flatter its offshore profile for waves to transmit the drift by the longshore energy available for it.

9. Groins have no influence on the erosion of the seabed beyond their tips and material collected between them is subject to wide dispersal due to rip currents generated during storms.

There are many more phenomena probably proven beyond reasonable doubt that can be put in the memory of the coastal computer, ready for any program to overcome a specific problem. It is these basic concepts that should be discussed at international conferences rather than solutions to given problems. Even if the latter are comprehensive enough to warrant presentation, the principles involved should take precedence over details of construction or design.

As Lee (1983) has stated: "Yet, it is clear that understanding the basic coastal processes was, and in the future will be, the key to good coastal engineering." Several quotations will be made in this chapter from the workshop conducted in Hawaii in 1982. It had 26 experts from around the world, formed into groups which concentrated on certain subjects deemed to require research attention. Summaries of their conclusions were presented without too much detail. Many topics presented in this chapter were not addressed, presumably because they were considered of little or no importance.

10.2 COSTS

Many modern activities suffer an exponential growth, such as power production. Marine structures for port development and coastal protection are also increasing in number, size, and hence cost. As new harbors, marinas, and fishing-craft havens are planned, more difficult sites must be utilized. Larger ships also demand deeper channels and larger basins. Breakwaters are running in ever deeper water and their extensions are resulting in greater interruption to sediment flow along shorelines, so causing erosion in downcoast areas. This attack on beaches comes at a time of increased demand for recreational and commercial facilities and hence conflicts of interest become severe.

Although structures in the form of breakwaters running out from the shore are obvious interrupters to littoral drift, the more insidious impediment is the channel dredged as an entrance to a port. As drafts of bulk carriers grow so these depressions in the seabed extend farther seaward. Any material caught in them from along the coast by wave action is removed and dumped out to sea for convenience. These channels thus become longer sinks for drift, to their outer limit which may be some kilometers offshore. This is one situation where littoral drift can cease absolutely, with consequent beach erosion for long distances downcoast.

Costs on coastlines can be divided into those necessary and those unnecessary. The former result from activities and structures that have been successfully planned to achieve a desired result. The latter emanate from mistakes made due to misunderstanding of the phenomenon involved or applying old staid procedures, even when these have proven to be failures elsewhere. Sometimes solutions in one place are adopted when the environmental conditions are different at another. There is a conservatism in the engineering profession, verging on a state of stubbornness, that will not consider an inventive scheme, no matter how well documented and rationalized.

10.3 NEED FOR REAPPRAISAL

Forgetting the mistakes made in the past, it is the future that must be considered when the question arises as to whether progress is being made in the right direction. There have been murmurs of discontent over certain types of shoreline structures and management (Edge et al. 1982, Hsu and Silvester 1989), but the most vociferous have emanated from environmentalists and geologists. As noted in Appendix B of Lee (1983): "Most geologists are against the present shoreline management policy of using stabilization techniques." But it is not fair to generalize in this manner because most geologists are not interested in the coast, and the few who are may be antagonistic. Unfortunately, few of these protagonists have received adequate training in wave action and fluid flow and hence cannot appreciate alternative solutions when they become available.

It is no good burying heads in the beach sand, or what is left of it, and deciding to leave the coastline to natural processes, as is proposed by some states in America. It is better to observe Nature and see where help can be provided to achieve results efficiently and economically. Officers of these government agencies should have their antennae out seeking messages on new methods and materials and pursue their relevance to the limit of financial and research resources available.

The present decision to ban further marine structures probably came from one eminent scientist, who had the ear of a powerful politician, with little knowledge of economics of the local situation or of the State as a whole. Such a drastic proposal should have a national or even an international enquiry at the highest level. To organize a conference or workshop to which worldwide experts are subsidized to attend would cost a fraction of the destruction suffered on a section of coast during a storm. The belief that all experts in any field are born and educated in one country is promoted more in the engineering profession than in the arts and sciences.

Coastal-defense structures in the past few years have been noteworthy for their magnitude (bigger is better), be it groins, seawalls, or breakwaters. Papers describing these mammoth features contain intricate details of concrete members but generally omit information or reasoning in respect to wave climate, sediment supply, or ultimate possible reshaping of the coast. With a surfeit of funds engineers can provide an answer to any question, but the point is "are the right questions being asked." Such questions might be:

1. Why is the movement of longshore drift not uniform throughout the year, as envisaged by calculations of energy flux from some average wave at some assumed obliquity to the coast?
2. Why does the shoreline want to reorient itself continually to the crests of incoming waves?
3. Can Nature be assisted in accomplishing tasks in shoreline stabilization?

Engineers need more flights of fancy, as is done in a *think tank* situation. There is a great need for *lateral thinking* instead of the old established *littoral thinking* in the coastal field (Silvester and Hsu 1991). The belief that Nature works uniformly, the "river of sand" concept, needs a drastic reassessment (Tanner 1987). Study of weather patterns should have proved that energy is applied to the sea in concentrated doses. The resulting storm waves are applied to the coast in batches while the swell emerging from distant storm centers is spread relatively uniformly along the continental margins. The interaction of these two systems effects a pulsational movement of material along the coast.

10.4 WAVES

As noted in Sections 2.3.3 and 7.2, short-crested waves are very effective in removing and transporting sand on the seabed and therefore knowledge of their characteristics is paramount if information is to be applied correctly. The work of Hsu (1979) and Hsu et al. (1979, 1980) has enhanced this expertise greatly, but its application in model studies and prototype monitoring needs much more attention. Breakwaters and other structures which are angled to the persistent swell should be physically modeled on large enough scale with a mobile bed, in order that the resulting scour can be observed over the full area of reflected wave action. Even if this swell arrives essentially normal to the structure, any flume tests should not be conducted with fixed beds. The extra cost of such verification is minimal compared to the costs of not undertaking it.

The reduction, due to diffraction, in reflected waves as they propagate seaward should be determined either analytically or by modeling, to find out their sediment transporting capability. All mathematical solutions so far have assumed equal heights of incident and reflected components. There is a need for kinematics of water-particle motions to be ascertained for differences in these energy levels. Prototype measurements of scour would also assist in the potential for pore pressure buildup and subsidence of the breakwater face.

Another topic in wave phenomena is that of mass transport, particularly that occurring near the seabed, where the bulk of material is in suspension or pseudo-suspension. Some researchers have given attention to this, but only for oblique reflection against vertical walls where reflected waves almost equal in height those of the incident waves. This net movement is two or three times greater than for the incoming progressive waves for complete short-crested waves, but could be different for a partial system.

This topic received attention in Lee (1983): "Even for the relatively fundamental case of regular waves propagating over a horizontal seabed, further research is needed to quantify the mass transport characteristics for the complete range of wave conditions and for both smooth and rough seabeds." This working group also noted the need to investigate the wind induced currents at the surface, which cannot be studied in models. They also discussed differences between prototype and model in the form of Reynolds number, three-dimensional effects, and sediment suspension. "Attempts should be made to conduct experiments which more closely simulate field conditions so that discrepancies between laboratory and field data can be reduced."

10.5 BEACH PROCESSES

This topic was discussed in Chapter 3, where several new ideas were introduced, for which further research is warranted. Looking at the region beyond the breaker zone, the oscillatory nature of the water particles, either in progressive or partially formed short-crested waves, results in ripple formation on the bed. These stabilize after some time from a constant wave input but will change in spacing when height or period changes, which is continual for swell waves. While assuming new crest positions the bed is more disturbed, so resulting in greater transport. The same occurs if the ripples are reoriented by a different approach direction of the waves. This topic is alluded to in Lee (1983): "Ripples which appear very commonly in the shallow water region play an important role in sediment transport."

Within the surf zone, where the bulk of littoral drift occurs, the interaction of storm and swell waves has received little attention. The transverse motion of material becomes significant because what was previously buried in the berm formed by previous swell is now within the reach of a new batch of swell. If this is oblique to the coast the returning bar material not only moves toward the beach but also along it in a massive pulse (Smith and Gordon 1980). Further verification of this pulsational movement is required by measuring drift volumes against some structure, or at the extremity of a spit, while profiles are taken of the offshore bar until it reaches the fully restored berm. This monitoring entails measurements daily for perhaps a week, since on oceanic shorelines, particularly western coasts, the complete refurbishment may take only a few days after the cessation of the storm.

This topic received attention in Lee (1983) as follows: "Additionally, the working group recognized the urgent need for a rapidly deployable, beach survey system. Such a system would be very useful in measuring shoreline elevation and profile changes on a routine basis, but it would be particularly desirable for documentation of large beach changes as soon as possible after severe or catastrophic storms."

It is well known that the offshore profile of a swell-built beach assumes a parabolic shape, but the constants and exponents in such equations have not been readily related to the wave climate of the area. Profiles are also dependent on the general size of sediment in the area so that the picture can become complicated. But these profiles determine how much sediment will be required to form a bar large enough to break incoming storm waves and so limit the erosion that can take place in any storm. This topic was discussed in Section 3.4.3.

The volume and characteristics of the bar formed by certain storm wave characteristics is discussed in the same section and also in Section 3.4.4. Much more data are required to confirm or dispute these relationships, so that the work of the authors referenced therein, conducting mainly flume studies, can be proven in some way in the field. This applies also to the volume of berm erosion that has been equated to that of the bar. It appeared correct for the large wave tank tests, but field conditions involve three-dimensional effects. This research will involve much manpower, available at short notice when a storm is imminent.

The topic of marine cliff erosion has received scant attention, both by geologists and engineers. It is mainly descriptive and presents data on existing conditions. Little discussion has taken place on stabilization procedures, even though it is admitted that a fronting beach will stop further degradation. There is a need to investigate the economics of placing structures such as headlands, in order that dynamic or static equilibrium bays can be maintained seaward of these steep but fragmenting scarps. This is more development than research but authorities need to be alerted to new possibilities.

10.6 GEOMORPHOLOGY OF COASTS

This is the subject of Chapter 4 in which three main issues were discussed, namely, crenulate shaped bays, physiographic units, and barrier beaches. These phenomena require much more data collection to verify the hypotheses put forward or to prove alternatives. The crenulate shaped bay, for example, in static equilibrium can be predicted for a known wave obliquity (Hsu et al. 1989a, 1989b, 1989c). More prototype bays need to be measured, which can be accepted as stable from the sand supply conditions, to refine the general relationships derived by Hsu and Evans (1989).

The use of bay shapes, as well as sea and swell data, in deriving directions of net sediment movement around continental margins has been undertaken by only one of these authors. The physiographic units so demarcated have been disputed on only one or two occasions for very small sections of coast. Other workers have derived drift directions which appear to change within small segments of the shoreline, not realizing that similar transport would have taken place over thousands of years. This implies that if adjacent coasts have apparent net movements toward each other then a larger shoal or a sandy protuberance should be the result, which has not always been the case.

The need to understand coastal changes over long periods was alluded to by Lee (1983): "In order to develop predictive capacity for describing beach evolution, it is important to understand fully the overall phenomenon by studying both short-term and long-term changes in the field." Reference is also made to the need for comprehensive studies to complement the ad hoc investigations of particular problems. "Basic research can be carried on in the spirit of mutual cooperation and competition between individual research groups, but experience has shown that many field projects are best performed through a fully coordinated and cooperative framework combining many specialists. This is because isolated field measurements have limited value in comparison to synoptic and simultaneous observations of the meteorological conditions, wave and current conditions, and geomorphology of the area, as in the case of a sand transport experiment, for example."

Barrier beaches or barrier islands are very dynamic areas which attract tourists and leisure seekers who want to occupy these coastal plains almost to the waterline. Their stabilization by some means other than the traditional requires attention, for which headland control is strongly suggested. These could be tried full scale with pilot structures which could be installed quite cheaply and removed if necessary. The mode of construction of these spits, which turn into mammoth barrier beaches, needs more attention. Their suggested origin in Section 4.4 should be proven or disputed by measurements as suggested in Section 10.5 above.

10.7 COASTAL DEFENSE

Erosion taking place on one section of coast and the concomitant accretion occurring elsewhere is due to the longer-term supply conditions, even though the less frequent lateral movement due to storm action is a significant part of this picture. This denudation is not uniform due to humps of sand traveling along the coast which cause fluctuations in the waterline over periods of months. Researchers taking spot measurements in time and place need to employ a more macroscopic view of the situation.

Hydrographic surveys are required at reasonably frequent intervals after any structure is placed on the coast, to see whether it is achieving the desired results. When these checks have been made the papers reporting these developments should be written, even though the achievements may be negative. The profession will learn more from these longer assessments than just descriptions of construction details.

10.7.1 Dune Stabilization

The accumulation of dunes along a coast indicates steady accretion or at least a steady state of sand supply. This is not to say that at times of storms with excessively high sea levels, perhaps due to storm surge, waves will not attack the foredune or first major dune. However, this is very transient as the resulting beach berm will be wider than usual. Coastal engineers need to ascertain how often these severe conditions can occur along any section of coast.

Means have been provided in Chapter 3 to determine the berm recession for a given storm wave input. These relationships are based solely on large wave tank tests with a slight reference to prototype conditions. There is a need for field and model results to be correlated, which involves measuring variables that are compatible. The varying sea conditions on the coast are difficult to compare with the constant monochromatic or even spectral waves utilized in flume tests.

Although dunes are accepted as God's given protection of shorelines, or certainly as a buffer against fluctuations, they are often protected as soon as they commence to disappear. Vegetation is introduced as one means of preventing their collapse into the sea. The root system causes the sea face to stand up more vertically, but this magnifies the reflection of breaking or broken waves that reach them. Researchers need to make field comparisons between erosive capability of natural and well-vegetated dunes under similar wave conditions. Does one type stand up better than another?

10.7.2 Previous Defense Solutions

As with the design of precast concrete blocks for armoring breakwaters, there have been many stabilization procedures adopted because others have been found wanting. Along many coasts seawalls have been installed, followed by a field of groins, and these are then encompassed by offshore breakwaters constructed further out to sea. But still the erosion continues of the very narrow beaches left, if any. Engineers have relied on bigness without fully explaining to themselves or others why the previous solution did not meet their expectations.

The above and other items will be discussed in turn, outlining the misconceptions of the past, as the current authors see them, with suggestions for research of these and new concepts. While it is normal to omit the phrase "I think" when writing professionally, the reader could insert it here, in order that they can come up with their own conclusions "I agree" or "I disagree," but including worthwhile reasons for such fixations. The outcome may be a novel solution better than any presented previously, and that is progress.

Seawalls.   The tests on seawalls or stone revetments have generally been conducted to determine the stability of their various components and maybe to reduce wave reflection. They are normally in flumes which immediately assumes the major waves arrive normal to the structure, which seldom is the case for extended periods. The more disturbing fact is that these model structures rest on a fixed bed so that the interaction of the sandy floor and the reflecting face is not examined. Even though the persistent swell and storm waves may arrive essentially normal to a structure, seabed erosion can occur relatively close to it and hence aid in its subsidence. This action is magnified many times should the more persistent swell waves arrive obliquely, as discussed in Chapters 5 and 7.

This indicates that all testing programs on seawalls and breakwaters should include model basins in which waves from a predominant direction, either oblique or normal to

526   What Direction Coastal Engineering   Chap. 10

the structure, should be tested with mobile beds. These must be to a scale where reflected waves are maintained over sufficient area where any erosion may influence the structure or adjacent beaches. This may involve the use of light-weight sediment because the model waves must be able to transport material at all points that prototype waves can move sand of the size indigenous to the area. Such models will be more expensive than flume tests, but they are necessary if an overall picture is to be obtained.

The complex water-particle motions in the resulting short-crested wave system generated by oblique reflection cause spatial differences in bed erosion for a given wave period. However, if these are monochromatic swell waves they should be varied in period since the mounds constructed parallel to the reflecting face will change in position. This implies that the complete floor can have material removed. As noted in Section 3.2.2, swell waves vary in height and period continually as various sections of the storm wave spectrum arrive. This does not necessarily mean the need to use spectral components of even a narrow spectrum simultaneously. They could be used consecutively, which helps in the assessment of the worst wave period that produces the scouring.

The reflected waves along the orthogonals from the wall limits diffract, so taking their toll of the wave energy between these bounding orthogonals, which ultimately becomes insufficient to add to that of the incident waves which are still traversing the offshore zone. The orientation of the net sediment movement is dictated by the balance of energy of these two wave components, which is difficult to determine with the present state of theory. The diffracting waves in the downcoast region will also suffer refraction toward the coast and so maintain a short-crested wave system offshore from the beaches beyond the extremity of the wall. This expedites sand movement so that the bed is deepened and very quickly the shoreline is recessed, as discussed in Chapter 3. All these possibilities must be kept in mind by engineers investigating the influence of seawalls on the coast.

As noted in Section 7.3, the short-crested wave crests arriving at the downcoast tip of marine structures can generate large scale vortices, which penetrate to the floor. They scour large holes in this region which can cause subsidence of the structure, generally blamed on tidal currents. This phenomenon, like most regarding oblique wave reflection, requires much more research, particularly in prototype situations where water-particle orbits and sand size are compatible. Research is also required into bed protection measures, since oblique reflection is a fact of life that has to be dealt with. This topic was discussed in Section 8.3, where geotextiles filled with grout were seen to be preferable to stone filling. All these aspects need attention.

Groins.   As noted previously, one way to reduce littoral drift is to orient the beach parallel to the incoming wave crests. Groins achieve this to a certain degree, dependent on their spacing and length. However, they are generally taken out only to the wave breaker line of the original beach, perhaps under storm wave conditions but generally less. Beyond their tips denudation continues if sand supply is insufficient to cope with downcoast transmission. This steepens the offshore profiles, which ultimately demands more material for the construction of the protective offshore bar. Erosion of the beaches between the groins is inevitable.

But as noted in Section 5.3.2, storm waves generate rip currents along the exposed downcoast faces of groins which disperse the berm material well out to sea, some of which remains there to make up for the continual erosion taking place there. Filling of the groin field is then required, which means that the downcoast region is devoid of sediment and so erosion ensues. All these facts must be considered when designing a groin system. There are locations that are almost stable where a groin can widen a beach for some useful purpose. But where erosion is rampant groins will not help but, in fact, exacerbate the problem. Lee (1983) states: "The present theory of groins is often inadequate because the important effect of wave reflection is not included." Modeling of groins requires storm wave input from many directions as well as the swell from some predominant direction.

Offshore breakwaters. These structures have received a great impetus from their proliferation throughout Japan, from the pressure of one engineer in a powerful position to promote them. However, they have been copied even though he was honest enough to report fully on their deficiencies. They are expensive to construct offshore, generally from precast concrete armor units, and are virtually a revetment along the full coastline, except for the small gaps in between units that run parallel to the original shoreline.

As noted in Section 5.3.3, the salients formed on their leeside can be lost due to the strong rip currents formed during storms, requiring extra defenses at the gaps between structures. Although these authors have provided useful relationships for the dimensions and shapes of these sand protuberances (see Section 6.2), they do not recommend this solution except for exceptional circumstances. However, for their application further research is necessary for the case of multiple breakwaters since prototype conditions can vary greatly. Lee (1983) has summarized the situation as follows: "More study is needed of submerged and detached breakwaters."

Perched beaches. Very little research has been reported on these structures, which purport to overcome the problem of the infrequent transient erosion of the berm. As stated previously, this cannot be overcome by man. They act to obviate littoral drift which they do not accomplish very well, but have the extra expense of maintenance due to the erosion beyond the offshore structure which results in structural subsidence. Perhaps with normal wave approach, for which most of the model tests are conducted, a perched beach can widen a working beach, but to overcome erosion due to oblique wave approach they cannot be recommended. Any research should involve the beaches both upcoast and downcoast of the installation and not flume studies.

Beach renourishment. This more recent development has come about due to the ability of man to deposit large volumes of material on a beach within the storm built profile. As recalled from Section 5.3.5 this added material is removed more swiftly than that on natural berms and generally has to be replaced at 2 or 3 yearly intervals. The variables involved are many but the most important one, of the number of storms experienced each year, does not seem to be taken into account. As stated by Lee (1983): "Artificial beach nourishment is done usually at great expense, but the result is seldom monitored for important data on this subject."

The failures of these projects has been emphasized in this book in order to alert engineers to the side effects that are omitted from most papers reporting them. These are generally written soon after completion so that long term effects are not exposed. Those that have included studies over several years always concluded that the expenditure in maintenance was much greater than expected. Although the procedure acknowledges that the beach is the best means of coastal defense, there would appear to be other methods that are preferable, which may include reclamation.

The best conditions for utilizing beach renourishment is where a stable beach already exists or where littoral drift is practically zero. Widening of a beach to preserve existing infrastructure can then experience a new constant waterline more seaward than the original. But to overcome erosion it is an expensive pill, to take away the pain and not solve the original disease of lack of sand supply. The side effects of the drug must be remembered, of causing siltation in harbors and river mouths downcoast.

Further research should include a demarcation of the most important variables and a consistency between observers. Comparisons should be made on one beach with successive refills to see which wave conditions remove them more swiftly. A design for other sites might then employ these data, although every location has its own unique wave climate, tidal current, and storm surge conditions. Much thought is required before this solution is employed because it has many drawbacks.

10.7.3 Headland Control

Research on this concept as a means for stabilizing shorelines has extended over 30 years with little acceptance by the profession. This may have been due to the initial use of the logarithmic spiral as the preferred geometric shape of the bays formed between headlands, which was difficult to apply in practice. It has since been found wanting in accuracy, especially for the downcoast segment which is straighter (Section 4.2.2). The more applicable parabolic shape has been derived by these authors in the form of equations, graphs, and tables, which should make this solution more acceptable, as in Section 4.2.3 (Hsu et al. 1989a, 1989b, 1989c, Hsu and Evans 1989).

Data from model tests and prototype bays known to be in static equilibrium have provided the dimensionless variables employed in the predictive relationships. Although being empirical they are extremely useful. Mathematical derivations based on wave energy, refraction, and diffraction could confirm the equations derived, but in the end it is observation of Nature that must be respected.

The few prototype applications have shown the practicality and economy of utilizing headlands to stabilize a coast, but more would be useful to promote their use. As can be gauged from this tome, these authors believe that headland control is the preferred solution for beach erosion as it copies Nature so well. Although the relationships can only predict the stable shape of bay, this knowledge is useful in locating more permanent structures on the coast where future erosion will not put them in jeopardy. Even though a bay may be in dynamic equilibrium at present, where sediment input equals sediment outlay, the future may not be so bright. Even today the littoral drift can fluctuate so much that bays can erode back to their limiting indentation temporarily.

Existing bays need to be tested for stability, in order to predict their long-term trends. When upcoast supply conditions change, due to longshore drift impediments being implemented, the final bay shape can be predicted. These new bays between headlands installed on the coast should be designed for complete stability, even though existing drift may remain the same for years. This applies to straight, bayed, or convex coasts alike, with full knowledge of the possible wave approach directions.

Man-made islands for chemical complexes, nuclear power stations, airports, factory sites, or ports are becoming more feasible. These are likely to be surrounded by revetment walls which will reflect waves with scouring consequences and large maintenance costs. The alternatives of headlands and stable bays are worthwhile considering for such situations, since many islands around the world have bayed shorelines that have been stable for thousands of years at least. Any authorities planning this type of development, no matter what size, should give consideration to this type of stabilization, as discussed in Section 6.11. The same area as a rectangular or oval shape could be achieved with the different plan dictated by the several bays around its periphery.

10.7.4 Cohesive Soils

Muddy coasts provide many problems because of their mobility and their inability to support heavy structures. This may mean using floating supports for membranes made of strong fabrics, although these are likely to be stolen by locals for use in their dwellings. Components such as bamboo could be used which are readily available in the region to obviate this theft. Another method of overcoming erosion problems on muddy coasts is to reclaim the area with sand. This can be expensive and needs much care in preventing the mass turning over and so losing the sand to the bottom. But, if this can be overcome by simultaneously draining the water from the mud, the new shoreline should be normal to the orthogonals of waves that can reach the shoreline (Pinchin and Nairn 1986). This therefore involves a bayed shape, for waves to arrive at right angles to the whole periphery.

Much work is required on the attenuation of waves through suspended mud and also the mass transport as far as waves can penetrate. Exchange of information is required between nations with good knowledge of cohesive soils, such as China, and other countries with less experience. The theory is fairly well developed but the experimental and field experience is more limited. There are vast expanses of such shorelines which are at present not used commercially because of the difficult structural conditions, but they may become more in demand as minerals are sought in such areas, for example.

This view was expressed in Lee (1983): "Very little is known about the dynamics of muddy coasts, so there is great need to study cohesive materials, a study which requires expertise in physical chemistry and biology. Muddy coasts occupy a large percentage of the world's shoreline, but compared to the work on sandy coasts, relatively few investigations of them have been made."

10.7.5 Shingle Beaches

Great advances have recently been made in predicting characteristics of these pebble structures, but based mainly on large scale flume studies. More research effort is required in getting field data and relating it to the wave climate. Such shorelines are very much more stable than sandy beaches because longshore drift is less. Any bays formed between headlands by pebbles are likely to be in static equilibrium as the coarse material will form a waterline parallel to the wave crests of the predominant incident waves. There is no doubt that shingle beaches will be used as defense mechanisms just as berm breakwaters are being considered.

10.8 EFFECTS OF MARITIME STRUCTURES

The chief effect to be considered here is that of scour in front and downcoast of the structure which may be a breakwater, groin, or plain seawall. This topic was alluded to in Section 10.7 above, so that little elaboration is necessary. Many breakwaters are angled to the coast so that oblique reflection often occurs. The large scale scouring that ensues prompted Japanese engineers to embark on a large scale research program, which has been enlarged on in Chapter 7.

Although the kinematics of short-crested waves formed by angled waves of equal height and period is well known, those for components with different height and period are not. Neither theory nor practical measurement of water-particle orbits in this complex situation are available. The next complication of introducing sediment suspensions and bed load needs attention. These data may help explain some of the complex bed forms that are observed on the coast and around structures at different times. However, this topic was not mentioned specifically in Lee (1983).

10.9 ALTERNATIVES TO NORMAL BREAKWATERS

Breakwaters generally comprise a rubble core, with larger stone on the face as an underlayer, which is overlain by armor units of very much larger size for two or more layers. It is stability of these more massive blocks that has been the concern of engineers because they have to withstand the full brunt of the storm waves. Many designs have appeared for these precast concrete units indicating that any particular shape has not provided the ideal solution. They are meant to dissipate wave energy besides prevent removal of the underlayer or core material.

Little thought has been given to the sandy base of the structure which is subject to wave action as much as its rock constituents. The partial standing waves or short-crested system set up by reflection are conducive to scour. These steep waves also build up pore pressure in bed material adjacent to the face which can result in liquefaction. This may cause the collapse of the small stone mound supporting armor units farther up the face which allows the armor units to slide, so permitting wave attack on the underlayer and even core (Silvester and Hsu 1989).

One way of overcoming this scour potential is to have the waves break properly, which can be achieved by using smaller stone which is shaped into a beach form. This concept of the *berm breakwater* does not quite achieve the desired result because the removed material deposits at the foot at almost the angle of repose for the rock. This is very steep and can still reflect the lower water column motions almost 100%. Partial standing waves are therefore still possible.

Another way to obviate reflection is to use a pure sand barrier as the breakwater. The requisite in this case is to prevent longshore movement of the material, else it would quickly be lost to the structure. This is where headlands and bay features can play a part, because as noted throughout this book, predictable stable bays will have no net littoral drift. Prebreaking on the beaches overcomes much of the spray generation that occurs with normal breakwaters.

A new type of construction provides large and more stable units throughout the breakwater and overcomes stability problems of large blocks resting on steep seaward slopes. This utilizes geotextiles in the form of skins into which mortar is pumped to form elongated units of large dimensions. These are cast in situ so that each has intimate contact with its neighbor below it. This prevents rocking so that the mixture strength can be very modest.

Where floating structures are demanded, such as in muddy conditions, the preferred version is a horizontal plate held a small distance below the water surface. This causes reflection and breaking and hence wave attenuation on the leeward side. It could also serve as a temporary headland when a bay is being formed out to its stable shape, after which a fixed structure is installed.

10.9.1 Berm Breakwater

This topic has received attention at the workshop conducted by the National Research Council of Canada in 1987, of which the conclusions were summarized by Anglin et al. (1987). The items considered in the roundtable discussion of research needs were:

1. **Design procedures:** It was agreed that hydraulic modeling was preferred either of actual site considerations or of tests on various sized stone and theory.

2. **Long-term stability:** These structures can be designed for static or dynamic stability. With the latter, maintenance of the surface layer may be required due to its degradation from continual wave motion. The former may prove too expensive, so that a balance must be made between initial and maintenance costs.

3. **Porosity and permeability:** Porosity appeared constant between 40% to 45% even after consolidation. There is a need to understand permeability since some long-standing berm breakwaters were considered impermeable.

4. **Profile development:** Further work was required to understand the role of stone size relative to wave conditions, stone gradation density and shape, permeability and porosity of the armor layer, and water level variations.

5. Oblique wave attack and roundheads: Oblique wave attack can transport material along the breakwater just like sand along a beach. Movement of stones at roundheads can be severe and needs attention.

6. Scour: "This topic was not discussed in detail. Scour might not be of importance to these unconventional structures, due to reduced wave reflections caused by the higher permeability of the berm structure resulting in lower shear stresses at the base of the structure."

7. Field measurements: There is a general lack of these, but they are important to evaluate scale effects in modeling. The observations needed are waves and water levels, pore water pressures, phreatic surfaces, water velocities within the breakwater, and breakwater profiles.

8. Model testing: Both conventional and unconventional designs are based on small scale models, which require true sea states in shallow water, rock durability, and wave dissipation in porous media.

9. Consolidation: This is a natural process that occurs under wave attack. It does not appear to affect the performance of the breakwater.

10. Rock durability: Deterioration due to abrasion and fracturing may have significant influence on performance. Normal techniques used in geology may not apply to wave action.

11. Construction: This mainly referred to the fine sands included in the quarry run material used. The resulting permeability may not matter much.

These authors would question conclusion (6) above, which has been quoted directly. The base of the "beach" profile shaped by the waves becomes the angle of repose of smaller stones included, which is very steep. The water column bounding this face can be a large proportion of the depth, especially for structures in deeper water, so that reflection can be substantial. The partially reflected waves when oblique can produce short-crested waves whose water-particle vortices can reach to the bottom and so cause scour, which can result in subsidence of the structure. This certainly needs attention in any monitoring program, besides measurements of reflection from realistic overall profiles.

10.9.2 Barrier Beach Breakwater

As noted in Section 8.2, these types of breakwater have been constructed successfully. They employ both sand and pebble beaches which are very mobile. The need therefore is to stabilize the resulting beaches in the places where they are required. This involves the use of headlands which form bays either in static or dynamic equilibrium, but preferably the former. At least they must be designed for this condition for safety even though they remain in dynamic equilibrium for years.

The headlands should be of walking stick shape with a quadrant at the downcoast end turned seaward and the main arm parallel to the accreted beach in front of it. This allows for smaller size overall and in the armor units to be used. At the tip of this quadrant a submerged groin may be required to prevent material, forming a bar during storms, from traveling into the next embayment past the tip of the breakwater. The last headland in a series may be inclined to the swell waves in order that sediment readily bypasses the breakwater across to the downcoast beach, as discussed in Section 9.8.

There is a need to investigate the advantages and disadvantages of this type of construction and compare costs with conventional breakwaters, both for installation and maintenance. There are many benefits in having land area around the seaward boundary of a harbor so formed, by a breakwater which also reduces spray generation. The cost of rubble material is growing with the ever-increasing demand for it and the expense in transportation. With this new type of structure material is available from dredging spoil associated with the port development. Further research is required into logistics and methods of constructing the headlands at an appropriate time in the construction. The spacing and seaward location of headlands can have an infinite variety, but some will prove more economical.

10.9.3 Geotextiles

The use of skins to contain cement mortar is a relatively new development (Porraz et al. 1977) so that design engineers and contractors need some education in the logistics and applications of this technology. If these large units are employed in wrong structures they will be found wanting, as are the alternative rubble-mound versions, so this point must be kept in mind. For coast stabilization headlands and stable bays are recommended, with the structures shaped and oriented as discussed in Section 6.1.

It has been recommended that these sausage-like units have relatively long lengths compared to their widths and heights, in order that they have intimate contact with several units beneath them (Silvester 1986). Filling should not take place to make circular sections as tension in the fabric will be too great, oblate spheroidal shape is preferred. But, these units can be made any size by sewing lengths of material lengthwise to provide any circumference desired. The only limitation is the need to fill them within about 3 hours, before initial setting takes place. This time restraint can be modified with experience and the use of admixtures to delay setting. Perhaps the use of detergents as suggested for pumping requirements may achieve this, which can be tested in the laboratory.

The suggested insertion of one layer at right angles to that immediately below it, and $45°$ to the axis of the structure, provides only their ends in contact with the water or waves. The weight of each nonrocking unit is such that these authors can not envisage storm waves, of the largest magnitude so far recorded, ever removing them from the body of the breakwater. Those at the summit, that may suffer the forces of overtopping water, are the most precarious, but these forces can be determined and the size and shape designed to give stability with an adequate safety factor. No tests so far have been conducted on the dimensions and weight of units for a given wave energy, so this needs attention.

Since the structures are more rectangular in shape than their rubble-mound alternative, they will reflect waves more effectively, unless spaces are provided between units to reduce forces and reflection. Research is required into these two aspects, together with wave transmission on the leeside of the structure. The generation of short-crested waves by oblique incidence has to be dealt with by protecting the seabed by similar types of units, perhaps more mattress-like than of sausage shape. The thickness of these covering units requires research as the forces to remove what is virtually a concrete slab from intimate contact with the bed have not been analyzed. They need to be as thin as possible for economic reasons. Such units need to be taken seaward far enough for the reflecting waves to be reduced sufficiently by diffraction for them to have little influence in the short-crested system.

Study is required not only in models but also in prototype situations to determine stability of units and of structures as a whole. The seabed needs monitoring to find the limits of scouring for given wave inputs, which is difficult in mobile bed models due to scale effects in sediment motion. But if the same investment is made in this novel approach as has been given to traditional structures, this more economical type of structure would soon be accepted.

10.9.4 Submerged Platform Breakwaters

Breakwaters floating at the sea surface have been examined in models and in one or two prototype cases, but have been found expensive and not able to withstand storm conditions. This region of the water column suffers the greatest turbulence and forces due to waves, so that anchoring requirements become severe as well as bending forces within the structure. Wave attenuation, particularly for long wave lengths, is not adequate so that wave transmission is still relatively strong.

The alternative presented in Section 8.4, of a horizontal plate supported either from floats or anchors at the bed, achieves better wave reduction with vastly decreased forces being applied to the breakwater. Research is required into materials and forces to make design more flexible. There are many applications for their use, with the main one concerning sedimentation being that of mobile headlands to aid in the formation of stable bays prior to a fixed structure being inserted.

10.10 BYPASSING MECHANISMS

The cost of the normal method of transferring littoral drift from one side of an inlet to another, namely dredging, has caused coastal engineers to consider alternative means. These include weir jetties, jet pumping, and sand fluidization, which are only at an experimental stage even though some have been installed in the field. Another two have been proposed in this book, those of natural reflection from specially oriented structures and accretion with headlands upcoast of the entrance. These all will be outlined with a view to suggesting research needs for each.

10.10.1 Dredging

This mode of bypassing is very expensive, running into millions of dollars annually around the world. It is generally carried out when shoaling occurs in navigable channels, or only when a difficulty has arisen and not as a precursor to the problem. The demand in the past few decades for deeper access to ports has magnified this expense, for which larger and more sophisticated dredges have been devised.

The equipment so invented and produced has been economical to the contractors using it even though the costs in power to operate it are excessive. Perhaps the use of the water-jet-pump principle (Silvester and Vongvisessomjai 1971) could make dredging more flexible, together with extra pumping stations in long delivery lines. This vacuum-cleaner effect using seawater as the driving fluid can be transferred from the domestic scene to the large scale sediment problem. In the end it is all a matter of economics and not a saving of fuel.

10.10.2 Weir Jetties

In many cases the dredging of littoral drift upcoast from a jetty controlling the flow from an inlet is dangerous and difficult due to unpredictable storm conditions or excessive swell. Exposure to the open sea will never be the choice of a contractor, who would like protected waters in which to operate. This is why the weir jetty was devised, which is a reduction in height of this structure adjacent to the shoreline to just below low water mark, for a limited length. The seaward segment still remains above this level to minimize overtopping. The cost of these structures is excessive because dismantling of the weir section subsequent to original construction to full height must take place.

The action purported to happen is that littoral drift, believed to be concentrated in the surf zone of the swell-built profile, is carried over the weir into sheltered water, from which it is dredged to the downcoast beach at any convenient frequency. This concept considers littoral drift like the "river of sand" which accepts that it is uniform throughout the year and is effected by moderate waves acting on the long duration swell-built profile of the beach. As noted already, it is in this condition that drift is negligible since most of the sand on the beach is tied up in the berm and is not available for transport.

As noted in Section 3.5 littoral drift is pulsational, with the bulk of movement occurring over a much wider surf zone during the return of the offshore bar formed during a storm some days before. Thus, as far as the weir jetties are concerned, sand is deposited against its upcoast face, not at the weir section but seaward of it. The pulse of sand quickly forms a beach beyond the limits of the weir so that its action of transmitting sediment over the working section is thwarted. In the prototype cases so far reported this action has not occurred, but publication has taken place soon after construction, so that a fully fledged storm may not have occurred prior to it. Any monitoring of sediment should take cognizance of the above possibility and the profession informed. Research is required into the location of beaches in respect to the weirs.

It is rather anomalous that weir jetties were conceived from the observation of Hodges (1955) at Hillsboro reef which was parallel to the shore and then subsequent structures were angled to the coast, with consequent cancellation of the concept. The deflection seaward of the littoral current caused the bulk of the drift to be transported offshore rather than over the weir. Much thought needs to be put into this phenomenon in order to save millions of dollars in full scale empirical methods.

10.10.3 Jet Pumping

Again to overcome the exigencies of open sea dredging, one land based pseudo-dredging proposal has included pumping sand from the seafloor by jet pumps, utilizing the water-jet-pump principle. These are spaced some distance apart and dig holes into which pumps lower themselves and are in continuous contact with the apexes of the conical depressions. Littoral drift pours into these large depressions and may at times bury them, which does not prevent them from working, since plain seawater is used as the driving force.

The shapes and proportions of the jet pumps reported to date have not been the most efficient (Silvester and Mueller 1968) so that further research is necessary. Another factor detracting from their use is the debris, such as metal cans, larger stones and shells, that can be caught in the throats of these pumps. The excessive power demand requires that they be turned off during periods of little or no drift. Knowledge of the pulsative nature of longshore movement is important in this respect. Very few applications have been reported, and they soon after construction. Thus, their long-term effectiveness has not been proven, although the concept has merit. The installation costs are moderate compared to the hiring of dredgers, if available at the time required, and downtime during stormy periods.

10.10.4 Sand Fluidization

This principle is somewhat similar to that of the jet pump except that sand is virtually placed in suspension for currents to carry it away from the area. If they are activated during the ebb flow in a river mouth the material is carried seaward, there to be acted upon by the waves. It is mainly utilized for river application where waves cannot effect their mass transport bed current which is only upstream.

10.10.5 Wave Reflection

As suggested in Section 9.8 the natural swell waves, arriving from essentially one direction, can be utilized to transmit sediment past an inlet or harbor entrance and beyond the opening to the downcoast beach. This it accomplishes without forming shoals since the short-crested wave system so formed can transport drift adequately at greater depths than the progressive incident waves. This energy is available continually on an oceanic margin and is increased commensurately with the volume of littoral drift.

Beyond the downcoast end of the reflecting structure a shoal may form due to this accelerated supply of material, but in the course of time could be dispersed to the adjacent shoreline. The zone near the tip of the structure is maintained at navigable depth for reasonably sized craft such as fishing boats. Even deeper channels could have sediment passed through them because of the vortices formed in the short-crested system.

It was also shown in the same section that craft can pass through these complex waves with less vertical motion than through normal swell waves if their speed is appropriate to the celerity of the island crests traveling parallel to the reflecting wall. An alternative is to travel along the wall between the passage of these crests. Some education of skippers may be necessary, but this is a small price to pay for the elimination of shoals.

More research is required to find the limiting depth of scour seaward and downcoast of such a feature. There are existing breakwaters and other structures receiving oblique swell which could provide this information, which is preferable to model studies that suffer scale effects with sediment problems.

There will be a need with any new design to protect the bed in the vicinity of the reflecting wall in order to safeguard it against subsidence. This topic has been discussed in the relevant section. The use of geotextile units is recommended because of the high reflectivity, but caissons floated out and sunk could be an economic alternative. However, for bed protection mattress-type flat units are recommended since they prevent vortices from reaching the sandy floor. Further research on this concept is required in models to confirm the action of swift sediment transmission and some pilot structures to test it at full scale. The same topics require attention as for all structures where oblique reflection is present.

10.10.6 Upcoast Accretion

This alternative has been noted in Section 9.9, where littoral drift is intercepted upcoast from the entrance by the installation of headlands progressively. Each forms a tombolo and an almost triangular zone of land is accreted, after which a new headland is placed to continue the action. This prevention of deposition at the mouth of the waterway can be carried on for years, or even decades, with the construction costs spread over this time. When all the coast under the jurisdiction of the authority has been so accreted, a second or third row of headlands can be inserted.

The cost of structures and administration can be offset by the lease or sale of the valuable land so won from the sea. It is known to be stable because of the presence of the headlands when the greatest indentation can be predicted. Downcoast of the inlet may have to be treated similarly, which is also the case if a channel is dredged and the spoil dumped out to sea. The problem must be viewed overall and not just at the entrance. Research is needed into the littoral drift rate in order to determine the volumes to be intercepted and hence the period over which shoaling at the inlet opening can be prevented. Such calculations should take into account the number of storms expected annually, if this is great the larger the annual drift rate.

10.11 DATA COLLECTION

It is seldom that a consistent long-term policy of data collection exists on any coast unless it is already highly developed and possibly suffering degradation. This is due to the large expense and lack of infrastructure to pay for it. Even wave recording may be instituted at the last minute for any large scale development, thus reducing the effectiveness of the data collected. But even some of the information gathered might be queried. The effort put into wave refraction diagrams is at times rather pointless. Thus, in all this expensive exercise of gathering field information the first question to be asked is "Why do it?"

On land, data collection is relatively easy but is not the case in the medium of the sea, where verification of theory under field conditions is extremely difficult. For this reason the phenomena need to be better understood, to show that variables being studied are known to be paramount in any problem. The two important inputs to transport of sediments are waves and currents. The former produce oscillatory and even elliptical motions of the water particles, which vary spatially both in the vertical and horizontal planes. Simultaneous measurement over the whole realm of interest is not possible economically and hence theory, either basic or empirical in nature, must be used in the selection of critical locations such as incipient breaking waves.

The second input, of currents, may be tidal in origin, due to wind stress, or possibly long-period oscillations occasioned by internal waves or large scale vortices. The importance of these sources of motion must be determined for any site. Again, extremes must be extrapolated in some manner, as are the winds that generate the waves. The work of meteorologists, oceanographers, and engineers requires integration, so that data are gathered not only for their own sake, but also for the sake of the appliers of this information. The mass transport of water near the bed due to waves plus ocean currents at the same level must be determined by measurements and theory. Wave duration is a relevant factor in sedimentary processes. The probability of short-crested systems should be considered from the wave sources and propagation principles across the continental shelf. Reflection from any structure to be inserted must be taken into account.

In any program of research or data gathering, a worker may commence with some preconceived idea or hypothesis regarding the importance of one variable. This can influence the choice and number of recorders to be used. For example, currents may be accepted as paramount when waves are more influential. Inconsistencies in sediment tracer movement and current direction have been unexplainable because wave direction and wave period were not recorded.

Before any program of data collection is commenced, it is as well to sit and contemplate the complete problem. This first analysis should not only take in the specific area of study but also zones upcoast and downcoast, particularly the former. Assess where the sediment has been coming from for the past few years and even the past few hundred years. What changes in this input have occurred naturally or are due to man? Where is the energy available to move it and how can this fluctuate over what periods of time? When ready, a report should be prepared for self or superior, outlining the phenomena as seen with suggestions for further information, be it theory, field data, or model investigations.

Lines of investigation will suggest themselves, but these can change as the problem is brought more into focus. The need is to keep an open mind, which may mean stepping back in the face of brighter suggestions, even from a younger colleague, but always rationalize procedures fully. There is much to be learned of this natural medium and nobody should be considered a clown for not knowing some of the facts. Do not take any statements on face value, even from the most eminent mouth or pen, for so many wrong solutions have been repeated along the shorelines of the world.

Reports presented to governmental and local authorities on beach erosion by consultants invariably contain pages and pages of data on sediment characteristics, beach profiles, and wave refraction, to name a few. When the reports are analyzed it is found that much of this information is not alluded to in the conclusions drawn or the recommendations made for remedial measures. The client has paid for the gathering and hence it warrants inclusion as appendices but serves effectively only as padding. It is a case of a little forethought determining the absolute need for this expensive information.

The coastal engineer should consider environmental issues continually. In *Effective Use of the Sea* (1966), previously alluded to, President Lyndon B. Johnson is quoted from his message on Natural Beauty as follows: "Our conservation must not be just classic protection and development, but a creative conservation, of restoration and innovation." The activities of environmental protection authorities the world over must be examined in this light, because as the report states: "Continuing population growth combined with increased dependence on the sea for food and recreation means that modification of marine environments will not only continue, but will drastically increase." Engineers must promote their picture of being enhancers of Nature rather than destroyers, as promulgated by ecologists at every turn.

REFERENCES

ANGLIN, C. D., DEAN, K. B., and WILLIS, D. H. 1987. Unconventional rubble-mound breakwaters—concerns. *Proc. Workshop on Berm Breakwaters: Unconventional Rubble-Mound Breakwaters, ASCE*, 270–78.

EDGE, B., BAIRD, J. M., CALDWELL, J. M., FAIRWEATHER, V., MAGOON, O. T., and TREADWELL, D. D. 1982. *Failure of the Breakwater at Port Sines, Portugal*. Coastal Eng. Res. Foundation, American Society of Civil Engineers.

Effective Use of the Sea. 1966. Report of the Panel on Oceanography, President's Science Advisory Committee, The White House, Washington, D.C.

HODGES, T. K. 1955. Sand bypassing at Hillsboro Inlet, Florida. *Bull. Beach Erosion Board*, U.S. Corps of Engrs., Vol. 9 (2).

HSU. J. R. C. 1979. Short-crested water waves. *Ph.D. thesis*, University of Western Australia.

HSU, J. R. C., and SILVESTER, R. 1989. Comparison of various defense measures. *Proc. 9th Austral. Conf. Coastal & Ocean Eng.*, 145–48.

HSU, J. R. C., and EVANS, C. 1989. Parabolic bay shapes and their applications. *Proc. Instn. Civil Engrs.* 87: 557-70.

HSU, J. R. C., SILVESTER, R., and XIA, Y. M. 1989a. Generalities on static equilibrium bays. *Coastal Eng. 12*: 353–69.

———. 1989b. Static equilibrium bays: new relationships. *J. Waterway, Port, Coastal & Ocean Eng., ASCE 115*(3): 285–98.

———. 1989c. Applications of headland control. *J. Waterway, Port, Coastal & Ocean Eng., ASCE 115*(3): 299–310.

Hsu, J. R. C., Silvester, R., and Tsuchiya, Y. 1980. Boundary-layer velocities and mass transport in short-crested waves. *J. Fluid Mech. 99*: 321–42.

Hsu, J. R. C., Tsuchiya, Y., and Silvester, R. 1979. Third-order approximation to short-crested waves. *J. Fluid Mech. 90*: 179–96.

Lee, T. T. 1983. *Research Needs and Facilities Requirements*. Proc. Workshop on Shallow Water Ocean Eng. Res., Honolulu, J.K.K. Look Lab. Rep. No. 56.

Pinchin, B. M., and Nairn, R. B. 1986. The use of numerical models for the design of artificial beaches to protect cohesive shores. *Proc. Symp. on Cohesive Shores*, Nat. Res. Council Canada, 196–209.

Porraz, M., Maza, J. A., and Munoz, M. L. 1977. Low cost structures using operational design systems. *Proc. Coastal Sediments '77, ASCE*, 672–85.

Silvester, R. 1986. Use of grout-filled sausages in coastal structures. *J. Waterway, Port, Coastal and Ocean Eng., ASCE 112*(1): 95–114.

Silvester, R., and Hsu, J. R. C. 1989. Sines revisited. *J. Waterway, Port, Coastal and Ocean Eng., ASCE 115*(3): 327–44.

———. 1991. New and old ideas in coastal sedimentation. Review in *Aquatic Sciences 4*(4): 375–410.

Silvester, R., and Mueller, N. H. G. 1968. Designing data for the liquid to liquid jet pump. *J. Hydraulic Res. 6*: 129–62.

Silvester, R., and Vongvisessomjai, S. 1971. Characteristics of jet-pump with liquids of different density. *Proc. 3rd World Dredging Conf.*, 293–315.

Smith, A. W., and Gordon, A. D. 1980. Secondary sand transport mechanisms. *Proc. 17th Inter. Conf. Coastal Eng., ASCE 2*: 1122–39.

Tanner, W. F. 1987. The beach: where is the 'river of sand.' *J. Coastal Res. 3*: 377–86.

Appendix

TABLE OF FUNCTIONS OF d/L_o

d/L_o	d/L	$2\pi d/L$	tanh $2\pi d/L$	sinh $2\pi d/L$	cosh $2\pi d/L$	$4\pi d/L$	sinh $4\pi d/L$	cosh $4\pi d/L$	H/H_o'
0	0	0	0	0	1	0	0	1	
.0001000	.003990	.02507	.02506	.02507	1.0003	.05014	.05016	1.001	4.467
.0002000	.005643	.03546	.03544	.03547	1.0006	.07091	.07097	1.003	3.757
.0003000	.006912	.04343	.04340	.04344	1.0009	.08686	.08697	1.004	3.395
.0004000	.007982	.05015	.05011	.05018	1.0013	.1003	.1005	1.005	3.160
.0005000	.008925	.05608	.05602	.05611	1.0016	.1122	.1124	1.006	2.989
.0006000	.009778	.06144	.06136	.06148	1.0019	.1229	.1232	1.008	2.856
.0007000	.01056	.06637	.06627	.06642	1.0022	.1327	.1331	1.009	2.749
.0008000	.01129	.07096	.07084	.07102	1.0025	.1419	.1424	1.010	2.659
.0009000	.01198	.07527	.07513	.07534	1.0038	.1505	.1511	1.011	2.582
.001000	.01263	.07935	.07918	.07943	1.0032	.1587	.1594	1.013	2.515
.001100	.01325	.08323	.08304	.08333	1.0035	.1665	.1672	1.014	2.456
.001200	.01384	.08694	.08672	.08705	1.0038	.1739	.1748	1.015	2.404
.001300	.01440	.09050	.09026	.09063	1.0041	.1810	.1820	1.016	2.357
.001400	.01495	.09393	.09365	.09407	1.0044	.1879	.1890	1.018	2.314
.001500	.01548	.09723	.09693	.09739	1.0047	.1945	.1957	1.019	2.275
.001600	.01598	.1004	.1001	.1006	1.0051	.2009	.2022	1.020	2.239
.001700	.01648	.1035	.1032	.1037	1.0054	.2071	.2086	1.022	2.205
.001800	.01696	.1066	.1062	.1068	1.0057	.2131	.2147	1.023	2.174
.001900	.01743	.1095	.1091	.1097	1.0060	.2190	.2207	1.024	2.145
.002000	.01788	.1123	.1119	.1125	1.0063	.2247	.2266	1.025	2.119
.002100	.01832	.1151	.1146	.1154	1.0066	.2303	.2323	1.027	2.094
.002200	.01876	.1178	.1173	.1181	1.0069	.2357	.2379	1.028	2.070
.002300	.01918	.1205	.1199	.1208	1.0073	.2410	.2433	1.029	2.047
.002400	.01959	.1231	.1225	.1234	1.0076	.2462	.2487	1.031	2.025
.002500	.02000	.1257	.1250	.1260	1.0079	.2513	.2540	1.032	2.005
.002600	.02040	.1282	.1275	.1285	1.0082	.2563	.2592	1.033	1.986
.002700	.02079	.1306	.1299	.1310	1.0085	.2612	.2642	1.034	1.967
.002800	.02117	.1330	.1323	.1334	1.0089	.2661	.2692	1.036	1.950
.002900	.02155	.1354	.1346	.1358	1.0092	.2708	.2741	1.037	1.933
.003000	.02192	.1377	.1369	.1382	1.0095	.2755	.2790	1.038	1.917
.003100	.02228	.1400	.1391	.1405	1.0098	.2800	.2837	1.040	1.902
.003200	.02264	.1423	.1413	.1427	1.0101	.2845	.2884	1.041	1.887
.003300	.02300	.1445	.1435	.1449	1.0104	.2890	.2930	1.042	1.873
.003400	.02335	.1467	.1456	.1472	1.0108	.2934	.2976	1.043	1.860
.003500	.02369	.1488	.1477	.1494	1.0111	.2977	.3021	1.045	1.847
.003600	.02403	.1510	.1498	.1515	1.0114	.3020	.3065	1.046	1.834
.003700	.02436	.1531	.1519	.1537	1.0117	.3061	.3109	1.047	1.822
.003800	.02469	.1551	.1539	.1558	1.0212	.3103	.3153	1.049	1.810
.003900	.02502	.1572	.1559	.1579	1.0124	.3144	.3196	1.050	1.799

TABLE OF FUNCTIONS OF d/L_o

d/L_o	d/L	$2\pi d/L$	tanh $2\pi d/L$	sinh $2\pi d/L$	cosh $2\pi d/L$	$4\pi d/L$	sinh $4\pi d/L$	cosh $4\pi d/L$	H/H_o'
.004000	.02534	.1592	.1579	.1599	1.0127	.3184	.3238	1.051	1.788
.004100	.02566	.1612	.1598	.1619	1.0130	.3224	.3280	1.052	1.777
.004200	.02597	.1632	.1617	.1639	1.0133	.3263	.3322	1.054	1.767
.004300	.02628	.1651	.1636	.1659	1.0137	.3302	.3362	1.055	1.756
.004400	.02659	.1671	.1655	.1678	1.0140	.3341	.3403	1.056	1.746
.004500	.02689	.1690	.1674	.1698	1.0143	.3380	.3444	1.058	1.737
.004600	.02719	.1708	.1692	.1717	1.0146	.3417	.3483	1.059	1.727
.004700	.02749	.1727	.1710	.1736	1.0149	.3454	.3523	1.060	1.718
.004800	.02778	.1745	.1728	.1754	1.0153	.3491	.3562	1.062	1.709
.004900	.02807	.1764	.1746	.1773	1.0156	.3527	.3601	1.063	1.701
.005000	.02836	.1782	.1764	.1791	1.0159	.3564	.3640	1.064	1.692
.005100	.02864	.1800	.1781	.1809	1.0162	.3599	.3678	1.066	1.684
.005200	.02893	.1818	.1798	.1827	1.0166	.3635	.3715	1.067	1.676
.005300	.02921	.1835	.1815	.1845	1.0169	.3670	.3753	1.068	1.669
.005400	.02948	.1852	.1832	.1863	1.0172	.3705	.3790	1.069	1.662
.005500	.02976	.1870	.1848	.1880	1.0175	.3739	.3827	1.071	1.654
.005600	.03003	.1887	.1865	.1898	1.0178	.3774	.3864	1.072	1.647
.005700	.03030	.1904	.1881	.1915	1.0182	.3808	.3900	1.073	1.640
.005800	.03057	.1921	.1897	.1932	1.0185	.3841	.3937	1.075	1.633
.005900	.03083	.1937	.1913	.1949	1.0188	.3875	.3972	1.076	1.626
.006000	.03110	.1954	.1929	.1967	1.0192	.3908	.4008	1.077	1.620
.006100	.03136	.1970	.1945	.1983	1.0195	.3941	.4044	1.079	1.614
.006200	.03162	.1987	.1961	.2000	1.0198	.3973	.4079	1.080	1.607
.006300	.03188	.2003	.1976	.2016	1.0201	.4006	.4114	1.081	1.601
.006400	.03213	.2019	.1992	.2033	1.0205	.4038	.4148	1.083	1.595
.006500	.03238	.2035	.2007	.2049	1.0208	.4070	.4183	1.084	1.589
.006600	.03264	.2051	.2022	.2065	1.0211	.4101	.4217	1.085	1.583
.006700	.03289	.2066	.2037	.2081	1.0214	.4133	.4251	1.087	1.578
.006800	.03313	.2082	.2052	.2097	1.0217	.4164	.4285	1.088	1.572
.006900	.03338	.2097	.2067	.2113	1.0221	.4195	.4319	1.089	1.567
.007000	.03362	.2113	.2082	.2128	1.0224	.4225	.4352	1.091	1.561
.007100	.03387	.2128	.2096	.2144	1.0227	.4256	.4386	1.092	1.556
.007200	.03411	.2143	.2111	.2160	1.0231	.4286	.4419	1.093	1.551
.007300	.03435	.2158	.2125	.2175	1.0234	.4316	.4452	1.095	1.546
.007400	.03459	.2173	.2139	.2190	1.0237	.4346	.4484	1.096	1.541
.007500	.03482	.2188	.2154	.2205	1.0240	.4376	.4517	1.097	1.536
.007600	.03506	.2203	.2168	.2221	1.0244	.4406	.4549	1.099	1.531
.007700	.03529	.2218	.2182	.2236	1.0247	.4435	.4582	1.100	1.526
.007800	.03552	.2232	.2196	.2251	1.0250	.4464	.4614	1.101	1.521
.007900	.03576	.2247	.2209	.2265	1.0253	.4493	.4646	1.103	1.517

TABLE OF FUNCTIONS OF d/L_o

d/L_o	d/L	$2\pi d/L$	tanh $2\pi d/L$	sinh $2\pi d/L$	cosh $2\pi d/L$	$4\pi d/L$	sinh $4\pi d/L$	cosh $4\pi d/L$	H/H_o'
.008000	.03598	.2261	.2223	.2280	1.0257	.4522	.4678	1.104	1.512
.008100	.03621	.2275	.2237	.2295	1.0260	.4551	.4709	1.105	1.508
.008200	.03644	.2290	.2250	.2310	1.0263	.4579	.4741	1.107	1.503
.008300	.03666	.2304	.2264	.2324	1.0266	.4607	.4772	1.108	1.499
.008400	.03689	.2318	.2277	.2338	1.0270	.4636	.4803	1.109	1.495
.008500	.03711	.2332	.2290	.2353	1.0273	.4664	.4834	1.111	1.491
.008600	.03733	.2346	.2303	.2367	1.0276	.4691	.4865	1.112	1.487
.008700	.03755	.2360	.2317	.2381	1.0280	.4719	.4896	1.113	1.482
.008800	.03777	.2373	.2330	.2396	1.0283	.4747	.4927	1.115	1.478
.008900	.03799	.2387	.2343	.2410	1.0286	.4774	.4957	1.116	1.474
.009000	.03821	.2401	.2356	.2424	1.0290	.4801	.4988	1.118	1.471
.009100	.03842	.2414	.2368	.2438	1.0293	.4828	.5018	1.119	1.467
.009200	.03864	.2428	.2381	.2452	1.0296	.4855	.5049	1.120	1.463
.009300	.03885	.2441	.2394	.2465	1.0299	.4882	.5079	1.122	1.459
.009400	.03906	.2455	.2407	.2479	1.0303	.4909	.5109	1.123	1.456
.009500	.03928	.2468	.2419	.2493	1.0306	.4936	.5138	1.124	1.452
.009600	.03949	.2481	.2431	.2507	1.0309	.4962	.5168	1.126	1.448
.009700	.03970	.2494	.2443	.2520	1.0313	.4988	.5198	1.127	1.445
.009800	.03990	.2507	.2456	.2534	1.0316	.5104	.5227	1.128	1.442
.009900	.04011	.2520	.2468	.2547	1.0319	.5040	.5257	1.130	1.438
.01000	.04032	.2533	.2480	.2560	1.0322	.5066	.5286	1.131	1.435
.01100	.04233	.2660	.2598	.2691	1.0356	.5319	.5574	1.145	1.403
.01200	.04426	.2781	.2711	.2817	1.0389	.5562	.5853	1.159	1.375
.01300	.04612	.2898	.2820	.2938	1.0423	.5795	.6125	1.173	1.350
.01400	.04791	.3010	.2924	.3056	1.0456	.6020	.6391	1.187	1.327
.01500	.04964	.3119	.3022	.3170	1.0490	.6238	.6651	1.201	1.307
.01600	.05132	.3225	.3117	.3281	1.0524	.6450	.6906	1.215	1.288
.01700	.05296	.3328	.3209	.3389	1.0559	.6655	.7158	1.230	1.271
.01800	.05455	.3428	.3298	.3495	1.0593	.6856	.7405	1.244	1.255
.01900	.05611	.3525	.3386	.3599	1.0628	.0751	.7650	1.259	1.240
.02000	.05763	.3621	.3470	.3701	1.0663	.7242	.7891	1.274	1.226
.02100	.05912	.3714	.3552	.3800	1.0698	.7429	.8131	1.289	1.213
.02200	.06057	.3806	.3632	.3898	1.0733	.7612	.8368	1.304	1.201
.02300	.06200	.3896	.3710	.3995	1.0768	.7791	.8603	1.319	1.189
.02400	.06340	.3984	.3786	.4090	1.0804	.7967	.8837	1.335	1.178
.02500	.06478	.4070	.3860	.4184	1.0840	.8140	.9069	1.350	1.168
.02600	.06613	.4155	.3932	.4276	1.0876	.8310	.9310	1.366	1.159
.02700	.06747	.4229	.4002	.4367	1.0912	.8478	.9530	1.381	1.150
.02800	.06878	.4322	.4071	.4457	1.0949	.8643	.9760	1.397	1.141
.02900	.07007	.4403	.4138	.4546	1.0985	.8805	.9988	1.413	1.133

TABLE OF FUNCTIONS OF d/L_o

d/L_o	d/L	$2\pi d/L$	tanh $2\pi d/L$	sinh $2\pi d/L$	cosh $2\pi d/L$	$4\pi d/L$	sinh $4\pi d/L$	cosh $4\pi d/L$	H/H'_o
.03000	.07135	.4483	.4205	.4634	1.1021	.8966	1.022	1.430	1.125
.03100	.07260	.4562	.4269	.4721	1.1059	.9124	1.044	1.446	1.118
.03200	.07385	.4640	.4333	.4808	1.1096	.9280	1.067	1.462	1.111
.03300	.07507	.4717	.4395	.4894	1.1133	.9434	1.090	1.479	1.104
.03400	.07630	.4794	.4457	.4980	1.1171	.9588	1.113	1.496	1.098
.03500	.07748	.4868	.4517	.5064	1.1209	.9737	1.135	1.513	1.092
.03600	.07867	.4943	.4577	.5147	1.1247	.9886	1.158	1.530	1.086
.03700	.07984	.5017	.4635	.5230	1.1285	1.0033	1.180	1.547	1.080
.03800	.08100	.5090	.4691	.5312	1.1324	1.018	1.203	1.564	1.075
.03900	.08215	.5162	.4747	.5394	1.1362	1.032	1.226	1.582	1.069
.04000	.08329	.5233	.4802	.5475	1.1401	1.047	1.248	1.600	1.064
.04100	.08442	.5304	.4857	.5556	1.1440	1.061	1.271	1.617	1.059
.04200	.08553	.5374	.4911	.5637	1.1479	1.075	1.294	1.636	1.055
.04300	.08664	.5444	.4964	.5717	1.1518	1.089	1.317	1.654	1.050
.04400	.08774	.5513	.5015	.5796	1.1558	1.103	1.340	1.672	1.046
.04500	.08883	.5581	.5066	.5876	1.1599	1.116	1.363	1.691	1.042
.04600	.08991	.5649	.5116	.5954	1.1639	1.130	1.386	1.709	1.038
.04700	.09098	.5717	.5166	.6033	1.1679	1.143	1.409	1.728	1.034
.04800	.09205	.5784	.5215	.6111	1.1720	1.157	1.433	1.747	1.030
.04900	.09311	.5850	.5263	.6189	1.1760	1.170	1.456	1.766	1.026
.05000	.09416	.5916	.5310	.6267	1.1802	1.183	1.479	1.786	1.023
.05100	.09520	.5981	.5357	.6344	1.1843	1.196	1.503	1.805	1.019
.05200	.09623	.6046	.5403	.6421	1.1884	1.209	1.526	1.825	1.016
.05300	.09726	.6111	.5449	.6499	1.1926	1.222	1.550	1.845	1.013
.05400	.09829	.6176	.5494	.6575	1.1968	1.235	1.574	1.865	1.010
.05500	.09930	.6239	.5538	.6652	1.2011	1.248	1.598	1.885	1.007
.05600	.1003	.6303	.5582	.6729	1.2053	1.261	1.622	1.906	1.004
.05700	.1013	.6366	.5626	.6805	1.2096	1.273	1.646	1.926	1.001
.05800	.1023	.6428	.5668	.6880	1.2138	1.286	1.670	1.947	.9985
.59000	.1033	.6491	.5711	.6956	1.2181	1.298	1.695	1.968	.9958
.06000	.1043	.6553	.5753	.7033	1.2225	1.311	1.719	1.989	.9932
.06100	.1053	.6616	.5794	.7110	1.2270	1.3231	1.744	2.011	.9907
.06200	.1063	.6678	.5834	.7187	1.2315	1.336	1.770	2.033	.9883
.06300	.1073	.6739	.5874	.7256	1.2355	1.348	1.795	2.055	.9860
.06400	.1082	.6799	.5914	.7335	1.2402	1.360	1.819	2.076	.9837
.06500	.1092	.6860	.5954	.7411	1.2447	1.372	1.845	2.098	.9815
.06600	.1101	.6920	.5993	.7486	1.2492	1.384	1.870	2.121	.9793
.06700	.1111	.6981	.6031	.7561	1.2537	1.396	1.896	2.144	.9772
.06800	.1120	.7037	.6069	.7633	1.2580	1.408	1.921	2.166	.9752
.06900	.1130	.7099	.6160	.7711	1.2628	1.420	1.948	2.189	.9732

TABLE OF FUNCTIONS OF d/L_o

d/L_o	d/L	$2\pi d/L$	tanh $2\pi d/L$	sinh $2\pi d/L$	cosh $2\pi d/L$	$4\pi d/L$	sinh $4\pi d/L$	cosh $4\pi d/L$	H/H_o'
.07000	.1139	.7157	.6144	.7783	1.2672	1.432	1.974	2.213	.9713
.07100	.1149	.7219	.6181	.7863	1.2721	1.444	2.000	2.236	.9694
.07200	.1158	.7277	.6217	.7937	1.2767	1.455	2.026	2.260	.9676
.07300	.1168	.7336	.6252	.8011	1.2813	1.467	2.053	2.284	.9658
.07400	.1177	.7395	.6289	.8088	1.2861	1.479	2.080	2.308	.9641
.07500	.1186	.7453	.6324	.8162	1.2908	1.490	2.107	2.332	.9624
.07600	.1195	.7511	.6359	.8237	1.2956	1.502	2.135	2.357	.9607
.07700	.1205	.7569	.6392	.8312	1.3004	1.514	2.162	2.382	.9591
.07800	.1214	.7625	.6427	.8386	1.3051	1.525	2.189	2.407	.9576
.07900	.1223	.7683	.6460	.8462	1.3100	1.537	2.217	2.432	.9562
.08000	.1232	.7741	.6493	.8538	1.3149	1.548	2.245	2.458	.9548
.08100	.1241	.7799	.6526	.8614	1.3198	1.560	2.274	2.484	.9534
.08200	.1251	.7854	.6558	.8687	1.3246	1.571	2.303	2.511	.9520
.08300	.1259	.7911	.6590	.8762	1.3295	1.583	2.331	2.537	.9506
.08400	.1268	.7967	.6622	.8837	1.3345	1.594	2.360	2.563	.9493
.08500	.1277	.8026	.6655	.8915	1.3397	1.605	2.389	2.590	.9481
.08600	.1286	.8080	.6685	.8989	1.3446	1.616	2.418	2.617	.9469
.08700	.1295	.8137	.6716	.9064	1.3497	1.628	2.448	2.644	.9457
.08800	.1304	.8193	.6747	.9141	1.3548	1.639	2.478	2.672	.9445
.08900	.1313	.8250	.6778	.9218	1.3600	1.650	2.508	2.700	.9433
.09000	.1322	.8306	.6808	.9295	1.3653	1.661	2.538	2.728	.9422
.09100	.1331	.8363	.6838	.9372	1.3706	1.672	2.568	2.756	.9411
.09200	.1340	.8420	.6868	.9450	1.3759	1.684	2.599	2.785	.9401
.09300	.1349	.8474	.6897	.9525	1.3810	1.695	2.630	2.814	.9391
.09400	.1357	.8528	.6925	.9600	1.3862	1.706	2.662	2.843	.9381
.09500	.1366	.8583	.6953	.9677	1.3917	1.717	2.693	2.873	.9371
.09600	.1375	.8639	.6982	.9755	1.3970	1.728	2.726	2.903	.9362
.09700	.1384	.8694	.7011	.9832	1.4023	1.739	2.757	2.933	.9353
.09800	.1392	.8749	.7039	.9908	1.4077	1.750	2.790	2.963	.9344
.09900	.1401	.8803	.7066	.9985	1.4131	1.761	2.822	2.994	.9335
.1000	.1410	.8858	.7093	1.006	1.4187	1.772	2.855	3.025	.9327
.1010	.1419	.8913	.7120	1.014	1.4242	1.783	2.888	3.057	.9319
.1020	.1427	.8967	.7147	1.022	1.4297	1.793	2.922	3.088	.9311
.1030	.1436	.9023	.7173	1.030	1.4354	1.805	2.956	3.121	.9304
.1040	.1445	.9076	.7200	1.037	1.4410	1.815	2.990	3.153	.9297
.1050	.1453	.9130	.7226	1.045	1.4465	1.826	3.024	3.185	.9290
.1060	.1462	.9184	.7252	1.053	1.4523	1.837	3.059	3.218	.9282
.1070	.1470	.9239	.7277	1.061	1.4580	1.848	3.094	3.251	.9276
.1080	.1479	.9293	.7303	1.069	1.4638	1.858	3.128	3.284	.9269
.1090	.1488	.9343	.7327	1.076	1.4692	1.869	3.164	3.319	.9263

TABLE OF FUNCTIONS OF d/L_o

d/L_o	d/L	$2\pi d/L$	tanh $2\pi d/L$	sinh $2\pi d/L$	cosh $2\pi d/L$	$4\pi d/L$	sinh $4\pi d/L$	cosh $4\pi d/L$	H/H_o'
.1100	.1496	.9400	.7352	1.085	1.4752	1.880	3.201	3.353	.9257
.1110	.1505	.9456	.7377	1.093	1.4814	1.891	3.237	3.388	.9251
.1120	.1513	.9508	.7402	1.101	1.4871	1.902	3.274	3.423	.9245
.1130	.1522	.9563	.7426	1.109	1.4932	1.913	3.312	3.459	.9239
.1140	.1530	.9616	.7450	1.117	1.4990	1.923	3.348	3.494	.9234
.1150	.1539	.9670	.7474	1.125	1.5051	1.934	3.385	3.530	.9228
.1160	.1547	.9720	.7497	1.133	1.5108	1.944	3.423	3.566	.9223
.1170	.1556	.9775	.7520	1.141	1.5171	1.955	3.462	3.603	.9218
.1180	.1564	.9827	.7543	1.149	1.5230	1.966	3.501	3.641	.9214
.1190	.1573	.9882	.7566	1.157	1.5293	1.977	3.540	3.678	.9209
.1200	.1581	.9936	.7589	1.165	1.5356	1.987	3.579	3.716	.9204
.1210	.1590	.9989	.7612	1.174	1.5418	1.998	3.620	3.755	.9200
.1220	.1598	1.004	.7634	1.182	1.5479	2.008	3.659	3.793	.9196
.1230	.1607	1.010	.7656	1.190	1.5546	2.019	3.699	3.832	.9192
.1240	.1615	1.015	.7678	1.198	1.5605	2.030	3.740	3.871	.9189
.1250	.1624	1.020	.7700	1.207	1.5674	2.041	3.782	3.912	.9186
.1260	.1632	1.025	.7721	1.215	1.5734	2.051	3.824	3.952	.9182
.1270	.1640	1.030	.7742	1.223	1.5795	2.061	3.865	3.992	.9178
.1280	.1649	1.036	.7763	1.231	1.5862	2.072	3.907	4.033	.9175
.1290	.1657	1.041	.7783	1.240	1.5927	2.082	3.950	4.074	.9172
.1300	.1665	1.046	.7804	1.248	1.5990	2.093	3.992	4.115	.9169
.1310	.1674	1.052	.7824	1.257	1.6060	2.104	4.036	4.158	.9166
.1320	.1682	1.057	.7844	1.265	1.6124	2.114	4.080	4.201	.9164
.1330	.1691	1.062	.7865	1.273	1.6191	2.125	4.125	4.245	.9161
.1340	.1699	1.068	.7885	1.282	1.6260	2.135	4.169	4.288	.9158
.1350	.1708	1.073	.7905	1.291	1.633	2.146	4.217	4.334	.9156
.1360	.1716	1.078	.7925	1.300	1.640	2.156	4.262	4.378	.9154
.1370	.1724	1.084	.7945	1.308	1.647	2.167	4.309	4.423	.9152
.1380	.1733	1.089	.7964	1.317	1.654	2.177	4.355	4.468	.9150
.1390	.1741	1.094	.7983	1.326	1.660	2.188	4.402	4.514	.9148
.1400	.1749	1.099	.8002	1.334	1.667	2.198	4.450	4.561	.9146
.1410	.1758	1.105	.8021	1.343	1.675	2.209	4.498	4.607	.9144
.1420	.1766	1.110	.8039	1.352	1.681	2.219	4.456	4.654	.9142
.1430	.1774	1.115	.8057	1.360	1.688	2.230	4.595	4.663	.9141
.1440	.1783	1.120	.8076	1.369	1.696	2.240	4.644	4.751	.9140
.1450	.1791	1.125	.8094	1.378	1.703	2.251	4.695	4.800	.9139
.1460	.1800	1.131	.8112	1.388	1.710	2.261	4.746	4.850	.9137
.1470	.1808	1.136	.8131	1.397	1.718	2.272	4.798	4.901	.9136
.1480	.1816	1.141	.8149	1.405	1.725	2.282	4.847	4.951	.9135
.1490	.1825	1.146	.8166	1.415	1.732	2.293	4.901	5.001	.9134

TABLE OF FUNCTIONS OF d/L_o

d/L_o	d/L	$2\pi d/L$	tanh $2\pi d/L$	sinh $2\pi d/L$	cosh $2\pi d/L$	$4\pi d/L$	sinh $4\pi d/L$	cosh $4\pi d/L$	H/H_o'
.1500	.1833	1.152	.8183	1.424	1.740	2.303	4.954	5.054	.9133
.1510	.1841	1.157	.8200	1.433	1.747	2.314	5.007	5.106	.9133
.1520	.1850	1.162	.8217	1.442	1.755	2.324	5.061	5.159	.9132
.1530	.1858	1.167	.8234	1.451	1.762	2.335	5.115	5.212	.9132
.1540	.1866	1.173	.8250	1.460	1.770	2.345	5.169	5.265	.9132
.1550	.1875	1.178	.8267	1.469	1.777	2.356	5.225	5.320	.9131
.1560	.1883	1.183	.8284	1.479	1.785	2.366	5.283	5.376	.9130
.1570	.1891	1.188	.8301	1.488	1.793	2.377	5.339	5.432	.9129
.1580	.1900	1.194	.8317	1.498	1.801	2.387	5.398	5.490	.9130
.1590	.1908	1.199	.8333	1.507	1.809	2.398	5.454	5.544	.9130
.1600	.1917	1.204	.8349	1.517	1.817	2.408	5.513	5.603	.9130
.1610	.1925	1.209	.8365	1.527	1.825	2.419	5.571	5.660	.9130
.1620	.1933	1.215	.8381	1.536	1.833	2.429	5.630	5.718	.9130
.1630	.1941	1.220	.8396	1.546	1.841	2.440	5.690	5.777	.9130
.1640	.1950	1.225	.8411	1.555	1.849	2.450	5.751	5.837	.9130
.1650	.1958	1.230	.8427	1.565	1.857	2.461	5.813	5.898	.9131
.1660	.1966	1.235	.8442	1.574	1.865	2.471	5.874	5.959	.9132
.1670	.1975	1.240	.8457	1.584	1.873	2.482	5.938	6.021	.9132
.1680	.1983	1.246	.8472	1.594	1.882	2.492	6.003	6.085	.9133
.1690	.1992	1.251	.8486	1.604	1.890	2.503	6.066	6.148	.9133
.1700	.2000	1.257	.8501	1.614	1.899	2.513	6.130	6.212	.9134
.1710	.2008	1.262	.8515	1.624	1.907	2.523	6.197	6.275	.9135
.1720	.2017	1.267	.8529	1.634	1.915	2.534	6.262	6.342	.9136
.1730	.2025	1.272	.8544	1.644	1.924	2.544	6.329	6.407	.9137
.1740	.2033	1.277	.8558	1.654	1.933	2.555	6.395	6.473	.9138
.1750	.2042	1.282	.8572	1.664	1.941	2.565	6.465	6.541	.9139
.1760	.2050	1.288	.8585	1.675	1.951	2.576	6.534	6.610	.9140
.1770	.2058	1.293	.8600	1.685	1.959	3.586	6.603	6.679	.9141
.1780	.2066	1.298	.8614	1.695	1.968	2.597	6.672	6.747	.9142
.1790	.2075	1.304	.8627	1.706	1.977	2.607	6.744	6.818	.9144
.1800	.2083	1.309	.8640	1.716	1.986	2.618	6.818	6.891	.9145
.1810	.2092	1.314	.8653	1.727	1.995	2.629	6.890	6.963	.9146
.1820	.2100	1.320	.8666	1.737	2.004	2.639	6.963	7.035	.9148
.1830	.2108	1.325	.8680	1.748	2.013	2.650	7.038	7.109	.9149
.1840	.2117	1.330	.8693	1.758	2.022	2.660	7.113	7.183	.9150
.1850	.2125	1.335	.8706	1.769	2.032	2.671	7.191	7.260	.9152
.1860	.2134	1.341	.8718	1.780	2.041	2.681	7.267	7.336	.9154
.1870	.2142	1.346	.8731	1.791	2.051	2.692	7.345	7.412	.9155
.1880	.2150	1.351	.8743	1.801	2.060	2.702	7.421	7.488	.9157
.1890	.2159	1.356	.8755	1.812	2.070	2.712	7.500	7.566	.9159

TABLE OF FUNCTIONS OF d/L_o

d/L_o	d/L	$2\pi d/L$	tanh $2\pi d/L$	sinh $2\pi d/L$	cosh $2\pi d/L$	$4\pi d/L$	sinh $4\pi d/L$	cosh $4\pi d/L$	H/H_o'
.1900	.2167	1.362	.8767	1.823	2.079	2.723	7.581	7.647	.9161
.1910	.2176	1.367	.8779	1.834	2.089	2.734	7.663	7.728	.9163
.1920	.2184	1.372	.8791	1.845	2.099	2.744	7.746	7.810	.9165
.1930	.2192	1.377	.8803	1.856	2.108	2.755	7.827	7.891	.9167
.1940	.2201	1.383	.8815	1.867	2.118	2.765	7.911	7.974	.9169
.1950	.2209	1.388	.8827	1.879	2.128	2.776	7.996	8.059	.9170
.1960	.2218	1.393	.8839	1.890	2.138	2.787	8.083	8.145	.9172
.1970	.2226	1.399	.8850	1.901	2.148	2.797	8.167	8.228	.9174
.1980	.2234	1.404	.8862	1.913	2.158	2.808	8.256	8.316	.9176
.1990	.2243	1.409	.8873	1.924	2.169	2.819	8.346	8.406	.9179
.2000	.2251	1.414	.8884	1.935	2.178	2.829	8.436	8.495	.9181
.2010	.2260	1.420	.8895	1.947	2.189	2.804	8.524	8.583	.9183
.2020	.2268	1.425	.8906	1.959	2.199	2.850	8.616	8.674	.9186
.2030	.2277	1.430	.8917	1.970	2.210	2.861	8.708	8.766	.9188
.2040	.2285	1.436	.8928	1.982	2.220	2.872	8.803	8.860	.9190
.2050	.2293	1.441	.8939	1.994	2.231	2.882	8.897	8.953	.9193
.2060	.2302	1.446	.8950	2.006	2.242	2.893	8.994	9.050	.9195
.2070	.2310	1.451	.8960	2.017	2.252	2.903	9.090	9.144	.9197
.2080	.2319	1.457	.8971	2.030	2.263	2.914	9.187	9.240	.9200
.2090	.2328	1.462	.8981	2.042	2.274	2.925	9.288	9.342	.9202
.2100	.2336	1.468	.8991	2.055	2.285	2.936	9.389	9.442	.9205
.2110	.2344	1.473	.9001	2.066	2.295	2.946	9.490	9.542	.9207
.2120	.2353	1.479	.9011	2.079	2.307	2.957	9.590	9.642	.9210
.2130	.2361	1.484	.9021	2.091	2.318	2.967	9.693	9.744	.9213
.2140	.2370	1.489	.9031	2.103	2.329	2.978	9.796	9.847	.9215
.2150	.2378	1.494	.9041	2.115	2.340	2.989	9.902	9.952	.9218
.2160	.2387	1.500	.9051	2.128	2.351	2.999	10.01	10.06	.9221
.2170	.2395	1.506	.9061	2.142	2.364	3.010	10.12	10.17	.9223
.2180	.2404	1.511	.9070	2.154	2.375	3.021	10.23	10.28	.9226
.2190	.2412	1.516	.9079	2.166	2.386	3.031	10.34	10.38	.9228
.2200	.2421	1.521	.9088	2.178	2.397	3.042	10.45	10.50	.9231
.2210	.2429	1.526	.9097	2.192	2.409	3.052	10.56	10.61	.9234
.2220	.2438	1.532	.9107	2.204	2.421	3.063	10.68	10.72	.9236
.2230	.2446	1.537	.9116	2.218	2.433	3.074	10.79	10.84	.9239
.2240	.2455	1.542	.9125	2.230	2.444	3.085	10.91	10.95	.9242
.2250	.2463	1.548	.9134	1.244	2.457	3.095	11.02	11.07	.9245
.2260	.2472	1.553	.9143	2.257	2.469	3.106	11.15	11.19	.9248
.2270	.2481	1.559	.9152	2.271	2.481	3.117	11.27	11.31	.9251
.2280	.2489	1.564	.9161	2.284	2.493	3.128	11.39	11.44	.9254
.2290	.2498	1.569	.9170	2.297	2.506	3.138	11.51	11.56	.9258

TABLE OF FUNCTIONS OF d/L_o

d/L_o	d/L	$2\pi d/L$	tanh $2\pi d/L$	sinh $2\pi d/L$	cosh $2\pi d/L$	$4\pi d/L$	sinh $4\pi d/L$	cosh $4\pi d/L$	H/H_o'
.2300	.2506	1.575	.9178	2.311	2.518	3.149	11.64	11.68	.9261
.2310	.2515	1.580	.9186	2.325	2.531	3.160	11.77	11.81	.9264
.2320	.2523	1.585	.9194	2.338	2.543	3.171	11.90	11.93	.9267
.2330	.2532	1.591	.9203	2.352	2.556	3.182	12.03	12.07	.9270
.2340	.2540	1.596	.9211	2.366	2.569	3.192	12.15	12.19	.9273
.2350	.2549	1.602	.9219	2.380	2.581	3.203	12.29	12.33	.9276
.2360	.2558	1.607	.9227	2.393	2.594	3.214	12.43	12.47	.9279
.2370	.2566	1.612	.9235	2.408	2.607	3.225	12.55	12.59	.9282
.2380	.2575	1.618	.9243	2.422	2.620	3.236	12.69	12.73	.9285
.2390	.2584	1.623	.9251	2.436	2.634	3.247	12.83	12.87	.9288
.2400	.2592	1.629	.9259	2.450	2.647	3.257	12.97	13.01	.9291
.2410	.2601	1.634	.9267	2.464	2.660	3.268	13.11	13.15	.9294
.2420	.2610	1.640	.9275	2.480	2.674	3.279	13.26	13.30	.9298
.2430	.2618	1.645	.9282	2.494	2.687	3.290	13.40	13.44	.9301
.2440	.2627	1.650	.9289	2.508	2.700	3.301	13.55	13.59	.9304
.2450	.2635	1.656	.9296	2.523	2.714	3.312	13.70	13.73	.9307
.2460	.2644	1.661	.9304	2.538	2.728	3.323	13.85	13.88	.9310
.2470	.2653	1.667	.9311	2.553	2.742	3.334	14.00	14.04	.9314
.2480	.2661	1.672	.9318	2.568	2.755	3.344	14.15	14.19	.9317
.2490	.2670	1.678	.9325	2.583	2.770	3.355	14.31	14.35	.9320
.2500	.2679	1.683	.9332	2.599	2.784	3.367	14.47	14.51	.9323
.2510	.2687	1.689	.9339	2.614	2.798	3.377	14.62	14.66	.9327
.2520	.2696	1.694	.9346	2.629	2.813	3.388	14.79	14.82	.9330
.2530	.2705	1.700	.9353	2.645	2.828	3.399	14.95	14.99	.9333
.2540	.2714	1.705	.9360	2.660	2.842	3.410	15.12	15.15	.9336
.2550	.2722	1.711	.9367	2.676	2.856	3.421	15.29	15.32	.9340
.2560	.2731	1.716	.9374	2.691	2.871	3.432	15.45	15.49	.9343
.2570	.2740	1.722	.9381	2.707	2.886	3.443	15.63	15.66	.9346
.2580	.2749	1.727	.9388	2.723	2.901	3.454	15.80	15.83	.9349
.2590	.2757	1.732	.9394	2.739	2.916	3.465	15.97	16.00	.9353
.2600	.2766	1.738	.9400	2.755	2.931	3.476	16.15	16.18	.9356
.2610	.2775	1.744	.9406	2.772	2.946	3.487	16.33	16.36	.9360
.2620	.2784	1.749	.9412	2.788	2.962	3.498	16.51	16.54	.9363
.2630	.2792	1.755	.9418	2.804	2.977	3.509	16.69	16.73	.9367
.2640	.2801	1.760	.9425	2.820	2.992	3.520	16.88	16.91	.9370
.2650	.2810	1.766	.9431	2.837	3.008	3.531	17.07	17.10	.9373
.2660	.2819	1.771	.9437	2.853	3.023	3.542	17.26	17.28	.9377
.2670	.2827	1.776	.9443	2.870	3.039	3.553	17.45	17.45	.9380
.2680	.2836	1.782	.9449	2.886	3.055	3.564	17.64	17.67	.9383
.2690	.2845	1.788	.9455	2.904	3.071	3.575	17.84	17.87	.9386

TABLE OF FUNCTIONS OF d/L_o

d/L_o	d/L	$2\pi d/L$	tanh $2\pi d/L$	sinh $2\pi d/L$	cosh $2\pi d/L$	$4\pi d/L$	sinh $4\pi d/L$	cosh $4\pi d/L$	H/H_o'
.2700	.2854	1.793	.9461	2.921	3.088	3.587	18.04	18.07	.9390
.2710	.2863	1.799	.9467	2.938	3.104	3.598	18.24	18.27	.9393
.2720	.2872	1.804	.9473	2.956	3.120	3.610	18.46	18.49	.9396
.2730	.2880	1.810	.9478	2.973	3.136	3.620	18.65	18.67	.9400
.2740	.2889	1.815	.9484	2.990	3.153	3.631	18.86	18.89	.9403
.2750	.2898	1.821	.9490	3.008	3.170	3.642	19.07	19.10	.9406
.2760	.2907	1.826	.9495	3.025	3.186	3.653	19.28	19.30	.9410
.2770	.2916	1.832	.9500	3.043	3.203	3.664	19.49	19.51	.9413
.2780	.2924	1.837	.9505	3.061	3.220	3.675	19.71	19.74	.9416
.2790	.2933	1.843	.9511	3.079	3.237	3.686	19.93	19.96	.9420
.2800	.2942	1.849	.9516	3.097	3.254	3.697	20.16	20.18	.9423
.2810	.2951	1.854	.9521	3.115	3.272	3.709	20.39	20.41	.9426
.2820	.2960	1.860	.9526	3.133	3.289	3.720	20.62	20.64	.9430
.2830	.2969	1.866	.9532	3.152	3.307	3.731	20.85	20.87	.9433
.2840	.2978	1.871	.9537	3.171	3.325	3.742	21.09	21.11	.9436
.2850	.2987	1.877	.9542	3.190	3.343	3.754	21.33	21.35	.9440
.2860	.2996	1.882	.9547	3.209	3.361	3.765	21.57	21.59	.9443
.2870	.3005	1.888	.9552	3.228	3.379	3.776	21.82	21.84	.9446
.2880	.3014	1.893	.9557	3.246	3.396	3.787	22.05	22.07	.9449
.2890	.3022	1.899	.9562	3.264	3.414	3.798	22.30	22.32	.9452
.2900	.3031	1.905	.9567	3.284	3.433	3.809	22.54	22.57	.9456
.2910	.3040	1.910	.9572	3.303	3.451	3.821	22.81	22.83	.9459
.2920	.3049	1.916	.9577	3.323	3.471	3.832	23.07	23.09	.9463
.2930	.3058	1.922	.9581	3.343	3.490	3.843	23.33	23.35	.9466
.2940	.3067	1.927	.9585	3.362	3.508	3.855	23.60	23.62	.9469
.2950	.3076	1.933	.9590	3.382	3.527	3.866	23.86	23.88	.9473
.2960	.3085	1.938	.9594	3.402	3.546	3.877	24.12	24.15	.9476
.2970	.3094	1.944	.9599	3.422	3.565	3.888	24.40	24.42	.9480
.2980	.3103	1.950	.9603	3.442	3.585	3.900	24.68	24.70	.9483
.2990	.3112	1.955	.9607	3.462	3.604	3.911	24.96	24.98	.9486
.3000	.3121	1.961	.9611	3.483	3.624	3.922	25.24	25.26	.9490
.3010	.3130	1.967	.9616	3.503	3.643	3.933	25.53	25.55	.9493
.3020	.3139	1.972	.9620	3.524	3.663	3.945	25.82	25.83	.9496
.3030	.3148	1.978	.9624	3.545	3.683	3.956	26.12	26.14	.9499
.3040	.3157	1.984	.9629	3.566	3.703	3.968	26.42	26.44	.9502
.3050	.3166	1.989	.9633	3.587	3.724	3.979	26.72	26.74	.9505
.3060	.3175	1.995	.9637	3.609	3.745	3.990	27.02	27.04	.9509
.3070	.3184	2.001	.9641	3.630	3.765	4.002	27.33	27.35	.9512
.3080	.3193	2.007	.9645	3.651	3.786	4.013	27.65	27.66	.9515
.3090	.3202	2.012	.9649	3.673	3.806	4.024	27.96	27.98	.9518

TABLE OF FUNCTIONS OF d/L_o

d/L_o	d/L	$2\pi d/L$	tanh $2\pi d/L$	sinh $2\pi d/L$	cosh $2\pi d/L$	$4\pi d/L$	sinh $4\pi d/L$	cosh $4\pi d/L$	H/H_o'
.3100	.3211	2.018	.9653	3.694	3.827	4.036	28.28	28.30	.9522
.3110	.3220	2.023	.9656	3.716	3.848	4.047	28.60	28.62	.9525
.3120	.3230	2.029	.9660	3.738	3.870	4.058	28.93	28.95	.9528
.3130	.3239	2.035	.9664	3.760	3.891	4.070	29.27	29.28	.9531
.3140	.3248	2.041	.9668	3.782	3.912	4.081	29.60	29.62	.9535
.3150	.3257	2.046	.9672	3.805	3.934	4.093	29.94	29.96	.9538
.3160	.3266	2.052	.9676	3.828	3.956	4.104	30.29	30.31	.9541
.3170	.3275	2.058	.9679	3.851	3.978	4.116	30.64	30.65	.9544
.3180	.3284	2.063	.9682	3.873	4.000	4.127	30.99	31.00	.9547
.3190	.3294	2.069	.9686	3.896	4.022	4.139	31.35	31.37	.9550
.3200	.3302	2.075	.9690	3.919	4.045	4.150	31.71	31.72	.9553
.3210	.3311	2.081	.9693	3.943	4.068	4.161	32.07	32.08	.9556
.3220	.3321	2.086	.9696	3.966	4.090	4.173	32.44	32.46	.9559
.3230	.3330	2.092	.9700	3.990	4.114	4.185	32.83	32.84	.9562
.3240	.3339	2.098	.9703	4.014	4.136	4.196	33.20	33.22	.9565
.3250	.3349	2.104	.9707	4.038	4.160	4.208	33.60	33.61	.9568
.3260	.3357	2.110	.9710	4.061	4.183	4.219	33.97	33.99	.9571
.3270	.3367	2.115	.9713	4.085	4.206	4.231	34.37	34.38	.9574
.3280	.3376	2.121	.9717	4.110	4.230	4.242	34.77	34.79	.9577
.3290	.3385	2.127	.9720	4.135	4.254	4.254	35.18	35.19	.9580
.3300	.3394	2.133	.9723	4.159	4.277	4.265	35.58	35.59	.9583
.3310	.3403	2.138	.9726	4.184	4.301	4.277	35.99	36.00	.9586
.3320	.3413	2.144	.9729	4.209	4.326	4.288	36.42	36.43	.9589
.3330	.3422	2.150	.9732	4.234	4.350	4.300	36.84	36.85	.9592
.3340	.3431	2.156	.9735	4.259	4.375	4.311	37.25	37.27	.9595
.3350	.3440	2.161	.9738	4.284	4.399	4.323	37.70	37.72	.9598
.3360	.3449	2.167	.9741	4.310	4.424	4.335	38.14	38.15	.9601
.3370	.3459	2.173	.9744	4.336	4.450	4.346	38.59	38.60	.9604
.3380	.3468	2.179	.9747	4.361	4.474	4.358	39.02	39.04	.9607
.3390	.3477	2.185	.9750	4.388	4.500	4.369	39.48	39.49	.9610
.3400	.3468	2.190	.9753	4.413	4.525	4.381	39.95	39.96	.9613
.3410	.3495	2.196	.9756	4.439	4.550	4.392	40.40	40.41	.9615
.3420	.3504	2.202	.9758	4.466	4.576	4.404	40.87	40.89	.9618
.3430	.3514	2.208	.9761	4.492	4.602	4.416	41.36	41.37	.9621
.3440	.3523	2.214	.9764	4.521	4.630	4.427	41.85	41.84	.9623
.3450	.3532	2.220	.9767	4.547	4.656	4.439	42.33	42.34	.9626
.3460	.3542	2.225	.9769	4.575	4.682	4.451	42.83	42.84	.9629
.3470	.3551	2.231	.9772	4.602	4.709	4.462	43.34	43.35	.9632
.3480	.3560	2.237	.9775	4.629	4.736	4.474	43.85	43.86	.9635
.3490	.3570	2.243	.9777	4.657	4.763	4.486	44.37	44.40	.9638

TABLE OF FUNCTIONS OF d/L_o

d/L_o	d/L	$2\pi d/L$	tanh $2\pi d/L$	sinh $2\pi d/L$	cosh $2\pi d/L$	$4\pi d/L$	sinh $4\pi d/L$	cosh $4\pi d/L$	H/H_o'
.3500	.3579	2.249	.9780	4.685	4.791	4.498	44.89	44.80	.9640
.3510	.3588	2.255	.9782	4.713	4.818	4.509	45.42	45.43	.9643
.3520	.3598	2.260	.9785	4.741	4.845	4.521	45.95	45.96	.9646
.3530	.3607	2.266	.9787	4.770	4.873	4.533	46.50	46.51	.9648
.3540	.3616	2.272	.9790	4.798	4.901	4.544	47.03	47.04	.9651
.3550	.3625	2.278	.9792	4.827	4.929	4.556	47.59	47.60	.9654
.3560	.3635	2.284	.9795	4.856	4.957	4.568	48.15	48.16	.9657
.3570	.3644	2.290	.9797	4.885	4.987	4.579	48.72	48.73	.9659
.3580	.3653	2.296	.9799	4.914	5.015	4.591	49.29	49.30	.9662
.3590	.3663	2.301	.9801	4.944	5.044	4.603	49.88	49.89	.9665
.3600	.3672	2.307	.9804	4.974	5.072	4.615	50.47	50.48	.9667
.3610	.3682	2.313	.9806	5.004	5.103	4.627	51.08	51.09	.9670
.3620	.3691	2.319	.9808	5.034	5.132	4.638	51.67	51.67	.9673
.3630	.3700	2.325	.9811	5.063	5.161	4.650	52.27	52.28	.9675
.3640	.3709	2.331	.9813	5.094	5.191	4.661	52.89	52.90	.9677
.3650	.3719	2.337	.9815	5.124	5.221	4.673	53.52	53.53	.9680
.3660	.3728	2.342	.9817	5.155	5.251	4.685	54.15	54.16	.9683
.3670	.3737	2.348	.9819	5.186	5.281	4.697	54.78	54.79	.9686
.3680	.3747	2.354	.9821	5.217	5.312	4.708	55.42	55.43	.9688
.3390	.3756	2.360	.9823	5.248	5.343	4.720	56.09	56.10	.9690
.3700	.3766	2.366	.9825	5.280	5.374	4.732	56.76	56.77	.9693
.3710	.3775	2.372	.9827	5.312	5.406	4.744	57.43	54.44	.9696
.3720	.3785	2.378	.9830	5.345	5.438	4.756	58.13	58.14	.9698
.3730	.3794	2.384	.9832	5.377	5.469	4.768	58.82	58.83	.9700
.3740	.3804	2.390	.9834	5.410	5.502	4.780	59.52	59.53	.9702
.3750	.3813	2.396	.9835	5.443	5.534	4.792	60.24	60.25	.9705
.3760	.3822	2.402	.9837	5.475	5.566	4.803	60.95	60.95	.9707
.3770	.3832	2.408	.9839	5.508	5.598	4.815	61.68	61.68	.9709
.3780	.3841	2.413	.9841	5.541	5.631	4.827	62.41	62.42	.9712
.3790	.3850	2.419	.9843	5.572	5.661	4.838	63.13	63.14	.9714
.3800	.3860	2.425	.9845	5.609	5.697	4.851	63.90	63.91	.9717
.3810	.3869	2.431	.9847	5.643	5.731	4.862	64.66	64.67	.9719
.3820	.3879	2.437	.9848	5.677	5.765	4.875	65.45	65.46	.9721
.3830	.3888	2.443	.9850	5.712	5.798	4.885	66.20	66.21	.9724
.3840	.3898	2.449	.9852	5.746	5.833	4.898	67.00	67.01	.9726
.3850	.3907	2.455	.9854	5.780	5.866	4.910	67.80	67.81	.9728
.3860	.3917	2.461	.9855	5.814	5.900	4.922	68.61	68.62	.9730
.3870	.3926	2.467	.9857	5.850	5.935	4.934	69.45	69.46	.9732
.3880	.3936	2.473	.9859	5.886	5.970	4.946	70.28	70.29	.9735
.3890	.3945	2.479	.9860	5.921	6.005	4.958	71.12	71.13	.9737

TABLE OF FUNCTIONS OF d/L_o

d/L_o	d/L	$2\pi d/L$	tanh $2\pi d/L$	sinh $2\pi d/L$	cosh $2\pi d/L$	$4\pi d/L$	sinh $4\pi d/L$	cosh $4\pi d/L$	H/H_o'
.3900	.3955	2.485	.9862	5.957	6.040	4.970	71.97	71.98	.9739
.3910	.3964	2.491	.9864	5.993	6.076	4.982	72.85	72.86	.9741
.3920	.3974	2.497	.9865	6.029	6.112	4.993	73.72	73.72	.9743
.3930	.3983	2.503	.9867	6.066	6.148	5.005	74.59	74.59	.9745
.3940	.3993	2.509	.9869	6.103	6.185	5.017	75.48	75.48	.9748
.3950	.4002	2.515	.9870	6.140	6.221	5.029	76.40	76.40	.9750
.3960	.4012	2.521	.9872	6.177	6.258	5.041	77.32	77.32	.9752
.3970	.4021	2.527	.9873	6.215	6.295	5.053	78.24	78.24	.9754
.3980	.4031	2.532	.9874	6.252	6.332	5.065	79.19	79.19	.9756
.3990	.4040	2.538	.9876	6.290	6.369	5.077	80.13	80.13	.9758
.4000	.4050	2.544	.9877	6.329	6.407	5.089	81.12	81.12	.9761
.4010	.4059	2.550	.9879	6.367	6.445	5.101	82.08	82.08	.9763
.4020	.4069	2.556	.9880	6.406	6.483	5.113	83.06	83.06	.9765
.4030	.4078	2.562	.9882	6.444	6.521	5.125	84.07	84.07	.9766
.4040	.4088	2.568	.9883	6.484	6.561	5.137	85.11	85.11	.9768
.4050	.4098	2.575	.9885	6.525	6.601	5.149	86.14	86.14	.9777
.4060	.4107	2.581	.9886	6.564	6.640	5.161	87.17	87.17	.9772
.4070	.4116	2.586	.9887	6.603	6.679	5.173	88.20	88.20	.9774
.4080	.4126	2.592	.9889	6.644	6.718	5.185	89.28	89.28	.9776
.4090	.4136	2.598	.9890	6.684	6.758	5.197	90.39	90.39	.9778
.4100	.4145	2.604	.9891	6.725	6.799	5.209	91.44	91.44	.9780
.4110	.4155	2.610	.9892	6.766	6.839	5.221	92.55	92.55	.9782
.4120	.4164	2.616	.9894	6.806	6.879	5.233	93.67	93.67	.9784
.4130	.4174	2.623	.9895	6.849	6.921	5.245	94.83	94.83	.9786
.4140	.4183	2.629	.9896	6.890	6.963	5.257	95.96	95.96	.9788
.4150	.4193	2.635	.9898	6.932	7.004	5.269	97.13	97.13	.9790
.4160	.4203	2.641	.9899	6.974	7.046	5.281	98.30	98.30	.9792
.4170	.4212	2.647	.9900	7.018	7.088	5.294	99.52	99.52	.9794
.4180	.4222	2.653	.9901	7.060	7.130	5.305	100.7	100.7	.9795
.4190	.4231	2.659	.9902	7.102	7.173	5.317	101.9	101.9	.9797
.4200	.4241	2.665	.9904	7.146	7.215	5.329	103.1	103.1	.9798
.4210	.4251	2.671	.9905	7.190	7.259	5.341	104.4	104.4	.9800
.4220	.4260	2.677	.9906	7.234	7.303	5.353	105.7	105.7	.9802
.4230	.4270	2.683	.9907	7.279	7.349	5.366	107.0	107.0	.9804
.4240	.4280	2.689	.9908	7.325	7.392	5.378	108.3	108.3	.9806
.4250	.4289	2.695	.9909	7.371	7.438	5.390	109.7	109.7	.9808
.4260	.4298	2.701	.9910	7.412	7.479	5.402	110.9	110.9	.9810
.4270	.4308	2.707	.9911	7.457	7.524	5.414	112.2	112.2	.9811
.4280	.4318	2.713	.9912	7.503	7.570	5.426	113.6	113.6	.9812
.4290	.4328	2.719	.9913	7.550	7.616	5.438	115.0	115.0	.9814

TABLE OF FUNCTIONS OF d/L_o

d/L_o	d/L	$2\pi d/L$	tanh $2\pi d/L$	sinh $2\pi d/L$	cosh $2\pi d/L$	$4\pi d/L$	sinh $4\pi d/L$	cosh $4\pi d/L$	H/H_o'
.4300	.4337	2.725	.9914	7.595	7.661	5.450	116.4	116.4	.9816
.4310	.4347	2.731	.9915	7.642	7.707	5.462	117.8	117.8	.9818
.4320	.4356	2.737	.9916	7.688	7.753	5.474	119.2	119.2	.9819
.4330	.4366	2.743	.9917	7.735	7.800	5.486	120.7	120.7	.9821
.4340	.4376	2.749	.9918	7.783	7.847	5.499	122.2	122.2	.9823
.4350	.4385	2.755	.9919	7.831	7.895	5.511	123.7	123.7	.9824
.4360	.4395	2.762	.9920	7.880	7.943	5.523	125.2	125.2	.9826
.4370	.4405	2.768	.9921	7.922	7.991	5.535	126.7	126.7	.9828
.4380	.4414	2.774	.9922	7.975	8.035	5.547	128.3	128.3	.9829
.4390	.4424	2.780	.9923	8.026	8.088	5.560	129.9	129.9	.9830
.4400	.4434	2.786	.9924	8.075	8.136	5.572	131.4	131.4	.9832
.4410	.4443	2.792	.9925	8.124	8.185	5.584	133.0	133.0	.9833
.4420	.4453	2.798	.9926	8.175	8.236	5.596	134.7	134.7	.9835
.4430	.4463	2.804	.9927	8.228	8.285	5.608	136.3	136.3	.9836
.4440	.4472	2.810	.9928	8.274	8.334	5.620	137.9	137.9	.9838
.4450	.4482	2.816	.9929	8.326	8.387	5.632	139.6	139.6	.9839
.4460	.4492	2.822	.9930	8.379	8.438	5.644	141.4	141.4	.9841
.4470	.4501	2.828	.9930	8.427	8.486	5.657	143.1	143.1	.9843
.4480	.4511	2.834	.9931	8.481	8.540	5.669	144.8	144.8	.9844
.4490	.4521	2.840	.9932	8.532	8.590	5.681	146.6	146.6	.9846
.4500	.4531	2.847	.9933	8.585	8.643	5.693	148.4	148.4	.9847
.4510	.4540	2.853	.9934	8.638	8.695	5.705	150.2	150.2	.9848
.4520	.4550	2.859	.9935	8.693	8.750	5.717	152.1	152.1	.9849
.4530	.4560	2.865	.9935	8.747	8.804	5.730	154.0	154.0	.9851
.4540	.4569	2.871	.9936	8.797	8.854	5.742	155.9	155.9	.9852
.4550	.4579	2.877	.9937	8.853	8.910	5.754	157.7	157.7	.9853
.4560	.4589	2.883	.9938	8.910	8.965	5.766	159.7	159.7	.9855
.4570	.4599	2.890	.9938	8.965	9.021	5.779	161.7	161.7	.9857
.4580	.4608	2.896	.9939	9.016	9.072	5.791	163.6	163.6	.9858
.4590	.4618	2.902	.9940	9.074	9.129	5.803	165.6	165.6	.9859
.4600	.4628	2.908	.9941	9.132	9.186	5.815	167.7	167.7	.9860
.4610	.4637	2.914	.9941	9.183	9.238	5.827	169.7	169.7	.9862
.4620	.4647	2.920	.9942	9.242	9.296	5.840	171.8	171.8	.9863
.4630	.4657	2.926	.9943	9.301	9.354	5.852	173.9	173.9	.9864
.4640	.4666	2.932	.9944	9.353	9.406	5.864	176.0	176.0	.9865
.4650	.4676	2.938	.9944	9.413	9.466	5.876	178.2	178.2	.9867
.4660	.4686	2.944	.9945	9.472	9.525	5.888	180.4	180.4	.9868
.4670	.4695	2.951	.9946	9.533	9.585	5.900	182.6	182.6	.9869
.4680	.4705	2.957	.9946	9.586	9.638	5.912	184.8	184.8	.9871
.4690	.4715	2.963	.9947	9.647	9.699	5.925	187.2	187.2	.9872

TABLE OF FUNCTIONS OF d/L_o

d/L_o	d/L	$2\pi d/L$	tanh $2\pi d/L$	sinh $2\pi d/L$	cosh $2\pi d/L$	$4\pi d/L$	sinh $4\pi d/L$	cosh $4\pi d/L$	H/H_o'
.4700	.4725	2.969	.9947	9.709	9.760	5.937	189.5	189.5	.9873
.4710	.4735	2.975	.9948	9.770	9.821	5.949	191.8	191.8	.9874
.4720	.4744	2.981	.9949	9.826	9.877	5.962	194.2	194.2	.9875
.4730	.4754	2.987	.9949	9.888	9.938	5.974	196.5	196.5	.9876
.4740	.4764	2.993	.9950	9.951	10.00	5.986	199.0	199.0	.9877
.4750	.4774	2.999	.9951	10.01	10.07	5.999	201.4	201.4	.9878
.4760	.4783	3.005	.9951	10.07	10.12	6.011	203.9	203.9	.9880
.4770	.4793	3.012	.9952	10.13	10.18	6.023	206.5	206.5	.9881
.4780	.4803	3.018	.9952	10.20	10.25	6.036	209.0	209.0	.9882
.4790	.4813	3.024	.9953	10.26	10.31	6.048	211.7	211.7	.9883
.4800	.4822	3.030	.9953	10.32	10.37	6.060	214.2	214.2	.9885
.4810	.4832	3.036	.9954	10.39	10.43	6.072	216.8	216.8	.9886
.4820	.4842	3.042	.9955	10.45	10.50	6.085	219.5	219.5	.9887
.4830	.4852	3.049	.9955	10.52	10.57	6.097	222.2	222.2	.9888
.4840	.4862	3.055	.9956	10.59	10.63	6.109	225.0	225.0	.9889
.4850	.4871	3.061	.9956	10.65	10.69	6.121	228.3	228.3	.9890
.4860	.4881	3.067	.9957	10.71	10.76	6.134	230.6	230.6	.9891
.4870	.4891	3.073	.9957	10.78	10.83	6.146	233.5	233.5	.9892
.4880	.4901	3.079	.9958	10.85	10.90	6.159	236.4	236.4	.9893
.4890	.4911	3.086	.9958	10.92	10.96	6.171	239.6	239.6	.9895
.4900	.4920	3.092	.9959	10.99	11.03	6.183	242.3	242.3	.9896
.4910	.4930	3.098	.9959	11.05	11.09	6.195	245.2	245.2	.9897
.4920	.4940	3.104	.9960	11.12	11.16	6.208	248.3	248.3	.9898
.4930	.4950	3.110	.9960	11.19	11.24	6.220	251.3	251.3	.9899
.4940	.4960	3.117	.9961	11.26	11.31	6.232	254.5	254.5	.9899
.4950	.4969	3.122	.9961	11.32	11.37	6.245	257.6	257.6	.9900
.4960	.4979	3.128	.9962	11.40	11.44	6.257	260.8	260.8	.9901
.4970	.4989	3.135	.9962	11.47	11.51	6.269	264.0	264.0	.9902
.4980	.4999	3.141	.9963	11.54	11.59	6.283	267.3	267.3	.9903
.4990	.5009	3.147	.9963	11.61	11.65	6.294	270.6	270.6	.9904
.5000	.5018	3.153	.9964	11.68	11.72	6.306	274.0	274.0	.9905
.5010	.5028	3.159	.9964	11.75	11.80	6.319	277.5	277.5	.9906
.5020	.5038	3.166	.9964	11.83	11.87	6.331	280.8	280.8	.9907
.5030	.5048	3.172	.9965	11.91	11.95	6.343	284.3	284.3	.9908
.5040	.5058	3.178	.9965	11.98	12.02	6.356	287.9	287.9	.9909
.5050	.5067	3.184	.9966	12.05	12.09	6.368	291.4	291.4	.9909
.5060	.5077	3.190	.9966	12.12	12.16	6.380	295.0	295.0	.9910
.5070	.5087	3.196	.9967	12.20	12.24	6.393	298.7	298.7	.9911
.5080	.5097	3.203	.9967	12.28	12.12	6.405	302.4	302.4	.9912
.5090	.5107	3.209	.9968	12.35	12.19	6.417	306.2	306.2	.9913

TABLE OF FUNCTIONS OF d/L_o

d/L_o	d/L	$2\pi d/L$	tanh $2\pi d/L$	sinh $2\pi d/L$	cosh $2\pi d/L$	$4\pi d/L$	sinh $4\pi d/L$	cosh $4\pi d/L$	H/H_o'
.5100	.5117	3.215	.9968	12.43	12.47	6.430	310.0	310.0	.9914
.5110	.5126	3.221	.9968	12.50	12.54	6.442	313.8	313.8	.9915
.5120	.5136	3.227	.9969	12.58	12.62	6.454	317.7	317.7	.9915
.5130	.5146	3.233	.9969	12.66	12.70	6.467	321.7	321.7	.9916
.5140	.5156	3.240	.9970	12.74	12.78	6.479	325.7	325.7	.9917
.5150	.5166	3.246	.9970	12.82	12.86	6.491	329.7	329.7	.9918
.5160	.5176	3.252	.9970	12.90	12.94	6.504	333.8	333.8	.9919
.5170	.5185	3.258	.9971	12.98	11.02	6.516	337.9	337.9	.9919
.5180	.5195	3.264	.9971	13.06	13.10	6.529	342.2	342.2	.9920
.5190	.5205	3.270	.9971	13.14	13.18	6.541	346.4	346.4	.9921
.5200	.5215	3.277	.9972	13.22	13.26	6.553	350.7	350.7	.9922
.5210	.5225	3.283	.9972	13.31	13.35	6.566	355.1	355.1	.9923
.5220	.5235	3.289	.9972	13.39	13.43	6.578	359.6	359.6	.9924
.5230	.5244	3.295	.9973	13.47	13.51	6.590	364.0	364.0	.9924
.5240	.5254	3.301	.9973	13.55	13.59	6.603	368.5	368.5	.9925
.5250	.5264	3.308	.9973	13.64	13.68	6.615	373.1	373.1	.9926
.5260	.5274	3.314	.9974	13.73	13.76	6.628	377.8	377.8	.9927
.5270	.5284	3.320	.9974	13.81	13.85	6.640	382.5	382.5	.9927
.5280	.5294	3.326	.9974	13.90	13.94	6.652	387.3	387.3	.9928
.5290	.5304	3.333	.9975	13.99	14.02	6.665	392.2	392.2	.9929
.5300	.5314	3.339	.9975	14.07	14.10	6.677	397.0	397.0	.9930
.5310	.5323	3.345	.9975	14.16	14.19	6.690	402.0	402.0	.9931
.5320	.5333	3.351	.9976	14.25	14.28	6.702	406.9	406.9	.9931
.5330	.5343	3.357	.9976	14.34	14.37	6.714	412.0	412.0	.9932
.5340	.5353	3.363	.9976	14.43	14.46	6.727	417.2	417.2	.9933
.5350	.5363	3.370	.9976	14.52	14.55	6.639	422.4	422.4	.9933
.5360	.5373	3.376	.9977	14.61	14.64	6.752	427.7	427.7	.9934
.5370	.5383	3.382	.9977	14.70	14.73	6.764	433.1	433.1	.9935
.5380	.5393	3.388	.9977	14.79	14.82	6.776	438.5	438.5	.9935
.5390	.5402	3.394	.9977	14.88	14.91	6.789	444.0	444.0	.9936
.5400	.5412	3.401	.9978	14.97	15.01	6.801	449.5	449.5	.9936
.5410	.5422	3.407	.9978	15.07	15.10	6.814	455.1	455.1	.9937
.5420	.5432	3.413	.9978	15.16	15.19	6.826	460.7	460.7	.9938
.5430	.5442	3.419	.9979	15.25	15.29	6.838	466.4	466.4	.9938
.5440	.5452	3.426	.9979	15.35	15.38	6.851	472.2	472.2	.9939
.5450	.5461	3.432	.9979	15.45	15.48	6.861	478.1	478.1	.9940
.5460	.5471	3.438	.9979	15.54	15.58	6.876	484.3	484.3	.9941
.5470	.5481	3.444	.9980	15.64	15.67	6.888	490.3	490.3	.9941
.5480	.5491	3.450	.9980	15.74	15.77	6.901	496.4	496.4	.9942
.5490	.5501	3.456	.9980	15.84	15.87	6.913	502.5	502.5	.9942

TABLE OF FUNCTIONS OF d/L_o

d/L_o	d/L	$2\pi d/L$	tanh $2\pi d/L$	sinh $2\pi d/L$	cosh $2\pi d/L$	$4\pi d/L$	sinh $4\pi d/L$	cosh $4\pi d/L$	H/H_o'
.5500	.5511	3.463	.9980	15.94	15.97	6.925	508.7	508.7	.9942
.5510	.5521	3.469	.9981	16.04	16.07	6.917	515.0	515.0	.9942
.5520	.5531	3.475	.9981	16.14	16.17	6.950	521.6	521.6	.9943
.5530	.5541	3.481	.9981	16.24	16.27	6.962	528.1	528.1	.9944
.5540	.5551	3.488	.9981	16.34	16.37	6.975	534.8	534.8	.9944
.5550	.5560	3.494	.9982	16.44	16.47	6.987	541.4	541.4	.9945
.5560	.5570	3.500	.9982	16.54	16.57	7.000	548.1	548.1	.9945
.5570	.5580	3.506	.9982	16.65	16.68	7.012	554.9	554.9	.9946
.5580	.5590	3.512	.9982	16.75	16.78	7.025	562.0	562.0	.9947
.5590	.5600	3.519	.9982	16.85	16.88	7.037	569.1	569.1	.9947
.5600	.5610	3.525	.9983	16.96	16.99	7.050	576.1	576.1	.9947
.5610	.5620	3.531	.9983	17.06	17.09	7.062	583.3	583.3	.9948
.5620	.5630	3.537	.9983	17.17	17.20	7.074	590.7	590.7	.9949
.5630	.5640	3.543	.9983	17.28	17.31	7.087	598.0	598.0	.9949
.5640	.5649	3.550	.9984	17.38	17.41	7.099	605.0	605.0	.9950
.5650	.5659	3.556	.9984	17.49	17.52	7.112	613.2	613.2	.9950
.5660	.5669	3.562	.9984	17.60	17.63	7.124	620.8	620.8	.9951
.5670	.5679	3.568	.9984	17.71	17.74	7.136	628.5	628.5	.9951
.5680	.5689	3.575	.9984	17.82	17.85	7.149	636.4	636.4	.9952
.5690	.5699	3.581	.9985	17.94	17.97	7.161	644.3	644.3	.9952
.5700	.5709	3.587	.9985	18.05	18.08	7.174	652.4	652.4	.9953
.5710	.5719	3.593	.9985	18.16	18.19	7.186	660.5	660.5	.9953
.5720	.5729	3.600	.9985	18.28	18.31	7.199	668.8	668.8	.9954
.5730	.5738	3.606	.9985	18.39	18.42	7.211	677.2	677.2	.9954
.5740	.5748	3.612	.9985	18.50	18.53	7.224	685.6	685.6	.9955
.5750	.5758	3.618	.9986	18.62	18.64	7.236	694.3	694.3	.9955
.5760	.5768	3.624	.9986	18.73	18.76	7.249	703.2	703.2	.9956
.5770	.5778	3.630	.9986	18.85	18.88	7.261	711.9	711.9	.9956
.5780	.5788	3.637	.9986	18.97	19.00	7.274	720.8	720.8	.9957
.5790	.5798	3.643	.9986	19.09	19.12	7.286	729.9	729.9	.9957
.5800	.5808	3.649	.9987	19.21	19.24	7.298	739.0	739.0	.9957
.5810	.5818	3.656	.9987	19.33	19.36	7.311	748.1	748.1	.9955
.5820	.5828	3.662	.9987	19.45	19.48	7.323	757.5	757.5	.9958
.5830	.5838	3.668	.9987	19.58	19.60	7.336	767.0	767.0	.9959
.5840	.5848	3.674	.9987	19.70	19.73	7.348	776.7	776.7	.9959
.5850	.5858	3.680	.9987	19.81	19.84	7.361	786.5	786.5	.9960
.5860	.5867	3.686	.9987	19.94	19.96	7.373	796.4	796.4	.9960
.5870	.5877	3.693	.9988	20.06	20.09	7.386	806.5	806.5	.9960
.5880	.5887	3.699	.9988	20.19	20.21	7.398	816.5	816.5	.9961
.5890	.5897	3.705	.9988	20.32	20.34	7.411	826.7	826.7	.9961

TABLE OF FUNCTIONS OF d/L_o

d/L_o	d/L	$2\pi d/L$	tanh $2\pi d/L$	sinh $2\pi d/L$	cosh $2\pi d/L$	$4\pi d/L$	sinh $4\pi d/L$	cosh $4\pi d/L$	H/H'_o
.5900	.5907	3.712	.9988	20.45	20.47	7.423	837.1	837.1	.9962
.5910	.5917	3.718	.9988	20.57	20.60	7.436	847.6	847.6	.9962
.5920	.5927	3.724	.9988	20.70	20.73	7.448	858.2	858.2	.9963
.5930	.5937	3.730	.9989	20.83	20.86	7.460	868.9	868.9	.9963
.5940	.5947	3.737	.9989	20.97	20.99	7.473	879.8	879.8	.9963
.5950	.5957	3.743	.9989	21.10	21.12	7.485	890.8	890.8	.9964
.5960	.5967	3.749	.9989	21.23	21.25	7.498	901.9	901.9	.9964
.5970	.5977	3.755	.9989	21.35	21.37	7.510	913.4	913.4	.9964
.5980	.5987	3.761	.9989	21.49	21.51	7.523	925.0	925.0	.9965
.5990	.5996	3.767	.9989	21.62	21.64	7.535	936.5	936.5	.9965
.6000	.6006	3.774	.9990	21.76	21.78	7.548	948.1	948.1	.9965
.6100	.6106	3.836	.9991	23.17	23.19	7.673	1,074	1,074	.9969
.6200	.6205	3.899	.9992	24.66	24.68	7.798	1,217	1,217	.9972
.6300	.6305	3.961	.9993	26.25	26.27	7.923	1,379	1,379	.9975
.6400	.6404	4.024	.9994	27.95	27.97	8.048	1,527	1,527	.9977
.6500	.6504	4.086	.9994	29.75	29.77	8.173	1,771	1,771	.9980
.6600	.6603	4.149	.9995	31.68	31.69	8.298	2,008	2,008	.9982
.6700	.6703	4.212	.9996	33.73	33.74	8.423	2,275	2,275	.9983
.6800	.6803	4.274	.9996	35.90	35.92	8.548	2,579	2,579	.9985
.6900	.6902	4.337	.9997	38.23	38.24	8.674	2,923	2,923	.9987
.7000	.7002	4.400	.9997	40.71	40.72	8.799	3,314	3,314	.9988
.7100	.7102	4.462	.9997	43.34	43.35	8.925	3,757	3,757	.9989
.7200	.7202	4.525	.9998	46.14	46.15	9.050	4,258	4,258	.9990
.7300	.7302	4.588	.9998	49.13	49.14	9.175	4,828	4,828	.9991
.7400	.7401	4.650	.9998	52.31	52.32	9.301	5,473	5,473	.9992
.7500	.7501	4.713	.9998	55.70	55.71	9.426	6,204	6,204	.9993
.7600	.7601	4.776	.9999	59.30	59.31	9.552	7,034	7,034	.9994
.7700	.7701	4.839	.9999	63.15	63.16	9.677	7,976	7,976	.9995
.7800	.7801	4.902	.9999	67.24	67.25	9.803	9,042	9,042	.9996
.7900	.7901	4.964	.9999	71.60	71.60	9.929	10,250	10,250	.9996
.8000	.8001	5.027	.9999	76.24	76.24	10.05	11,620	11,620	.9996
.8100	.8101	5.090	.9999	81.19	81.19	10.18	13,180	13,180	.9996
.8200	.8201	5.153	.9999	86.44	86.44	10.31	14,940	14,940	.9997
.8300	.8301	5.215	.9999	92.05	92.05	10.43	17,340	17,340	.9997
.8400	.8400	5.278	1.000	98.01	98.01	10.56	19,210	19,210	.9997
.8500	.8500	5.341	1.000	104.4	104.4	10.68	21,780	21,780	.9998
.8600	.8600	5.404	1.000	111.1	111.1	10.81	24,690	24,690	.9998
.8700	.8700	5.467	1.000	118.3	118.3	10.93	28,000	28,000	.9998
.8800	.8800	5.529	1.000	126.0	126.0	11.06	31,750	31,750	.9998
.8900	.8900	5.592	1.000	134.2	134.2	11.18	36,000	36,000	.9998

TABLE OF FUNCTIONS OF d/L_o

d/L_o	d/L	$2\pi d/L$	tanh $2\pi d/L$	sinh $2\pi d/L$	cosh $2\pi d/L$	$4\pi d/L$	sinh $4\pi d/L$	cosh $4\pi d/L$	H/H_o'
.9000	.9000	5.655	1.000	142.9	142.9	11.31	40,810	40,810	.9999
.9100	.9100	5.718	1.000	152.1	152.1	11.44	46,280	46,280	.9999
.9200	.9200	5.781	1.000	162.0	162.0	11.56	52,470	52,470	.9999
.9300	.9300	5.844	1.000	172.5	172.5	11.69	59,500	59,500	.9999
.9400	.9400	5.906	1.000	183.7	183.7	11.81	67,470	67,470	.9999
.9500	.9500	5.969	1.000	195.6	195.6	11.94	76,490	76,490	.9999
.9600	.9600	6.032	1.000	203.5	203.5	12.06	86,740	86,740	.9999
.9700	.9700	6.095	1.000	222.8	222.8	12.19	98,350	98,350	.9999
.9800	.9800	6.158	1.000	236.1	236.1	12.32	111,500	111,500	.9999
.9900	.9900	6.220	1.000	251.4	251.4	12.44	126,500	126,500	1.000
1.000	1.000	6.283	1.000	267.7	267.7	12.57	143,400	143,400	1.000

Index